Centrosomes in Development and Disease

Edited by Erich A. Nigg

Related Titles

R. A. Meyers (Ed.)

Encyclopedia of Molecular Cell Biology and Molecular Medicine, 2nd Edition

2004

ISBN 3-527-30542-4

M. Schliwa (Ed.)

Molecular Motors

2003

ISBN 3-527-30594-7

J. Crocker, P. G. Murray (Eds.)

Molecular Biology in Cellular Pathology

2003

ISBN 0-470-84475-2

Y. Yawata

Cell Membrane

2003

ISBN 3-527-30463-0

A. Lockshin, Z. Zakeri (Eds.)

When Cells Die II: A Comprehensive Evaluation of Apoptosis and Programmed Cell Death

2003

ISBN 0-471-21947-9

F. M. Ausubel et al. (Eds.)

Short Protocols in Molecular Biology, 5th Edition

2002

ISBN 0-471-25092-9

Centrosomes in Development and Disease

Edited by Erich A. Nigg

WILEY-VCH

WILEY-VCH Verlag GmbH & Co. KGaA

Editor:

Prof. Dr. Erich A. Nigg
Max-Planck-Institute of Biochemistry
Department of Cell Biology
Am Klopferspitz 18a
82152 Martinsried
Germany

Cover Illustration
The cover illustration is based on an immunofluorescence picture showing a mitotic mammalian cell (BSC-1) with a monoastral spindle (courtesy of Dr. Thomas Mayer, Max-Planck-Institute of Biochemistry, Martinsried, Germany). Centrosomes are shown in yellow, spindle microtubules in green and chromosomes in blue. In this cell, centrosome separation was blocked by treatment with monastrol, a small molecule inhibitor of the centrosome-associated kinesin-related motor Eg5.

Library of Congress Card No.: applied for

British Library Cataloguing-in-Publication Data:
A catalogue record for this book is available from the British Library.

Bibliographic information published by Die Deutsche Bibliothek
Die Deutsche Bibliothek lists this publication in the Deutsche Nationalbibliografie; detailed bibliographic data is available in the Internet at <http://dnb.ddb.de>.

Printed in the Federal Republic of Germany.
Printed on acid-free paper.

Typesetting hagedorn kommunikation, Viernheim
Printing betz-druck gmbh, Darmstadt
Bookbinding Grossbuchbinderei J. Schäffer GmbH & Co. KG, Grünstadt

ISBN 3-527-30980-2

Contents

Centrosomes in Development and Disease. Edited by Erich A. Nigg

ISBN 3-527-30980-2

Preface

Much like the smile on Mona Lisa's face: beautiful and mysterious...

Ever since the centrosome was discovered more than a hundred years ago, many aspects of its structure, function and reproduction have been shrouded by mystery. However, new information is now rapidly leading to a better understanding of this fascinating organelle, particularly with regard to its role in reproduction, development and disease. The centrosome is a tiny organelle intimately involved with the organization of the microtubule cytoskeleton. Hence, it governs most microtubule-related functions, including intracellular transport, cell motility and polarity, as well as the segregation of chromosomes during cell division. Importantly, the centrioles – cylindrical structures embedded within the animal centrosome – are evolutionarily related to basal bodies. These in turn give rise to cilia and flagella which perform key functions not only in specialized epithelia and motile gametes, but also in many unicellular organisms, including parasites. Thus, wherever centrioles/basal bodies have been conserved in evolution, they are indispensable for cell cycle progression, cell motility or sensory perception. Likewise, the spindle pole body (SPB) of yeast, a microtubule organizing center (MTOC) functionally analogous to the centrosome, is essential for cell viability.

Many of the fundamental problems in centrosome biology, notably its mode of reproduction and its relevance to human development and cancer, were already introduced by Theodor Boveri (1862-1915), the eminent scientist who pioneered the study of centrosomes at the end of the 19th century. However, the centrosome had proven refractory to molecular analysis for decades, largely due to its low abundance and small size. Thanks to modern techniques and the application of complementary research strategies to several distinct organisms, answers to long-standing questions about the centrosome (and related microtubule-organizing centers) are now beginning to emerge. In particular, forward and reverse genetics, mass spectrometry-based proteomics approaches, and the combination of live-cell imaging and laser microsurgery have yielded important new information on the composition of the centrosome, its duplication and its role in the cell division cycle. These results also set the stage for new enquiries into the role of the centrosome in the etiology of cancer and other human diseases, its impact on stem cell biology,

Centrosomes in Development and Disease. Edited by Erich A. Nigg

ISBN 3-527-30980-2

human reproduction and infertility, and last but not least, its relevance to the propagation of intracellular parasites. From this perspective, I hope that this book will serve as a rich source of information for a wide audience, experienced centrosome-researchers and newcomers alike.

My sincere thanks go to all authors for contributing excellent, comprehensive and authoritative chapters, to Ms Alison Dalfovo for expert secretarial assistance and to Dr. Andreas Sendtko and his colleagues at Wiley-VCH for a very pleasant collaboration throughout the preparation of this book.

Erich A. Nigg
Martinsried, June 2004

List of Contributors

David A. Agard
Department of Biochemistry
& Biophysics
University of California, San Francisco
600 16th Street, San Francisco,
CA 94143-2240
USA

Jens S. Andersen
Center for Experimental
BioInformatics (CEBI)
University of Southern Denmark
Campusvej 55
5230 Odense M
Denmark

Juliette Azimzadeh
Institut Curie
UMR 144-CNRS
26, rue d'Ulm
75248 Paris Cedex 05
France

Michel Bornens
Institut Curie
UMR 144-CNRS
26, rue d'Ulm
75248 Paris Cedex 05
France

Laura Briggs
Sir William Dunn School of Pathology
University of Oxford
South Parks Road
Oxford OX1 3RE
UK

William R. Brinkley
Graduate School of Biomedical Sciences
Baylor College of Medicine
One Baylor Plaza-N204
Houston, TX 77030
USA

Stefan Duensing
Molecular Virology Program
University of Pittsburgh Cancer Institute
Hillman Cancer Center
Research Pavilion Suite 1–8
5117 Centre Avenue
Pittsburgh, PA 15213
USA

Susan K. Dutcher
Department of Genetics
Washington University School
of Medicine
875 McDonnell Medical Sciences
Building
St. Louis, MO 63110
USA

Centrosomes in Development and Disease. Edited by Erich A. Nigg

ISBN 3-527-30980-2

Andrew M. Fry
Department of Biochemistry
University of Leicester
University Road
Leicester LE1 7RH
UK

Joseph G. Gall
Department of Embryology
Carnegie Institution of Washington
115 West University Parkway
Baltimore, MD 21210
USA

Thea M. Goepfert
Department of Molecular
and Cellular Biology
Baylor College of Medicine
One Baylor Plaza-117AA
Houston, TX 77030
USA

Keith Gull
Sir William Dunn School of Pathology
University of Oxford
South Parks Road
Oxford OX1 3RE
UK

Rebecca S. Hames
Department of Biochemistry
University of Leicester
University Road
Leicester LE1 7RH
UK

Anthony A. Hyman
Max Planck Institute of
Molecular Cell Biology and Genetics
Pfotenhauerstrasse 108
01307 Dresden
Germany

Alexey Khodjakov
Wadsworth Center
New York State Department of Health
P. O. Box 509
Albany, NY 12201-0509
USA

Matthias Mann
Center for Experimental BioInformatics
(CEBI)
University of Southern Denmark
Campusvej 55
5230 Odense M
Denmark

Kazuhiro Mizumoto
Department of Surgery and Oncology
Graduate School of Medical Sciences
Kyushu University
3-1-1 Maidashi, Higashi-ku
Fukuoka 812-8582
Japan

Mette M. Mogensen
School of Biological Sciences
University of East Anglia
Norwich NR4 7TJ
UK

Michelle Moritz
Department of Biochemistry
& Biophysics
600 16th Street
University of California, San Francisco
San Francisco, CA 94143-2240
USA

Thomas Müller-Reichert
Max Planck Institute of Molecular Cell
Biology and Genetics
Pfotenhauerstrasse 108
01307 Dresden
Germany

Karl Münger
Pathology Department
Harvard Medical School
77 Avenue Louis Pasteur
NRB, Room 0958c
Boston, MA 02115-5727
USA

Christopher S. Navara
Departments of Obstetrics, Gynecology and Reproductive Sciences
Pittsburgh Development Center and University of Pittsburgh
204 Craft Avenue
Pittsburgh, PA 15213
USA

Erich A. Nigg
Department of Cell Biology
Max-Planck-Institute of Biochemistry
Am Klopferspitz 18a
82152 Martinsried
Germany

Nurhan Özlü
Max Planck Institute of Molecular Cell Biology and Genetics
Pfotenhauerstrasse 108
01307 Dresden
Germany

Laurence Pelletier
Max Planck Institute of Molecular Cell Biology and Genetics
Pfotenhauerstrasse 108
01307 Dresden
Germany

Jordan W. Raff
Wellcome/CR UK Institute of Cancer and Developmental Biology
University of Cambridge
Tennis Court Road
Cambridge CB2 1QR
UK

Luke M. Rice
Department of Biochemistry & Biophysics
University of California, San Francisco
600 16th Street
San Francisco, CA 94143-2240
USA

Conly L. Rieder
Wadsworth Center
New York State Department of Health
P. O. Box 509
Albany, NY 12201-0509
USA

Norihiro Sato
Department of Surgery and Oncology
Graduate School of Medical Sciences
Kyushu University
3-1-1 Maidashi, Higashi-ku
Fukuoka 812-8582
Japan

Niki Scaplehorn
Cell Motility Group
Cancer Research UK
Lincoln's Inn Fields Laboratories
44 Lincoln's Inn Fields
London WC2A 3PX
UK

G. P. Schatten
Pittsburgh Development Center
Magee Womens Research Institute
University of Pittsburgh School of Medicine
Departments of Obstetrics, Gynecology and Reproductive Sciences
204 Craft Avenue
Pittsburgh, PA 15213
USA

Anne-Lore Schlaitz
Max Planck Institute of Molecular Cell Biology and Genetics
Pfotenhauerstrasse 108
01307 Dresden
Germany

Ody C. M. Sibon
Department of Radiation and Stress Cell Biology
University of Groningen
Ant. Deusinglaan 1
9713 AV Groningen
The Netherlands

C. R. Simerly
Departments of Obstetrics, Gynecology and Reproductive Sciences
Pittsburgh Development Center and University of Pittsburgh
204 Craft Avenue
Pittsburgh, PA 15213
USA

Greenfield Sluder
Department of Cell Biology
University of Massachusetts Medical School
377 Plantation Street
Worcester, MA 01605
USA

Martin Srayko
Max Planck Institute of Molecular Cell Biology and Genetics
Pfotenhauerstrasse 108
01307 Dresden
Germany

Tim Stearns
Department of Biological Sciences
Stanford University
337 Campus Drive
Stanford, CA 94305-5020
USA

Masao Tanaka
Department of Surgery and Oncology
Graduate School of Medical Sciences
Kyushu University
3-1-1 Maidashi, Higashi-ku
Fukuoka 812-8582
Japan

William E. Theurkauf
Program in Cell Dynamics and Program in Molecular Medicine
University of Massachusetts Medical School
373 Plantation Street
Worcester, MA 01605
USA

Sue Vaughan
Sir William Dunn School of Pathology
University of Oxford
South Parks Road
Oxford OX1 3RE
UK

Suzanne van Kreeveld Naone
Molecular, Cellular and Developmental Biology
University of Colorado at Boulder
Porter Biosciences, 347 UCB
Boulder, CO 80309-0347
USA

Michael Way
Cell Motility Group
Cancer Research UK
Lincoln's Inn Fields Laboratories
44 Lincoln's Inn Fields
London WC2A 3PX
UK

Christopher J. Wilkinson
Department of Cell Biology
Max-Planck-Institute of Biochemistry
Am Klopferspitz 18a
82152 Martinsried
Germany

Mark Winey
Molecular, Cellular and Developmental Biology
University of Colorado at Boulder
Porter Biosciences, 347 UCB
Boulder, CO 80309-0347
USA

Color Plates

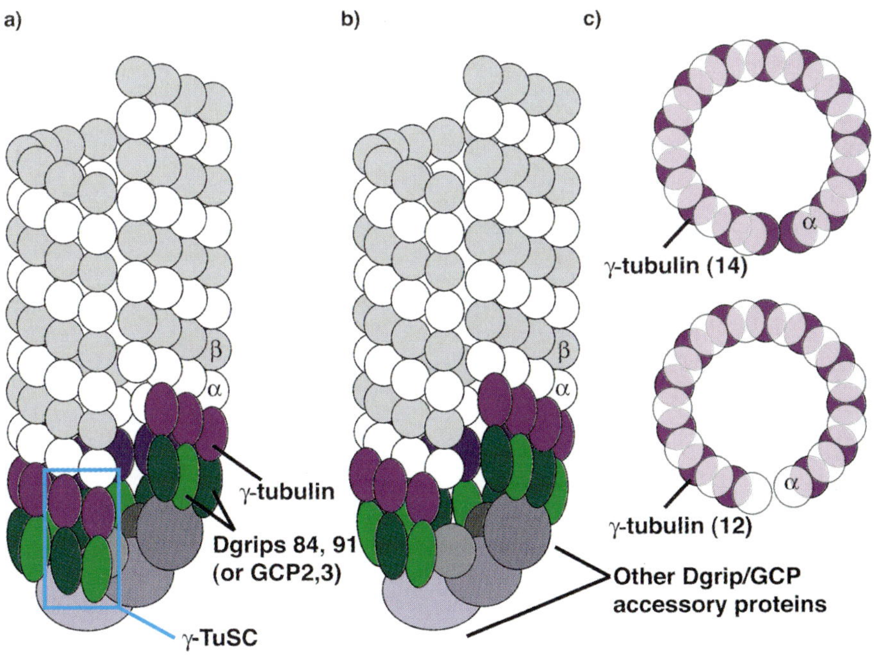

Figure 3.5 Modified template model of γTuRC-mediated microtubule nucleation. (a) The original template model proposed that γ-tubulins bind to α-tubulins at the minus ends of protofilaments similarly to longitudinal α/β-tubulin binding within a protofilament (reviewed in [15, 17]). (b) The modified template model takes into account physical properties of γ-tubulin and the mechanism of γ-tubulin-mediated microtubule nucleation by proposing that γ-tubulin binds between protofilaments [26]. A γTuRC containing 12 γ-tubulins is shown associated with the microtubule, but a 14-γ-tubulin γTuRC could also be accommodated. (c) Cross-sectional views illustrating the proposed binding sites for γ-tubulins between the α-tubulins at the minus end of each protofilament. This mode of binding provides an explanation for how a γTuRC containing an even number of γ-tubulins could template a 13-protofilament microtubule, the most common architecture observed *in vivo*.

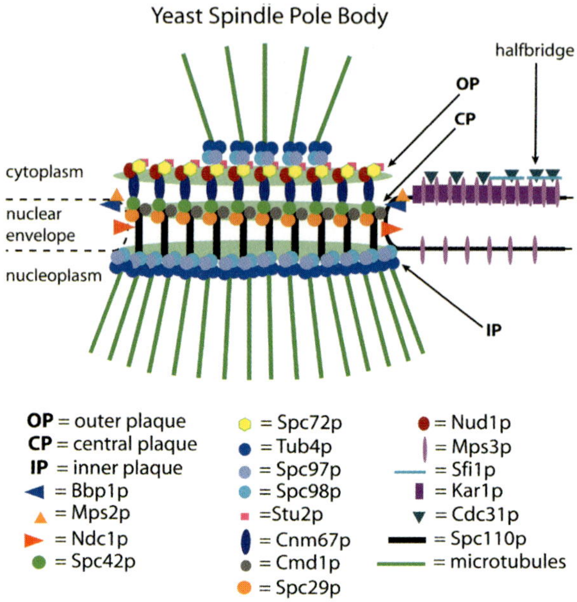

Figure 4.2 Yeast Spindle Pole Body. Shown here is a schematic of the organization of most of the components described in Table 4.1.

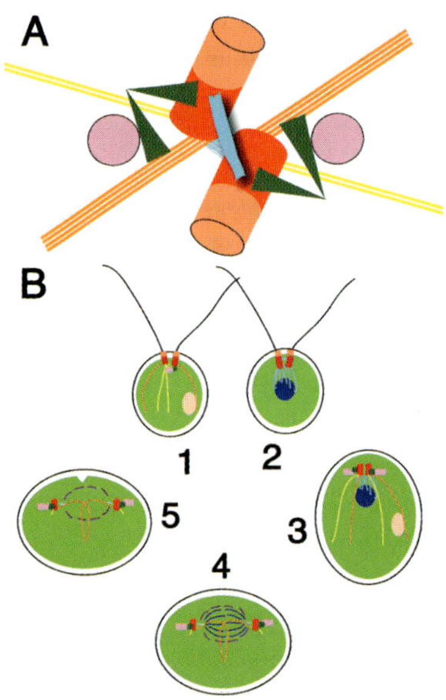

Figure 5.5 The three fiber systems of the basal body complex. (A) The mature basal bodies are shown in red, the transition zones in peach and the probasal bodies are shown in pink. The rootlet microtubules have four microtubules (orange) or two microtubules (yellow) and attach at specific triplet microtubules of the basal body. The distal (solid) and proximal (striped) striated fibers are shown in light blue. They connect the two mature basal bodies at the two ends. The lateral fibers are shown in green. They connect the mature basal body to its daughter probasal body across the rootlet microtubules. (B) Changes in the fiber systems during the cell cycle. 1, During interphase the basal bodies and transition zones are continuous with the flagella. The rootlet microtubules are adjacent to the plasma membrane. One of the four-membered rootlet microtubules lie adjacent to the eyespot (rose). 2, Another view of interphase cells illustrates that the basal bodies are connected to the nucleus and to each other by centrin fibers. 3, At preprophase, the flagella are lost. The probasal bodies elongate. The distal and proximal striated fibers are lost. 4, The two-membered rootlet microtubules shorten. The centrioles (without transition zones) are found at the poles of the spindle. The four-membered rootlet microtubules arc over the spindle. The eyespot is disassembled. 5, Cytokinesis is initiated at one end of the cell. This will be followed by extension of the two-membered rootlet microtubules, the striated fibers, and assembly of new rootlet microtubules and of a new eyespot in association with the new four-membered rootlet microtubules.

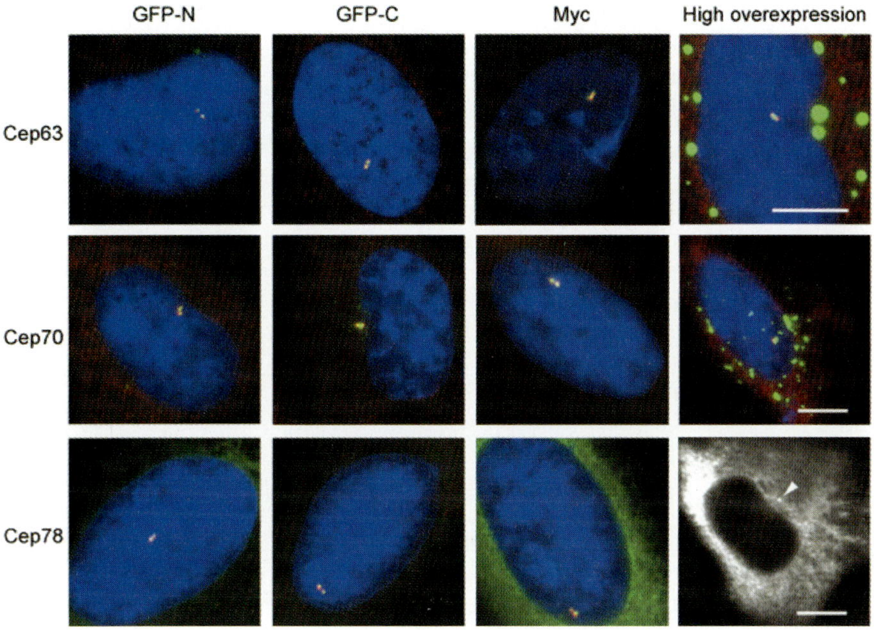

Figure 7.2 A selection of differently tagged, novel centrosome proteins. Rows from top to bottom show Cep63, Cep70 and Cep78. Columns from left to right show N-terminal GFP, C-terminal GFP and N-terminal myc-tagged proteins, respectively. The most right-hand column shows the results of very high overexpression of these proteins (tagged at the N-terminus with GFP), generating large aggregates or a high cytoplasmic background. Green, ectopically expressed centrosomal proteins; red, γ-tubulin; blue, DNA (DAPI). The arrowhead points to the position of the centrosome. Scale bars, 10 μm; panels in the three left columns are to the same scale as the top right panel.

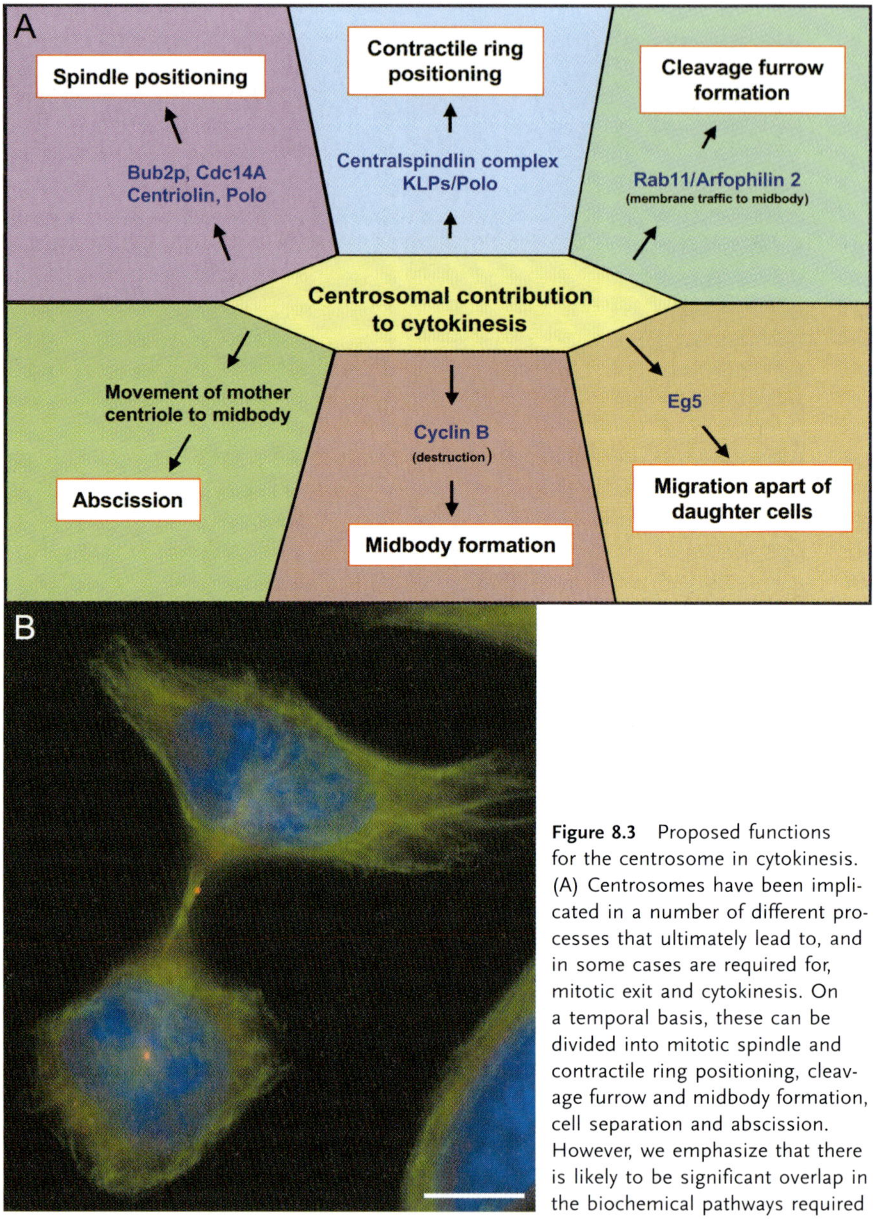

Figure 8.3 Proposed functions for the centrosome in cytokinesis. (A) Centrosomes have been implicated in a number of different processes that ultimately lead to, and in some cases are required for, mitotic exit and cytokinesis. On a temporal basis, these can be divided into mitotic spindle and contractile ring positioning, cleavage furrow and midbody formation, cell separation and abscission. However, we emphasize that there is likely to be significant overlap in the biochemical pathways required for each of these endpoints. Examples of proteins that localize to mitotic centrosomes and are implicated in these pathways are indicated in dark blue. (B) One of the most intriguing questions relating to the role of the centrosome in cytokinesis is why the mother centriole migrates towards the midbody prior to cell abscission. HeLa cells are shown following methanol fixation and staining with antibodies against α-tubulin (green) and γ-tubulin (red). DNA is stained with Hoechst 33258 (blue). Scale bar, 10 μm.

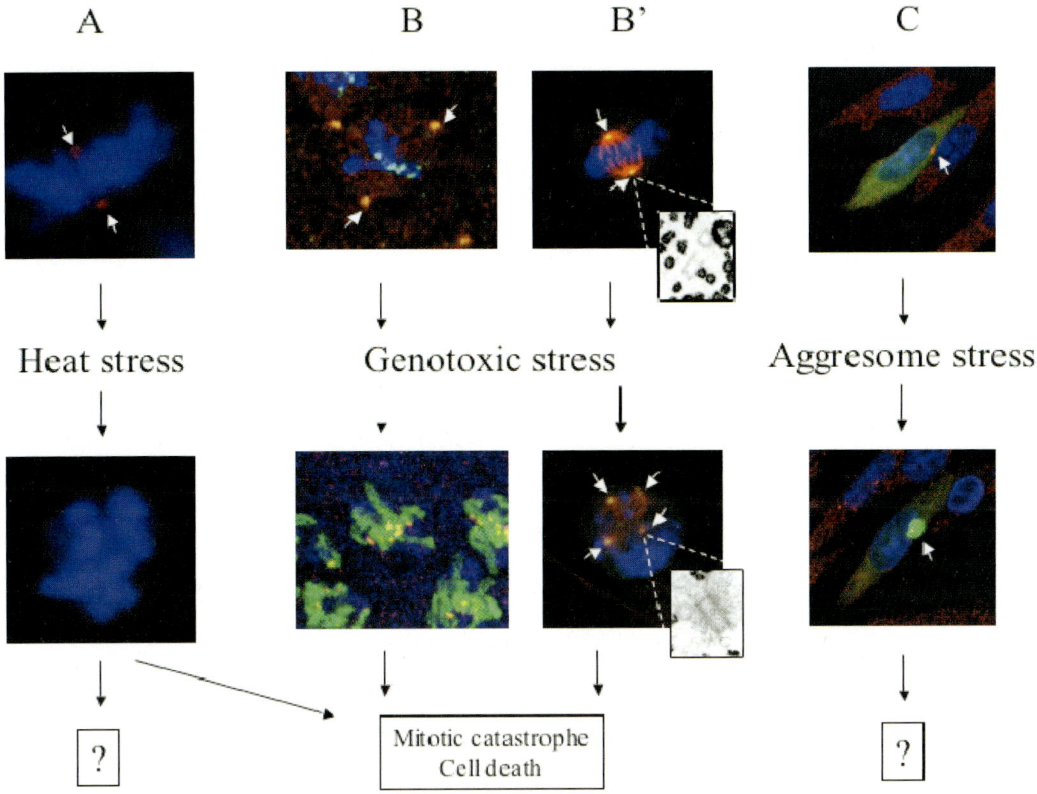

Figure 11.1 Centrosome alterations in response to heat, genotoxic and aggresome stress. In diverse systems, γ-tubulin (red) localizes to centrosomes at the mitotic spindle poles (A, B, B) and close to interphase nuclei (C). In Chinese hamster ovary (CHO) cells, heat stress (A) triggers loss of γ-tubulin localization to the poles (courtesy of H. Hut) while genotoxic stress (B) leads to mitotic centrosome fragmentation. Electron microscopic examination demonstrates that the centrosome fragments contain single centrioles (insets). In response to heat shock and genotoxic stress, centrosome disruption is associated with failures of mitotic division and mitotic catastrophe. In *Drosophila* embryos, genotoxic stress also leads to dissociation of γ-tubulin from the spindle poles (B) and mitotic catastrophe. Over-expression of a mutant form of GFP taggered the Huntingtin protein (green) in hamster cells (C), leads to aggresome formation around interphase centrosomes (courtesy of F. Salomons and M. Rujano). The significance of aggresome formation is not known, but this structure may contribute to neurodegeneration in a number of pathological conditions. In all panels, γ-tubulin is in red and DNA is in blue. In B, the kinetochore marker MeiS332 is in green. In C the Huntingtin-GFP protein is in green.

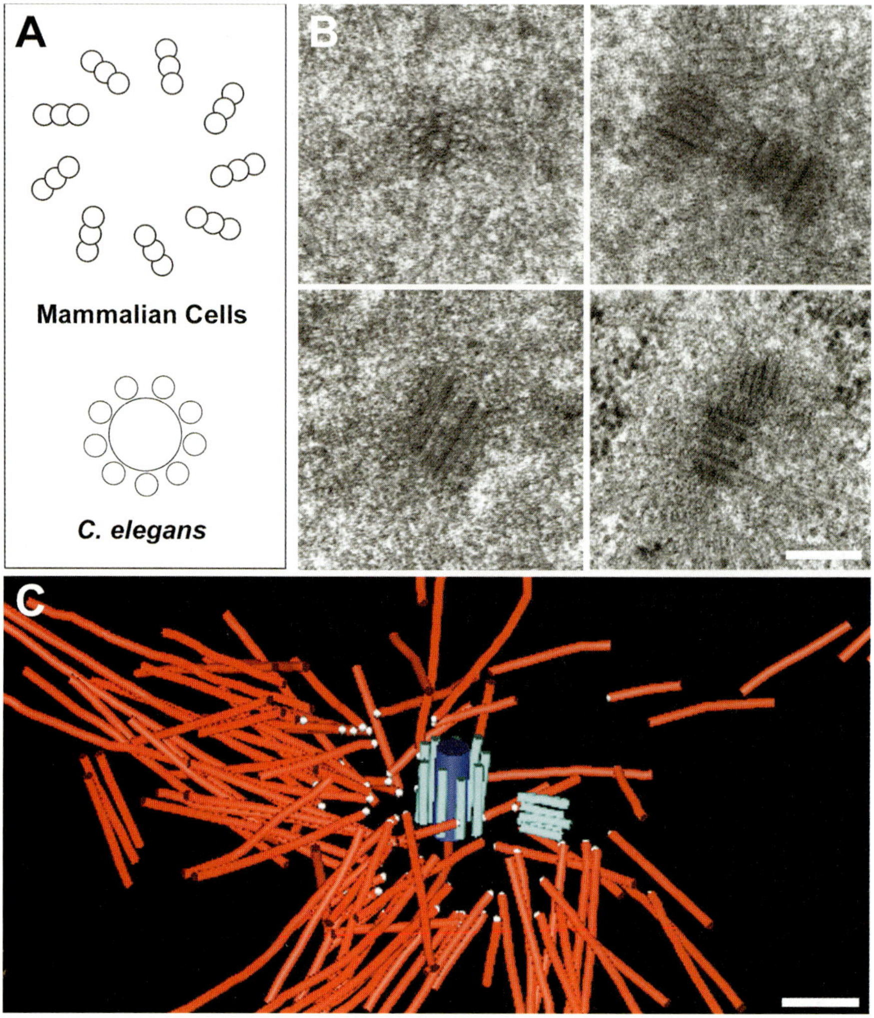

Figure 12.2 The ultrastructure of the *C. elegans* centrosome. (A) Schematic representation of the triplet structure of centrioles found in mammalian cells (top) and the singlet structure observed in *C. elegans* (bottom). (B) Electron micrographs of wild-type centrioles in cross-section and longitudinal orientation (left) and wild-type centriole pairs in orthogonal orientation (right). (C) 3-D model of a centriole pair during prometaphase derived from a tomographic reconstruction. Microtubules (red) are organized mainly around one centriole (blue), referred to as the mother centriole. Note that the minus ends of the microtubules do not come in contact with this centriole. Scale bars = 250 nm.

Figure 12.3 PCM recruitment and spindle assembly in *C. elegans*. Early embryos at different stages of the cell cycle were fixed and labeled for DNA (blue), microtubules (green) and γ-tubulin (red). Z-stacks through entire embryos were acquired, the images deconvolved and shown as two-dimensional projections. Scale bar = 10 μm. The anterior is to the left in all the images. (a) An acentrosomal meiotic spindle can be observed soon after fertilization (arrow). At this stage the centrosome contributed by the sperm has yet to separate. (b) At the beginning of pronuclear migration, the sperm-derived centrosome has separated and recruited some γ-tubulin therefore increasing the amount of microtubules it is able to nucleate. (c) At the time when the pronu- ▶

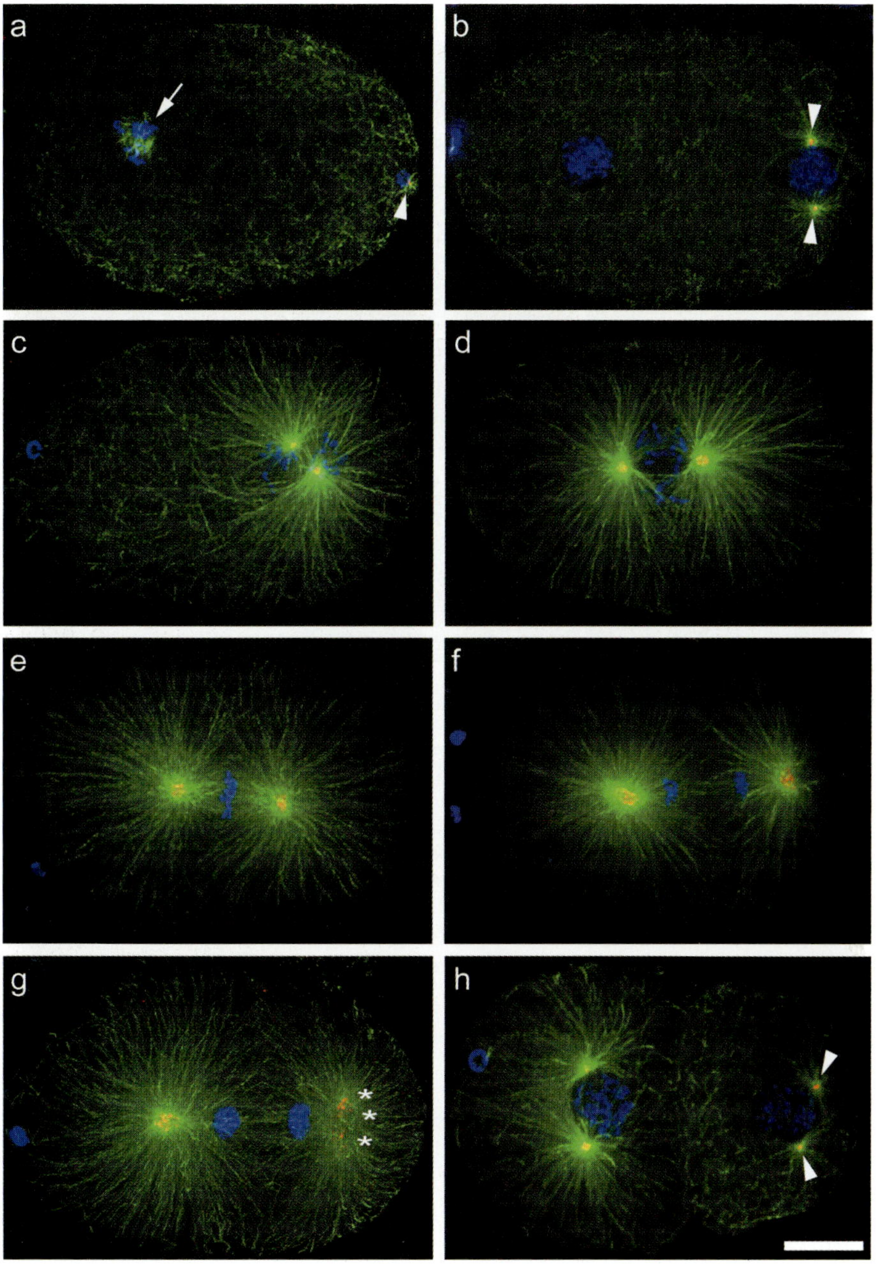

cleus–centrosome complex begins rotating, the DNA is condensed, the nuclear envelope broken down, and the levels of γ-tubulin and the nucleating capacity begin increasing rapidly, a process termed centrosome maturation. (d) After rotation, spindle assembly begins. (e) Late metaphase spindle with aligned chromosomes. (f) Late anaphase. (g) During telophase, the nuclear envelope re-forms and the posterior centrosome adopts a flattened morphology. At this time the γ-tubulin staining becomes more diffuse. (h) After cytokinesis, the centrosomes separate again in preparation for the next cellular division.

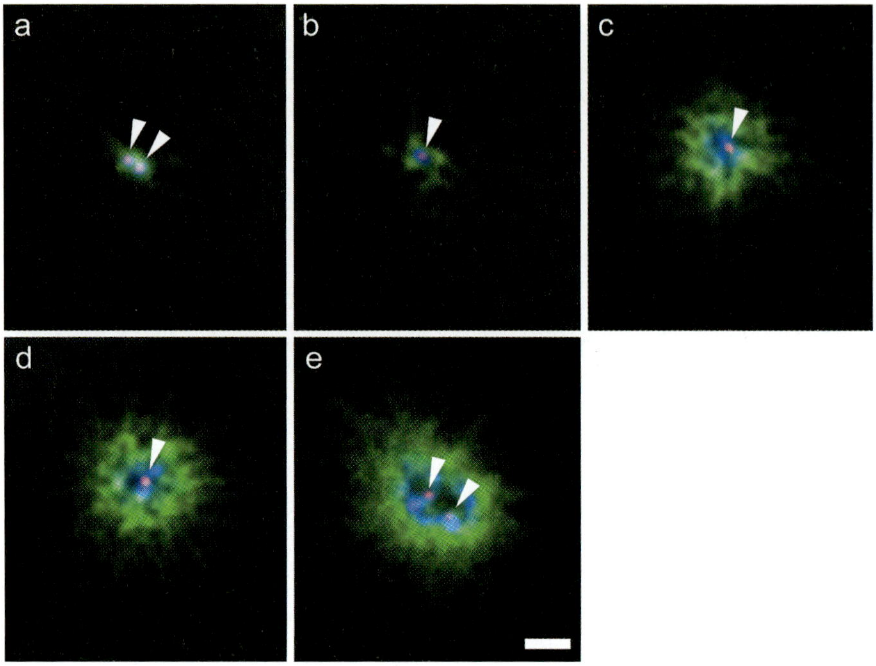

Figure 12.4 Layers of the *C. elegans* centrosome. Early embryos at different stages of the cell cycle were fixed and triple labeled for γ-tubulin (light blue), AIR-1 (green) and the centriolar protein SAS-4 (red). Z-stacks through entire embryos were acquired and the images deconvolved. Individual images or two-dimensional projections of two sections are shown. (a) Centrosome during female meiosis. Note the presence of two SAS-4 positive structures, indicating that the two centrioles contributed by the sperm have separated. (b) One pole after centrosome separation. (c) One pole during the rotation of the pronucleus–centrosome complex. (d) One pole during metaphase. (e) One pole during telophase. Note that at this stage two SAS-4 positive structures can be seen at each pole, suggesting that centriole duplication has occurred.

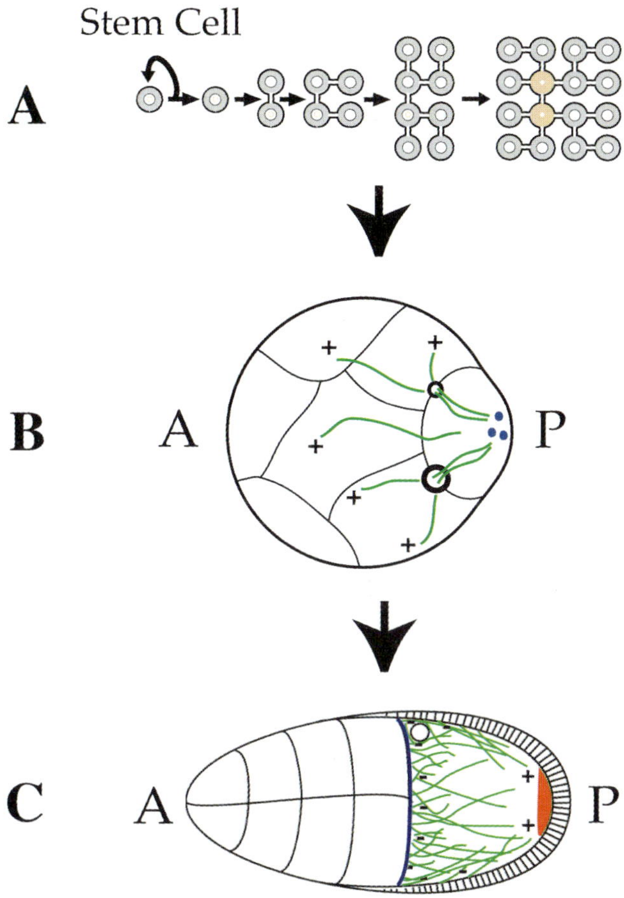

Figure 13.1 A schematic summary of oogenesis in *Drosophila*. (A) The division of a stem cell gives rise to another stem cell and a cystoblast that goes through four rounds of mitosis to generate a cyst of 16 cells which remain interconnected by ring canals. Two cells (orange) contain four ring canals, and one of these invariably becomes positioned at the posterior of the cyst and becomes the oocyte, while the other cells become nurse cells. (B) After mitosis is finished, the centrioles (blue) of the nurse cells lose all PCM markers and migrate into the oocyte. As the oocyte enlarges, the centrioles move to the posterior and an MTOC is assembled in this region, although it is not clear that the centrioles are required to form this MTOC. The microtubules (green) extend away from the MTOC and spread through the ring canals to the nurse cells (only two ring canals are shown here). (C) At later stages of oogenesis, the centrioles disappear, and the posterior MTOC is disassembled. The minus ends of the microtubules now associate with a diffuse region spread along the anterior cortex. The exact organization of microtubules at this stage of oogenesis is controversial (see, for example [151]) and long microtubules, like those depicted here, are rarely visible at this stage of oogenesis. A microtubule plus end-directed kinesin, however, can accumulate at the posterior pole (red), while a minus end-directed kinesin can accumulate at the anterior cortex (blue) suggesting that the microtubules have an overall polarity [152]. In this figure, anterior is on the left and posterior on the right, as indicated. This figure was adapted from a figure provided by Daniel St Johnston.

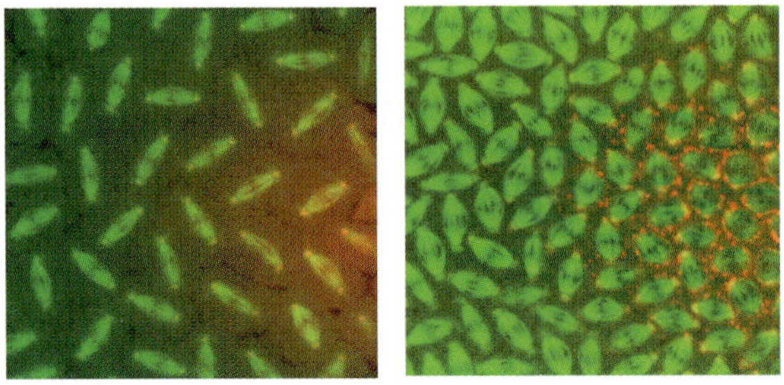

Figure 13.2 Perturbing centrosome function by antibody injection in syncytial embryos. Embryos that express a tubulin-GFP fusion protein were injected with Texas Red-labeled anti-CP190 (A) or anti-D-TACC (B) antibodies and imaged on a confocal microscope. The anti-CP190 antibodies bind to the centrosomes closest to the injection site. The behavior of the microtubules, however, is not perturbed, suggesting that CP190 is not involved in regulating centrosome or microtubule behavior (see Section 13.3.4). The anti-D-TACC antibodies precipitate the endogenous D-TACC protein and form large lumps in the cytoplasm, effectively depleting D-TACC from the spindles around the injection site. As a result, the spindles closest to the injection site are noticeably shorter than those away from the injection site, suggesting that D-TACC is involved in stabilizing spindle microtubules. Scale bar = 10 μm.

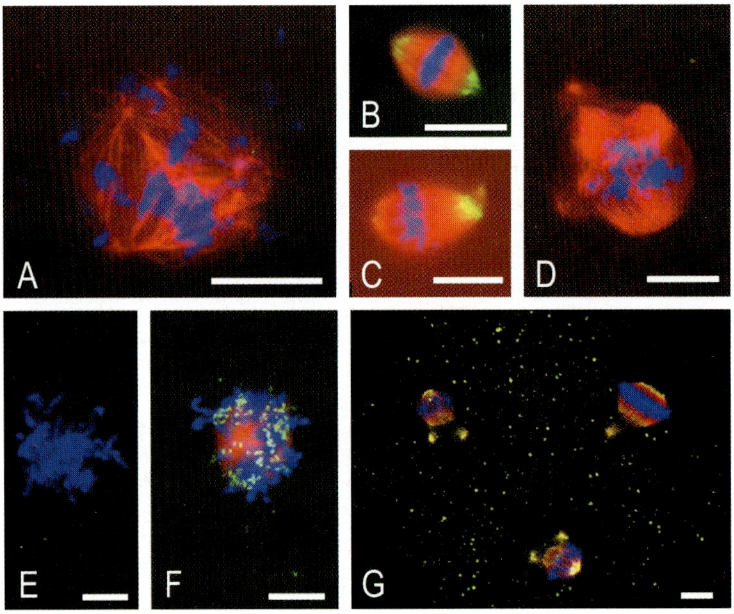

Figure 14.5 Faulty mitotic spindles produce aneuploid embryos after primate nuclear transfer. (A) Defective NT mitotic spindle with misaligned chromosomes. Centrosomal NuMA at meiosis (B) and mitosis (C), but not in mitotic spindles after NT (D). The centrosomal kinesin HSET is also missing after NT (E), but not centromeric Eg5 (F). Bipolar mitotic spindles with aligned chromosomes and centrosomal NuMA after NT into fertilized eggs (G). Blue, DNA; red, B-tubulin; green, NuMA in B, C, D, and G; HSET in E; and Eg5 in F. Reprinted with permission from Simerly et al. [91]. Bars = 10 μm.

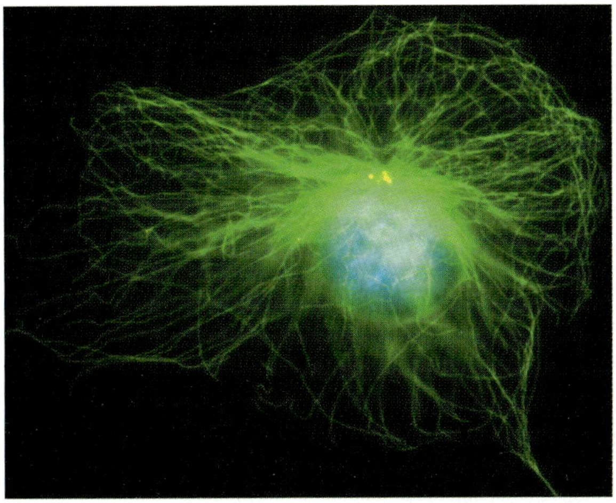

Figure 15.1 Cell with a radial array of microtubules focused on a juxta-nuclear centrosome. A wide field fluorescent image of a cell triple labeled for microtubules with an antibody to α-tubulin (green), for the centrosome with an antibody to γ-tubulin (red or yellow where co-localized with α-tubulin) and for the nucleus with DAPI (blue). Images by courtesy of Gemma Bellett.

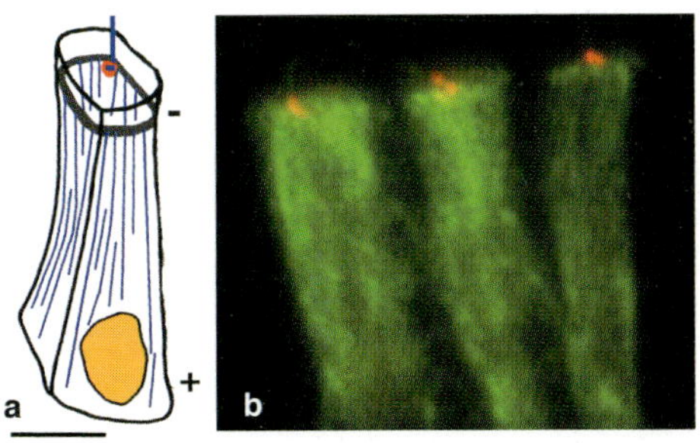

Figure 15.2 Apico-basal microtubule arrays in cochlear inner pillar epithelial cells. (a) Schematic 3-D representation of the microtubule organization (blue) in an inner pillar cell during assembly of the apico-basal array. The centrosome (red with centriole and basal body/primary cilium in blue) is located at the apex and most of the microtubule minus-ends are anchored at a peripheral apical ring (grey). The nucleus is shown in yellow. Adapted from [52]. (b) Wide field fluorescent image of the apical halves of three inner pillar cells showing an apico-basal array of microtubules (labeled with an antibody to α-tubulin in green) and γ-tubulin (red) concentrated at the apical centrosome. Images by courtesy of Gemma Bellett. Bar = 10 μm.

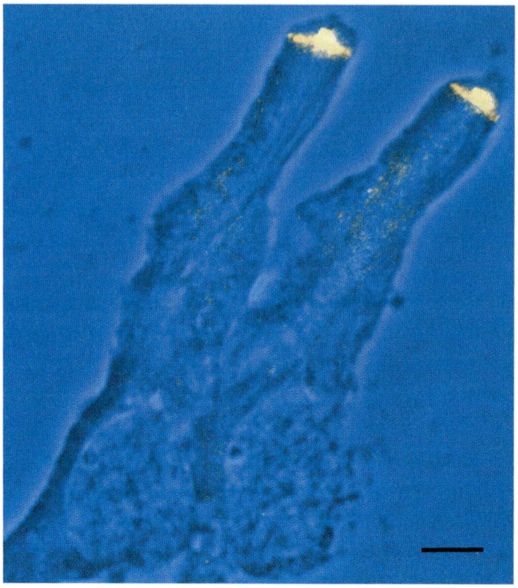

Figure 15.4 Ninein localization in inner pillar cells. Projection of confocal optical sections through two isolated inner pillar cells at a stage when some 3000 microtubule minus-ends are concentrated at the apical sites. The pillar cells have been labeled with an antibody for ninein (yellow). Ninein is concentrated at the centrosome and at the apical sites but ninein speckles are also evident within the apical half of the cytoplasm. From [52]. Bar = 5 μm.

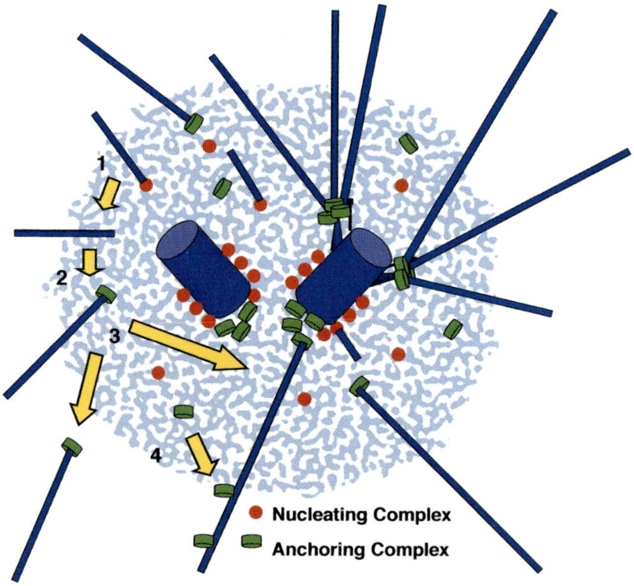

Figure 15.5 Two centrosomal complexes: a nucleating and an anchoring complex. Schematic diagram showing the organization of nucleating (γ-TURC) and anchoring (ninein) complexes within the centrosome based on recent findings. The possible fates of a microtubule nucleated by a γ-TURC and centrosomal anchoring complexes are outlined in 1–4. (1) Microtubule release from the γ-TURC following nucleation; (2) microtubule minus-end capping by a capping/anchoring complex; (3) microtubule release from the centrosome or firm anchorage within the PCM closely associated with the mother centriole; (4) release of anchoring complexes from the centrosome and their transport along a microtubule.

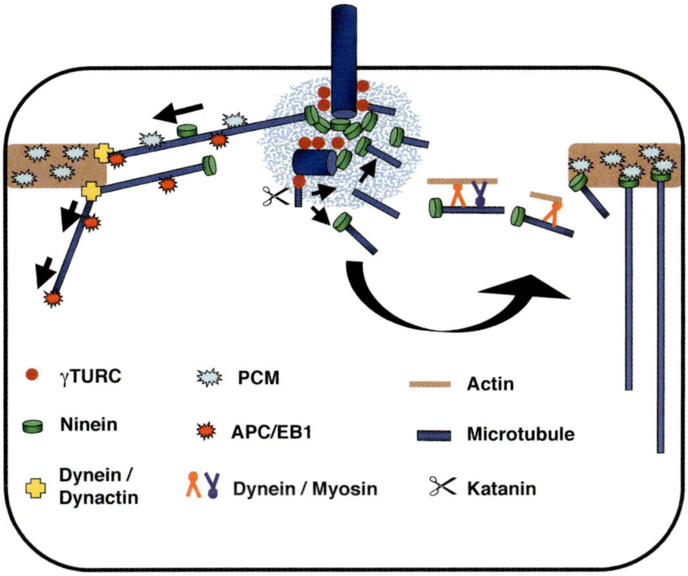

Figure 15.6 Release and capture. Two variants of the release and capture mechanism are illustrated to show the assembly of non-centrosomal apico-basal microtubule arrays in polarized epithelial cells. The right-hand side shows the classic release and capture version involving the release of relatively short microtubules and their subsequent translocation and capture at apical sites, whereas the left-hand side shows the modified version involving microtubule plus-end capture, release, translocation and minus-end capture. The two variant models are not mutually exclusive.

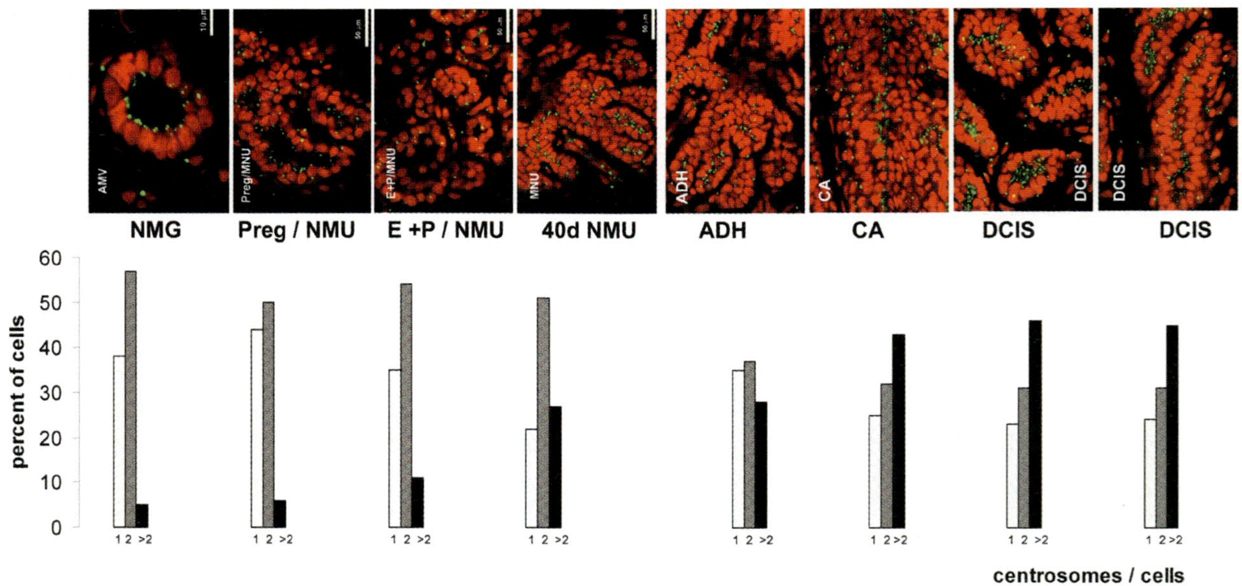

Figure 16.3 Centrosome amplification correlates with tumorigenesis in rat mammary glands. Confocal microscopy optical sections of tissue from different regimens and centrosome plots. NMG, normal mammary gland; Preg/NMU, pregnancy and involution prior to NMU; E + P/NMU, hormone treatment prior to NMU; 40 days NMU: 40 days post treatment with NMU. Pre-malignant lesions, e. g. ADH, atypical ductal hyperplasia; CA, cancer; DCIS, ductal carcinoma *in situ*.

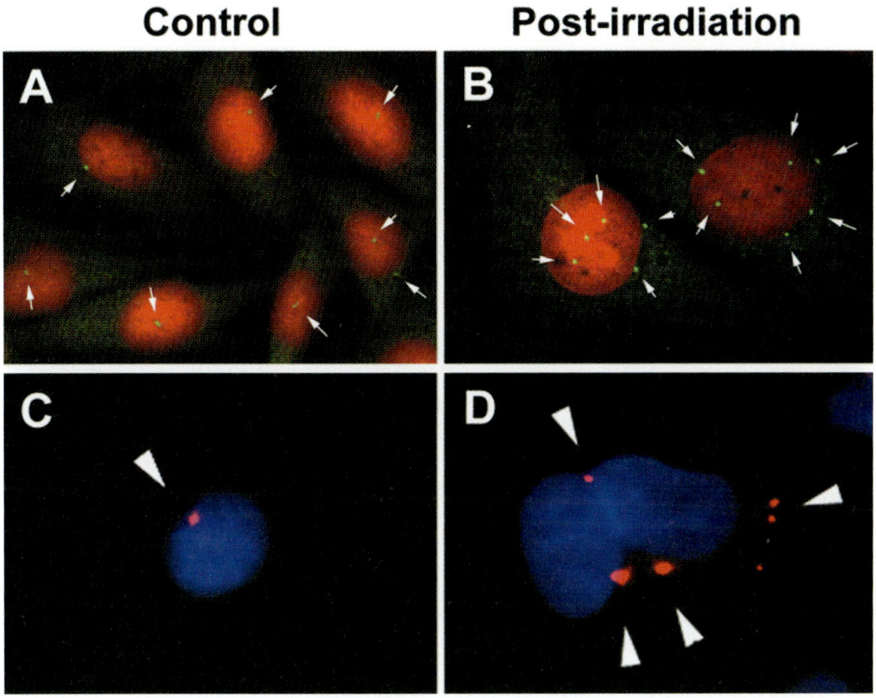

Figure 17.1 Ionizing radiation induces numerical aberrations of centrosomes in human tumor cells. Cells were stained with an antibody to γ-tubulin (green, A and B) or pericentrin (red, C and D) for centrosomes and counterstained with propidium iodide (red, A and B) or Hoechst 33258 (blue, C and D) for nuclear labeling. U2-OS osteosarcoma cells before (A) and 48 h after (B) γ-radiation at 10 Gy. Mia PaCa2 pancreatic cancer cells before (C) and 48 h after (D) γ-radiation at 10 Gy.

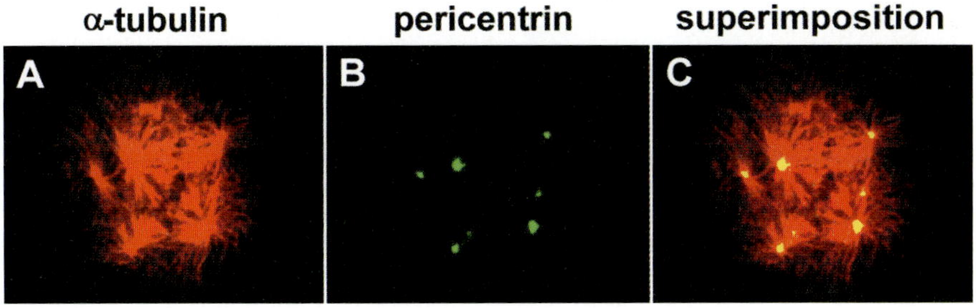

Figure 17.2 Aberrant mitotic cell with multipolar spindles after γ-radiation at 10 Gy. Cells were double-stained with α-tubulin (red) and pericentrin (green) antibodies.

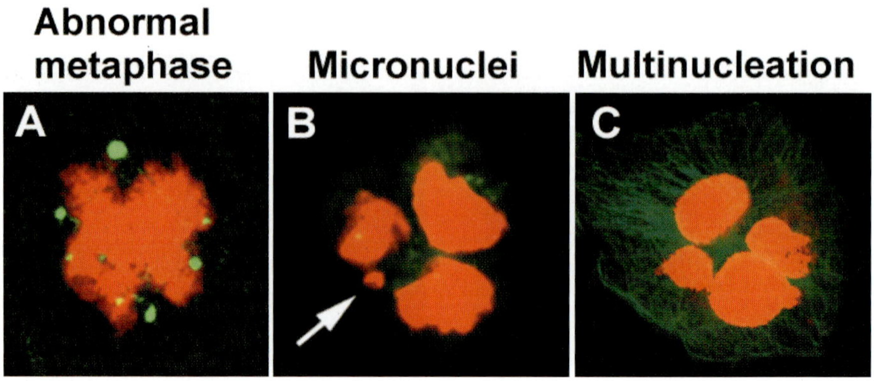

Figure 17.3 MiaPaCa2 pancreatic cancer cells after 10 Gy radiation showing abnormal mitosis (metaphase) with multiple centrosomes (green, pericentrin labeling) dispersed chaotically throughout the condensed chromosomes (A), followed by the appearance of the micro- (B) and multinucleated (C) phenotype, characteristic of mitotic cell death.

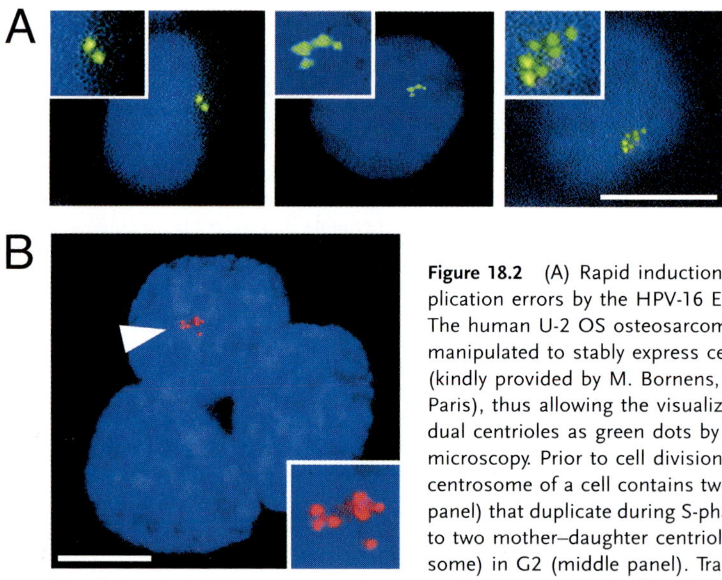

Figure 18.2 (A) Rapid induction of centriole duplication errors by the HPV-16 E7 oncoprotein. The human U-2 OS osteosarcoma cell line was manipulated to stably express centrin-GFP (kindly provided by M. Bornens, Institut Curie, Paris), thus allowing the visualization of individual centrioles as green dots by fluorescence microscopy. Prior to cell division, the normal centrosome of a cell contains two centrioles (left panel) that duplicate during S-phase and give rise to two mother–daughter centriole pairs (diplosome) in G2 (middle panel). Transient overexpression of the HPV-16 E7 oncoprotein increases the proportion of cells with uncontrolled centriole duplication (right panel). Cells were evaluated at 48 h post transfection; no drug treatment was used to induce S-phase arrest. Note the differences in fluorescence intensity between centrioles. Younger centrioles have been shown to contain less centrin than older centrioles supporting the notion that HPV E7 triggers abnormal daughter centriole formation. Nuclei stained with DAPI. Scale bar indicates 10 μm. (B) Centrosome accumulation in HPV-16 E6-expressing cells. An example of a primary human keratinocyte expressing HPV-16 E6 oncoprotein that has accumulated multiple centrosomes in parallel with nuclear atypia. Centrosomes were detected by immunofluorescence microscopy for γ-tubulin, a pericentriolar marker. The abnormal cell nucleus (stained with DAPI) is lobulated and dramatically enlarged. This nuclear morphology suggests a division failure, indicating that DNA replication may be uncoupled from cell division. Scale bar indicates 10 μm.

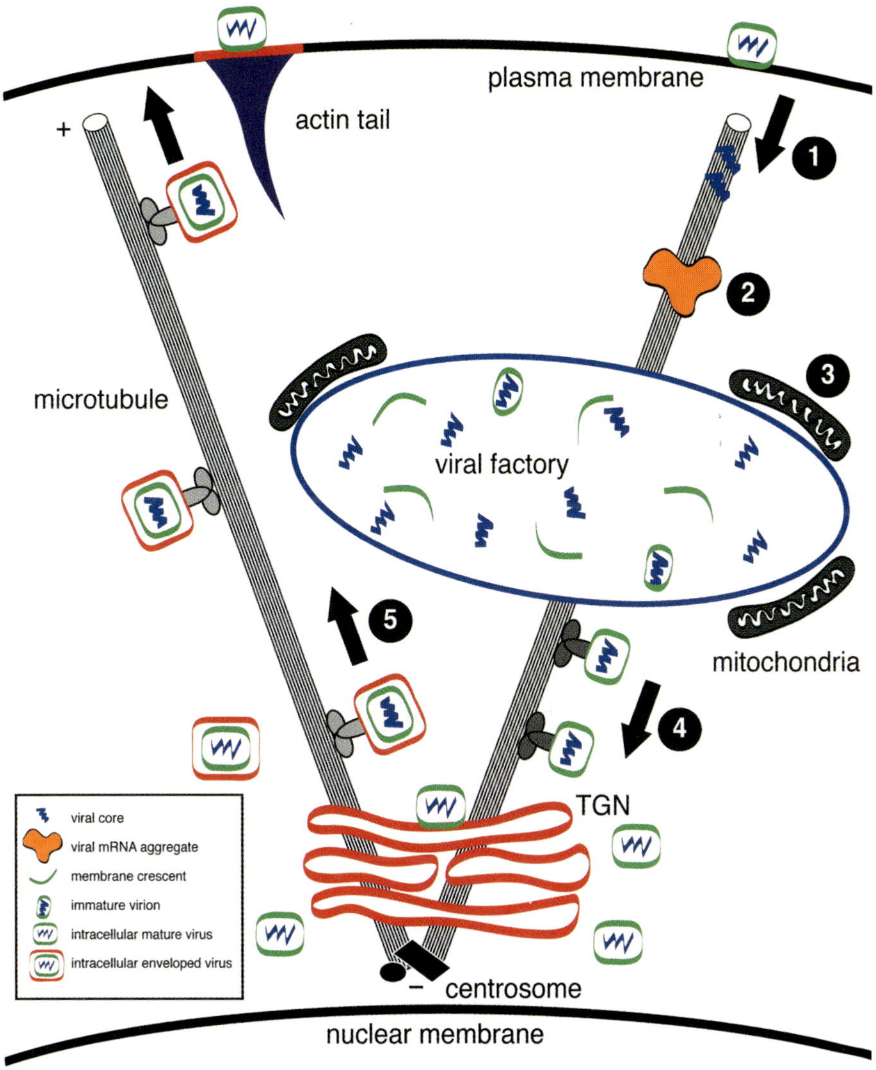

Figure 19.3 Interactions between vaccinia and microtubule cytoskeleton. Microtubules are thought to be involved in many steps of the vaccinia virus life cycle. (1) Incoming viral cores appear to associate with microtubules. (2) Viral mRNAs are organized into granules in a microtubule-dependent process. (3) Mitochondria retract from the cell periphery to surround the newly formed viral-factory region. (4) Accumulation of IMV particles at the centrosome and TGN requires intact microtubules. (5) IEV move from the TGN to the cell periphery along microtubules before leaving the cell and stimulating actin-tail formation.

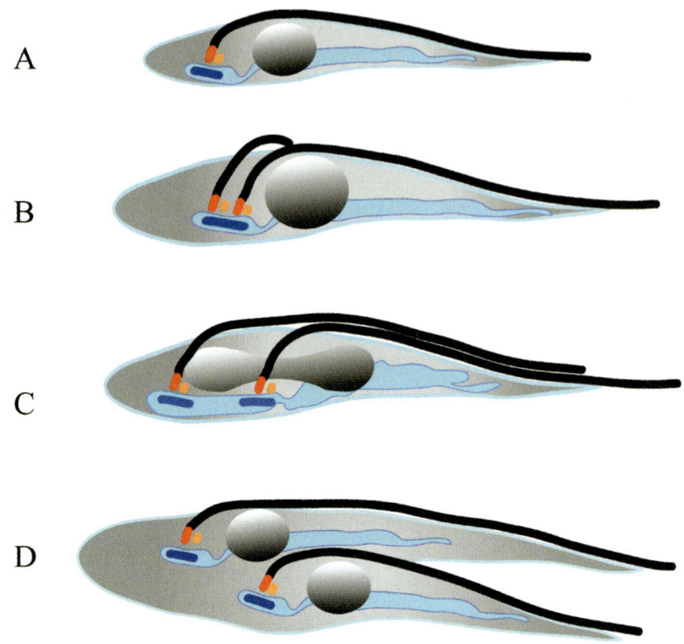

Figure 20.1 A schematic representation of the cell division cycle of pro-cyclic forms of *T. brucei*. (A) The G1 cell contains a single flagellum with a basal body and pro-basal body. The basal bodies are connected via the Tripartite Attachment Complex to the kinetoplast (dark blue) inside the mitochondrion. (B) The pro-basal body matures, nucleates a new flagellum and two new pro-basal bodies are formed. (C) Mitosis occurs across the axis of the old flagellum. (D) Cytokinesis occurs from the anterior end of the trypanosome.

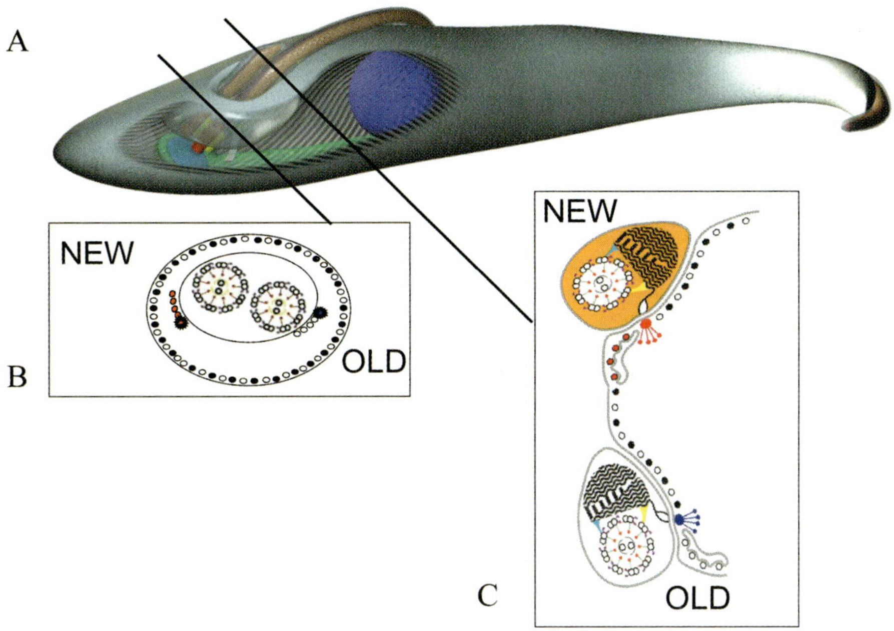

Figure 20.2 (A) A diagram of an early stage in the division of a pro-cyclic *T. brucei* [13] depicting the three-dimensional arrangement of the flagella and flagellar pocket in a cell soon after the initiation of the formation of a new flagellum. The specialized set of four FAZ microtubules (white) are nucleated near the basal bodies. The cartoon shows two theoretical slices through the trypanosome at the level of the flagellar pocket and at a position through the cell body more to the anterior of the cell. The diagrams (B) and (C) illustrate the semi-conservative inheritance of the sub-pellicular microtubules (shown as black/white for old/new) and the formation and conservative inheritance of a completely new flagellum (orange) and completely new FAZ filament and four microtubules (both red).

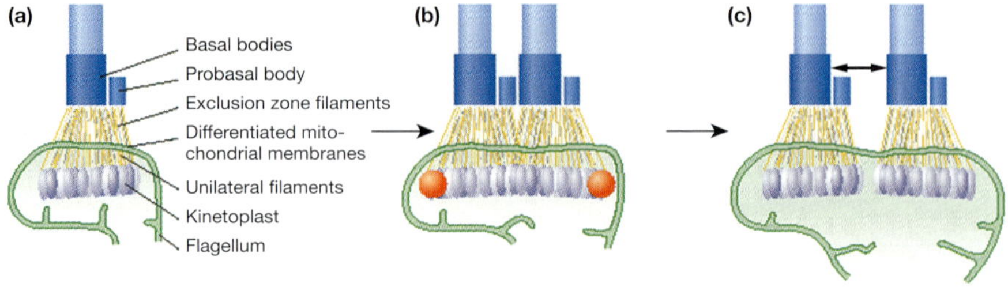

Figure 20.6 A schematic diagram of the tripartite attachment complex in Trypanosomes. Panel (a) illustrates the basal bodies, kinetoplast and the components of the TAC (exclusion zone filaments, differentiated mitochondrial membranes and unilateral filaments) in a trypanosome in G1 of the cell cycle. In this period there is a single flagellum, a basal body and a pro-basal body. Panel (b) shows the organization of the S-phase TAC. When the cell enters S-phase discrete fibrous lobes appear at the poles of the kinetoplast, the pro-basal body matures into a basal body and subtends the new flagellum and two new pro-basal bodies are formed. Two nascent TAC complexes are discernable at this period of the cell cycle. Panel (c) shows the period where movement apart of the flagella basal bodies segregates the replicated kinetoplast DNA. Note that the position and orientation of the basal bodies have been idealized in this two-dimensional diagram (from [58]).

Part I
Microtubule Organization and Dynamics

Centrosomes in Development and Disease. Edited by Erich A. Nigg

ISBN 3-527-30980-2

1
Early Studies on Centrioles and Centrosomes

Joseph G. Gall

1.1
Introduction

From its discovery in 1876 through the early decades of the 20th century, the centriole occupied the attention of many investigators, who showed that it played a key role in organizing fibrillar structures in the cell, including the spindle and asters during mitosis, cytoplasmic filaments in interphase cells, and cilia and flagella in everything from the tracheal epithelium of mammals to the flagellated sperms of lower plants. In a broad sense, the function of the centriole as an organizer or controller of fibrillar structures was established at the outset, long before any hint of its molecular composition or even its fine structure was known. Centriole replication also attracted attention from the very beginning. Because of its regular doubling during the cell cycle in animals, the centriole was first thought to be an autonomous, self-replicating structure with properties similar to those of chromosomes. As evidence against this view began to accumulate, interest in centrioles waned, reaching a low point in the middle of the 20th century. A resurgence of interest accompanied the introduction of electron microscopy into cell biology in the 1950s, which at last revealed the fine structure of the centriole and established even more clearly its relationship to cilia, flagella, and other microtubule-based structures. But little progress was made on the problem of centriole replication, and still nothing was known about its molecular composition. The modern era, encompassing roughly the last 20 years, began with the introduction of specific immunofluorescent staining, and has picked up ever-increasing speed with improvements in microscopy, especially confocal microscopy and the use of GFP-tagged proteins in living cells. Once again, attention is focused on the replication problem, this time backed by the full force of modern molecular and microscopical techniques. In this introductory chapter I will review the history of centriole research during the past 125 years, concentrating heavily on the problem of centriole replication, which has dominated so much of the thinking about this fascinating cell organelle.

Centrosomes in Development and Disease. Edited by Erich A. Nigg

ISBN 3-527-30980-2

1.2 Pioneering Studies

In 1876 Édouard Van Beneden, then a 30-year-old Professor of Zoology at Liège, provided the first tentative glimpse of the centrosome [1][1]. His paper was not about centrosomes or even about mitosis, which had not yet been accurately described. Instead, it concerned the anatomy and development of an obscure group of parasites, the Dicyemidae, which live in the kidneys of squids and octopuses. Van Beneden noticed that something unusual happened to the nucleus during the first few cleavages of the egg. He made drawings of what he saw, the details of which are still difficult to interpret. Nevertheless, at the poles of what we can now recognize as the mitotic spindle he drew a small dot or circle, and labeled it the polar corpuscle (Figure 1.1). Seven years later, while examining fertilization in eggs of the nematode *Ascaris megalocephala* (now called *Parascaris equorum*), he published much more accurate and detailed observations on centrosomes [2]. But again his major interest lay elsewhere, for in this momentous 375-page mono-

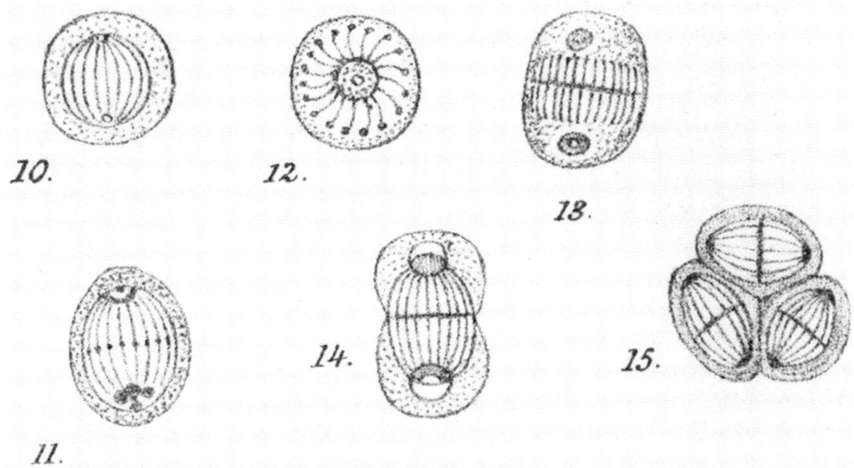

Figure 1.1 Van Beneden's drawings of mitosis during the first cleavage division of the mesozoan *Dicyemella*. At the poles of the spindle Van Beneden drew a small dot or circle, which he called the polar corpuscle. Van Beneden's original paper was published in the *Bulletin of the Royal Belgian Academy* [1]. The figures reproduced here were taken from a reprinting of Van Beneden's plate that appeared in 1877 in the *Quarterly Journal of Microscopical Science*, vol 17 (new series), plate 10. To save space I have rearranged the images so that Figures 10 and 11 are next to Figures 12–15.

[1] It was Boveri, not Van Beneden, who introduced the terms centrosome [4] and centriole [5]. As originally conceived, *centriole* referred to the tiny granule at the very center of the astral configuration, whereas *centrosome* included the larger area of differentiated cytoplasm around it – the pericentriolar material of recent authors. From the beginning, however, the terms were used interchangeably, and only from the context can one be sure what any particular author was referring to.

graph Van Beneden concentrated on the behavior of the chromosomes during meiosis and fertilization, laying the foundation for the chromosome theory of inheritance. In 1887 he published a second and much shorter paper on *Ascaris* in which he dealt with fertilization and the first two cleavage divisions [3]. Here at last he paid careful attention to the centrosomes, showing their now-familiar association with the poles of the mitotic spindle and, even more remarkably, their duplication at telophase and separation during early interphase (Figure 1.2). He speculated that centrosomes, which by then he called central corpuscles, were permanent self-replicating cell organelles that acted as the insertion site for the astral rays.

While his second *Ascaris* paper was being readied for publication, Van Beneden learned that Theodor Boveri at the Zoological Institute in Munich had independently submitted a manuscript on the same subject. Boveri was soon to become one of the dominant cell biologists of his time, but in 1887 he was a young postdoctoral fellow just beginning the studies on *Ascaris* that brought him worldwide acclaim. In the short paper he published in 1887 [6] and in a longer monograph that appeared in 1888 [4], Boveri came to two major conclusions about centrosomes. First, in agreement with Van Beneden, he suggested that the centrosome is a permanent cell organelle endowed with the property of self-replication. Second, he postulated that a major function of the sperm during fertilization was to supply a functional centrosome to the egg – the egg having lost its centrosome during the divisions leading to formation of the polar bodies. Boveri's theory of fertilization provided a backdrop for later studies by other investigators, but was gradually abandoned as the complexities of fertilization became apparent over the next 10–20 years (see the discussion in [7]). On the other hand, the idea that the centrosome is a permanent self-replicating organelle became widely accepted. Indeed, this hypothesis was so attractive on theoretical grounds and so cogently argued by its

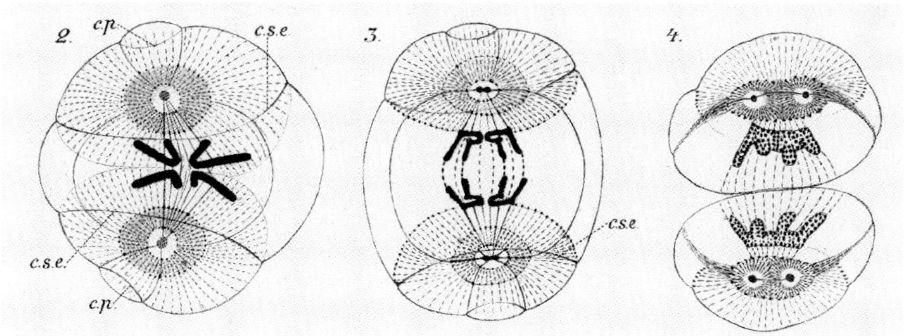

Figure 1.2 Van Beneden's drawings of the first cleavage division in the egg of the nematode worm *Ascaris megalocephala* (now *Parascaris equorum*). Van Beneden clearly depicted doubling of the centriole at anaphase and separation of the two resulting centrioles during early interphase. Presumably, doubling actually occurs prior to metaphase, but Van Beneden could not resolve individual centrioles until anaphase. These figures were taken from Van Beneden and Neyt [3], rearranged to bring Figure 4 next to Figures 2 and 3.

supporters that it persisted until quite recently, in the face of much contrary evidence.

Boveri's drawings of centrioles, chromosomes, and other features of the *Ascaris* egg are among the most elegant products of late 19th and early 20th century cytology. Fortunately, a number of Boveri's original microscope slides are still in existence, permitting us to examine the very material with which he worked (Figure 1.3).

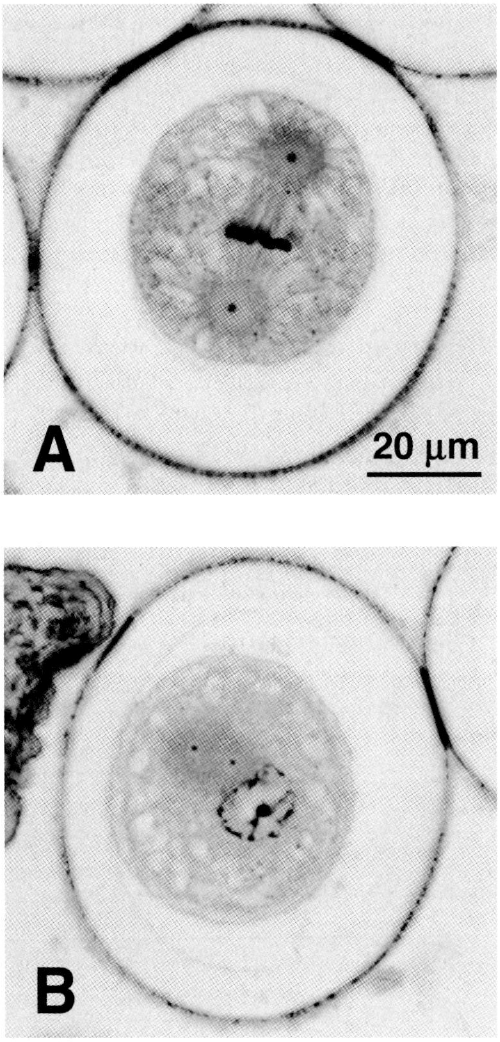

Figure 1.3 Photographs taken by the author from one of Boveri's own slides of *Ascaris* eggs. This slide, which was probably made before 1910, was almost certainly stained with Heidenhain's iron hematoxylin method. (A) Metaphase of the first cleavage division in *Ascaris megalocephala* (now *Parascaris equorum*). Compare with Van Beneden's Figure 2 (Figure 1.2). (B) End of the first cleavage division. Two centrioles lie next to one of the two nuclei formed by the first mitosis. Compare with Van Beneden's Figure 4 (Figure 1.2).

1.3
Self-replication versus *De Novo* Formation

In 1925 E. B. Wilson began his discussion of centrioles and centrosomes [7] with the comment, "...we must admit that there is a certain presumption in favor of the conclusion of Van Beneden, Boveri and their followers that the division-center (centriole) may be regarded as a permanent and autonomous cell-organ that arises only by the division of a preexisting body of the same kind". He went on to marshal the evidence for and against this model, but in the end he was swayed by those cases in which centrioles appear to arise *de novo*, and therefore cannot be permanent and autonomous. Wilson based his argument heavily on the formation of the so-called cytasters that appear in the cytoplasm when the eggs of various marine invertebrates are induced to develop parthenogenetically. These cytasters often contain distinct centrioles, which, of course, cannot have been derived from a sperm centriole. However, it could be imagined that the egg contains an undetected centriole that undergoes rapid division during induction of the cytasters. Wilson argued against this possibility, but it was ultimately the inability to rigorously exclude "undetected" or "invisible" centrioles that plagued all discussions about their possible *de novo* origin.

From an early date it was recognized that mitosis in somatic cells of mosses, ferns, and seed plants differs from the typical condition in animal cells. Specifically, centrioles and asters are missing from the poles of the spindle, which itself has a more rounded configuration than that in animals. It thus became clear that, even if centrioles were self-replicating and permanent organelles in some animals, they were not even present in most plants. However, the situation in plants was complicated by the existence of giant centrosome-like structures (blepharoplasts) at the poles of the spindle during the last divisions of the male gametophyte in mosses, ferns, cycads, and even *Ginkgo*. I will return to these interesting cases shortly, but first I will discuss the relationship between centrioles and the basal bodies of cilia and flagella.

1.4
Centrioles and Basal Bodies

Sperm formation was a favorite topic of investigation during the early years of the 20th century. All of the cell organelles – nucleus, mitochondria, Golgi apparatus, and centrioles – undergo striking changes during the transformation of the round spermatid into the elongated sperm with a single flagellum. It was relatively easy to follow the fate of the centriole from the end of the second meiotic division right through to the mature sperm, and to see that the centriole forms the basal body of the flagellum. Meves [8] even described a remarkable case in the moth *Pygaera* in which flagella grow out from the centrioles while they are still situated at the poles of the first meiotic spindle (Figure 1.4). This case showed conclusively that one and the same structure could function simultaneously as the basal body of

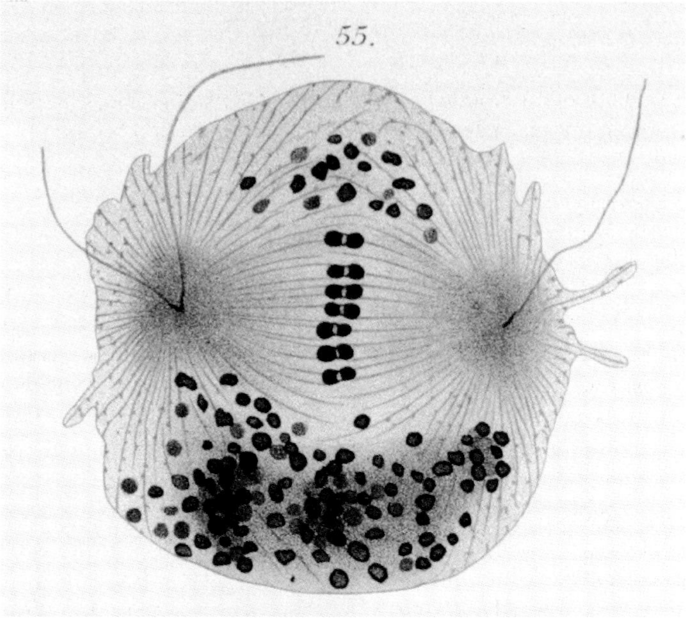

Figure 1.4 Metaphase of the first meiotic division in a primary spermatocyte of the moth *Pygaera*. Note that the centrioles serve simultaneously as the poles of the meiotic spindle and as basal bodies for the precociously formed flagella. Drawing created by Friedrich Meves [8].

a flagellum and as the pole of a spindle. At about the same time Henneguy [9] and Lenhossék [10] proposed that the basal bodies of cilia might also be identical to centrioles. They reached this conclusion from a study of ciliated epithelial cells, which apparently lacked the two centrioles that normally reside next to the nucleus during interphase. They suggested that these two centrioles migrate to the cell surface, where they multiply to form a cluster of centrioles. These multiple centrioles then line up at the surface of the cell and function as the basal bodies for the cilia.

Because centrioles and basal bodies were structures near the limit of resolution of the light microscope, lingering doubts remained about their structural identity. These doubts were completely overcome by electron microscopic observations made in the 1950s, which showed that basal bodies and centrioles are identical to each other in fine structure, and both are related to the axoneme of cilia and flagella [11–16]. The axoneme consists of nine doublet microtubules surrounding two central microtubules, whereas basal bodies and centrioles have nine triplets in their walls but lack the central microtubules.

1.5
Blepharoplasts

Some of the most striking cases of *de novo* centriole formation take place during spermiogenesis in lower plants, including mosses, ferns, fern allies, cycads and *Ginkgo*. All of these plants have flagellated sperms, sometimes with as few as one or two flagella but in many cases with 100 or more (Figure 1.5). The champion in this respect is the cycad sperm – a huge sphere some 250–300 μm in diameter – with up to 25000 flagella arranged along a spiral band. Each flagellum has a basal body, and the question is: where do these basal bodies come from? Light microscopic studies undertaken in the late 19th and early 20th century showed that they arise from the breakdown of a spherical structure, the blepharoplast, which resides in the cytoplasm of the spermatid [17–19] (Figure 1.5). The size of the blepharoplast varies from species to species more or less in proportion to the number of basal bodies that it will produce. Thus the blepharoplast of the cycads is a gigantic structure up to 25 μm in diameter. During the last gametophyte mitosis (which gives rise to the spermatids), two blepharoplasts occupy the poles of the mitotic spindle. These two blepharoplasts are produced by still smaller structures that can be found in the cytoplasm during the preceding interphase. However, most cells of the gametophyte completely lack anything that looks like a blepharoplast or centriole, and so the origin of the blepharoplast itself remains a mystery.

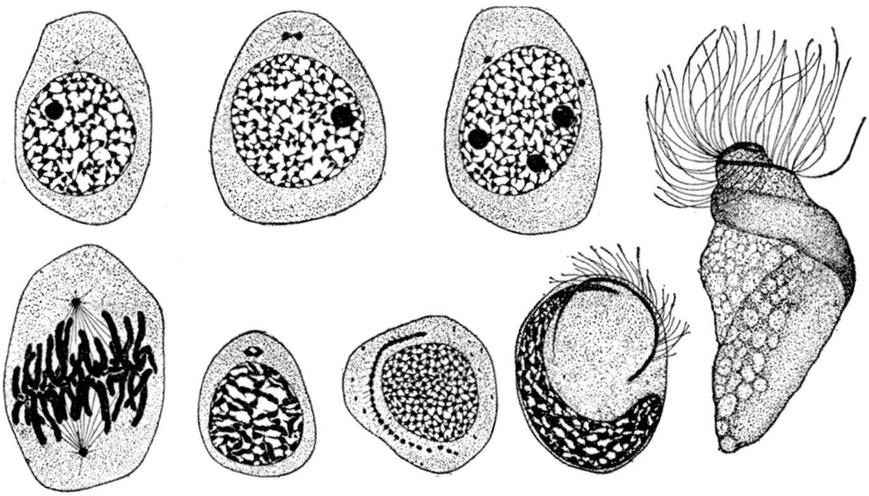

Figure 1.5 Stages in the formation of the flagellated sperm of the horsetail *Equisetum*. A small blepharoplast arises *de novo* in the cytoplasm during the interphase preceding the last gametophyte mitosis. This blepharoplast divides into two blepharoplasts that occupy the poles of last mitotic spindle. In the spermatid the blepharoplast breaks up into a row of centrioles which line up at the cell surface and create the multiple flagella of the sperm. The drawings are by Lester W. Sharp [18]. Sharp's original figures were rearranged to give the sequence shown here. Reproduced from L. Sharp, Introduction to Cytology, 1934, Figure 122, page 205. ©McGraw-Hill.

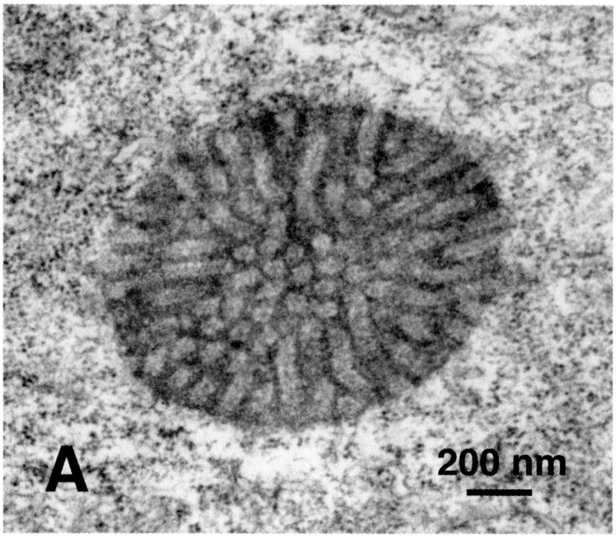

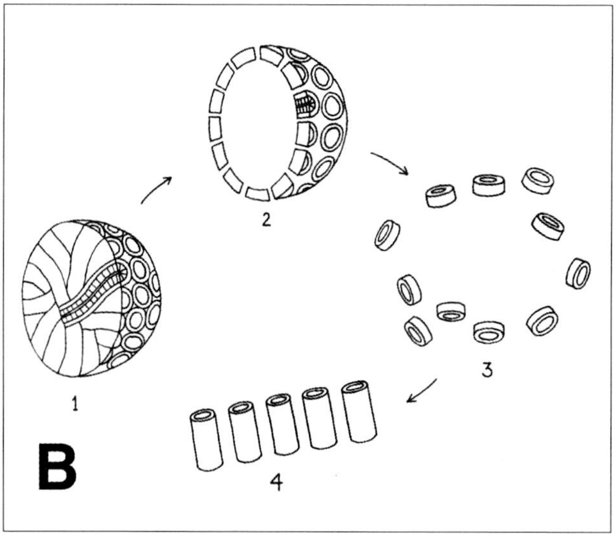

Figure 1.6 Blepharoplasts in the water fern *Marsilea*. (A) Electron micrograph of one of the two blepharoplasts in the cytoplasm of a haploid gametophyte cell. It consists of somewhat irregular tubules that display nine-fold symmetry in cross-section. One blepharoplast will occupy each of the poles of the last mitotic division (see Figure 1.5). It will then transform into a cluster of procentrioles, which elongate and function as the basal bodies for the multiple flagella of the motile sperm. (B) A diagram showing stages in the formation of basal bodies from the blepharoplast. The blepharoplast begins as a spherical mass of somewhat irregular tubules (1), each of which has nine-fold symmetry in cross-section. The blepharoplast eventually resolves into a cluster of procentrioles arranged on the surface of the sphere (2). The procentrioles lose their regular arrangement (3) and elongate to form full-length basal bodies (4). Reproduced from I. Mizukami and J. Gall [20], The Journal of Cell Biology, 1966, 29, Figures 7 and 14, pages 102 and 103. ©The Rockefeller University Press.

EM studies show that the earliest blepharoplasts consist of curious tubular structures, whose nine-fold symmetry in cross-section is clearly related to centrioles (Figure 1.6A). By the time the blepharoplast comes to occupy the pole of the last mitotic division, it has become hollow, and now consists of many short centrioles (procentrioles) covering the surface of a somewhat amorphous sphere [20, 21]. Once released, these procentrioles elongate into full-length centrioles which migrate to the surface of the cell, and function as basal bodies for the flagella (Figure 1.6B).

As in other cases of *de novo* origin of centrioles, the very earliest stages in formation of the blepharoplast remain obscure. Nevertheless, lower plants provide a clear case of *de novo* centriole formation, at least in the limited sense that the blepharoplast arises in the absence of preexisting centrioles.

1.6 The Search for DNA

Undaunted by the evidence for the *de novo* origin of centrioles, or perhaps unaware of it, proponents of self-replication hoped to find DNA in centrioles and basal bodies. In 1924 the German chemist Robert Feulgen introduced the first staining procedure that was specific for DNA in cytological preparations [22]. Feulgen's method involved mild acid hydrolysis of a tissue followed by staining with a decolorized solution of the dye basic fuchsin. It gave a deep magenta color that in most tissues was strictly limited to interphase nuclei and to the chromosomes during mitosis. Just 1 year after Feulgen published his cytochemical test for DNA, Bresslau and Scremin [23] applied the method to several species of the flagellated protozoan *Trypanosoma*, including *T. brucei*, the causative agent of sleeping sickness. They saw not only the expected staining of the nucleus but also a clearly positive reaction in a small body at the base of the flagellum, which was known in the protozoological literature as the kinetoplast. For many years it was assumed that the kinetoplast of *Trypanosoma* was identical to the basal body of other ciliated and flagellated cells. The kinetoplast appeared to provide strong support for the idea that basal bodies contain DNA and hence might be self-replicating structures like chromosomes. The truth turned out to be quite different. When trypanosomes were finally examined with the electron microscope in the 1950s, the kinetoplast was found to be a highly modified mitochondrion closely associated with the basal body of the flagellum, but distinct from it. Kinetoplast DNA is thus mitochondrial DNA – albeit a most unusual type of mitochondrial DNA [24].

Nevertheless, the search for basal body DNA continued. For many years the basal bodies of ciliated protozoa, such as *Tetrahymena* and *Paramecium*, provided fertile ground for speculation. The cilia in these protozoa are not randomly arrayed on the surface of the organism, but instead occur in precisely defined rows and patches, whose patterns provide key taxonomic characters on which individual species are recognized. When these organisms undergo cell division, the ciliary patterns are faithfully replicated in the offspring, starting with replication of the rows of basal bodies. In an influential book published in 1950 Lwoff argued that the basis of the faithful replication of the ciliary rows is self-replication of the basal bodies themselves [25]. This theme was pushed to its logical extreme in a study published in 1965 by Randall and Disbrey [26], who claimed that the basal bodies of *Tetrahymena* contain DNA. They provided two lines of evidence: staining with the fluorescent dye acridine orange and incorporation of [^{3}H]-labeled thymidine. Acridine orange binds to both RNA and DNA in cytological preparations, but its fluorescence is different in the two cases – more orange in combination with RNA and more greenish with DNA. Randall and Disbrey published an image of a *Tetrahymena* pellicle in which the rows of ciliary basal bodies appeared greenish. They stated that the fluorescence was removed by prior treatment with DNase. They also showed an autoradiograph of a pellicle isolated from an animal that had been incubated in [^{3}H]-thymidine. The silver grains in the emulsion appeared in rows roughly corresponding to the ciliary rows of the pellicle, suggesting that the basal bodies had incorporated the DNA precursor during replication. An alternative interpretation of the thymidine experiment was provided by the observations of Miller [27], who noted that rows of mitochondria lie under the ciliary rows and that these mitochondria incorporate a detectable amount of [^{3}H]-thymidine in experiments like those performed by Randall and Disbrey. The color shift with acridine orange staining thus remained the major evidence for DNA in the basal bodies of *Tetrahymena*.

The last major claim for DNA in basal bodies – in the flagellated alga *Chlamydomonas* – was made in 1989 by David Luck and his associates [28]. Their case rested on a combination of genetic data, *in situ* hybridization, and fluorescent staining, this time with the much more sensitive and highly specific DNA dye 4′,6-diamidino-2-phenylindole (DAPI). However, the putative basal body DNA itself was not isolated and characterized. Luck estimated that the amount of DNA in the basal bodies was 6–9 Mb – more than in the *E. coli* genome – an amount that could have been seen earlier, even by the Feulgen reaction. A detailed critique of the DNA data was subsequently published by Johnson and Rosenbaum [29, 30], who could find no evidence for DNA in the *Chlamydomonas* basal body.

Evidence for unique physical or chemical properties of the putative basal body DNA was notably lacking in all of these accounts. The first claims that mitochondria and chloroplasts contain DNA were also based on staining, but these were soon backed up by isolation of the DNA and demonstration of its unique physical properties and sequence specificity (reviewed in [31]). Any credible claim for basal body DNA should have met the same stringent criteria.

1.7 On to Self-assembly

Viewed in historical context, the search for DNA in basal bodies and centrioles was entirely logical. After all, there seemed to be four self-replicating organelles in the cell – chromosomes, mitochondria, chloroplasts, and centrioles – and three of them contained DNA. Moreover, the semi-conservative replication of the DNA molecule provided a stunning explanation for the duplication of chromosomes, and there was every hope that it might do the same for centrioles and basal bodies. In many ways the duplication of centrioles during the mitotic cell cycle (at least in animal cells) mirrors the duplication of chromosomes much more closely than it does the somewhat random multiplication of mitochondria and chloroplasts: chromosomes and centrioles both go from one to two to four copies in a synchronized fashion under tight cell-cycle control, as first seen by Van Beneden and Boveri. Nevertheless, there is now general consensus that centriole duplication differs fundamentally from chromosome duplication, and that the search for centriole DNA, although undoubtedly a necessary step in our understanding, was largely a diversion from the real issues.

What is the fundamental difference in mode of duplication? When a chromosome duplicates, the DNA replicates semi-conservatively, resulting in the formation of two essentially identical daughter chromosomes. However, when a centriole duplicates, the two products are not identical. Instead, the original centriole, the mother, remains intact and a new centriole, the daughter, is formed next to it. At first the daughter, termed a procentriole, is shorter than the mother, but eventually grows to full length. Curiously, the procentriole is oriented perpendicular to the mother and lies at the "old" end of the mother. Like so much else in centriole biology, these fundamental facts have now been demonstrated elegantly with molecular markers, but they were clearly presaged by observations dating back many decades. One of the earliest such observations is shown in Figure 1.7, taken from a study of spermatogenesis in the hagfish *Myxine*, published in 1905 by A. and K. Schreiner [32]. For some reason the centrioles in the hagfish spermatocytes are unusually long; hence, one can see not only two mature centrioles but also two shorter daughters, all four retaining orthogonal orientations with respect to each other. The increased resolution of the electron microscope showed that the daughter procentriole shares the same nine-fold symmetry as the mature centriole and

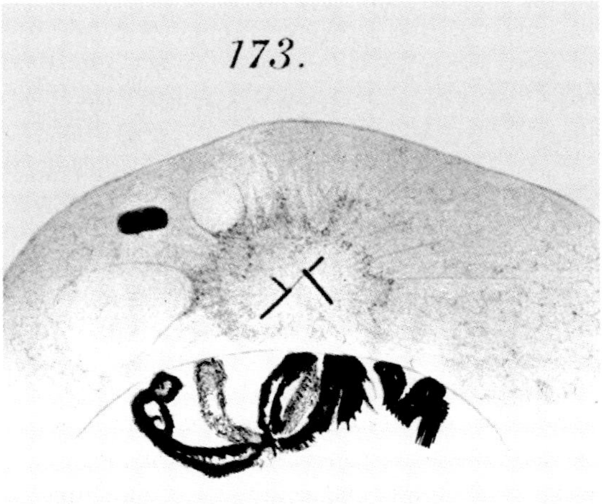

Figure 1.7 Drawing of replicating centrioles in a spermatocyte of the hagfish *Myxine*. Two short daughter centrioles are present at the bases of the two long mother centrioles. All four centrioles retain orthogonal relationships to each other. The drawing is by A. and K. Schreiner [32].

established the topological relationships between the mother and daughter [15, 20, 33].

So what can we learn from this brief history? First and foremost, centrioles and basal bodies are identical structures that *can* arise in the absence of preexisting centrioles or basal bodies. If we want to know how they replicate, we need to know what they are made of and how their components self-assemble. A good start has been made on this problem in recent years, and subsequent chapters in this book will bring us up to date on these issues. Although self-assembly must be at the heart of centriole replication, it is equally clear that old centrioles have a profound influence on the site at which new centrioles arise. The old centriole may simply serve as a place where precursors are concentrated, or it could be more actively involved as a catalyst or template for some steps that might otherwise be extremely slow or rate-limiting. In addition to assembly issues *per se*, there is clear evidence that in many animal cells centriole replication is tightly coordinated with the mitotic cycle – the very feature that so impressed and to a certain extent misled Van Beneden and Boveri. When centriole replication and the cell cycle are not properly coordinated, there are dire consequences for the cell, and this time Boveri correctly predicted that cancerous growth might be one of them [34].

After several ups and downs in popularity over more than a century of study, centrioles and centrosomes have once again moved to center stage in cell biology research. The remaining chapters in this book will summarize where we now stand and where we hope to be in the near future, in understanding these fascinating structures.

References

1. Van Beneden, É. **1875–6**. Recherches sur les Dicyémides, survivant actuels d'un embranchement des Mésozoaires. *Bull. Acad. R. Méd. Belg., 2me sér.* 41:1160–1205; 42:35–97.
2. Van Beneden, É. **1883**. Recherches sur la maturation de l'oeuf, la fécondation et la division cellulaire. *Arch. Biol.* 4:265–640.
3. Van Beneden, É., and A. Neyt. **1887**. Nouvelles recherches sur la fécondation et la division mitosique chez l'Ascaride mégalocéphale. *Bull. Acad. R. Méd. Belg., 3me sér.* 14:215–295.
4. Boveri, T. **1888**. Zellen-Studien II. Die Befruchtung und Teilung des Eies von Ascaris megalocephala. *Jena. Z. Naturwissen.* 22:685–882.
5. Boveri, T. **1895**. Ueber die Befruchtungs- und Entwickelungsfähigkeit kernloser Seeigeleier und über die Möglichkeit ihrer Bastardierung. *Arch. Entwicklungsmech. Org. (Wilhelm Roux)*. 2:394–443.
6. Boveri, T. **1887**. Ueber die Befruchtung der Eier von Ascaris megalocephala. *Sitz.-Ber. Ges. Morph. Phys. München.* 3.
7. Wilson, E. B. **1925**. *The Cell in Development and Heredity.* Macmillan, New York.
8. Meves, F. **1900**. Ueber den von la Valette St. George entdeckten Nebenkern (Mitochondrienkörper) der Samenzellen. *Arch. mikr. Anat.* 56:553–606.
9. Henneguy, L. F. **1897**. Sur les rapports des cils vibratiles avec les centrosomes. *Arch. Anat. Microsc.* 1:481–496.
10. Lenhossék, M. v. **1898**. Ueber Flimmerzellen. *Verh. Anat. Ges.* 12:106.

11. Fawcett, D. W., and K. R. Porter. **1954**. A study of the fine structure of ciliated epithelia. *J. Morphol.* 94:221–282.

12. Burgos, M. H., and D. W. Fawcett. **1955**. Studies on the fine structure of mammalian testis. I. Differentiation of the spermatids in the cat (*Felis domestica*). *J. Biophys. Biochem. Cytol.* 1:287–300.

13. de Harven, E., and W. Bernhard. **1956**. Étude au microscope électronique de l'ultrastructure du centriole chez les vertébrés. *Z. Zellforsch.* 45:378–398.

14. Bessis, M., J. Breton-Gorius, and J.-P. Thiéry. **1958**. Centriole, corps de Golgi et aster des leucocytes. Étude au microscope électronique. *Rev. Hematol.* 13:363–370.

15. Bernhard, W., and E. de Harven. **1960**. L'ultrastructure du centriole et d'autres éléments de l'appareil achromatique. *Fourth Internat. Conf. Electron Microscopy.* 2:217–227.

16. Gibbons, I. R., and A. V. Grimstone. **1960**. On flagellar structure in certain flagellates. *J. Biophys. Biochem. Cytol.* 7:697–716.

17. Webber, H. J. **1897**. Notes on the fecundation of *Zamia* and the pollen tube apparatus of *Ginkgo*. *Bot. Gaz.* 24:225–235.

18. Sharp, L. W. **1912**. Spermatogenesis in Equisetum. *Bot. Gaz.* 54:89–119.

19. Sharp, L. W. **1914**. Spermatogenesis in Marsilia. *Bot. Gaz.* 58:419–431.

20. Mizukami, I., and J. Gall, G. **1966**. Centriole replication II. Sperm formation in the fern, *Marsilea*, and the cycad, *Zamia*. *J. Cell Biol.* 29:97–111.

21. Hepler, P. K. **1976**. The blepharoplast of *Marsilea*: its *de novo* formation and spindle association. *J. Cell Sci.* 21:361–390.

22. Feulgen, R., and H. Rossenbeck. **1924**. Mikroskopisch-chemischer Nachweis einer Nucleinsäure vom Typus der Thymonucleinsäure und die darauf beruhende elektive Färbung von Zellkernen in mikroskopischen Präparaten. *Hoppe-Seyler's Z. physiol. Chem.* 135:203–248.

23. Bresslau, E., and L. Scremin. **1924**. Die Kerne der Trypanosomen und ihr Verhalten zur Nuclealreaktion. *Arch. Protist.* 48:509–516.

24. Shapiro, T. A., and P. T. Englund. **1995**. The structure and replication of kinetoplast DNA. *Annu. Rev. Microbiol.* 49:117–143.

25. Lwoff, A. 1950. *Problems of Morphogenesis in Ciliates.* Wiley and Sons, New York.

26. Randall, J., and C. Disbrey. **1965**. Evidence for the presence of DNA at basal body sites in *Tetrahymena pyriformis*. *Proc. R. Soc. Lond. B. Biol. Sci.* 162:473–491.

27. Miller, O. L. **1965**. A stable mitochondrial DNA in *Tetrahymena pyriformis*. *J. Exp. Zool.* 159:33–37.

28. Hall, J. L., Z. Ramanis, and D. J. L. Luck. **1989**. Basal body/centriolar DNA: molecular genetic studies in *Chlamydomonas*. *Cell* 59:121–132.

29. Johnson, K. A., and J. L. Rosenbaum. **1990**. The basal bodies of *Chlamydomonas reinhardtii* do not contain immunologically detectable DNA. *Cell* 62:615–619.

30. Johnson, K. A., and J. L. Rosenbaum. **1991**. Basal bodies and DNA. *Trends Cell Biol.* 1:145–149.

31. Gillham, N. W. **1994**. *Organelle Genes and Genomes.* Oxford University Press, New York, Oxford.

32. Schreiner, A., and K. Schreiner. **1905**. Ueber die Entwicklung der männlichen Geschlechtszellen von *Myxine glutinosa* (L). *Arch. Biol.* 21:315.

33. Gall, J. G. **1961**. Centriole replication. A study of spermatogenesis in the snail *Viviparus*. *J. Biophys. Biochem. Cytol.* 10:163–193.

34. Boveri, T. **1914**. Zur Frage der Entstehung maligner Tumoren. Gustav Fischer, Jena.

2
The Tubulin Superfamily

Tim Stearns

2.1
History

The properties of the microtubule cytoskeleton are a combination of the properties of microtubules themselves and the proteins that temporally and spatially control microtubule growth. Tubulin, the subunit of microtubules, was identified in the 1960s [1], and the genes corresponding to the two peptide components, α-tubulin and β-tubulin, which form a heterodimer of tubulin, were cloned in the 1970s [2]. By the 1980s it was clear that α-tubulin and β-tubulin were conserved from human to yeast [3, 4], and a large amount of work went into the study of tubulin and microtubule structure and function. This led to the discovery of the remarkable polymerization dynamics of tubulin [5], and ultimately to the determination of the three-dimensional structure of the α/β-tubulin heterodimer [6].

Although α-tubulin and β-tubulin are the major components of the microtubule cytoskeleton, it now clear that there is a tubulin superfamily and that all of the members of this superfamily play roles in microtubule cytoskeleton function. γ-Tubulin (gamma) was discovered in *Aspergillus nidulans*, a filamentous fungus, as a suppressor of a mutant β-tubulin [7]. Like α- and β-tubulin, γ-tubulin is conserved in all eukaryotes; more recently identified members of the tubulin superfamily appear to be more limited in distribution, and more specialized in their function. δ-Tubulin (delta) was discovered in *Chlamydomonas reinhardtii*, a motile green alga, as a flagellar assembly mutant [8]. ε-Tubulin (epsilon) and ζ-tubulin (zeta) were identified in database searches for tubulin homologs in mammals and in parasitic organisms, respectively [9, 10]. η-Tubulin (eta) was identified in *Paramecium tetraurelia*, a ciliated protozoan, as a basal body duplication mutant [11]. *Paramecium* may harbor the most extended tubulin superfamily of all, with at least two more uncharacterized tubulins in Genbank (AJ427480 and AJ427481, Dupuis-Williams, unpublished data). I will not discuss the bacterial tubulin ortholog, ftsZ, which has been reviewed recently [12].

Centrosomes in Development and Disease. Edited by Erich A. Nigg

ISBN 3-527-30980-2

2.2 Family Relations

The tubulin superfamily members differ both in how widely they are found, and in how well-conserved they are in the groups in which they are found. The α-, β- and γ-tubulin families are found in all eukaryotes examined. However, in all three families there are examples of organisms having multiple genes for each family. This is particularly common for the α-tubulin and β-tubulin families in higher eukaryotes, in which these tubulins are among the most abundant proteins in cells. In most cases the level of identity among family members within a species is greater than 90 %, thus identification as an α- β- or γ-tubulin is usually unambiguous. δ-Tubulin and ε-tubulin are less widely distributed, found in vertebrates, chordates, and many single-celled organisms, but not in worms, flies, or fungi. η-Tubulin and ζ-tubulin are even more narrowly distributed, being described only in single-celled organisms, excluding fungi. Since there is little available genome information for ciliates and eukaryotic parasites, it is not yet clear how widely distributed η- and ζ-tubulin are even within those groups.

Within a species, each of the tubulin families shares about 25–35 % identity with the other families. A phylogenetic tree of the human tubulin superfamily is shown in Figure 2.1A as an example of this relationship. In such trees, the lengths of the horizontal lines are a measure of similarity so that proteins separated by shorter lines are more similar. It is apparent that α- and β-tubulin are the most similar, but also that none of the tubulin families is grossly divergent. This level of similarity is consistent with all of the tubulins being similarly shaped proteins, and efforts have been made to use the α-/β-tubulin structure to predict the structure of γ-tubulin, δ-tubulin, and ε-tubulin [13].

Comparing family members between species reveals that the level of conservation is dramatically different between the families. Phylogenetic trees of α-, γ-, and δ-tubulin, from human, frog, *Chlamydomonas*, and trypanosome are shown in Figure 2.1B. Trees for β-tubulin are similar to those for α-tubulin, and trees for ε-tubulin are similar to those for δ-tubulin, and are not shown. These trees show that α-tubulin proteins are the most conserved and δ-tubulins the least con-

Figure 2.1 Rooted trees of tubulin superfamily comparisons. The length of the horizontal lines is proportional to the degree of similarity, and is indicated by the numbers on each line. Vertical lines are not relevant. Sequences were aligned with ClustalW and trees were plotted with njplot. (A) Comparison of human tubulin superfamily members. (B) Comparisons of α- (top), γ- (middle) and δ-tubulins (bottom) from human, frog (*Xenopus laevis*), *Chlamydomonas*, and trypanosome (*Trypanosoma brucei*). Sequences used in Figures 2.1 and 2.2:
hum_alpha NP_005992; hum_beta AAH05838; hum_gamma NP_001061;
hum_delta AAF09584; hum_epsilon AAF09585.
xen_alpha CAA30093; xen_beta S05968; xen_gamma UBXLG; xen_delta AAL27450;
xen_epsilon AAN77278.
chlamy_alpha AAN87017; chlamy_beta UBKM; chlamy_gamma AAB71841; chlamy_delta T07903;
chlamy_epsilon AAM23012.
tryp_alpha AAA30262; tryp_beta P04107; tryp_gamma CAA68866; tryp_delta AAF32301;
tryp_epsilon AAF32302.

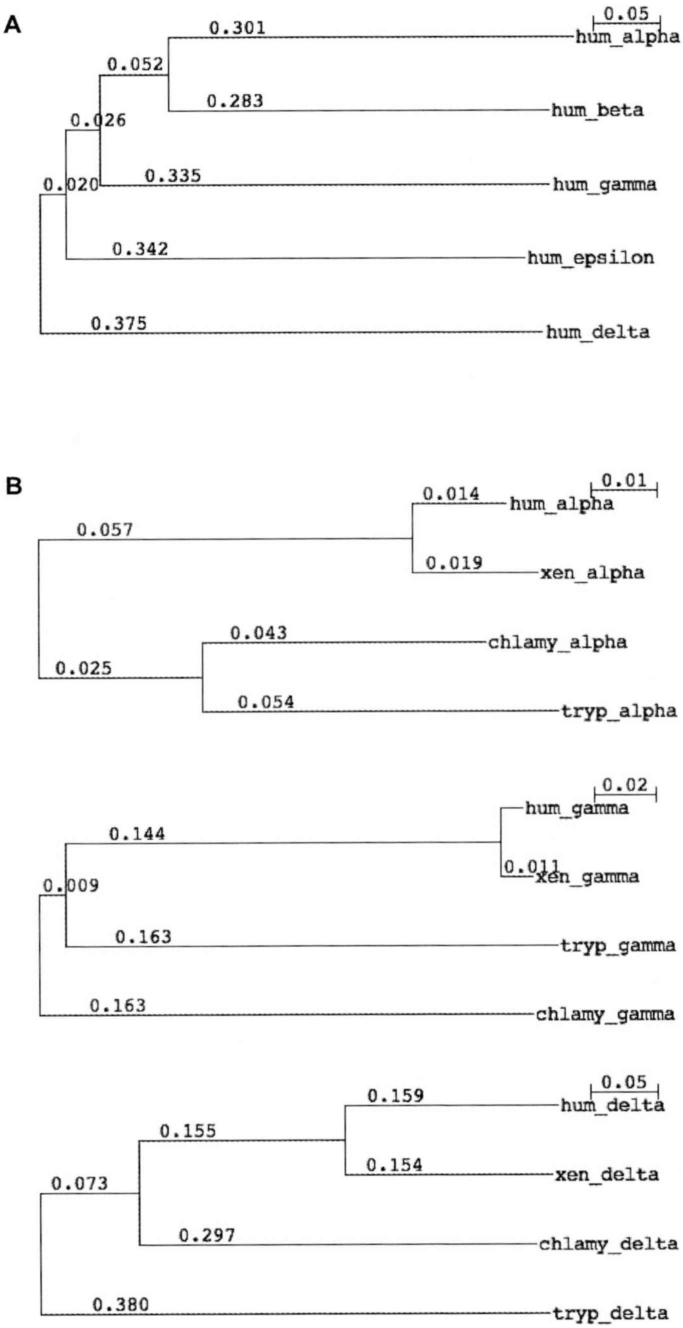
A
0.05
0.301
hum_alpha
0.052
0.283
hum_beta
0.026
0.020
0.335
hum_gamma
0.342
hum_epsilon
0.375
hum_delta
B
0.01
0.014
hum_alpha
0.057
0.019
xen_alpha
0.043
chlamy_alpha
0.025
0.054
tryp_alpha
0.02
hum_gamma
0.144
0.011
xen_gamma
0.009
0.163
tryp_gamma
0.163
chlamy_gamma
0.05
0.159
hum_delta
0.155
0.154
xen_delta
0.073
0.297
chlamy_delta
0.380
tryp_delta

served, with γ-tubulins in between. An unrooted tree that includes all of the sequences from these four organisms (except ζ-tubulin from trypanosome) makes this point most clearly (Figure 2.2). As examples of the most widely divergent proteins, human and trypanosome α-tubulins are 84% identical, human and trypansome γ-tubulins are 64% identical, and human and trypanosome δ-tubulins are only 28% identical. The δ-tubulin comparison reveals a general problem: the human and trypanosome sequences are within the range of homology that exists between different tubulin families, so how can one be sure that the trypanosome sequence is a δ-tubulin, and not a member of another tubulin family? There are some identifying sequence elements of the individual families [9, 13], but this

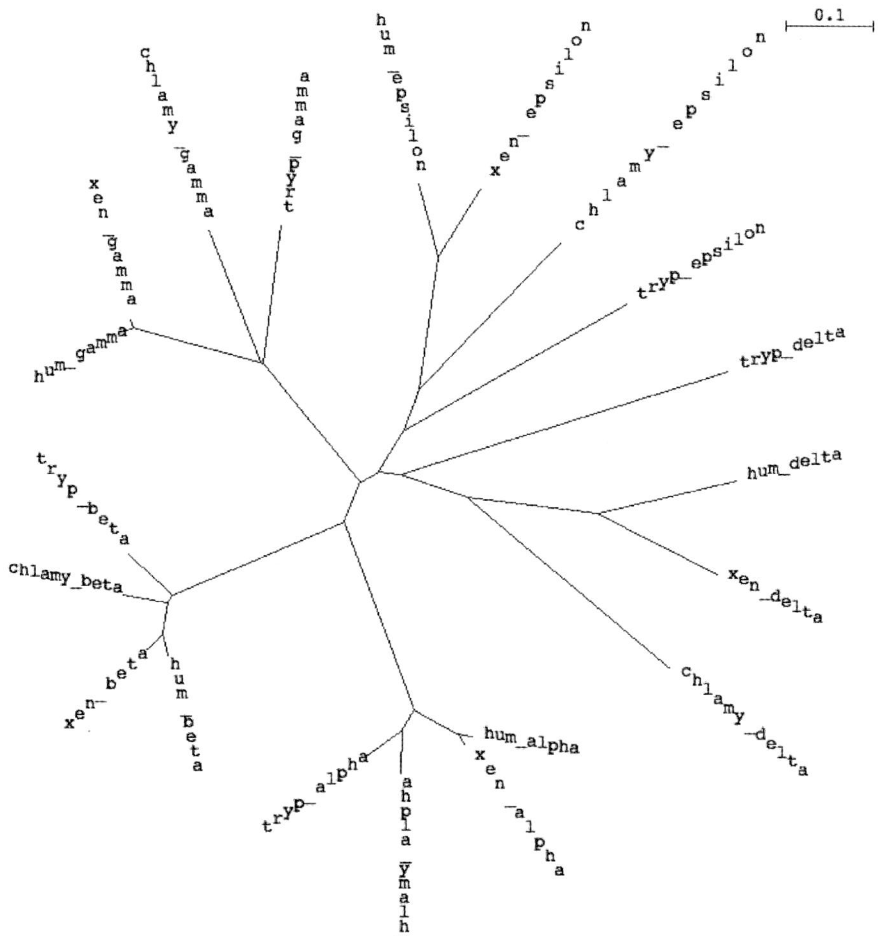

Figure 2.2 Unrooted tree derived from an alignment of all of the tubulin sequences listed in Figure 2.1. Path length between proteins is proportional to similarity. Sequences were aligned as in Figure 2.1, and the tree was plotted with unrooted.

issue will only be definitively resolved with more work on the proteins so that identity can be based on more than sequence alone.

2.3
Localization and Function

The number of tubulin superfamily members is surprising to those who have long toiled on the microtubule cytoskeleton – where were they all those years, where are they in the cell, and what do they do? The remarkable answer is that most or all of the new tubulin superfamily members are involved in the function of the microtubule organizing center. This information is summarized in Table 2.1.

Table 2.1 Summary of the tubulin superfamily.

Name	*Localization*	*Loss of function phenotype*	*Where found in eukaryotes*
α-Tubulin	Microtubules centrioles flagella	No microtubules essential	All
β-Tubulin	Microtubules, centrioles, flagella	No microtubules essential	All
γ-Tubulin	Centrosome	Defect in microtubule nucleation	All
δ-Tubulin	Centrioles	Defect in centriole microtubules, associated structures	Vertebrates, chordates, ciliated single-celled
ε-Tubulin	Sub-distal appendage of centrioles	Defect in centriole duplication, centriole microtubules	Vertebrates, chordates, ciliated single-celled
η-Tubulin	Not known	Defect in centriole microtubules	*Paramecium*
ζ-Tubulin	Not known	Not known	*Trypanosoma, Leishmania*

2.4
γ-Tubulin

Since its initial discovery through fungal genetics, γ-tubulin has been found in all eukaryotic groups, and, in organisms in which the experiment can be undertaken, is found to be essential for viability. Many eukaryotic groups, including fungi, higher plants and at least some invertebrate animals, only have α-, β-, and γ-tubulin, thus these define the necessary and sufficient set for constructing a basic microtubule cytoskeleton. γ-Tubulin is found in all species in a complex with at least two members of a related family of proteins called GCPs (Gamma Complex Protein) [14] or Grips (Gamma Ring Protein) [15]. The founding members of this family were Spc97p and Spc98p [16, 17] from the budding yeast *Saccharomyces cerevisiae*. Interestingly, budding yeast only has those two proteins whereas most organisms have at least five GCPs [18], and possibly one other type of subunit [19].

γ-Tubulin was shown to localize the microtubule organizing center in all cell types examined, which led to the idea that it might be involved in microtubule nucleation (reviewed in [20]). But the breakthrough for understanding γ-tubulin function came when the γ-tubulin complex was purified and found to have both a ring shape and a subunit structure similar to the minus ends of microtubules, and to be able to nucleate microtubule polymerization *in vitro* [21]. It is now widely accepted that this γ-tubulin ring complex is responsible, at least in part, for nucleating microtubule polymerization from the organizing center (see [22] for recent experiments), and that it does so by acting as a template for microtubule polymerization (see Chapter 3).

2.5
δ-Tubulin

Dutcher and Trabuco [8] showed that the *UNI3* gene in *Chlamydomonas* encoded a new tubulin, which they called δ-tubulin. *uni3-1* mutant cells have aberrant basal bodies, with doublet rather than triplet microtubles, and have defects in flagellar number. Detailed studies show that other basal body-associated structures are also defective in this mutant [23]. This phenotype is strikingly similar to that seen in *Paramecium* by silencing δ-tubulin [24], suggesting that δ-tubulin plays a role in the formation of the outer C tubule of the basal body. Interestingly, Dutcher and colleagues have shown that a mutation in α-tubulin can completely suppress defects in the flagellar number of *uni3-1* mutants, and partially suppress defects of the basal body structure [25].

δ-Tubulin was identified in mammals by database searching [9, 26]. The two groups that identified δ-tubulin in mammals differed on its localization in cells. Chang and Stearns [9] reported localization to the centrioles of the centrosome, whereas Smrzka et al. [26] reported localization to microtubule-associated structures in developing sperm, but not to the somatic centrosome. Given the phenotypes associated with loss of δ-tubulin function, and a preliminary report of δ-tubulin localization to the centrioles in *Chlamydomonas* [27], it seems likely that the centriolar localization in mammalian cells is correct. No functional experiments in vertebrate cells have been reported yet, so the role of δ-tubulin in cell types that do not make motile cilia is not yet clear.

2.6
ε-Tubulin

ε-Tubulin was discovered in the human genome database as a “beta tubulin-like” cDNA [9]. Like δ-tubulin, it has since been identified in other vertebrates, and single-celled organisms with cilia or flagella, but not in invertebrates, fungi or higher plants. In animal cells ε-tubulin localizes to the sub-distal appendages of the mature centriole [28]. The sub-distal appendages are only assembled on the newer of the two original centrioles at the G2/M transition, so ε-tubulin is asymmetric with respect to the centrioles prior to G2/M. This staining pattern is similar to that reported for ninein [29], and an increasing list of proteins (for example, see [30]). Experiments in which ε-tubulin was depleted from a frog egg extract centrosome duplication assay showed that it is required for centriole formation, but not for microtubule nucleation or mitotic aster formation [28].

Dutcher et al. [31] showed that the product of the *BLD2* gene is the *Chlamydomonas* ortholog of ε-tubulin, that ε-tubulin localized to basal bodies, and that the mutant was defective for forming the doublet and triplet microtubules of the basal bodies. Similar experiments with gene silencing of ε-tubulin in *Paramecium* by Dupuis-Williams et al. [32] also showed defects in basal body microtubule formation.

2.7 Other Members of the Fold

Again illustrating the power of *Paramecium* for identifying components required for centriole/basal body function, η-tubulin was identified as the product of the *SM19* gene [11]. The *sm19-1* mutant blocks basal body duplication and results in delocalization of γ-tubulin. In parallel with the case for *uni3-1* in *Chlamyomonas,* mutations in *SM19* can be suppressed by mutations in β-tubulin [33]. Although there are many possible explanations for this result, it does raise the possibility that η-tubulin interacts with β-tubulin during the formation of basal bodies. It has not yet been possible to determine how conserved this role might be because η-tubulin has not yet been found in other organisms. Database searches of *Trypanosoma* and *Leishmania* genome sequence yielded another unique tubulin superfamily member, ζ-tubulin [10], about which there is no functional information. Lastly, ESTs from *Xenopus laevis* that were originally cited as being representatives of the η-tubulin and ζ-tubulin families [27] actually belong to a single cDNA that appears to encode a tubulin unlike any other (Genbank accession AAN52369, Keith Gull, unpublished data). There are several ESTs corresponding to this cDNA, so it is likely that it is expressed.

In summary, the tubulin superfamily consists of the subunits that make up the microtubules, but also closely related proteins that are all likely to be involved in the assembly and function of the centrosome, and of the microtubules of the centriole/basal body. It seems likely that the reason that large swaths of eukaryotic groups do not have some of these family members is that they either lack basal bodies and flagella, or have variations on the conserved theme. Likely all of the conserved tubulin families have already been identified in the available genome sequences, but there is still much room for "black sheep" to have developed in organisms that specialize in the ciliated lifestyle.

References

1. Weisenberg, R. C., Borisy, G. G., and Taylor, E. W. **1968**. The colchicine-binding protein of mammalian brain and its relation to microtubules. Biochemistry *7*, 4466–4479.
2. Cleveland, D. W., Kirschner, M. W., and Cowan, N. J. **1978**. Isolation of separate mRNAs for alpha- and beta-tubulin and characterization of the corresponding *in vitro* translation products. Cell *15*, 1021–1031.
3. Neff, N. F., Thomas, J. H., Grisafi, P., and Botstein, D. **1983**. Isolation of the beta-tubulin gene from yeast and demonstration of its essential function *in vivo*. Cell *33*, 211–219.
4. Schatz, P. J., Pillus, L., Grisafi, P., Solomon, F., and Botstein, D. **1986**. Two functional alpha-tubulin genes of the yeast *Saccharomyces cerevisiae* encode divergent proteins. Mol Cell Biol *6*, 3711–3721.
5. Mitchison, T., and Kirschner, M. **1984**. Dynamic instability of microtubule growth. Nature *312*, 237–242.
6. Nogales, E., Wolf, S. G., and Downing, K. H. **1998**. Structure of the alpha beta tubulin dimer by electron crystallography. Nature *391*, 199–203.
7. Oakley, C. E., and Oakley, B. R. **1989**. Identification of gamma-tubulin, a new member of the tubulin superfamily encoded by mipA gene of *Aspergillus nidulans*. Nature *338*, 662–664.
8. Dutcher, S. K., and Trabuco, E. C. **1998**. The *UNI3* gene is required for assembly of basal bodies of Chlamydomonas and encodes delta-tubulin, a new member of the tubulin superfamily. Mol Biol Cell *9*, 1293–1308.
9. Chang, P., and Stearns, T. **2000**. Delta-tubulin and epsilon-tubulin: two new human centrosomal tubulins reveal new aspects of centrosome structure and function. Nature Cell Biol *2*, 30–35.
10. Vaughan, S., Attwood, T., Navarro, M., Scott, V., McKean, P., and Gull, K. **2000**. New tubulins in protozoal parasites. Curr Biol *10*, R258–R259.
11. Ruiz, F., Krzywicka, A., Klotz, C., Keller, A., Cohen, J., Koll, F., Balavoine, G., and Beisson, J. **2000**. The *SM19* gene, required for duplication of basal bodies in Paramecium, encodes a novel tubulin, eta-tubulin. Curr Biol *10*, 1451–1454.
12. Romberg, L., and Levin, P. A. **2003**. Assembly dynamics of the bacterial cell division protein FTSZ: poised at the edge of stability. Annu Rev Microbiol *57*, 125–154.
13. Inclan, Y. F., and Nogales, E. **2001**. Structural models for the self-assembly and microtubule interactions of gamma-, delta- and epsilon-tubulin. J Cell Sci *114*, 413–422.
14. Murphy, S. M., Urbani, L., and Stearns, T. **1998**. The mammalian gamma-tubulin complex contains homologues of the yeast spindle pole body components spc97p and spc98p. J Cell Biol *141*, 663–674.
15. Gunawardane, R. N., Martin, O. C., Cao, K., Zhang, L., Dej, K., Iwamatsu, A., and Zheng, Y. **2000**. Characterization and reconstitution of Drosophila gamma-tubulin ring complex subunits. J Cell Biol *151*, 1513–1524.
16. Geissler, S., Pereira, G., Spang, A., Knop, M., Soues, S., Kilmartin, J., and Schiebel, E. **1996**. The spindle pole body component Spc98p interacts with the gamma-tubulin-like Tub4p of *Saccharomyces cerevisiae* at the sites of microtubule attachment. Embo J *15*, 3899–3911.
17. Knop, M., Pereira, G., Geissler, S., Grein, K., and Schiebel, E. **1997**. The spindle pole body component Spc97p interacts with the gamma-tubulin of *Saccharomyces cerevisiae* and functions in microtubule organization and spindle pole body duplication. Embo J *16*, 1550–1564.
18. Murphy, S. M., Preble, A. M., Patel, U. K., O'Connell, K. L., Dias, D. P., Moritz, M., Agard, D., Stults, J. T., and Stearns, T. **2001**. GCP5 and GCP6: two new members of the human gamma-

tubulin complex. Mol Biol Cell *12*, 3340–3352.

19. Gunawardane, R. N., Martin, O. C., and Zheng, Y. (2003). Characterization of a new gammaTuRC subunit with WD repeats. Mol Biol Cell *14*, 1017–1026.

20. Job, D., Valiron, O., and Oakley, B. **2003**. Microtubule nucleation. Curr Opin Cell Biol *15*, 111–117.

21. Zheng, Y., Wong, M. L., Alberts, B., and Mitchison, T. **1995**. Nucleation of microtubule assembly by a gamma-tubulin-containing ring complex. Nature *378*, 578–583.

22. Hannak, E., Oegema, K., Kirkham, M., Gonczy, P., Habermann, B., and Hyman, A. A. **2002**. The kinetically dominant assembly pathway for centrosomal asters in *Caenorhabditis elegans* is gamma-tubulin dependent. J Cell Biol *157*, 591–602.

23. O'Toole, E. T., Giddings, T. H., McIntosh, J. R., and Dutcher, S. K. **2003**. Three-dimensional organization of basal bodies from wild-type and delta-tubulin deletion strains of *Chlamydomonas reinhardtii*. Mol Biol Cell *14*, 2999–3012.

24. Garreau de Loubresse, N., Ruiz, F., Beisson, J., and Klotz, C. **2001**. Role of delta-tubulin and the C-tubule in assembly of Paramecium basal bodies. BMC Cell Biol *2*, 4.

25. Fromherz, S., Giddings, T. H., Jr., Gomez-Ospina, N., and Dutcher, S. K. **2004**. Mutations in alpha-tubulin promote basal body maturation and flagellar assembly in the absence of delta-tubulin. J Cell Sci *117*, 303–314.

26. Smrzka, O. W., Delgehyr, N., and Bornens, M. **2000**. Tissue-specific expression and subcellular localisation of mammalian delta-tubulin. Curr Biol *10*, 413–416.

27. Dutcher, S. K. **2001**. The tubulin fraternity: alpha to eta. Curr Opin Cell Biol *13*, 49–54.

28. Chang, P., Giddings, T. H., Jr., Winey, M., and Stearns, T. **2003**. Epsilon-tubulin is required for centriole duplication and microtubule organization. Nature Cell Biol *5*, 71–76.

29. Mogensen, M. M., Malik, A., Piel, M., Bouckson-Castaing, V., and Bornens, M. **2000**. Microtubule minus-end anchorage at centrosomal and non-centrosomal sites: the role of ninein. J Cell Sci *113* (Pt 17), 3013–3023.

30. Louie, R. K., Bahmanyar, S., Siemers, K. A., Votin, V., Chang, P., Stearns, T., Nelson, W. J., and Barth, A. I. **2004**. Adenomatous polyposis coli and EB1 localize in close proximity of the mother centriole and EB1 is a functional component of centrosomes. J Cell Sci *117*, 1117–1128.

31. Dutcher, S. K., Morrissette, N. S., Preble, A. M., Rackley, C., and Stanga, J. **2002**. Epsilon-tubulin is an essential component of the centriole. Mol Biol Cell *13*, 3859–3869.

32. Dupuis-Williams, P., Fleury-Aubusson, A., de Loubresse, N. G., Geoffroy, H., Vayssie, L., Galvani, A., Espigat, A., and Rossier, J. **2002**. Functional role of epsilon-tubulin in the assembly of the centriolar microtubule scaffold. J Cell Biol *158*, 1183–1193.

33. Ruiz, F., Dupuis-Williams, P., Klotz, C., Forquignon, F., Bergdoll, M., Beisson, J., and Koll, F. **2004**. Genetic evidence for interaction between eta- and beta-tubulins. Eukaryot Cell *3*, 212–220.

3
Microtubule Nucleation

Michelle Moritz, Luke M. Rice and David A. Agard

3.1
Introduction

The microtubule cytoskeleton provides a vital framework for the polarization of cells, the movement of vesicle traffic, and the segregation of chromosomes. Microtubules are highly dynamic, yet rigid, 25-nm diameter cylindrical polymers of α/β-tubulin that can extend for tens of microns inside the cell, and for millimeters *in vitro*. The assembly of microtubules begins with a thermodynamically unfavorable process in which small oligomers of α/β-tubulin heterodimers form. Rapid polymerization of the microtubule does not occur until a stable oligomer of a certain size is reached. This rate-limiting, stable oligomer is termed the nucleus, and its size varies with buffer conditions, ionic strength, pH and the presence of accessory proteins that facilitate nucleation. The nucleation-elongation behavior exhibited during microtubule assembly is therefore characterized by three phases: a slow phase in which nuclei form, followed by a rapid growth phase in which α/β-tubulin heterodimers elongate the polymer, and finally by a steady-state phase in which most of the α/β-tubulin is incorporated into polymers (Figure 3.1; reviewed in [1–3]). In addition to nucleation-elongation behavior, a microtubule assembly displays a striking phenomenon termed dynamic instability. This involves random switching between polymerization and depolymerization of some microtubules, the subunits of which are then incorporated into others. This results in a smaller number of longer microtubules at steady-state (reviewed in [1] and [4]).

Centrosomes in Development and Disease. Edited by Erich A. Nigg

ISBN 3-527-30980-2

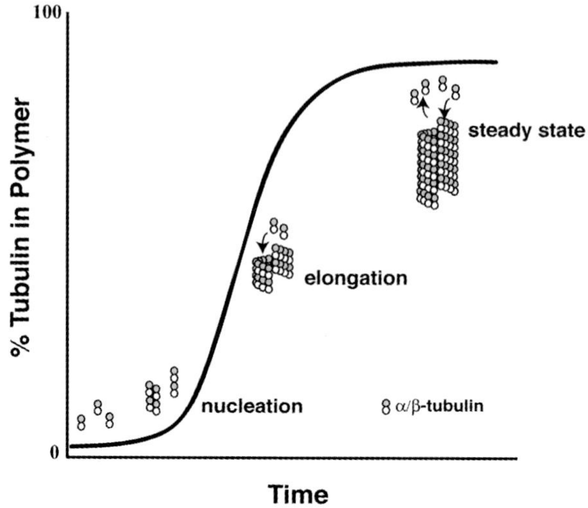

Figure 3.1 The three phases of microtubule assembly. First, α/β-tubulin heterodimers assemble slowly into small oligomers in a thermodynamically unfavorable process called nucleation, which gives rise to the characteristic lag seen in microtubule growth. Rapid elongation of microtubules occurs once rate-limiting oligomers, called nuclei, have formed. When most of the tubulin is incorporated into polymer, steady state is reached. (Figure adapted from [69]).

3.1.1
The Nucleation of Microtubules can occur Spontaneously *In Vitro*, but Requires γ-Tubulin *In Vivo*

The nucleation and polymerization of microtubules can occur spontaneously *in vitro* from pure tubulin in the presence of Mg^{2+} and GTP. However, *in vivo*, the process usually involves microtubule-organizing centers such as the centrosome of animal cells and the spindle pole body of yeasts, which result in a radial array of microtubules that are focused at their minus (slow growing) ends, with their fast-growing plus ends oriented outward (reviewed in [5]). Certain cell types employ other modes of producing microtubules. For example, in some animal cells containing centrosomes, non-centrosome-associated microtubules sometimes form [6]. In plants, microtubules do not grow out of one discrete site, but emanate from different points on the surface of the nucleus as well as on the cell cortex [7, 8]. Some meiotic cells also lack centrosomes yet produce microtubules [9].

The nucleating activity of the centrosome and spindle pole body is thought to depend on the tubulin isoform γ-tubulin (see Chapter 2). A variety of genetic, antibody interference, and biochemical studies have implicated this protein as the major factor in microtubule nucleation (reviewed in [10]). Studies in animal cells of microtubules that did not originate from the centrosome found that the microtubules still grew from small cytoplasmic foci containing γ-tubulin [6]. γ-Tubulin's role is less clear in plants, but it most likely nucleates the unfocused microtubules that are distributed around the cell. Plant γ-tubulin may also provide some other function, because it can also be found along the lengths of microtubules [7]. In meiotic cells lacking centrosomes, it may be that soluble γ-tubulin complexes initiate microtubule growth, and then microtubule motors and cross-linking proteins organize the microtubules into arrays such as the meiotic spindle. Microtubule-associated proteins such as TPX2 and XMAP215 are also important for this chromatin-based microtubule formation (see below, and [9, 11–14]). In all cell types studied to date, however, γ-tubulin appears to be required for the efficient formation of at least a subset of microtubules.

The biochemical characterization of γ-tubulin from cell extracts led to the discovery that γ-tubulin is found in two major, highly conserved protein complexes inside cells. The most abundant complex in most cell types is the ~ 2.2-MDa γ-tubulin ring complex (γ-TuRC), which is thought to consist of 12 or 14 γ-tubulin molecules and at least six additional proteins. A second complex of ~ 280 kDa is called the γ-tubulin small complex (γ-TuSC) in Drosophila or the Tub4 complex in *S. cerevisiae*. It is a heterotetramer, consisting of two copies of γ-tubulin and one copy each of two accessory proteins, Dgrip84/Spc97/GCP2 (nomenclature from Drosophila/*S. cerevisiae*/human) and Dgrip91/Spc98/GCP3. In higher eukaryotes, this complex is a subunit of the γ-TuRC (reviewed in [15–17]).

Most of the accessory proteins of these complexes share homology within and between species and it has been proposed that these regions allow interactions between the accessory proteins or with γ-tubulin [18–21]. In the electron microscope, the γ-TuRC displays an open ring structure of ~ 25 nm diameter (Figure 3.2) [22–24]. Due to the repeating subunits visible around the open rim of the ring, as well as the stoichiometry of the proteins in the complex, it is thought that this is where the γ-tubulins reside. The accessory proteins would then compose the closed face, or "cap", of the ring [24]. The role of these accessory proteins, however, is not known. It is clear that γ-tubulin can nucleate microtubules on its own *in vitro* (see Section 3.2 and [25, 26]), so the accessory proteins may play a regulatory role, or a purely structural role such as maintaining the open-ring shape of the γ-TuRC and/or attaching the γ-tubulins to the centrosome. *In vitro* studies comparing the nucleating activities of the γ-TuSC and γ-TuRC revealed that the large complex is a better nucleator, suggesting that the full complement of accessory proteins enhances the activity [23].

3.1.2
Models for the Mechanism of γ-TuRC/Tub4 Complex-mediated Microtubule Nucleation

The dimensions and physical appearance of the γ-TuRC by itself or when in complex with microtubules, as well as its ability to functionally cap the minus ends of microtubules suggests a "templating" model for microtubule nucleation (Figure 3.2). This model proposes that the γ-tubulins on the open face of the ring interact laterally with one another, in reference to the interactions between α/β-tubulins in adjacent protofilaments of the microtubule. This ring of γ-tubulins is proposed to facilitate the assembly of α/β-tubulins, forming the minus end of the microtubule, which then grows rapidly and takes on the 13-protofilament architec-

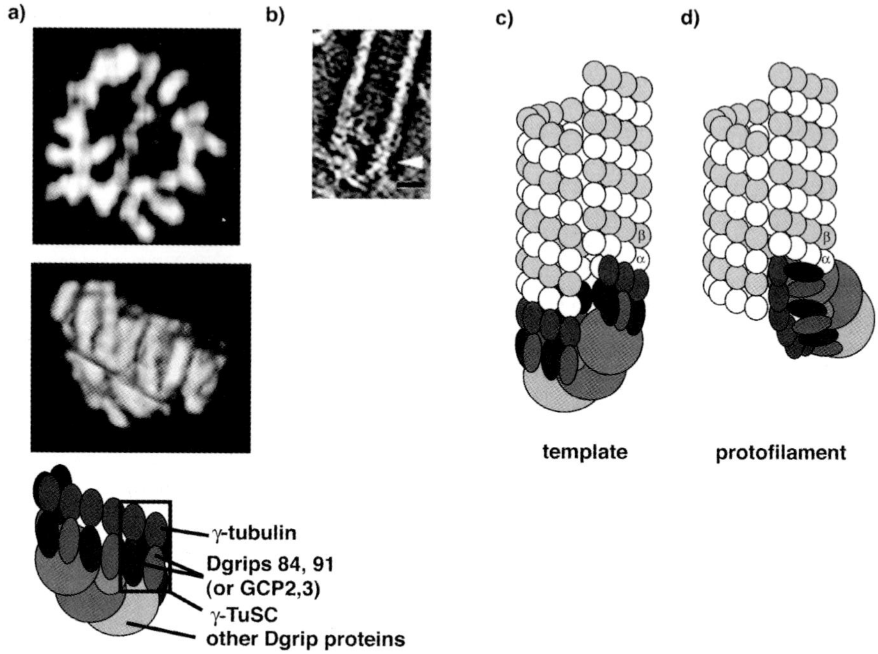

Figure 3.2 Structure of the γTuRC by electron microscopy. (a) Top: image from electron tomography of isolated, negative-stained Drosophila γTuRC showing a modular, capped open ring structure. Centre: side view of the γTuRC ring revealing repeating subunits around the ring wall and the asymmetric, globular cap on one face of the ring. Bottom: model of the subunit organization of the γTuRC structure based on electron microscopy and the known stoichiometry of γTuRC components. (b) Image from electron tomography of a microtubule grown from an isolated Drosophila γTuRC. White arrow indicates the cap (presumably γTuRC) at the end of the microtubule. (c) Template model for microtubule nucleation by the γTuRC. γ-Tubulins are proposed to interact laterally with each other and to bind to α-tubulins at the minus end of the microtubule. (d) Protofilament model for microtubule nucleation by the γTuRC. γ-Tubulins are proposed to interact longitudinally with each other and to promote formation of the first protofilament of the microtubule (reviewed in [15, 17]). Figures reproduced with permission from Moritz et al. [24].

ture dictated by the structure of the γ-TuRC. An alternative "protofilament" model proposes that the γ-tubulins in the ring interact longitudinally, in a manner similar to α/β-tubulins in the same protofilament, and that they provide enough stability to form the first protofilament of the microtubule (Figure 3.2; reviewed in [15, 17]. These models differ radically in their assumptions of how the γ-tubulins interact with one another, which carries over into how the γ-tubulins interact with α/β-tubulin and facilitate nucleation.

Inferences about γ-tubulin structure from the known atomic structures of α/β-tubulin suggest that γ-tubulin is likely to interact longitudinally with α- or β-tubulin, and may interact laterally. The preliminary information that could be gleaned from these structural comparisons did not discriminate between the template or protofilament models, although they may slightly favor the former [27]. Interestingly, in a study of γ-tubulin overexpression *in vivo*, γ-tubulin formed novel tubular structures [28]. Moreover, a detailed analysis of pure human γ-tubulin *in vitro* showed that γ-tubulin displays interaction and assembly characteristics that are unlike those of α/β-tubulin [26]. These results suggest that it may be misleading to infer γ-tubulin's behavior from comparisons with α/β-tubulin. For example, in the *in vitro* study, under physiological conditions γ-tubulin existed as oligomers that could only be converted to monomers at KCl concentrations of 0.5 M and above. At low ionic strength and at low concentration γ-tubulin forms a tetramer, while at high concentrations it oligomerizes further into filaments and tubules that are insensitive to cold, independent of GTP, and do not display the protofilament architecture typical of microtubules [26]. Furthermore, studies of the interaction of γ-tubulin peptides with α/β-tubulin *in vitro* revealed evidence for unique, lateral binding properties unlike those of α/β-tubulin [29]. The functional significance of these regions of γ-tubulin *in vivo* is supported by mutagenesis studies [30]. These results indicate that the biochemical properties of γ-tubulin are notably distinct from those of α/β-tubulin and begin to suggest that the template and protofilament models are at best simplifications of the actual mechanism of γ-tubulin-mediated microtubule nucleation.

How might the relatively small Tub4 complex of *S. cerevisiae* fit into this picture of microtubule nucleation? The heterotetrameric Tub4 complex is the predominant γ-tubulin complex that can be isolated from *S. cerevisiae* and there is little evidence to suggest that it forms a large complex similar to the γ-TuRC *in vivo* [31, 32]. A weak microtubule-nucleating activity has been observed for the Tub4 complex *in vitro* [32], and this may be sufficient to produce the relatively small number of microtubules needed in yeast cells. It is possible, however, that the Tub4 complex forms a larger aggregate at the spindle pole body that is functionally similar to the γ-TuRC. The capped appearance of microtubules at the spindle pole body in the electron microscope [33, 34] is very similar to that of microtubules grown from isolated γ-TuRCs [24, 35, 36] and from centrosomes [24, 37], suggesting that a larger complex may assemble.

3.2
Kinetic Models of the Mechanism of Microtubule Nucleation

Microtubule assembly from α/β-tubulin heterodimers follows a nucleation-elongation mechanism, in which efficient polymer elongation depends on the intrinsically inefficient formation of oligomers of α/β-tubulin heterodimers. The largest, rate-limiting oligomer constitutes an important species, known as the nucleus. While the molecular details of microtubule assembly remain poorly understood, the past 20 years of study has provided important clues about the mechanism of this process.

Microtubule assembly reactions are commonly monitored by following the increase in light scattering via absorbance at 350 nm, which allows time-dependent measurement of the extent of polymerization. A set of polymerization curves measured over a range of initial protein concentration contains a wealth of information about the underlying mechanism. Voter and Erickson published such a set of microtubule assembly curves in 1984 [38], and these data retain benchmark status in the search for a complete understanding of microtubule nucleation. Initial attempts to fit the assembly data with a kinetic model originally developed for actin polymerization failed [38]. This kinetic model, formulated by Oosawa, assumes that the critical nucleus for actin is formed in a single, rate-limiting step [39, 40]. The fact that the Oosawa model was unable to fit the microtubule assembly data demonstrated that microtubule assembly was fundamentally more complex than that of actin, and made it clear that a more complicated kinetic mechanism was required to accurately describe the microtubule assembly data. None of the models proposed at the time however, were able to provide convincing fits of the kinetic data [38]. A detailed molecular model of the organization of the nucleus and how it assembled from α/β-tubulin subunits remained elusive.

A key step in understanding microtubule assembly (and any other biological self-assembly) is the ability to specify the concentration dependence of the assembly reactions. In a model-independent analysis of the Voter and Erickson data, Flyvbjerg and colleagues demonstrated that the overall concentration dependence of the assembly reactions could be determined from two parameters extracted directly from the assembly curves [41–43]. The assembly curves displayed an important property, scaling, which, simply put, means that all the curves are fundamentally the same shape, independent of tubulin concentration and time (Figure 3.3). Scaling was demonstrated in a two-step process. First, the amplitude of each assembly curve was normalized by dividing the amplitude at each time-point by the final pla-

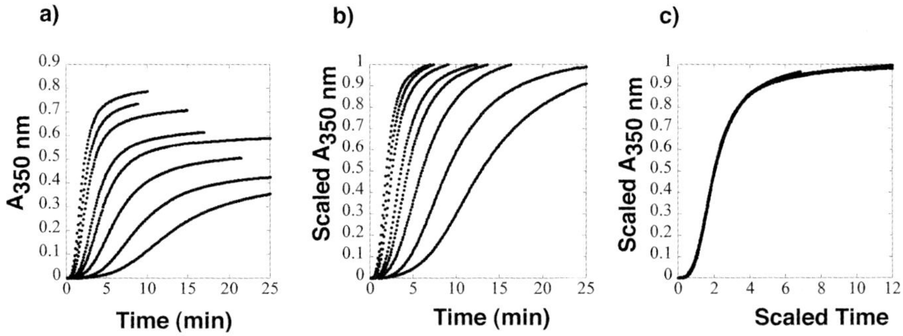

Figure 3.3 Illustration of scaling behavior displayed by microtubule assembly curves. (a) Microtubule assembly can be followed over a range of tubulin concentrations by turbidity at A_{350nm}. Increasing concentrations of α/β-tubulin give higher values of A_{350nm} at steady state, and have faster assembly reactions. (b) The first step in the scaling process is to normalize the amplitudes (*y*-axis) of each curve, dividing the value of the assembly curve at each time point by the final value obtained by that curve at steady-state. (c) The second step in the scaling process is to normalize the time series (*x*-axis) for each curve. This is accomplished by dividing all the time points for a given curve by $T_{0.1}$, the time at which that curve reached 10% of its steady-state amplitude. The assembly curves scale, indicating that the same mechanism of microtubule nucleation is used at all tubulin concentrations. This places strict limitations on the possible kinetic models that can be used to describe the assembly process [42, 43].

teau value for that curve. Second, the time of each assembly curve was normalized by dividing the time of each time-point by the time required to reach 10% assembly (the so-called characteristic time $T_{0.1}$) for that curve. After these two operations, all the transformed assembly curves were remarkably similar. Flyvbjerg and colleagues showed that both the amplitude and the characteristic time of each curve were simply related to the initial α/β-tubulin concentration. This implied that one single assembly mechanism was operating over the entire concentration range. In other words, once the initial α/β-tubulin concentration was specified, one single function of concentration and time would describe all the assembly curves [41–43].

How was the overall concentration dependence of microtubule assembly determined? Flyvbjerg and colleagues noted that, at all initial α/β-tubulin concentrations, the assembly curves displayed early growth proportional to the fourth power of time (t^4). This suggested that there were four rate-limiting steps (including elongation) in the assembly mechanism. It remained to be determined how many α/β-tubulin subunits participated at each step. Flyvbjerg et al. demonstrated that the characteristic time $T_{0.1}$ was related to the inverse cube of the initial α/β-tubulin concentration, which was consistent with all rate-limiting intermediates being composed of multiples of 3 α/β-tubulin subunits. The mechanism that they fit convincingly to the microtubule assembly data could be described as a progression of increasingly large rate-limiting intermediates (hexamer, then nonamer) leading up to the nucleus (dodecamer), which then undergoes elongation by repeated monomer addition [41–43].

Pure γ-tubulin exhibits microtubule-nucleating activity *in vitro* [25]. Our ability to purify large quantities of recombinant human γ-tubulin provided us the opportunity to apply the Flyvbjerg method of analysis to our own microtubule assembly data obtained in the presence or absence of this nucleator over a broad range of α/β-tubulin concentrations (Figure 3.4) [26]. Surprisingly, we found that γ-tubulin-mediated microtubule nucleation proceeded by different mechanisms, depending on the concentration and oligomerization state of γ-tubulin. At low nanomolar concentrations, where γ-tubulin forms a tetramer, it simply accelerated the rate of nucleus formation, apparently by stabilizing a pre-nucleus intermediate. It did not decrease the nucleus size. However, at higher concentrations, where γ-tubulin can form large filaments and arrays, it dramatically facilitated microtubule nucleation by decreasing the size of the nucleus. The oligomerization-state dependence of the mechanism and the biochemical properties of these oligomers suggest new working models for how γ-tubulins in the γ-TuRC may be interacting with α/β-tubulins in a microtubule (Figure 3.5) [26].

Fits of the Flyvbjerg model to the microtubule assembly data with and without added γ-tubulin provide rate constants that predict the efficiency with which each of the rate-limiting intermediates form. Comparing the rate constants for the γ-tubulin tetramer-nucleated data to those of the buffer control indicated that formation of the first intermediate (the hexamer of α/β-tubulins) was unaffected; instead, γ-tubulin favored formation of a later intermediate (a nonamer of α/β-tubulins), most likely by specifically binding it. Based on thermodynamic arguments and the hexamer precursor, the most likely configuration of this nonamer inter-

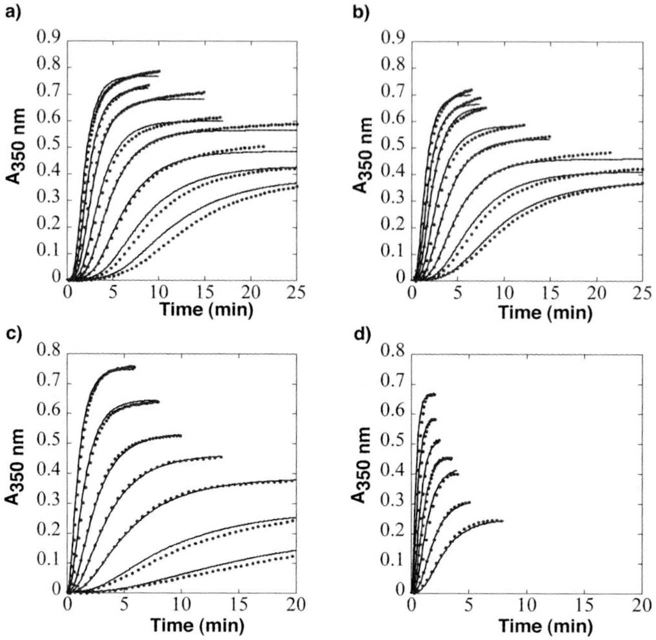

Figure 3.4 Analysis of γ-tubulin-mediated microtubule assembly data using the Flyvbjerg method. [26, 42, 43]. In all four panels, microtubule assembly data measured in the spectrophotometer is displayed as filled grey circles, and the kinetic models that were fit to the data are shown as solid lines. (a) Assembly of α/β-tubulin at a range of concentrations, no γ-tubulin added. (b) Assembly of identical concentrations of α/β-tubulin as in (a), but with 100 nM pure human γ-tubulin added to the reactions. γ-Tubulin forms tetramers under these conditions, and there is a small but measurable increase in the rate at which microtubules assemble, indicating that γ-tubulin tetramers have microtubule nucleating activity. (c) Assembly of α/β-tubulin at a range of concentrations, no γ-tubulin added. (d) Assembly of identical concentrations of α/β-tubulin as in (c), but with 670 nM pure human γ-tubulin added to the reactions. At these high concentrations, γ-tubulin forms larger oligomers that have a dramatic effect on the rate of microtubule assembly. The differences between (a) and (c) result from slightly different buffers used in the two different experiments. In all cases, fits to the data obtained using the Flyvbjerg model are very good, and therefore provide an important framework for understanding γ-tubulin function in microtubule nucleation [26].

mediate is a sheet of three-by-three heterodimers. One way to explain how this nonamer intermediate is specifically bound by the γ-tubulin tetramer while the hexamer is not is to propose that the γ-tubulin tetramer binds *between* the three short protofilaments [26].

Interestingly, when considered in terms of the entire γTuRC, this mode of binding provides a logical explanation for how this nucleating complex can produce microtubules with the 13 protofilaments most often seen in vivo (Figure 3.5b,c). Historically, this fact has presented a puzzle because studies of the stoichiometry of proteins in the Tub4 complex in yeast suggest that it contains *two* copies of γ-tubulin and one copy each of the accessory proteins Spc97 (Dgrip84/GCP2) and Spc98

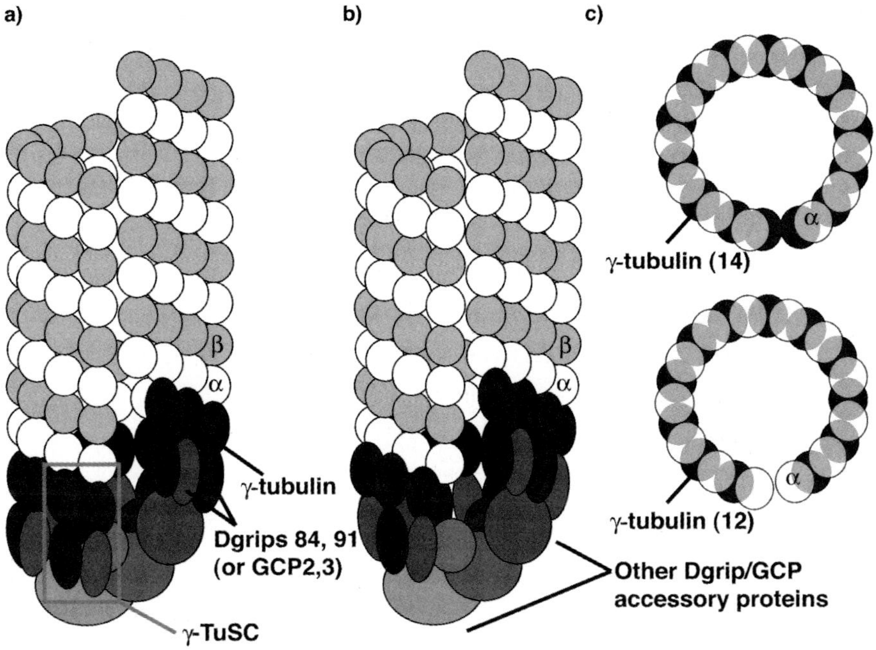

Figure 3.5 Modified template model of γTuRC-mediated microtubule nucleation. (a) The original template model proposed that γ-tubulins bind to α-tubulins at the minus ends of protofilaments similarly to longitudinal α/β-tubulin binding within a protofilament (reviewed in [15, 17]). (b) The modified template model takes into account physical properties of γ-tubulin and the mechanism of γ-tubulin-mediated microtubule nucleation by proposing that γ-tubulin binds between protofilaments [26]. A γTuRC containing 12 γ-tubulins is shown associated with the microtubule, but a 14-γ-tubulin γTuRC could also be accommodated. (c) Cross-sectional views illustrating the proposed binding sites for γ-tubulins between the α-tubulins at the minus end of each protofilament. This mode of binding provides an explanation for how a γTuRC containing an even number of γ-tubulins could template a 13-protofilament microtubule, the most common architecture observed *in vivo* (see Color Plates page XXIII).

(Dgrip91/GCP3) [31]. By analogy and some physical evidence, the corresponding γTuSC subunit of the γTuRC that is found in higher eukaryotes is thought to have the same composition [18, 23]. The presumed γTuSC subunits visible in the wall of the γTuRC ring in the electron microscope also appear to be bi-lobed structures (Figure 3.2a) [24]. Unfortunately, our current electron microscopy images of the γTuRC have allowed only a rough count of the subunits that constitute the ring wall. However, taking into account all of this evidence, the γTuRC is most likely to contain an even number (12 or 14) of γ-tubulins. Thus, placing the γ-tubulins *between* the protofilaments provides a template that would produce 13-protofilament microtubules. It should be possible to test this model by higher-resolution imaging and cross-linking studies of the binding of γ-tubulin to α/β-tubulin.

The kinetic model derived by Flyvbjerg and colleagues represents an important advance in our understanding of microtubule assembly, and provides a necessary framework for dissecting the mode of action of microtubule nucleating proteins such as γ-tubulin and its complexes. Nevertheless, the insight provided by the Flyvbjerg analysis is limited in several key aspects. What is the role of GTP hydrolysis by α/β- and/or γ-tubulin in microtubule nucleation? What is the subunit organization of the various rate-limiting intermediates? Why are they rate limiting? The answers to these and other questions will come from a more detailed understanding of microtubule assembly based in part on the known structure of the α/β-tubulin heterodimers and of the microtubule lattice. A detailed molecular model combining kinetic data with subunit and polymer structure has recently been published to describe actin assembly [44], so an analogous model for microtubules should be forthcoming in the near future.

3.3
The Involvement of Non-γ-TuRC Proteins in Microtubule Nucleation

There are fundamental differences between the interphase and mitotic microtubule arrays in cells, suggesting that there is cell cycle-dependent regulation of γTuRC localization and activity. In interphase, the microtubules are relatively long and stable, and most grow from a single organizing center such as the centrosome. In mitosis, more γTuRC is recruited to the centrosome (a phenomenon termed centrosome maturation) resulting in the nucleation of a larger number of microtubules that are also relatively short and highly dynamic as they form the bipolar array of the spindle [45, 46]. A complicated network of interacting regulatory proteins such as the small GTPase Ran, Aurora A kinase, Polo-like kinase, as well as other cell cycle-control proteins has been found to control centrosome maturation, microtubule nucleation and spindle assembly during mitosis and meiosis [11, 47–55]. Only recently have we begun to understand some of the details of these intricate pathways, which are described in more detail in other chapters in this book.

What is intriguing from the standpoint of microtubule nucleation is that certain non-γTuRC proteins are emerging that may contribute to this activity. For example, a target of Polo-like kinase, ninein-like protein (Nlp), has been shown to associate with γ-tubulin and the γTuRC protein, GCP4. Nlp's regulated association is important for interphase microtubule nucleation [54]. Some other non-γTuRC proteins implicated in microtubule nucleation fall under the control of the Ran signaling pathway. Ran was originally described as an abundant GTPase that is required for transport of proteins and RNA in and out of the nucleus, but more recently it has been found to play an additional important role in controlling microtubule nucleation and organization during mitosis and meiosis (reviewed in [56–58]). So far, no direct connection between regulatory proteins and the γTuRC has been revealed. RanGTP has been implicated in the release of activated spindle assembly factors such as the microtubule-associated proteins TPX2, NuMA and XMAP215

[11, 14, 52, 55, 59, 60]. These studies have pointed to a requirement for distinct classes of microtubules that are regulated, and perhaps generated, differentially. For example, depletion of TPX2 had little effect on centrosomal microtubules, but was important for microtubules that form around chromatin during spindle assembly [12]. In addition, another layer of regulation has emerged in studies of TPX2, which was found to target Aurora A kinase to the spindle, and to stimulate its activity in a microtubule-dependent manner [55, 61–63].

Evidence from fluorescence microscopy assays of aster formation suggests that TPX2, XMAP215 and, surprisingly, the ribonucleotide reductase protein R1 have microtubule nucleating activity [12–14, 64]. Since TPX2 and XMAP215 interact with microtubules, it follows that they may play a role in nucleation. R1, on the other hand, is a subunit of an enzyme involved in DNA synthesis and as such would not be predicted to affect microtubule nucleation. It has been proposed that R1 is bifunctionally required for distinct cellular events, DNA synthesis and centrosome/spindle pole body activation. One possibility that should be explored further is that R1 plays a role in preventing spindle formation in the presence of DNA damage or replication blocks [64]. Indeed, a connection between DNA defects and centrosome activity mediated through checkpoint kinase 2 has been revealed in Drosophila [65, 66] (see Chapter 11). R1 may be similarly involved. Thus, the possibility has been raised that multiple nucleators contribute to the formation of microtubules and that specific nucleators may act on different populations of microtubules [14]. This idea is also supported by studies in which γ-tubulin was mutated or depleted by RNAi in *C. elegans,* which showed that only interphase microtubules failed to form [67]. Many other studies in which γ-tubulin activity was impaired have indicated that at least subsets of microtubules still polymerize (reviewed in [3]). These tantalizing results must be tempered by the limitations of fluorescence microscopy, a technique that does not allow the distinction between bundling and stabilization of pre-existing, small microtubule seeds from bona fide nucleation. However, this is an intriguing subject that should be investigated further using purified proteins and a kinetic assay such as the one described in Section 3.2.

3.4 Future Directions

The microtubule cytoskeleton is a remarkably complicated and dynamic cellular machine whose proper function relies on the coordinated activity of a large number of structural and regulatory proteins. Many proteins involved in producing and modifying microtubules for their various functions have now been identified and crude understandings of their purpose have been catalogued. The next steps will be to discern the molecular and kinetic mechanisms by which accessory proteins and nucleotide binding states modify the activity of microtubule nucleating proteins such as γ-tubulin. Clearly, we must explore the roles of GTP binding, hydrolysis and release in γ-tubulin/γTuRC function. In addition, the possibility that accessory proteins are required for GTP exchange on γ-tubulin should be pursued. γ-Tubulin is also phosphorylated in a cell cycle-dependent manner and this modification has an effect on microtubule number and organization [68]. This aspect of γ-tubulin regulation deserves a more detailed characterization in other cell types and *in vitro* with purified proteins. It will be important to discover the kinase(s) responsible for this phosphorylation. Complementary studies of γ-tubulin assembly properties have revealed γ-tubulin's unique characteristics among tubulins. We must now pursue further its interaction with α/β-tubulin and with accessory proteins. In conclusion, through the use of an effective method of kinetic analysis as described here, we are beginning to understand the basic molecular mechanism of γ-tubulin-mediated microtubule nucleation. We can now apply this method to understanding the contributions of accessory proteins in the γ-TuRC and beyond, as well as to understanding the mechanisms and roles of putative non-γ-tubulin microtubule nucleators.

Acknowledgments

We thank the Agard Laboratory centrosome group for useful input and discussion.

Our work is supported by the Howard Hughes Medical Institute and the National Institutes of Health. L. M. R. is a Paul Sigler/Ayouron Institute Fellow of the Helen Hay Whitney Foundation.

References

1. Bayley, P., K. Sharma and S. R. Martin. **1994**. Microtubule dynamics *in vitro*. In *Microtubules* vol. 13, J. S. Hyams and C. W. Lloyd (eds). New York: Wiley-Liss, 111–137.
2. Nogales, E. **2000**. Structural insights into microtubule function. *Ann Rev Biochem* 69: 277–302.
3. Job, D., O. Valiron and B. Oakley. **2003**. Microtubule nucleation. *Curr Opin Cell Biol* 15: 111–117.
4. Desai, A. and T. J. Mitchison. 1997. Microtubule polymerization dynamics. *Ann Rev Cell Dev Biol* 13: 83–117.
5. Kellogg, D. R., M. Moritz and B. M. Alberts. **1994**. The centrosome and cellular organization. *Annu Rev Biochem* 63: 639–674.
6. Tulu, U. S., N. M. Rusan and P. Wadsworth. **2003**. Peripheral, non-centrosome-associated microtubules contribute to spindle formation in centrosome-containing cells. *Curr Biol* 13: 1894–1899.
7. Chan, J., G. M. Calder, J. H. Doonan and C. W. Lloyd. **2003**. EB1 reveals mobile microtubule nucleation sites in Arabidopsis. *Nature Cell Biol* 5: 967–971.
8. Shaw, S. L., R. Kamyar and D. W. Ehrhardt. **2003**. Sustained microtubule treadmilling in Arabidopsis cortical arrays. *Science* 300: 1715–1718.
9. Becker, B. E., S. J. Romney and D. L. Gard. **2003**. XMAP215, XKCM1, NuMA, and cytoplasmic dynein are required for the assembly and organization of the transient microtubule array during the maturation of Xenopus oocytes. *Dev Biol* 261: 488–505.
10. Oakley, B. R. 2000. γ-Tubulin. In *The Centrosome in Cell Replication and Early*

Development, vol 49. R. E. Palazzo and G. P. Schatten (Eds). San Diego: Academic Press, 27–54.
11. Gruss, O. J., R. E. Carazo-Salas, C. A. Schatz, G. Guarguaglini, J. Kast, M. Wilm, N. Le Bot, I. Vernos, E. Karsenti and I. W. Mattaj. **2001**. Ran induces spindle assembly by reversing the inhibitory effect of importin alpha on TPX2 activity. *Cell* 104: 83–93.
12. Gruss, O. J., M. Wittmann, H. Yokoyama, R. Pepperkok, T. Kufer, H. Sillje, E. Karsenti, I. W. Mattaj and I. Vernos. **2002**. Chromosome-induced microtubule assembly mediated by TPX2 is required for spindle formation in HeLa cells. *Nature Cell Biol* 4: 871–879.
13. Popov, A. V., F. Severin and E. Karsenti. **2002**. XMAP215 is required for the microtubule-nucleating activity of centrosomes. *Curr Biol* 12: 1326–1330.
14. Schatz, C. A., R. Santarella, A. Hoenger, E. Karsenti, I. W. Mattaj, O. J. Gruss and R. E. Carazo-Salas. **2003**. Importin α-regulated nucleation of microtubules by TPX2. *EMBO J.* 22: 2060–2070.
15. Wiese, C. and Y. Zheng. **1999**. γ-Tubulin complexes and their interaction with microtubule-organizing centers. *Curr Opin Struct Biol* 9: 250–259.
16. Schiebel, E. **2000**. γ-Tubulin complexes: binding to the centrosome, regulation and microtubule nucleation. *Curr Opin Cell Biol* 12: 113–118.
17. Moritz, M. and D. A. Agard. **2001**. γ-Tubulin complexes and microtubule nucleation. *Curr Opin Struct Biol* 11: 174–181.
18. Murphy, S. M., L. Urbani and T. Stearns. **1998**. The mammalian γ-tubulin complex contains homologues of the yeast spindle pole body components Spc97p and Spc98p. *J Cell Biol* 141: 663–674.
19. Tassin, A. M., C. Celati, M. Moudjou and M. Bornens. **1998**. Characterization of the human homologue of the yeast spc98p and its association with γ-tubulin. *J Cell Biol* 141: 689–701.
20. Gunawardane, R. N., O. C. Martin, K. Cao, L. Zhang, K. Dej, A. Iwamatsu and Y. Zheng. **2000**. Characterization and reconstitution of Drosophila γ-tubulin ring complex subunits. *J Cell Biol* 151: 1513–1524.
21. Murphy, S. M., A. M. Preble, U. K. Patel, K. L. O'Connell, D. P. Dias, M. Moritz, D. Agard, J. T. Stults and T. Stearns. **2001**. GCP5 and GCP6: two new members of the human γ-tubulin complex. *Mol Biol Cell* 12: 3340–3352.
22. Zheng, Y., M. L. Wong, B. Alberts and T. Mitchison. **1995**. Nucleation of microtubule assembly by a γ-tubulin-containing ring complex. *Nature* 378: 578–583.
23. Oegema, K., C. Wiese, O. C. Martin, R. A. Milligan, A. Iwamatsu, T. J. Mitchison and Y. Zheng. **1999**. Characterization of two related Drosophila γ-tubulin complexes that differ in their ability to nucleate microtubules. *J Cell Biol* 144: 721–733.
24. Moritz, M., M. B. Braunfeld, V. Guenebaut, J. Heuser and D. A. Agard. **2000**. Structure of the γ-tubulin ring complex: a template for microtubule nucleation. *Nature Cell Biol* 2: 365–370.
25. Leguy, R., R. Melki, D. Pantaloni and M. F. Carlier. **2000**. Monomeric γ-tubulin nucleates microtubules. *J Biol Chem* 275: 21975–21980.
26. Moritz, M., L. M. Rice, H. Aldaz, T. Stearns and D. Agard. **2004**. Mechanisms of microtubule nucleation by human γ-tubulin (Public Library of Science, in press).
27. Inclan, Y. F. and E. Nogales. **2001**. Structural models for the self-assembly and microtubule interactions of γ-, δ- and ε-tubulin. *J Cell Sci* 114: 413–422.
28. Shu, H. B. and H. C. Joshi. **1995**. γ-Tubulin can both nucleate microtubule assembly and self-assemble into novel tubular structures in mammalian cells. *J Cell Biol* 130: 1137–1147.
29. Llanos, R., V. Chevrier, M. Ronjat, P. Meurer-Grob, P. Martinez, R. Frank, M. Bornens, R. H. Wade, J. Wehland and D. Job. **1999**. Tubulin binding sites on γ-tubulin: identification and molecular characterization. *Biochemistry* 38: 15712–15720.
30. Jung, M. K., N. Prigozhina, C. E. Oakley, E. Nogales and B. R. Oakley. **2001**. Alanine-scanning mutagenesis of

Aspergillus γ-tubulin yields diverse and novel phenotypes. *Mol Biol Cell* 12: 2119–2136.

31. Knop, M. and E. Schiebel. **1997**. Spc98p and Spc97p of the yeast γ-tubulin complex mediate binding to the spindle pole body via their interaction with Spc110p. *EMBO J.* 16: 6985–6995.

32. Vinh, D. B., J. W. Kern, W. O. Hancock, J. Howard and T. N. Davis. **2002**. Reconstitution and characterization of budding yeast γ-tubulin complex. *Mol Biol Cell* 13: 1144–1157.

33. Byers, B., K. Shriver and L. Goetsch. **1978**. The role of spindle pole bodies and modified microtubule ends in the initiation of microtubule assembly in *Saccharomyces cerevisiae. J Cell Sci* 30: 331–352.

34. Bullitt, E., M. P. Rout, J. V. Kilmartin and C. W. Akey. **1997**. The yeast spindle pole body is assembled around a central crystal of Spc42p. *Cell* 89: 1077–1086.

35. Keating, T. J. and G. G. Borisy. **2000**. Immunostructural evidence for the template mechanism of microtubule nucleation. *Nature Cell Biol* 2: 352–357.

36. Wiese, C. and Y. Zheng. **2000**. A new function for the γ-tubulin ring complex as a microtubule minus-end cap. *Nature Cell Biol* 2: 358–364.

37. Moritz, M., M. B. Braunfeld, J. W. Sedat, B. Alberts and D. A. Agard. **1995**. Microtubule nucleation by γ-tubulin-containing rings in the centrosome. *Nature* 378: 638–640.

38. Voter, W. A. and H. P. Erickson. **1984**. The kinetics of microtubule assembly. *J Biol Chem* 259: 10430–10438.

39. Oosawa, F. and M. Kasai. **1962**. A theory of linear and helical aggregations of macromolecules. *J Mol Biol* 4: 10–21.

40. Oosawa, F. and S. Asakura. **1975**. *Thermodynamics of the Polymerization of Protein*. New York: Academic Press.

41. Fygenson, D. K., H. Flyvbjerg, K. Sneppen, A. Libchaber and S. Leibler. **1995**. Spontaneous nucleation of microtubules. *Phys Rev E* 51: 5058–5063.

42. Flyvbjerg, H., E. Jobs and S. Leibler. **1996**. Kinetics of self-assembling microtubules: an "inverse problem" in biochemistry. *Proc Natl Acad Sci USA* 93: 5975–5979.

43. Flyvbjerg, H. and E. Jobs. **1997**. Microtubule dynamics. II. Kinetics of self assembly. *Phys Rev E* 56: 7083–7099.

44. Sept, D. and J. A. McCammon. 2001. Thermodynamics and kinetics of actin filament nucleation. *Biophys J* 81: 667–674.

45. Khodjakov, A. and C. L. Rieder. **1999**. The sudden recruitment of γ-tubulin to the centrosome at the onset of mitosis and its dynamic exchange throughout the cell cycle, do not require microtubules. *J Cell Biol* 146: 585–596.

46. Palazzo, R. E., J. M. Vogel, B. J. Schnackenberg, D. R. Hull and X. Wu. **2000**. Centrosome maturation. *Curr Top Dev Biol* 49: 449–470.

47. Nigg, E. A. 1998. Polo-like kinases: positive regulators of cell division from start to finish. *Curr Opin Cell Biol* 10: 776–783.

48. Ohba, T., M. Nakamura, H. Nishitani and T. Nishimoto. **1999**. Self-organization of microtubule asters induced in Xenopus egg extracts by GTP-bound Ran. *Science* 284: 1356–1358.

49. Fleig, U., S. S. Salus, I. Karig and S. Sazer. **2000**. The fission yeast ran GTPase is required for microtubule integrity. *J Cell Biol* 151: 1101–1111.

50. Carazo-Salas, R. E., O. J. Gruss, I. W. Mattaj and E. Karsenti. **2001**. Ran-GTP coordinates regulation of microtubule nucleation and dynamics during mitotic-spindle assembly. *Nature Cell Biol* 3: 228–234.

51. Hannak, E., M. Kirkham, A. A. Hyman and K. Oegema. **2001**. Aurora-A kinase is required for centrosome maturation in *Caenorhabditis elegans. J Cell Biol* 155: 1109–1116.

52. Nachury, M. V., T. J. Maresca, W. C. Salmon, C. M. Waterman-Storer, R. Heald and K. Weis. **2001**. Importin β is a mitotic target of the small GTPase Ran in spindle assembly. *Cell* 104: 95–106.

53. Berdnik, D. and J. A. Knoblich. **2002**. Drosophila Aurora-A is required for

centrosome maturation and actin-dependent asymmetric protein localization during mitosis. *Curr Biol* 12: 640–647.

54. Casenghi, M., P. Meraldi, U. Weinhart, P. I. Duncan, R. Korner and E. A. Nigg. **2003**. Polo-like kinase 1 regulates Nlp, a centrosome protein involved in microtubule nucleation. *Dev Cell* 5: 113–125.

55. Kahana, J. A. and D. W. Cleveland. **1999**. Beyond nuclear transport. RanGTP as a determinant of spindle assembly. *J Cell Biol* 146: 1205–1210.

56. Tsai, M. Y., C. Wiese, K. Cao, O. Martin, P. Donovan, J. Ruderman, C. Prigent and Y. Zheng. **2003**. A Ran signalling pathway mediated by the mitotic kinase Aurora A in spindle assembly. *Nature Cell Biol* 5: 242–248.

57. Dasso, M. **2001**. Running on Ran: nuclear transport and the mitotic spindle. *Cell* 104: 321–324.

58. Walczak, C. E. **2001**. Ran hits the ground running. *Nature Cell Biol* 3: E69–E70.

59. Wilde, A. and Y. Zheng. **1999**. Stimulation of microtubule aster formation and spindle assembly by the small GTPase Ran. *Science* 284: 1359–1362.

60. Wiese, C., A. Wilde, M. S. Moore, S. A. Adam, A. Merdes and Y. Zheng. **2001**. Role of importin-β in coupling Ran to downstream targets in microtubule assembly. *Science* 291: 653–656.

61. Kufer, T. A., H. H. Sillje, R. Korner, O. J. Gruss, P. Meraldi and E. A. Nigg. **2002**. Human TPX2 is required for targeting Aurora-A kinase to the spindle. *J Cell Biol* 158: 617–623.

62. Eyers, P. A. and J. L. Maller. **2003**. Regulating the regulators: Aurora A activation and mitosis. *Cell Cycle* 2: 287–289.

63. Kufer, T. A., E. A. Nigg and H. H. Sillje. **2003**. Regulation of Aurora-A kinase on the mitotic spindle. *Chromosoma* 112: 159–163.

64. Takada, S., T. Shibata, Y. Hiraoka and H. Masuda. **2000**. Identification of ribonucleotide reductase protein R1 as an activator of microtubule nucleation in Xenopus egg mitotic extracts. *Mol Biol Cell* 11: 4173–4187.

65. Sibon, O. C., A. Kelkar, W. Lemstra and W. E. Theurkauf. **2000**. DNA-replication/DNA-damage-dependent centrosome inactivation in Drosophila embryos. *Nature Cell Biol* 2: 90–95.

66. Takada, S., A. Kelkar and W. E. Theurkauf. **2003**. Drosophila checkpoint kinase 2 couples centrosome function and spindle assembly to genomic integrity. *Cell* 113: 87–99.

67. Hannak, E., K. Oegema, M. Kirkham, P. Gonczy, B. Habermann and A. A. Hyman. **2002**. The kinetically dominant assembly pathway for centrosomal asters in *Caenorhabditis elegans* is γ-tubulin dependent. *J Cell Biol* 157: 591–602.

68. Vogel, J., B. Drapkin, J. Oomen, D. Beach, K. Bloom and M. Snyder. **2001**. Phosphorylation of γ-tubulin regulates microtubule organization in budding yeast. *Dev Cell* 1: 621–631.

69. Alberts, B., D. Bray, J. Lewis, M. Raff, K. Roberts and J. D. Watson. **1994**. The cytoskeleton. In *Molecular Biology of the Cell*. New York: Garland, 805.

4
The Budding Yeast Spindle Pole Body: A Centrosome Analog

Suzanne van Kreeveld Naone and Mark Winey

4.1
Introduction

Microtubules are nucleated and organized by a diverse group of organelles known as microtubule organizing centers (MTOCs). These organelles are responsible for non-random arrays of microtubules that perform such functions as ciliary and flagellar movement, maintenance of cell shape and form, intracellular and axoplasmic transport, anchorage of cell surface receptors and chromosome movement. The latter task is performed by a subset of MTOCs that function as spindle poles, orienting the bipolar spindle during mitosis in the cell cycle and during meiosis in the formation of germ cells.

To form bipolar spindles, the spindle pole organelle must be duplicated once and only once in preparation for each round of chromosome segregation, be it mitosis or one of the meiotic segregations. After duplication, the resulting spindle pole organelles move to opposite sides of the nucleus to organize the microtubules of the spindle, which function to capture and move one set of chromosomes into each daughter cell along with one spindle pole. Correct arrangement of a bipolar spindle ensures proper segregation of DNA to each daughter cell, and failure to do so results in aneuploid cells, which is a hallmark of many human cancers.

Two spindle pole organelles that have been studied in greater detail are the budding yeast spindle pole body (SPB) and the vertebrate cell centrosome. SPBs and centrosomes share analogous functions of organizing the bipolar spindle, as well as the more recently studied signaling roles that affect cell cycle progression. To accomplish these analogous functions, SPBs and centrosomes appear to share many homologous components despite the significant morphological differences between the organelles. Furthermore, the organelles may have similar mechanisms by which they regulate duplication. Due to the conservation of function between SPBs and centrosomes and because yeast is a genetically tractable organism amenable to biochemical and cell biological studies, many consider the SPB as a model for the study of centrosomes.

Centrosomes in Development and Disease. Edited by Erich A. Nigg

ISBN 3-527-30980-2

SPB structure in *Saccharomyces cerevisiae* has been studied in detail using such techniques as electron microscopy (EM) of SPBs *in situ* and of isolated SPBs, cryoEM of isolated SPBs, and high voltage electron tomography of SPBs *in situ*. These studies reveal that the SPB is a multi-layered cylinder embedded in the nuclear envelope (Figure 4.1a), its basic structure comprising an outer, central and inner plaque [1–7]. The SPB is anchored in the nuclear envelope by hook-like appendages that emanate from the central plaque [7], and it nucleates cytoplasmic and nuclear microtubules from the outer and inner plaques, respectively. The overall dimensions of the organelle are 150 nm in thickness (outer plaque to inner plaque) with a diameter of 110 nm in haploid cells [3]. Interestingly, the diameter is larger in diploids (160 nm), increasing the surface area of the SPB and allowing for additional microtubule nucleation sites. The mass of a diploid SPB is between 1 and 1.5 gigadatton [6], and the mass of the SPB core is ~ 0.3 GDa [8]. Finally, on one side of the SPB is an electron-dense structure called the halfbridge, which is a modification of the nuclear envelope and the site at which SPB duplication is initiated.

As is apparent in Figure 4.1b, the centrosome is a quite dissimilar looking organelle that consists of two centrioles surrounded by an amorphous substance called the pericentriolar material (PCM). The centrioles are barrel-shaped structures com-

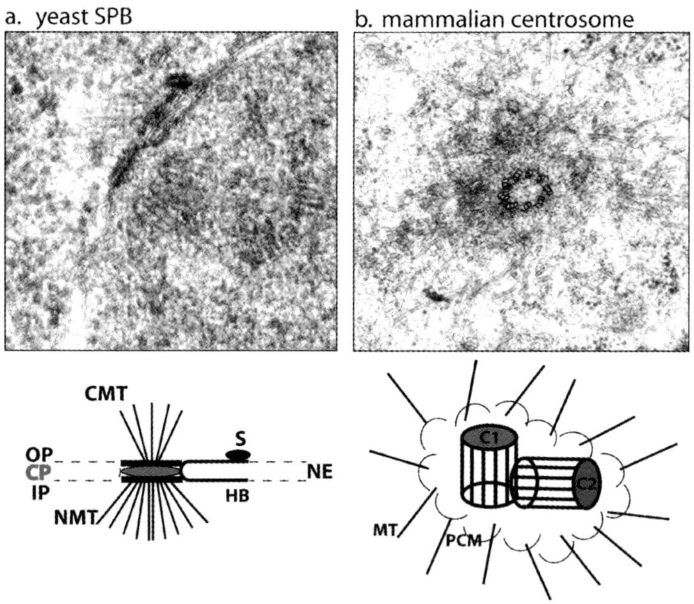

Figure 4.1 Spindle pole organelles. Electron micrographs of the (a) budding yeast SPB and the (b) mammalian centrosome (courtesy of Dr. Thomas Giddings, Jr.). The barrel of triplet microtubules can be clearly distinguished in the micrograph of the centrosome, which shows a centriole in cross-section. Schematics beneath the micrographs show the basic structures of each organelle (HB, halfbridge; OP, outer plaque; CP, central plaque; IP, inner plaque; S, satellite; CMT, cytoplasmic microtubule; NMT, nuclear microtubule; NE, nuclear envelope; C1, centriole 1; C2, centriole 2; PCM, pericentriolar material; MT, microtubule).

posed of polyglutamylated microtubule triplets, and they often exhibit a perpendicular orientation with respect to each other (Figure 4.1b). The centrosome is roughly 10-fold larger than the SPB and, unlike the SPB, resides in the cytoplasm. The structure of the centrosome is described in more detail elsewhere in this volume (see Chapters 7 and 9).

The nuclear envelope-embedded, multi-layered SPB is not an oddity of the budding yeast *Saccharomyces cerevisiae*, but is common in fungi. Similar organelles are also found in other organisms, sometimes under the name "spindle plaque", which was the original name of the SPB. Organisms in which spindle plaques have been documented include other fungi (*Schizosaccharomyces pombe, Aspergillus nidulans, Exophiala dermatitidis*), slime molds (*Dictyostelium discoideum*), ciliates (*Paramecium, Tetrahymena*) and protists (*Naegleria*). The fission yeast (*S. pombe*) spindle pole body is described, using electron microscopy, as an ellipsoid-shaped structure of amorphous internal structure with dimensions of 90 nm in thickness and 180 nm in diameter. This organelle is located near the nuclear envelope and is associated with a darkly-staining structure that is analogous to the halfbridge in budding yeast [9]. Unlike the budding yeast SPB, the fission yeast SPB localizes to the cytoplasm during most of interphase (where it also duplicates), but moves into the nuclear envelope to perform its role in mitosis [9]. Electron microscopy of the *Dictyostelium* SPB shows a layered core structure surrounded by a corona made up of dense nodules embedded in amorphous material (reviewed in [10]).

4.2
Molecular Composition of the Spindle Pole Body

Beyond the morphological characterization of the SPB, the molecular components have also been under study. A large variety of approaches over the years has yielded a list of at least 19 core components of the SPB, i. e. those proteins that contribute to the structure or function of the SPB and are found at the organelle throughout the cell cycle. This definition excludes a number of signaling molecules that transiently localize to the SPB during the cell cycle (i. e. Tem1p of the mitotic exit network), but they are discussed later, and also excludes proteins brought to the SPB by their association with microtubules (i. e. dynein). The techniques used to uncover SPB components include identifying the epitopes for monoclonal antibodies raised against cellular extracts enriched for SPBs (i. e. Spc110p, [11]), identifying proteins by mass spectrometric analysis of cellular extracts enriched for SPBs (several, e. g. Nud1p, [12]), immunoprecipitation approaches using known SPB components to find associated proteins (i. e. Sfi1p, [13]), and genomics approaches, either identifying yeast orthologs of centrosomal proteins (i. e. *TUB4* encodes gamma(γ)-tubulin, [14]) or localizing the gene products of unannotated ORFs using green fluorescent protein (GFP) gene fusions and looking for SPB localization (i. e. *CMN67*, [15]). Genetic studies also have contributed a number of components whose genes were identified by mutant alleles in screens designed to uncover phenotypes affecting SPB function in the cell cycle (i. e. Cdc31p, [16]), karyo-

Table 4.1 Core SPB components and their localization with respect to the SPB, their relevant SPB functions (MT, microtubule), their mammalian counterparts and relevant references.

Gene	*Localization*	*Function*	*Vertebrate homolog*	*Partial reference*
BBP1	Nuclear membrane at SPB	SPB duplication, structural component		23
CDC31	Halfbridge	SPB duplication	Centrin 3, 2 and 1	16, 89, 90
CMD1	Inner plaque	Nuclear MTs, structural component	Calmodulin 2, 1 and 3	24, 32, 33
CNM67	Outer plaque, satellite	Cytoplasmic MTs, structural component		15, 34
KAR1	Halfbridge	SPB duplication		17, 47, 48
MPS1	SPB	Protein kinase, SPB duplication	hMPS1, mMPS1, xMPS1	18, 26, 96, 97
MPS2	Nuclear membrane at SPB	SPB duplication		18, 23, 43
MPS3	Halfbridge	SPB duplication		19
NDC1	Nuclear membrane at SPB, nuclear pore	SPB duplication		44, 92
NUD1	Outer plaque, satellite	Cytoplasmic MTs, structural component	Centriolin	8, 35, 36
SFI1	Halfbridge	SPB duplication	hSfi1	13
SPC29	Inner plaque, satellite	SPB duplication, structural component		8, 28
SPC42	Central plaque, satellite	SPB duplication, structural component		6, 25, 26
SPC72	Outer plaque and halfbridge	Cytoplasmic MTs, structural component		71, 73
SPC97	Gamma-tubulin complex	MT nucleation and SPB duplication	TUBGCP2	22, 30, 59, 62
SPC98	Gamma-tubulin complex	MT nucleation	TUBGCP3	21, 22, 30, 61, 62
SPC110	Inner plaque	MT nucleation	Kendrin	11, 31–33
STU2	Outer plaque, microtubules	Cytoplasmic MTs	XMAP215	20, 37, 38, 42
TUB4	Gamma-tubulin complex	MT nucleation	TUBG1,2	14, 30, 57, 58

gamy (i. e. Kar1p, [17]), microtubule organization (i. e. Mps1p and Mps2p, [18]), or incorporation of GFP-tagged SPB components into SPBs (i. e. Mps3p, [19]). In addition, the use of genes encoding SPB components in two-hybrid screens (i. e. Spc72p, [20]) and the use of mutant alleles of these genes in suppressor or enhancer screens (i. e. Spc98p, [21]; Spc97p and Spc98p, [22]; Bbp1p, [23]) have also contributed components. Finally, studies examining genes with no apparent connection to the SPB have also uncovered components (i. e. Cmd1p, [24]). Table 4.1 includes the list of core SPB components that are found in either the layers of the SPB or in the adjacent half bridge.

4.2.1
The Central Plaque

The SPB can be viewed as being built around a crystalline array of the Spc42p protein in the central plaque region (Figure 4.2). Spc42p is a coiled coil protein that exists as dimers and forms an extended plaque when the protein is overexpressed in yeast cells [25]. Structural analysis of these isolated extended plaques (we call them "superplaques", [26]) by cryo-EM [6] revealed hexagonal crystalline packing of Spc42p; interestingly, electron tomography of intact SPBs *in situ* also suggests the presence of the Spc42p crystalline array [7]. Beyond the presumptive role of the coiled coil domains in assembly, the Spc42p protein must also be phosphorylated properly [8, 26]. Coiled coil domains are common in SPB and centrosomal components (see also Chapter 7). These domains have been shown to be protein

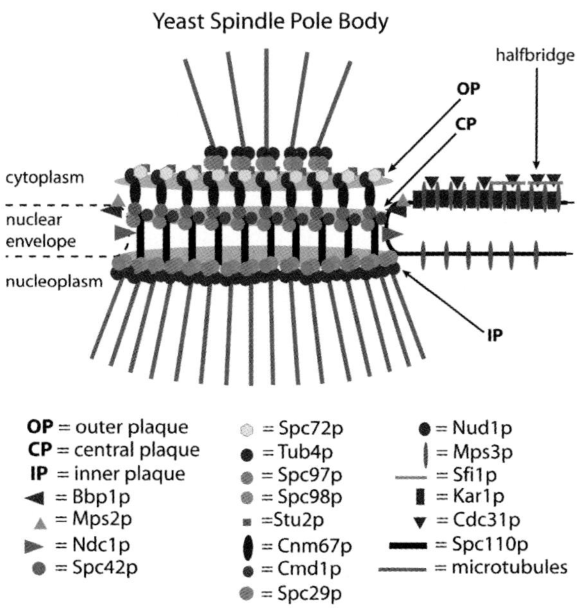

Figure 4.2 Yeast Spindle Pole Body. Shown here is a schematic of the organization of most of the components described in Table 4.1 (see Color Plates page XXIV.

interaction motifs and are found where a rigid structure is needed (reviewed in [27]). In fact, the SPB can be viewed as layers of coiled coil proteins, one on top of the next. In order to build a model for the SPB, it has been important to understand the orientation of each of the proteins in the organelle and to identify their binding partners. This has been accomplished largely through identifying binding partners by two hybrid and co-immunoprecipitation tactics and then mapping the interacting domains. This data has been used to infer how the organelle is organized and a model is presented in Figure 4.2. The validity of this model is being tested using fluorescence resonance energy transfer (FRET). These analyses utilize strains in which various SPB components are tagged with the GFP variants, yellow fluorescent protein (YFP) and cyan fluorescent protein (CFP); these differentially-tagged molecules will generate a FRET signal when they are in close proximity (as discussed in [19]).

4.2.2
The Inner Plaque

From the Spc42p array working towards the nucleus, there are the proteins Spc29p and Spc110p, both of which bind to the N-terminus of Spc42p indicating that the N-terminus of Spc42p points towards the nucleus and the C-terminus points toward the cytoplasm [8, 28]. Spc29p binds Spc42p and Spc110p, as well as other proteins discussed later, indicating that it holds these two coiled coil proteins together [28]. Spc110p binds Spc42p via its C-terminus, and this is probably indirectly via Spc29p binding [8]. The C-terminus of Spc110p also binds Cmd1p (calmodulin), but at a site different from that on Spc110p which binds Spc29p [29]. At its N-terminus, Spc110p binds the γ-tubulin complex, composed of Tub4p (γ-tubulin), Spc97p and Spc98p (discussed in detail below), via direct binding of Spc98p [22, 30]. This complex is responsible for microtubule nucleation and forms the inner plaque or the nuclear face of the SPB. The connection between the central plaque and the inner plaque formed by Spc110p encouraged the idea that this protein formed a strut that set the distance between these two plaques. This idea was proven true in a series of elegant experiments in which John Kilmartin lengthened or shortened the central coiled coil domain of Spc110p by gene engineering [31]. Using the assumption that this domain existed as an elongated structure, he could predict the change in distance that would result between the central and inner plaque, which was found to be accurate by EM analysis of the resultant SPBs. Spc110p is the sole length-determining strut holding the γ-tubulin complex onto the nuclear face of the SPB, and this observation has been informative in understanding the function of the human kendrin that also binds calmodulin and is thought to be in the pericentriolar material associating with γ-tubulin complexes [32, 33].

4.2.3
The Outer Plaque

The outer plaque is more complex, because more proteins are involved and important signaling functions occur here. From the Spc42p array working towards the cytoplasm, the organization is such that the C-terminus of Spc42p binds to the C-terminus of Cnm67p [8, 34]. Like the spacer function of Spc110p in the inner plaque, Cnm67p shows a similar function in the outer plaque, as described using engineered versions of the gene [34]. Next, the N-terminus of Cnm67p binds to the C-terminus of Nud1p [28]. Nud1p is the protein with which components of the mitotic exit pathway (MEN) associate as they coordinate the completion of mitosis and cytokinesis (discussed later), and a Nud1p domain appears to be conserved in a vertebrate ortholog, centriolin, that is found at centrosomes [35]. After this, Nud1p binds to Spc72p, preferentially to the phosphorylated form of Spc72p [36]. Lastly, the γ-tubulin complex of the inner plaque has the same composition as that found on the outer plaque, but on the outer plaque the γ-tubulin complexes bind to Spc72p, instead of Spc110p.

Another Spc72p binding partner is Stu2p [20]. Stu2p seems to have several functions, as mutations in this protein leads to fewer and less dynamic cytoplasmic microtubules, cell cycle arrest and failure to elongate the mitotic spindle [37, 38]. Although the exact function of Stu2p at SPBs is not fully understood, recent studies point to a role in microtubule plus-end destablization [37, 39]. There is a Stu2p-related family of proteins in vertebrates, XMAP215-like, that bind centrioles via an interaction with Spc72p-related proteins called transforming acidic coiled coils (TACCs) [40–42]. Overall, the major components of the entire SPB from γ-tubulin complexes of the inner plaque in the nucleus to the γ-tubulin complexes of the outer plaque in the cytoplasm are known and much regarding their arrangement and interactions has been elucidated.

4.2.4
Nuclear Membrane Factors

The multiple layers of the spindle pole body span the nuclear envelope resulting in access to both the nucleoplasm and cytoplasm. This is achieved by placing the spindle pole body in a fenestra in the nuclear envelope similar to nuclear pore complexes [5]. Recent electron tomographic analysis of SPBs revealed a "hook like" structure that may anchor the SPB in the nuclear envelope [7]. There are two known membrane proteins in the SPB, Mps2p and Ndc1p, that are thought to anchor the SPBs in the nuclear envelope [43, 44]. Mps2p has a single transmembrane-spanning segment and binds Bbp1p [23]. In turn, Bbp1p binds Spc29p of the central plaque providing a link from the membrane to the central plaque of the SPB [23]. Ndc1p has multiple transmembrane segments, but it is unknown with which other SPB components it interacts. The *S. pombe* ortholog of Ndc1p is Cut11, which has also been shown to localize at SPBs. In *cut11* mutants, the SPBs are observed in the nucleoplasm, suggesting that they are unable to interact at the nuclear envelope fenestra [45]. In addition, expression of budding yeast *NDC1* in *S. pombe* cut11 mutant strains leads to a partial rescue of this phenotype [45]. Both Cut11 and Ndc1p are also found at nuclear pore complexes and may be involved in anchoring these organelles in the nuclear envelope similar, to their function at the SPB.

4.2.5 The Halfbridge

Longitudinal sections through the SPB reveal a darkly staining (indicating protein) segment of the nuclear envelope on one side of the SPB called the halfbridge (Figure 4.1a). When viewed *en face*, the SPB is not completely round; the halfbridge interacts with the SPB along a straight edge in the otherwise round SPB [5]. The halfbridge is critical to SPB assembly and becomes a further modified "full bridge" during the assembly of new SPBs (discussed below). Four proteins are known to be in a complex at the halfbridge: Cdc31p, Kar1p, Mps3p and Sfi1p. *CDC31* is the centrin homolog in budding yeast and encodes a small calcium binding protein [46]. Centrin is widely conserved among eukaryotes. Cdc31p binds two different membrane proteins, Kar1p and Mps3p [19, 47, 48] and one internal repeat protein Sfi1p [13]. Both Kar1p and Mps3p have single transmembrane-spanning domains and may form a trimeric complex with Cdc31p [19]. Sfi1p binds multiple Cdc31p molecules at its repeat domains and is also conserved in humans [13]. These proteins are the only proteins found thus far at the halfbridge; however, there may be other components, particularly on the nuclear side of the halfbridge.

4.2.6 Structure Summary

There are a few additional proteins that have been localized to SPBs, but their exact location in the organelle has not been determined. Notable is the Mps1p protein kinase that is required during SPB duplication (see below). The kinase has been localized to SPBs, and it has been shown to bind Spc42p [26]. The binding to Spc42p could be in a cytoplasmic pool that has not yet assembled, and/or it could be in the SPB. Mps1p levels are very low in the cell, and it is expected that it would be a low stoichiometry component of the SPB. There may be similar proteins that have yet to be characterized (i. e. Spc105p, [12]).

Although the SPB is presented as a static structure of interlocked coiled coil proteins that make a large structure viewable in the EM, it is a dynamic organelle [49]. Adams and Kilmartin have reported that the size of the SPB changes during the cell cycle (reviewed in [50]). Furthermore, SPBs can incorporate ectopically expressed, GFP-tagged SPB components indicating some level of turnover of components [19]. The dynamic nature of the SPB has been studied carefully in strains in which Spc110p was tagged with GFP and fluorescence recovery after photobleaching (FRAP) was used to look at the rate of its recovery at the SPB [49]. About a 50% exchange of Spc110p occurs during G1/S, a stabilization follows during G2 (no ex-

exchange of Spc110p occurs during G1/S, a stabilization follows during G2 (no exchange), and more incorporation of Spc110p takes place during G2/M (no exchange). These data suggest that SPB assembly/duplication should be described in terms of growth and exchange of components, rather than in terms of conservative or dispersive assembly [49]. It may not be surprising that component turnover occurs. SPBs are absolutely essential for the cell, yet there is no report of their *de novo* formation as has been reported for centrioles in centrosomes [51]. The collective data in the field indicate that all SPBs are formed from existing ones as discussed below and that SPBs need to be maintained, which could include repairing damage. Component turnover in SPBs may serve both to affect the function of the organelle and to maintain its integrity.

4.3 Microtubule Nucleation

The most significant function of SPBs, like other spindle pole organelles, is the nucleation and organization of microtubules. Microtubules in budding yeast make up the structural framework of mitotic and meiotic spindles, whose role it is to move chromosomes during mitosis and meiosis [52], and to position the nucleus [53, 54]. The widely conserved γ-tubulin complex is responsible for the nucleation of microtubules from α/β tubulin dimers.

γ-Tubulin was discovered in *Aspergillus nidulans* [55, 56], and is a member of the tubulin superfamily, which is discussed in detail elsewhere in this volume (see Chapters 2 and 5). *TUB4*, the budding yeast gene that encodes γ-tubulin, was found in the genome project, and several groups have shown by genetic analysis that γ-tubulin is needed for microtubule nucleation [14, 57, 58]. Tub4p forms a stable 6S complex with two other proteins: Spc98p and Spc97p [21, 30, 59]. Each complex contains one molecule of Spc97p, one molecule of Spc98p and two or more molecules of Tub4p. A reconstituted γ-tubulin complex using recombinant proteins with the addition of several other recombinant SPB components exhibits limited ability to nucleate microtubules *in vitro* [60]. The 6S γ-tubulin complex is similar to complexes reported in other organisms, and the γ-tubulin binding proteins have been conserved and include vertebrate forms of Spc98p (HsSPC98 or hGCP3; [61, 62]) and Spc97p (hGCP2, [62]). In other systems, a larger complex, the γ-TURC with a diameter of 25 nm and a sediment coefficient of 26–32 S has been identified [63], but this complex does not seem to exist in yeast. Genetic analysis in yeast has shown that the proteins encoded by *SPC97* and *SPC98* both function in microtubule organization, but their specific roles, if any, in microtubule nucleation is not known.

S. cerevisiae exhibits a closed mitosis, with the SPB embedded in the nuclear envelope for the entire life cycle of these cells [64]. Because of its unique location, the SPB is capable of nucleating two subsets of microtubules: one in the cytoplasm and one in the nucleoplasm. As discussed before, the organelle has different tethering molecules for these two sets of microtubules. Spc98p and Spc97p of the γ-tubulin complex bind a protein of the inner plaque, Spc110p [30]. Specifically, Spc110p binds Spc98p more directly and Spc97p binding is dependent on Spc98p [22]. Similar to the components of the γ-tubulin complex, Spc110p is conserved and its ortholog, pericentrin/kendrin, is found at the pericentriolar material of centrosomes where it binds γ-tubulin complexes by its N-terminal region [22, 33, 65]. Mutations in *SPC110* or *CMD1* (which encodes a binding partner of Spc110p) cause defects in microtubule attachment and SPB structure [66, 67]. Spc110p is a phosphoprotein *in vivo*, and it is a substrate of the Mps1p kinase [68] and likely of the Cdc28p kinase [69], the major CDK in yeast. Proper Spc110p modification appears to enhance binding of the γ-tubulin kinase complex. Interestingly, Spc98p is also a phosphoprotein, and this modification is dependent on both the cell cycle and the Mps1p kinase [70]. Although γ-tubulin complexes are found on both faces of the SPB, phospho-Spc98p appears to be limited to the nuclear face [70].

Cytoplasmic microtubules are nucleated from sites both on the outer plaque of the SPB and from the bridge during SPB duplication. γ-Tubulin complexes are tethered via Spc72p to either the outer plaque, by Spc72p binding to Cnm67p, or to the halfbridge by Spc72p binding to the N-terminus of the halfbridge component Kar1p [71, 72]. Mutations in *SPC72* result in defects in spindle elongation, cytoplasmic microtubule organization and nuclear migration [71], as expected for defects in tethering of cytoplasmic microtubules. Spc72p and Cnm67p are not essential individually in some strain backgrounds, presumably because the other gene will supply sufficient cytoplasmic microtubules [15, 73, 74]; in fact, the double mutation is lethal [74]. Spc72p localization changes in a cell cycle-dependent manner, as does the phosphorylation pattern of Spc72p [8, 72]. It is likely that unphosphorylated Spc72p binds Kar1p in G1 and phosphorylated Spc72p binds Nud1p in S/G2 and M [36]. Spc72p plays only a minor role during vegetative growth; this localization and therefore recruitment of the γ-tubulin complex becomes essential during karyogamy [72]. The Tub4p binding domain on Spc72p has been mapped to the N-terminal 176 amino acids [71].

Finally, microtubule nucleation capacity is regulated on centrosomes, with nucleation capacity increasing during "maturation" in preparation for mitosis (see Chapters 8–10). The number of MTs nucleated by the SPB appears to be limited by the size of the SPB, on which the MTs are packed tightly on the nuclear face [7], and does not appear to change much during the cell cycle. The SPB does increase in size with ploidy as a result of fusion of SPBs during karyogamy, and the larger SPBs appear to accommodate the increased number of MTs necessary to capture the additional chromosomes during mitosis. There also appears to be regulation of the γ-tubulin complex by post-translational modification. *SPC97* is not controlled transcriptionally, but *SPC98* is repressed during stationary growth [75]. *TUB4*, on the other hand, is controlled transcriptionally throughout the cell cycle and is likely elevated after the G1/S transition [14]. Post-translational modification of Spc98p is discussed above, and a study by Vogel and Snyder [76] shows that the carboxy terminus of Tub4p is important for the re-organization of existing microtubules and the movement of the spindle into the bud. Tub4p also has been shown to be modified as part of its nucleation function [77]. Tub4p is a phosphoprotein *in vivo* and is phosphorylated at a C-terminal tyrosine; this phosphorylation seems to be most prominent in G1 of the cell cycle and is absent during mitosis. Phosphorylation of Tub4p at Tyr445 seems to be an important *in vivo* modification, as mutation of this residue caused a defect in the kinetics of microtubule dynamics and in their organization [77]. The mechanism by which Tub4p affects MT dynamics is not clear, particularly since yeast does not appear to have minus-end flux that would likely be controlled by γ-tubulin [78].

4.4 Assembly/Duplication of SPBs and Centrosomes

Two very important duplication events which occur once and only once each cell cycle, are the duplication of DNA and the duplication of the centrosome. The fidelity of these events ensures that each daughter cell receives exactly one copy of the duplicated genome. Regulation of centrosome duplication must be stringent, as failure of this event may result in either monopolar or multipolar spindles and aneuploid cells; both phenomena have been observed in several types of human cancer and other diseases (see Chapters 16–18). Finally, even though the yeast SPB and the vertebrate centrosome are structurally different, many of the components are conserved, and the manner by which they duplicate during each cell cycle is comparable (reviewed in [50]).

4.4.1 Electron Microscopic Description of Duplication

SPB duplication in budding yeast was first described, using electron microscopic techniques, as three discernible steps [3]. Early in G1, each cell starts out with one SPB and its associated halfbridge structure (Figure 4.3a). The first step of duplication is elongation of the halfbridge structure and the appearance of electron-dense material on the distal, cytoplasmic face of the halfbridge (Figure 4.3b). This material, called the satellite, is thought to be the nascent SPB and is composed of Spc42p, Spc29p, Nud1p and Cnm67p [8]. Spc42p, a central plaque component, and Spc29p, an inner plaque component that binds Spc42p, are both required for SPB duplication. Cnm67p and Nud1p are both outer plaque proteins that have not been shown to be required in duplication, but such a function could be masked by their cytoplasmic microtubule function [15, 8]. The placement of the satellite at the distal, cytoplasmic site on the halfbridge is consistent, but it is not known what specifies this assembly site. Sfi1p was reported recently to be localized at this site of satellite formation [13].

Second, a "duplication plaque" has been identified recently (Figure 4.3c), and is described as an intermediate between the satellite-bearing SPB and duplicated side-by-side SPBs. The duplication plaque is an expanded satellite that resembles the layered appearance of the SPB, and it can nucleate cytoplasmic microtu-

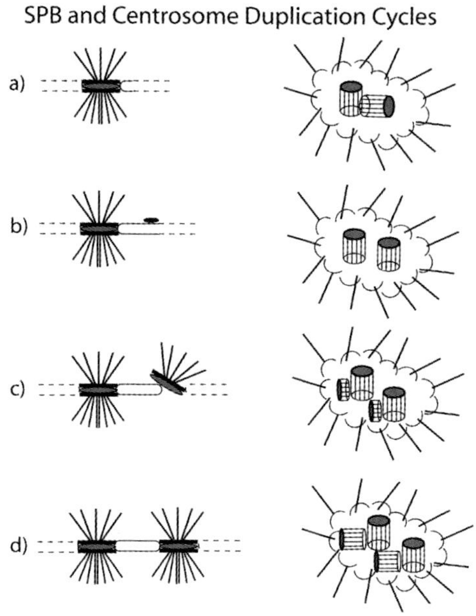

Figure 4.3 SPB and centrosome duplication cycles. Both processes start with (a) single organelles at the beginning of the cell cycle. First, (b) each undergoes the initial step of satellite formation and centriole splitting, respectively. Second, (c) both duplication intermediates continue to mature, and third, (d) fully duplicated organelles result. The newly assembled yeast SPB is inserted into the nuclear envelope, and the new centrioles have achieved full size.

bules [8], dependent on recruitment of components of the outer plaque, including Spc72p and the γ-tubulin complex (Spc97p, Spc98p and Tub4p). The duplication plaque appears to be the structure that is inserted into a pore in the nuclear envelope that may be formed by the fusion of the ends of the halfbridge and retraction of the halfbridge lipid bilayers [8, 50]. The third step revealed by Byers and Goetsch [3] was the duplicated side-by-side SPBs that are attached to each other by the complete bridge (Figure 4.3d). Later, these two duplicated SPBs will separate and move to opposite ends of the nucleus to form the bipolar spindle (not shown), each taking with them a halfbridge structure. While the mechanism of bridge scission is unknown, it is known that kinesin-like motor proteins and Cdc28p, with B-type cyclin, activity is required for the separation event [79–81].

There are obvious similarities between SPB duplication and the duplication of centrioles in centrosomes. In both cases, there is an existing organelle on which a new organelle is assembled at a specific site, either the formation of the satellite on the halfbridge or the formation of the procentriole at the base of the centriole that continues to lengthen and mature by acquiring modifications (i. e. distal and subdistal appendages) [82, 83]. A review by Adams and Kilmartin [50] offers a more extensive comparison of these processes. Both processes appear to be conservative, in which a new SPB organelle or centriole is made adjacent to an existing structure [84], but the analysis of Spc110p turnover indicates the duplication involves significant subunit turnover in the existing SPB [49]. Furthermore, both duplication events are close to the G1/S boundary, with both processes starting in G1 (satellite formation and centriole separation).

4.4.2
Cell Cycle Regulation of Duplication

Both SPB duplication and centrosome duplication are under the control of CDK kinases. The role of CDK2/cyclin A or E in centrosome duplication is treated elsewhere in this volume (see Chapter 9). In yeast, the SPB cycle is under the control of the cyclin-dependent kinase Cdc28p, together with an activating cyclin (Cln and Clb proteins) [85]. Cdc28p appears to play a role both in promoting SPB duplication and preventing reduplication of this organelle. Specifically in G1, the activity of Cln1, 2, and 3p are necessary for the initiation of SPB duplication. Low Cdc28p activity, from mutant phenotypes or mating factor treatment, leads to a satellite bearing SPB indicating that Cdc28p is needed to proceed beyond this step [4]. During S and G2/M, when the SPBs are maturing and moving apart to form the bipolar spindle, the activity of Clb5, 6 is necessary. The activity of the mitotic cyclins, Clb1–4, seems to be necessary to prevent reduplication [85]. Lastly, as cells exit mitosis, undergo cytokinesis and segregate both DNA and a single SPB into each daughter cell, Clb1–4 activity needs to be low, as this is thought to "license" the SPB to be able to duplicate in the next cell cycle [85].

4.4.3
Genetic Analysis of Duplication

Genetic analysis of SPB duplication has identified a number of genes required for the duplication of this organelle. All of the genes are essential and are studied primarily using conditional alleles. The duplication genes have been identified in genetic screens (i. e. [18, 19]) or in the course of the analysis of genes encoding components of the SPB (i. e. [23, 25]). The general phenotype of cells harboring these mutations is a mitotic arrest with a monopolar spindle instead of the normal bipolar spindle. This phenotype was first recognized for *cdc31-1* [16] and indicates that cells will proceed into the cell cycle growing a bud and replicating chromosomes in the absence of SPB duplication. The mutant cells arrest in mitosis, because monopolar spindles cannot capture chromosomes correctly and the spindle assembly checkpoint is activated [86]. The cells can eventually exit the arrest (i. e.[87, 88]), but how this happens is not understood. Finally, the observation that aberrant microtubule arrays seen in these mutants arise from a single, unduplicated SPB revealed that these SPBs exhibit different morphologies in different mutant strains (e. g. [18]). These different morphologies indicate that the genes act at different steps in SPB duplication, an interpretation that has been supported by execution-point experiments and epistasis tests (e. g. [18]). Such work has led to the identification of genes required for each of the steps of SPB duplication that had been identified previously by EM analysis.

As mentioned above, *CDC31* was the first gene known to be required for SPB duplication in yeast. The single SPB in *cdc31* mutants was observed to have little or no bridge structure, suggesting that Cdc31p is necessary for forming or maintaining the bridge and, therefore, would be required early in SPB duplication. Indeed, execution-point experiments showed that Cdc31p is not required after the formation of the satellite [18]. Furthermore, Cdc31p is found at the halfbridge [89]. Importantly, *CDC31* encodes the yeast homolog of centrin. Centrins are small, conserved calcium binding proteins that are found at MTOCs and are involved in the duplication of centrioles [90] and basal bodies [91]. Cdc31p is not a membrane protein, yet it is associated with the halfbridge. Cdc31p is tethered to the halfbridge by the membrane proteins Kar1p and Mps3p. Furthermore, Kar1p interacts with Mps3p indicating that there is a complex of these two proteins that binds Cdc31p. Another halfbridge binding partner, Sfi1p, has been discovered quite recently [13]. Much like those with *cdc31* mutations, cells containing temperature-sensitive mutations in *KAR1, MPS3*, and *SFI1* fail in SPB duplication at the restrictive temperature and have unduplicated SPBs with little or no halfbridge. How these proteins in various complexes act to initiate SPB duplication leading to the formation of the satellite is unknown, but is likely to be conserved in that centrin is required for the duplication of SPBs, centrioles and basal bodies. Moreover, Sfi1p is widely conserved [13].

Analysis of the SPB components Spc42p and Spc29p reveals that the two genes responsible for these components are also required for SPB duplication. These proteins are required after formation of the satellite, because when strains mutant in either gene are released from mating factor arrest (after satellite formation) at the restrictive temperature, they still fail in SPB duplication [25, 28]. These findings are well explained by the discovery that both these proteins are components of the satellite, and would be added to the satellite as the duplication plaque is formed. Interestingly, overexpression of Spc42p leads to a large lateral expansion of the central plaque that has been used for structural analysis and as the basis of assembly assays [25]. The protein kinase Mps1p, discussed further below, is also required for the maturation of the satellite, and its activity is required for proper assembly of Spc42p, most likely because Spc42p is a substrate of Mps1p [26]. The other satellite components Cnm67p and Nud1p have other cellular functions which complicate experiments to determine if they have roles in SPB duplication [15, 8].

The final step in SPB duplication, identified by mutation, is the insertion of the nascent SPB, or duplication plaque, into a pore in the nuclear envelope. Mutants in two membrane proteins of the nuclear envelope, Mps2p and Ndc1p, appear to fail in this step, giving rise to an unusual phenotype [18, 92]. In these mutants, two SPB-like organelles are observed, but only one of them has normal nuclear microtubules and is associated with the chromatin. The other SPB is on the cytoplasmic side of the nuclear envelope and only has microtubules in the cytoplasm, which will serve to move this defective SPB away from the other SPB. As it is moved, the defective SPB will only bring along the nuclear envelope but not the chromosomes. The defective SPB in these mutants appears to be quite similar to a duplication plaque that has not been inserted into the nuclear envelope. It is not known how these proteins function, but Mps2p is known to bind Bbp1p [23]. Mutations in *BBP1* have a similar phenotype to those of *MPS2* and *NDC1* mutants, suggesting that Bbp1p participates in the insertion event. Furthermore, Bbp1p binds Spc29p, suggesting a role for anchoring the SPB in the nuclear envelope [23]. Upon insertion into the pore in the nuclear envelope, the new SPB has access to components present in the nucleoplasm and can thus form the inner plaque. As mentioned, components of the inner plaque include Spc110p and its binding partner calmodulin which, in turn, bind the γ-tubulin complex (Spc97/98p and Tub4p). Mutations in genes encoding these components can show defects in the inner plaque or its complete absence [57, 66]. Completion of the inner plaque marks the completion of SPB duplication, resulting in duplicated side-by-side SPBs (Figure 4.3d).

Critical for SPB duplication is the Mps1p protein kinase. Mps1p is found at both SPBs and kinetochores, where it functions in both SPB duplication and in the spindle assembly checkpoint [18, 26, 86, 93]. The original allele of *MPS1*, *mps1-1*, exhibits unduplicated SPBs with a large halfbridge, indicating that the satellite was not formed or was not functional for duplication [18]. Execution-point experiments revealed that Mps1p is required upon release from mating factor arrest and the likely substrate for this requirement may be Spc42p, an Mps1p binding partner and *in vitro* substrate [26]. Further analysis of *MPS1* led to the identification of additional mutant alleles that appear to be defective earlier in SPB duplication similar to *CDC31/KAR1/MPS3* mutants (*mps1–8*, [26]) or later in duplication similar to *NDC1/MPS2/BBP1* mutants (*mps1–737*, [94]). These findings indicate that Mps1p is required at all known steps of SPB duplication either because it has multiple substrates and/or some substrate(s) needs multiple phosphorylation events to act at all steps during duplication. No other regulator of SPB duplication behaves this way. Mps1p protein kinase is conserved, and vertebrate orthologs have been shown to be involved in the spindle assembly checkpoint [95–97]. Localization of Mps1 at centrosomes [96, 98, 99] and Mps1 control of centrosome duplication [98, 99] has been controversial [97].

Several other important regulators of SPB duplication have been identified and include the proteosome subunit Pcs1p [100], the ubiquitin-like proteins Rad23p and Dsk2p [101], the chaperone Cdc37p [102] and the heat shock transcription factor Hsf1p [103]. Which steps in duplication require these genes are not known, but it is clear that a diverse collection of protein activities is required for SPB duplication. In many cases, the vertebrate orthologs of genes required for SPB duplication are clearly required for centrosome duplication. More detailed analysis of SPB duplication will yield insights into the control of this process, which will also shed light on the control of centrosome duplication.

4.5 Signaling Platform

In addition to their well-characterized role in organizing the bipolar spindle during mitosis, spindle pole organelles have more recently been implicated as having a role in mitotic exit, cytokinesis, progression into S phase and possibly in commitment to enter mitosis. The common theme for the role of the spindle pole organelle in these processes is that the organelles function as a scaffold for the signaling molecules in the given pathway.

The best characterized example of this scaffolding function of the budding yeast SPB is for the signaling molecules of the mitotic exit network (MEN). The most important role for the MEN is the destruction of B-type cyclins and the decrease of cyclin–CDK activity, which allows the cells to exit mitosis and undergo cytokinesis. The major MEN-regulated event responsible for CDK inhibition is release into the cytoplasm of the Cdc14p phosphatase from its inhibitory binding partner Net1p/Cfi1p in the nucleous [104]. In order to release Cdc14p from the nucleolus, the cell must align its mitotic spindle appropriately so that one SPB (the older one) is directed toward the bud and the other is directed back into the mother cell [105]. Several MEN components localize preferentially to the cytoplasmic side of the old SPB. A clever experiment using strains harboring a red fluorescent protein (RFP)-tagged Spc42p has shown that the old SPB preferentially goes to the daughter cell [104]. On this old SPB is found the GTPase Tem1p and the GAP complex consisting of the Bub2p/Bfa1p complex. This Bub2p/Bfa1p/Tem1p complex is tethered to the SPB via interactions with Nud1p. Nud1p is also responsible for the correct localization of the MEN components Mob1p/Dbf2p and Cdc14p to the SPB. Initially, Tem1p localizes to the older SPB in its inactive GDP-bound state. When the nucleus is moved to the bud neck between the mother and bud, Tem1p has access to its activating GEF Lte1p, which transforms it into the active GTP-bound state. Lte1p is spatially limited to the bud and, therefore, will only act on Tem1p as the SPB crosses into the bud. Activated Tem1p then acts through the protein kinases Cdc15p and the Dbf2p-Mob1p complex. Another MEN pathway component Cdc5p (the budding yeast polo kinase) is found at the SPB and has been shown to promote Dbf2p kinase activity and to activate Bub2p/Bfa1p. Once in the cytoplasm, Cdc14p acts directly to stabilize Sic1p (a CDK inhibitor), indirectly to increase levels of Sic1p (by stabilizing the Swi5p transcription factor), and directly to allow Cdh1p/Hct1p to bind and activate the APC to destroy B-type cyclins (reviewed in [106]).

An analogous pathway to the MEN may exist in vertebrates. Among the known mammalian orthologs of proteins in the MEN pathway are proteins with significant orthology to Cdc14p called hCdc14A and hCdc14B [107, 108]. Recently, hCdc14B was shown to localize to the nucleolus, whereas hCdc14A localizes to interphase centrosomes [108, 109]. Studies with hCdc14A also show a role in centrosome structure, mitotic spindle organization, chromosome segregation, karyokinesis, cytokinesis [108, 109], and activation of APC (via dephosphorylation of Cdh1p subunit, [107]). In addition to the Cdc14p ortholog in mammalian cells, there are similar proteins to Dbf2p and Mob1p in humans [110–112]. Although these orthologs have been identified, their molecular functions are unknown. In addition, a possible ortholog for the Bfa1p/Bub2p GAP complex has been found and is known as GAPCenA (GAP centrosome associated, [113]). Next, polo kinase family members (Cdc5p in budding yeast) have been known for some time to be localized to centrosomes or SPBs, where they have roles in maturation and in signaling (see Chapters 8 and 9). Similar to the MEN function of Cdc5p, the polo kinase in *Xenopus laevis* known as Plx1, is required for M-phase exit in egg extracts [114]. Lastly, a novel centriolar protein called centriolin has been described to have a Nud1p-like domain, and a defect in this protein points to a function in both cytokinesis and cell cycle progression [35]. Even though a complete analogous MEN pathway is not yet clear in higher eukaryotes, it has been shown that spindle misalignment in mammalian cells does delay cell cycle progression [115].

Like the yeast SPBs, centrosomes, or perhaps components associated with these organelles, have a role in cytokinesis as well. To illustrate this, removal of centrosomes [116, 117] results in cytokinesis failure and the cells remain binucleate or attached by an intercellular bridge. Consistent with a role for centrosomes in cytokinesis, live-cell imaging shows that the maternal centriole moves to the site of the intercellular bridge just before the bridge narrows and the microtubules depolymerize [118].

Fission yeast have a signaling pathway called the septation initiation network (SIN; analogous to MEN) wherein the importance of SPB localization of the signaling molecules was first recognized (reviewed in [106]). It has been shown that the Cdc14p ortholog in the SIN (Clp1/Flp1) can regulate CDK activity, as well as having a less well understood role in cytokinesis. Like in the MEN, Clp1/Flp1 is localized to SPBs and the nucleolus; however, it is released earlier in mitosis than in the MEN, and the SIN functions to maintain its localization out of the nucleolus. Unlike the MEN, the SIN is only responsible for cytokinesis/septation and not mitotic exit. The SIN is composed of the following components: Spg1 (GTPase), Cdc16-Byr4 (GAP complex), and protein kinases (Cdc7, Sid1, Sid2 and Plo1). Like in the MEN, Spg1 localizes to SPBs and is inhibited by Cdc16-Byr4. Localization of these MEN components to the SPB relies on the SPB components Sid4 and Cdc11. Once mitosis is initiated, the GAP complex leaves the SPBs and the kinase Cdc7 (Cdc15p ortholog) is recruited; it is thought that the GAP presence at SPBs prevents Cdc7 recruitment [119, 120]. The protein kinase Sid1, along with its associated factor Cdc14, localizes to the SPB once anaphase has begun, and this localization correlates with late anaphase events such as the reduction in CDK activity [121]. Unlike the Sid1 kinase, the Sid2 kinase, along with associated factor Mob1, localizes to SPBs throughout the cell cycle and requires activation at the SPB (likely via Cdc7 and Sid1/Cdc14) to move to the site of septation, the medial ring [122–125]. The polo kinase Plo1 is also required for septation, and it is likely to fulfill its function before the SIN is initiated [126–128].

A new role for SPBs in mitotic commitment has recently come forth in studies of *S. pombe* [129]. The progression into mitosis is regulated by the cyclin/Cdk kinase complex (Cdc2/Cdc13) [130]. Cdk activity is regulated by the activating phosphatase Cdc25 and by the inhibiting Wee1 kinase. Once Cdk is activated, there is a feedback loop that further activates Cdc25 and this is accomplished by phosphorylation of Cdc25 by Cdk, other kinases [131–133] and perhaps polo kinase [129]. It appears that a fission yeast SPB component Cut12, which localizes to the cytoplasmic domain of the SPB [134], can bind and activate the polo kinase Plo1 during mitotic commitment [129]. It will be interesting to see if this function is conserved in other organisms since polo kinases are so widely distributed.

The large number and variety of proteins found at centrosomes suggest that these organelles may play a part in a number of signaling pathways.

4.6 Developmental Alteration of SPB Function

Centrosomes can be specialized for different functions. During the cell cycle there is maturation in preparation for mitosis. Maternal and paternal centrosomes can be selectively controlled during fertilization by association with the centromatrix [135]. Centrosomes can be transformed to perform functions involving ciliary growth and organization (e. g. centrioles interconverting to basal bodies, see Chapters 1 and 5). Developmental specialization of spindle pole organelles can be studied in budding yeast, wherein the SPB acts in membrane organization. This work may be relevant to the membrane association of centrosomes, particularly the Golgi, as centrosomes may play a role in the organization of this organelle (reviewed in [136]).

During meiosis in yeast, the SPB is duplicated twice. The first duplication event gives rise to the poles of the meiosis I spindle, and the second duplication of the two meiosis I SPBs gives rise to the four SPBs needed to form the two meiosis II spindles. Both of the meiosis II spindles reside in the same nucleus that is partitioned into the four spores after meiosis II [5]. The SPBs play a critical role in the sporulation process after meiotic segregation. The outer plaque of the SPB changes to a very dark-staining structure on which the prospore membrane begins to form. The prospore membrane extends out from each of the SPBs and engulfs part of the nucleus and the cytoplasm that will be included in the spore. The spore wall is then assembled on the prospore membrane [2, 137–140]. This strategy appears to ensure that four spores each include an SPB and the chromosomes that are attached to it via the spindle microtubules.

The molecular basis of the specialized function of the meiosis II SPB has come to light recently through the identification of the SPB components that are specific to this special SPB. Some components were found by screening for genes that encoded predicted coiled coil proteins whose transcription increased during meiosis [141–143]. The genes recovered were tested for SPB localization by tagging with GFP and these were *MPC54* and *MPC70* (*SPO21*). Mpc54p and Mpc70p exhibit physical interactions with the SPB components Nud1p, Spc42p and Cnm67p, and localize only to meiotic SPBs [141]. During meiosis, SPBs contain the central plaque proteins Spc42p and the outer plaque proteins Nud1p and Cnm67p. However, Spc72p is no longer at the SPB, and it is replaced by the meiotic plaque containing the proteins Mpc54p and Mpc70p [141]. Mutants of either *MPC54* or *MPC70* exhibit a normal progression through meiosis, but they are unable to sporulate. More recently, Mpc54p and Mpc70p and additional components involved in prospore membrane and ascus formation, Ady3p, Ady4p and Spo74p, have been recovered in a deletion screen looking for genes with meiotic defects [144]. These were shown to localize to the SPB as well [145, 146].

The relatively simple yeast cell offers the opportunity to look at the modification of SPBs for specialized functions. It would be interesting if some of the molecules involved in this process of prospore formation were conserved for similar membrane management functions in other cell types. It remains to be seen in what other ways the SPBs may be modified for specific functions or at specific points in the organism's life cycle.

4.7
Parting Thoughts

Early on in the study of the budding yeast SPB, it was not clear if this unusual membrane embedded organelle would bear any resemblance to other MTOCs, particularly centrosomes, beyond their shared functions in microtubule nucleation. The significant differences in morphology and the lack of centrioles in yeast certainly lent credence to the hypothesis that SPBs were unique in the realm of MTOCs. In some respects, this is true. SPBs in yeast duplicate and function without homologs of tubulin isoforms such as delta or epsilon, and without apparent tubulin modification (both of which are present in vertebrate cells). Nonetheless, it has been clearly demonstrated that SPBs use conserved molecules both for microtubule nucleation and as signaling platforms, as well as having conserved molecules regulating their duplication events. This conservation means that the analysis of yeast SPBs has made, and will continue to make, significant contributions to the understanding of spindle pole organelles in general.

Acknowledgments

We thank Michele Jones and Chris Mattison for comments on the manuscript and all members of the Winey laboratory for valuable help. We are grateful to Tom Giddings for EM images of the SPB and centrosome.

This work was supported by funding from NIH: GM51312 to M. Winey and GM08759 to S. van Kreeveld Naone.

References

1. Robinow, C. F. and J. Marak. **1966**. A fiber apparatus in the nucleus of the yeast cell. *J Cell Biol* **29**, 129–150.
2. Moens, P. and E. Rapport. **1971**. Spindles, spindle plaques and meiosis in the yeast *Saccharomyces cerevisiae (hansen)*. *J Cell Biol.* **50**, 344–361.
3. Byers, B. and L. Goetsch. **1974**. Duplication of spindle plaques and integration of the yeast cell cycle. *Cold Spring Harb Symp Quant Biol.* **38**, 123–131.
4. Byers, B. and L. Goetsch. **1975**. Behavior of spindles and spindles plaques in the cell cycle and conjugation of *Saccharomyces cerevisiae*. *J Bacteriol* **124**, 511–523.
5. Byers, B. **1981**. Cytology of the yeast cell cycle. In *The Molecular Biology of the Yeast Saccharomyces: Life Cycle and Inheritance*, E.W.J. J. N. Strathern, and J. R. Broach, (Eds). Cold Spring Harbor Laboratory: Cold Spring Harbor, NY, 59–96.
6. Bullitt, E., M. Rout, J. Kilmartin, and C. Akey. **1997**. The yeast spindle pole body is assembled around a central crystal of Spc42p. *Cell* **89**, 1077–1086.
7. O'Toole, E., M. Winey, and J. R. McIntosh. **1999**. High-voltage Electron Tomography of spindle pole bodies and early mitotic spindles in the yeast *Saccharomyces cerevisiae*. *Mol Biol Cell* **10**, 2017–2031.
8. Adams, I. R. and J. V. Kilmartin. **1999**. Localization of core Spindle Pole Body (SPB) components during SPB duplication in *Saccharomyces cerevisiae*. *J Cell Biol* **145**, 809–823.
9. Ding, R., R. R. West, D. M. Morphew, B. R. Oakley, and J. R. McIntosh. **1997**. The spindle pole body of *Schizosaccharomyces pombe* enters and leaves the nuclear envelope as the cell cycle proceeds. *Mol Biol Cell* **8**, 1461–1479.
10. Graf, R., N. Brusis, C. Daunderer, U. Euteneuer, A. Hestermann, M. Schliwa, and M. Ueda. **2000**. Comparative structural, molecular, and functional aspects of the *Dictyostelium discoideum* centrosome. *Curr Top Dev Biol* **49**, 161–185.
11. Rout, M. and J. Kilmartin. **1990**. Components of the yeast spindle and spindle pole body. *J Cell Biol* **111**, 1913–1927.
12. Wigge, P. A., O. N. Jensen, S. Holmes, S. Soues, M. Mann, and J. V. Kilmartin. **1998**. Analysis of the *Saccharomyces* spindle pole by matrix-assisted laser desorption/ionization (MALDI) mass spectrometry. *J Cell Biol* **141**, 967–977.
13. Kilmartin, J. V. **2003**. Sfi1p has conserved centrin-binding sites and an essential function in budding yeast spindle pole body duplication. *J Cell Biol* **162**, 1211–1221.
14. Sobel, S. and M. Snyder. **1995**. A highly divergent γ-tubulin gene is essential for cell growth and proper microtubule organization in *Saccharomyces cerevisiae*. *J Cell Biol* **131**, 1775–1788.
15. Brachat, A., J. V. Kilmartin, A. Wach, and P. Philippsen. **1998**. *Saccharomyces cerevisiae* cells with defective spindle pole body outer plaques accomplish nuclear migration via half-bridge-organized microtubules. *Mol Biol Cell* **9**, 977–991.
16. Byers, B. **1981**. Multiple roles of the spindle pole bodies in the life cycle of *Saccharomyces cerevisiae*. In *Molecular Genetics in Yeast*, J. F. D. von Wettstein, M. Kielland-Brandt, and A. Stenderup (Eds). Alfred Benzon Symposium 16: Munksgaard, Copenhagen, 119–131.
17. Rose, M. and G. Fink. **1987**. *KAR1*, a gene required for function of both intranuclear and extranuclear microtubules in yeast. *Cell* **48**, 1047–1060.
18. Winey, M., L. Goetsch, P. Baum, and B. Byers. **1991**. *MPS1* and *MPS2*: Novel yeast genes defining distinct steps of spindle pole body duplication. *J Cell Biol* **114**, 745–754.
19. Jaspersen, S. L., T. H. Giddings, Jr., and M. Winey. **2002**. Mps3p is a novel component of the yeast spindle pole

body that interacts with the yeast centrin homologue Cdc31p. *J Cell Biol* **159**, 945–956.

20. Chen, X. P., H. Yin, and T. C. Huffaker. **1998**. The yeast spindle pole body component Spc72p interacts with Stu2p and is required for proper microtubule assembly. *J Cell Biol* **141**, 1169–1179.

21. Geissler, S., G. Pereira, A. Spang, M. Knop, S. Soues, J. V. Kilmartin, and E. Schiebel. **1996**. The spindle pole body component Spc98p interacts with the gamma-tubulin-like Tub4p of *Saccharomyces cerevisiae* at the sites of microtubule attachment. *EMBO J* **15**, 3899–3911.

22. Nguyen, T., D. B. Vinh, D. K. Crawford, and T. N. Davis. **1998**. A genetic analysis of interactions with Spc110p reveals distinct functions of Spc97p and Spc98p, components of the yeast gamma-tubulin complex. *Mol Biol Cell* **9**, 2201–2216.

23. Schramm, C., S. Elliott, A. Shevchenko, and E. Schiebel. **2000**. The Bbp1p/Mps2p complex connects the SPB to the nuclear envelope and is essential for SPB duplication. *EMBO J* **19**, 421–433.

24. Davis, T. N., M. S. Urdea, F. R. Masiarz, and J. Thorner. **1986**. Isolation of the yeast calmodulin gene: calmodulin is an essential protein. *Cell* **47**, 423–431.

25. Donaldson, A. D. and J. V. Kilmartin. **1996**. Spc42p: a phosphorylated component of the *S. cerevisiae* spindle pole body (SPB) with an essential function during SPB duplication. *J Cell Biol* **132**, 887–901.

26. Castillo, A. R., J. B. Meehl, G. Morgan, A. Schutz-Geschwender, and M. Winey. **2002**. The yeast protein kinase Mps1p is required for assembly of the integral spindle pole body component Spc42p. *J Cell Biol* **156**, 453–465.

27. Burkhard, P., J. Stetefeld, and S. V. Strelkov. **2001**. Coiled coils: a highly versatile protein folding motif. *Trends Cell Biol* **11**, 82–88.

28. Elliott, S., M. Knop, G. Schlenstedt, and E. Schiebel. **1999**. Spc29p is a component of the Spc110p subcomplex and is essential for spindle pole body duplication. *Proc Natl Acad Sci USA* **96**, 6205–6210.

29. Geiser, J., H. Sundberg, B. Chang, E. Muller, and T. N. Davis. **1993**. The essential mitotic target of calmodulin is the 110-Kilodalton component of the spindle pore body in *Saccharomyces cerevisiae. Mol Cell Biol* **13**, 7913–7924.

30. Knop, M. and E. Schiebel. **1997**. Spc98p and Spc97p of the yeast γ-tubulin complex mediate binding to the spinde pole body via their interaction with Spc110p. *EMBO (Eur Mol Biol Organ) J* **18**, 6985–6995.

31. Kilmartin, J., S. Dyos, D. Kershaw, and J. Finch. **1993**. A spacer protein in the *Saccharomyces cerevisiae* spindle pole body whose transcript is cell cycle-regulated. *J Cell Biol* **123**, 1175–1184.

32. Tassin, A. M., C. Celati, M. Paintrand, and M. Bornens. **1997**. Identification of an Spc110p-related protein in vertebrates. *J Cell Sci* **110**, 2533–2545.

33. Flory, M. R., M. J. Moser, R. J. Monnat, Jr., and T. N. Davis. **2000**. Identification of a human centrosomal calmodulin-binding protein that shares homology with pericentrin. *Proc Natl Acad Sci USA* **97**, 5919–5923.

34. Schaerer, F., G. Morgan, M. Winey, and P. Philippsen. **2001**. Cnm67p is a spacer protein of the *Saccharomyces cerevisiae* spindle pole body outer plaque. *Mol Biol Cell* **12**, 2519–2533.

35. Gromley, A., A. Jurczyk, J. Sillibourne, E. Halilovic, M. Mogensen, I. Groisman, M. Blomberg, and S. Doxsey. **2003**. A novel human protein of the maternal centriole is required for the final stages of cytokinesis and entry into S phase. *J Cell Biol* **161**, 535–545.

36. Gruneberg, U., K. Campbell, C. Simpson, J. Grindlay, and E. Schiebel. **2000**. Nud1p links astral microtubule organization and the control of exit from mitosis. *EMBO J* **19**, 6475–6488.

37. Kosco, K. A., C. G. Pearson, P. S. Maddox, P. J. Wang, I. R. Adams, E. D. Salmon, K. Bloom, and T. C. Huffaker. **2001**. Control of microtubule dynamics by Stu2p is essential for spindle orientation and metaphase chromosome alignment in yeast. *Mol Biol Cell* **12**, 2870–2880.

38. Severin, F., B. Habermann, T.C. Huffaker, and T. Hyman. **2001**. Stu2 promotes mitotic spindle elongation in anaphase. *J Cell Biol* **153**, 435–442.
39. Van Breugel, M., D. Drechsel, and A. Hyman. **2003**. Stu2p, the budding yeast member of the conserved Dis1/XMAP215 family of microtubule-associated proteins is a plus end-binding microtubule destabilizer. *J Cell Biol* **161**, 359–369.
40. Gergely, F., D. Kidd, K. Jeffers, J.G. Wakefield, and J.W. Raff. **2000**. D-TACC: a novel centrosomal protein required for normal spindle function in the early Drosophila embryo. *EMBO J* **19**, 241–252.
41. Lee, M.J., F. Gergely, K. Jeffers, S.Y. Peak-Chew, and J.W. Raff. **2001**. Msps/xmap215 interacts with the centrosomal protein D-TACC to regulate microtubule behaviour. *Nature Cell Biol* **3**, 643–649.
42. Usui, T., H. Maekawa, G. Pereira, and E. Schiebel. **2003**. The XMAP215 homologue Stu2 at yeast spindle pole bodies regulates microtubule dynamics and anchorage. *EMBO J* **22**, 4779–4793.
43. Munoz-Centeno, M.C., S. McBratney, A. Monterrosa, B. Byers, C. Mann, and M. Winey. **1999**. *Saccharomyces cerevisiae* MPS2 encodes a membrane protein localized at the spindle pole body and the nuclear envelope. *Mol Biol Cell* **10**, 2393–2406.
44. Chial, H.J., M.P. Rout, T.H. Giddings, and M. Winey. **1998**. *Saccharomyces cerevisiae* Ndc1p is a shared component of nuclear pore complexes and spindle pole bodies. *J Cell Biol* **143**, 1789–1800.
45. West, R.R., E.V. Vaisberg, R. Ding, P. Nurse, and J.R. McIntosh. **1998**. *cut11(+)*: A gene required for cell cycle-dependent spindle pole body anchoring in the nuclear envelope and bipolar spindle formation in *Schizosaccharomyces pombe*. *Mol Biol Cell* **9**, 2839–2855.
46. Baum, P., C. Furlong, and B. Byers. **1986**. Yeast gene required for spindle pole body duplication: homology of its product with Ca^{2+}-binding proteins. *Proc Natl Acad Sci USA* **83**, 5512–5516.
47. Biggins, S. and M. Rose. **1994**. Direct interaction between yeast spindle pole body components: Kar1p is required for Cdc31p localization to the spindle pole body. *J Cell Biol* **125**, 843–852.
48. Spang, A., I. Courtney, K. Grein, M. Matzner, and E. Schiebel. **1995**. The Cdc31p-binding protein Kar1p is a component of the half bridge of the yeast spindle pole body. *J Cell Biol* **128**, 863–877.
49. Yoder, T.J., C.G. Pearson, K. Bloom, and T.N. Davis. **2003**. The *Saccharomyces cerevisiae* spindle pole body is a dynamic structure. *Mol Biol Cell* **14**, 3494–3505.
50. Adams, I.R. and J.V. Kilmartin. **2000**. Spindle pole body duplication: a model for centrosome duplication? *Trends Cell Biol* **10**, 329–335.
51. Khodjakov, A., C.L. Rieder, G. Sluder, G. Cassels, O. Sibon, and C.L. Wang. **2002**. *De novo* formation of centrosomes in vertebrate cells arrested during S phase. *J Cell Biol* **158**, 1171–1181.
52. Jacobs, C.W., A.E. Adams, P.J. Szaniszlo, and J.R. Pringle. **1988**. Functions of microtubules in the *Saccharomyces cerevisiae* cell cycle. *J Cell Biol* **107**, 1409–1426.
53. Palmer, R.E., D.S. Sullivan, T. Huffaker, and D. Koshland. **1992**. Role of astral microtubules and actin in spindle orientation and migration in the budding yeast, *Saccharomyces cerevisiae*. *J Cell Biol* **119**, 583–593.
54. Sullivan, D.S. and T.C. Huffaker. **1992**. Astral microtubules are not required for anaphase B in *Saccharomyces cerevisiae*. *J Cell Biol* **119**, 379–388.
55. Weil, C.F., C.E. Oakley, and B.R. Oakley. **1986**. Isolation of mip (microtubule-interacting protein) mutations of *Aspergillus nidulans*. *Mol Cell Biol* **6**, 2963–2968.
56. Oakley, C.E. and B.R. Oakley. **1989**. Identification of gamma-tubulin, a new member of the tubulin superfamily encoded by *mipA* gene of *Aspergillus nidulans*. *Nature* **338**, 662–664.
57. Marschall, L., R. Jeng, J. Mulholland, and T. Stearns. **1996**. Analysis of Tub4p, a yeast gamma-tubulin-like

protein implications for microtubule-organizing center function. J. *Cell Biol* **134**, 443–454.

58. Spang, A., S. Geissler, K. Grein, and E. Schiebel. **1996**. Gamma-tubulin-like Tub4p of *Saccharomyces cerevisiae* is associated with the spindle pole body substructures that organize microtubules and is required for mitotic spindle formation. *J Cell Biol* **134**, 429–441.

59. Knop, M., G. Pereira, S. Geissler, K. Grein, and E. Schiebel. **1997**. The spindle pole body component Spc97p interacts with the gamma-tubulin of *Saccharomyces cerevisiae* and functions in microtubule organization and spindle pole body duplication. *EMBO J* **16**, 1550–1564.

60. Vinh, D. B., J. W. Kern, W. O. Hancock, J. Howard, and T. N. Davis. **2002**. Reconstitution and characterization of budding yeast gamma-tubulin complex. *Mol Biol Cell* **13**, 1144–1157.

61. Tassin, A. M., C. Celati, M. Moudjou, and M. Bornens. **1998**. Characterization of the human homologue of the yeast Spc98p and its association with gamma-tubulin. *J Cell Biol* **141**, 689–701.

62. Murphy, S. M., L. Urbani, and T. Stearns. **1998**. The mammalian gamma-tubulin complex contains homologues of the yeast spindle pole body components Spc97p and Spc98p. *J Cell Biol* **141**, 663–674.

63. Zheng, Y., M. Wong, B. Alberts, and T. Mitchison. **1995**. Nucleation of microtubule assembly by a γ-tubulin-containing ring complex. *Nature* **378**, 578–583.

64. Winey, M. and E. T. O'Toole. **2001**. The spindle cycle in budding yeast. *Nature Cell Biol* **3**, E23–E27.

65. Doxsey, S. **2001**. Re-evaluating centrosome function. *Nature Rev Mol Cell Biol* **2**, 688–698.

66. Sundberg, H., L. Goetsch, B. Byers, and T. N. Davis. **1996**. Role of calmodulin and Spc110p interaction in the proper assembly of spindle pole body components. *J Cell Biol* **133**, 111–124.

67. Sundberg, H. and T. Davis. **1997**. A mutational analysis identifies three functional regions of the spindle pole component Spc110p in *Saccharomyces cerevisiae. Mol Biol Cell* **8**, 2575–2590.

68. Friedman, D. B., J. W. Kern, B. J. Huneycutt, D. B. Vinh, D. K. Crawford, E. Steiner, D. Scheiltz, J. Yates, 3rd, K. A. Resing, N. G. Ahn, M. Winey, and T. N. Davis. **2001**. Yeast Mps1p phosphorylates the spindle pole component Spc110p in the N- terminal domain. *J Biol Chem* **276**, 17958–17967.

69. Friedman, D. B., H. A. Sundberg, E. Y. Huang, and T. N. Davis. **1996**. The 110-kD spindle pole body component of *Saccharomyces cerevisiae* is a phosphoprotein that is modified in a cell cycle-dependent manner. *J Cell Biol* **132**, 903–914.

70. Pereira, G., M. Knop, and E. Schiebel. **1998**. Spc98p directs the yeast gamma-tubulin complex into the nucleus and is subject to cell cycle-dependent phosphorylation on the nuclear side of the spindle pole body. *Mol Biol Cell* **9**, 775–793.

71. Knop, M. and E. Schiebel. **1998**. Receptors determine the cellular localization of a gamma-tubulin complex and thereby the site of microtubule formation. *EMBO J* **17**, 3952–3967.

72. Pereira, G., U. Gruneberg, M. Knop, and E. Schiebel. **1999**. Interaction of the yeast γ-tubulin complex-binding protein Spc72p with Kar1p is essential for microtubule function during karyogamy. *EMBO J* **18**, 4180–4195.

73. Soues, S. and I. R. Adams. **1998**. SPC72: a spindle pole component required for spindle orientation in the yeast *Saccharomyces cerevisiae. J Cell Sci* **111**, 2809–2818.

74. Hoepfner, D., F. Schaerer, A. Brachat, A. Wach, and P. Philippsen. **2002**. Reorientation of mispositioned spindles in short astral microtubule mutant spc72Δ is dependent on spindle pole body outer plaque and Kar3 motor protein. *Mol Biol Cell* **13**, 1366–1380.

75. Planta, R. J., A. J. Brown, J. L. Cadahia, M. E. Cerdan, M. de Jonge, M. E. Gent, A. Hayes, C. P. Kolen, L. J. Lombardia, M. Sefton, S. G. Oliver, J. Thevelein, H. Tournu, Y. J. van Delft, D. J. Verbart, and J. Winderickx. **1999**. Transcript analysis of 250 novel yeast genes from chromosome XIV. *Yeast* **15**, 329–350.

76. Vogel, J. and M. Snyder. **2000**. The carboxy terminus of Tub4p is required for gamma-tubulin function in budding yeast. *J Cell Sci* **113**, 3871–3882.

77. Vogel, J., B. Drapkin, J. Oomen, D. Beach, K. Bloom, and M. Snyder. **2001**. Phosphorylation of gamma-tubulin regulates microtubule organization in budding yeast. *Dev Cell* **1**, 621–631.

78. Maddox, P., E. Chin, A. Mallavarapu, E. Yeh, E. D. Salmon, and K. Bloom. **1999**. Microtubule dynamics from mating through the first zygotic division in the budding yeast *Saccharomyces cerevisiae*. *J Cell Biol* **144**, 977–987.

79. Fitch, I., C. Dahmann, U. Surana, A. Amon, K. Nasmyth, L. Goetsch, B. Byers, and B. Futcher. **1992**. Characterization of four B-type cyclin genes of the budding yeast *Saccharomyces cerevisiae*. *Mol Biol Cell* **3**, 805–818.

80. Saunders, W. **1993**. Mitotic spindle pole separation. *Trends Cell Biol* **3**, 432–437.

81. Lim, H. H., P. Y. Goh, and U. Surana. **1996**. Spindle pole body separation in *Saccharomyces cerevisiae* requires dephosphorylation of the tyrosine 19 residue of Cdc28. *Mol Cell Biol* **16**, 6385–6397.

82. Kuriyama, R. and G. Borisy. **1981**. Microtubule-nucleating activity of centrosomes in Chinese hamster ovary cells is independent of the centriole cycle but coupled to the mitotic cycle. *J Cell Biol* **91**, 822–826.

83. Lange, B. M.H., A. Bachi, M. Wilm, and C. Gonzalez. **2000**. Hsp90 is a core centrosomal component and is required at different stages of the centrosome cycle in Drosophila and vertebrates. *EMBO J* **19**, 1252–1262.

84. Kochanski, R. S. and G. G. Borisy. 1990. Mode of centriole duplication and distribution. *J Cell Biol* **110**, 1599–1605.

85. Hasse, S. B., M. Winey, and S. I. Reed. **2001**. Multi-step control of spindle pole body duplication by cyclin-dependent kinase. *Nature Cell Biol* **3**, 38–42.

86. Weiss, E. and M. Winey. **1996**. The *Saccharomyces cerevisiae* spindle pole body duplication gene MPS1 is part of a mitotic checkpoint. *J Cell Biol* **132**, 111–123.

87. Thomas, J. and D. Botstein. **1986**. A gene required for the separation of chromosomes on the spindle apparatus in yeast. *Cell* **44**, 65–76.

88. Chial, H. J. and M. Winey. **1999**. Mechanisms of genetic instability revealed by analysis of yeast spindle pole body duplication. *Biol Cell* **91**, 439–450.

89. Spang, A., I. Courtney, U. Fackler, M. Matzner, and E. Schiebel. **1993**. The calcium-binding protein division cycle 31 of *Saccharomyces cerevisiae* is a component of the half bridge of the spindle pole body. *J Cell Biol* **123**, 405–416.

90. Salisbury, J., Suino, R. Busby, and M. Springett. **2002**. Centrin-2 is required for centriole duplication in mammalian cells. *Curr Biol* **12**, 1287–1292.

91. Koblenz, B., J. Schoppmeier, A. Grunow, and K. F. Lechtreck. **2003**. Centrin deficiency in *Chlamydomonas* causes defects in basal body replication, segregation and maturation. *J Cell Sci* **116**, 2635–2646.

92. Winey, M., A. Hoyt, C. Chan, L. Goetsch, D. Botstein, and B. Byers. **1993**. NDC1: a nuclear periphery component required for yeast spindle pole body duplication. *J Cell Biol* **122**, 743–751.

93. Lauzé, E., B. Stoelcker, F. C. Luca, E. Weiss, A. Schutz, and M. Winey. **1995**. Yeast spindle pole duplication gene MPS1 encodes an essential dual specificity protein kinase. *EMBO (Eur Mol Biol Organ) J* **14**, 1655–1663.

94. Schutz, A. R. and M. Winey. **1998**. New alleles of the yeast MPS1 gene reveal multiple requirements in spindle pole body duplication. *Mol Biol Cell* **9**, 759–774.

95. Abrieu, A., L. Magnaghi-Jaulin, J. A. Kahana, M. Peter, A. Castro, S. Vigneron, T. Lorca, D. W. Cleveland, and J. C. Labbe. **2001**. Mps1 is a kinetochore-associated kinase essential for

the vertebrate mitotic checkpoint. *Cell* **106**, 83–93.

96. Liu, S. T., G. K. Chan, J. C. Hittle, G. Fujii, E. Lees, and T. J. Yen. **2003**. Human MPS1 kinase is required for mitotic arrest induced by the loss of CENP-E from kinetochores. *Mol Biol Cell* **14**, 1638–1651.

97. Stucke, V. M., H. H. Sillje, L. Arnaud, and E. A. Nigg. **2002**. Human Mps1 kinase is required for the spindle assembly checkpoint but not for centrosome duplication. *EMBO J* **21**, 1723–1732.

98. Fisk, H. A. and M. Winey. **2001**. The mouse Mps1p-like kinase regulates centrosome duplication. *Cell* **106**, 95–104.

99. Fisk, H. A., C. P. Mattison, and M. Winey. **2003**. The human Mps1 protein kinase is required for centrosome duplication and normal mitotic progression. *Proc Natl Acad Sci USA* **100**, 14875–14880.

100. McDonald, H. B. and B. Byers. **1997**. A proteasome cap subunit required for spindle pole body duplication in yeast. *J Cell Biol* **137**, 539–553.

101. Biggins, S., I. Ivanowska, and M. Rose. **1996**. Yeast ubiquitin-like genes are involved in duplication of the microtubule organizing center. *J Cell Biol*, **133**, 1331–1346.

102. Schutz, A. R., T. H. Giddings, E. Steiner, and M. Winey. **1997**. The yeast *CDC37* gene interacts with MPS1 and is required for proper execution of spindle pole body duplication. *J Cell Biol* **136**, 969–982.

103. Zarzov, P., H. Boucherie, and C. Mann. **1997**. A yeast heat shock transcription factor (Hsf1) mutant is defective in both Hsc82/Hsp82 synthesis and spindle pole body duplication. *J Cell Sci* **110**, 1879–1891.

104. Visintin, R., E. S. Hwang, and A. Amon. **1999**. Cfi 1 prevents premature exit from mitosis by anchoring Cdc 14 phosphatase in the nucleolus. *Nature* **398**, 818–823.

105. Pereira, G., T. U. Tanaka, K. Nasmyth, and E. Schiebel. **2001**. Modes of spindle pole body inheritance and segregation of the Bfa1p-Bub2p checkpoint protein complex. *EMBO J* **20**, 6359–6370.

106. Bardin, A. J. and A. Amon. 2001. Men and sin: what's the difference? *Nature Rev Mol Cell Biol* **2**, 815–826.

107. Bembenek, J. and H. Yu. **2001**. Regulation of the anaphase-promoting complex by the dual specificity phosphatase human Cdc14a. *J Biol Chem* **276**, 48237–48242.

108. Mailand, N., C. Lukas, B. K. Kaiser, P. K. Jackson, J. Bartek, and J. Lukas. **2002**. Deregulated human Cdc14A phosphatase disrupts centrosome separation and chromosome segregation. *Nature Cell Biol* **4**, 318–322.

109. Kaiser, B. K., Z. A. Zimmerman, H. Charbonneau, and P. K. Jackson. **2002**. Disruption of centrosome structure, chromosome segregation, and cytokinesis by misexpression of human Cdc14A phosphatase. *Mol Biol Cell* **13**, 2289–2300.

110. Nishiyama, Y., T. Hirota, T. Morisaki, T. Hara, T. Marumoto, S. Iida, K. Makino, H. Yamamoto, T. Hiraoka, N. Kitamura, and H. Saya. **1999**. A human homolog of Drosophila warts tumor suppressor, h-warts, localized to mitotic apparatus and specifically phosphorylated during mitosis. *FEBS Lett* **459**, 159–165.

111. Tao, W., S. Zhang, G. S. Turenchalk, R. A. Stewart, M. A. St John, W. Chen, and T. Xu. **1999**. Human homologue of the *Drosophila melanogaster* lats tumour suppressor modulates CDC2 activity. *Nature Genet* **21**, 177–181.

112. Moreno, C. S., W. S. Lane, and D. C. Pallas. **2001**. A mammalian homolog of yeast MOB1 is both a member and a putative substrate of striatin family-protein phosphatase 2A complexes. *J Biol Chem* **276**, 24253–24260.

113. Cuif, M. H., F. Possmayer, H. Zander, N. Bordes, F. Jollivet, A. Couedel-Courteille, I. Janoueix-Lerosey, G. Langsley, M. Bornens, and B. Goud. **1999**. Characterization of GAPCenA, a GTPase activating protein for Rab6, part of which associates with the centrosome. *EMBO J* **18**, 1772–1782.

114. Descombes, P. and E. A. Nigg. **1998**. The polo-like kinase Plx1 is required

for M phase exit and destruction of mitotic regulators in *Xenopus* egg extracts. *EMBO J* **17**, 1328–1335.
115. O'Connell, C. B. and Y. L. Wang. **2000**. Mammalian spindle orientation and position respond to changes in cell shape in a dynein-dependent fashion. *Mol Biol Cell* **11**, 1765–1774.
116. Hinchcliffe, E. H., F. J. Miller, M. Cham, A. Khodjakov, and G. Sluder. **2001**. Requirement of a centrosomal activity for cell cycle progression through G1 into S phase. *Science* **291**, 1547–1550.
117. Khodjakov, A. and C. L. Rieder. **2001**. Centrosomes enhance the fidelity of cytokinesis in vertebrates and are required for cell cycle progression. *J Cell Biol* **153**, 237–242.
118. Piel, M., J. Nordberg, U. Euteneuer, and M. Bornens. **2001**. Centrosome-dependent exit of cytokinesis in animal cells. *Science* **291**, 1550–1553.
119. Cerutti, L. and V. Simanis. **1999**. Asymmetry of the spindle pole bodies and spg1p GAP segregation during mitosis in fission yeast. *J Cell Sci* **112**, 2313–2321.
120. Li, C., K. A. Furge, Q. Cheng, and C. F. Albright. **2001**. Byr4 localizes to spindle-pole bodies in a cell cycle-regulated manner to control cdc7 localization and septation in fission yeast. *J Biol Chem* **275**, 14381–14387.
121. Guertin, D. A., L. Chang, F. Irshad, K. L. Gould, and D. McCollum **2000**. The role of the Sid1p kinase and Cdc14p in regulating the onset of cytokinesis in fission yeast. *EMBO J* **19**, 1803–1815.
122. Balasubramanian, M. K., D. McCollum, L. Chang, K. C. Wong, N. I. Naqvi, X. He, S. Sazer, and K. L. Gould. **1998**. Isolation and characterization of new fission yeast cytokinesis mutants. *Genetics* **149**, 1265–1275.
123. Sparks, C., M. Morphew, and D. McCollum. **1999**. Sid2p, a spindle pole body kinase that regulates the onset of cytokinesis. *J Cell Biol* **146**, 777–790.
124. Hou, M. C., J. Salek, and D. McCollum. **2000**. Mob1p interacts with the Sid2p kinase and is required for cytokinesis in fission yeast. *Curr Biol* **10**, 619–622.
125. Salimova, E., M. Sohrman, N. Fournier, and V. Simanis. **2000**. The *S. pombe* orthologue of the S. cerevisiae mob1 gene is essential and functions in signaling the onset of septum formation. *J Cell Sci* **113**, 1695–1704.
126. Ohkura, H., I. Hagan, and D. Glover. **1995**. The conserved *Schizosaccharomyces pombe* kinase plo1, required to form a bipolar spindle, the actin ring, and septum, can drive septum formation in G1 and G2 cells. *Genes Dev* **9**, 1059–1073.
127. Mulvihill, D., J. Petersen, H. Ohkura, D. Glover, and I. Hagan. **1999**. Plo1 kinase recruitment to the spindle pole body and its role in cell division in *Schizosaccharomyces pombe. Mol Biol Cell* **10**, 2271–2785.
128. Tanaka, K., J. Petersen, F. MacIver, D. P. Mulvihill, D. M. Glover, and I. M. Hagan. **2001**. The role of Plo1 kinase in mitotic commitment and septation in *Schizosaccharomyces pombe. EMBO J* **20**, 1259–1270.
129. MacIver, F. H., K. Tanaka, A. M. Robertson, and I.M Hagan. **2003**. Physical and functional interactions between polo kinase and the spindle pole component Cut12 regulate mitotic commitment in *S. pombe. Genes Dev* **17**, 1507–1523.
130. Nurse, P. **1990**. Universal control mechanism regulating onset of M-phase. *Nature* **344**, 503–508.
131. Strausfeld, U., J. C. Labbe, D. Fesquet, J. C. Cavadore, A. Picard, K. Sadhu, P. Russell, and M. Doree. **1991**. Dephosphorylation and activation of a p34cdc2/cyclin B complex in vitro by human CDC25 protein. *Nature* **351**, 242–245.
132. Izumi, T. and J. L. Maller. **1993**. Elimination of cdc2 phosphorylation sites in the cdc25 phosphatase blocks initiation of M-phase. *Mol Biol Cell* **4**, 1337–1350.
133. Izumi, T. and J. L. Maller. **1995**. Phosphorylation and activation of the *Xenopus* Cdc25 phosphatase in the absence of Cdc2 and Cdk2 kinase activity. *Mol Biol Cell* **6**, 215–226.
134. Bridge, A. J., M. Morphew, R. Barlett, and I. M. Hagan. **1998**. The fission yeast SPB component Cut12 links

bipolar spindle formation to mitotic control. *Genes Dev* **12**, 927–942.

135. Wu, X. and R. E. Palazzo. **1999**. Differential regulation of maternal vs. paternal centrosomes. *Proc Natl Acad Sci USA* **96**, 1397–1402.

136. Rios, R. M. and M. Bornens. **2003**. The Golgi apparatus at the cell centre. *Curr Opin Cell Biol* **15**, 60–66.

137. Lynn, R. R. and P. T. Magee. **1970**. Development of the spore wall during ascospore formation in *Saccharomyces cerevisiae*. *J Cell Biol* **44**, 688–692.

138. Peterson, J., R. Gray, and H. Ris. **1972**. Meiotic spindle plaques in *Saccharomyces cerevisiae*. *J Cell Biol* **53**, 837–841.

139. Zickler, D. and L. Olson. **1975**. The synaptonemal complex and the spindle plaque during meiosis in yeast. *Chromosoma* **50**, 1–23.

140. Moreno-Borchart, A. C. and M. Knop. **2003**. Prospore membrane formation: how budding yeast gets shaped in meiosis. *Microbiol Res* **158**, 83–90.

141. Knop, M. and K. Strasser. **2000**. Role of the spindle pole body of yeast in mediating assembly of the prospore membrane during meiosis. *EMBO J* **19**, 3657–3667.

142. Chu, S., J. DeRisi, M. Eisen, J. Mulholland, D. Botstein, P. O. Brown, and I. Herskowitz. **1998**. The transcriptional program of sporulation in budding yeast. *Science* **282**, 699–705.

143. Lupas, A., M. Van Dyke, and J. Stock. **1991**. Predicting coiled coils from protein sequences. *Science* **252**, 1162–1164.

144. Rabitsch, K. P., A. Toth, M. Galova, A. Schleiffer, G. Schaffner, E. Aigner, C. Rupp, A. M. Penkner, A. C. Moreno-Borchart, M. Primig, R. E. Esposito, F. Klein, M. Knop, and K. Nasmyth. **2001**. A screen for genes required for meiosis and spore formation based on whole-genome expression. *Curr Biol* **11**, 1001–1009.

145. Nickas, M. E. and A. M. Neiman. **2002**. Ady3p links spindle pole body function to spore wall synthesis in *Saccharomyces cerevisiae*. *Genetics* **160**, 1439–1450.

146. Nickas, M. E., C. Schwartz, and A. M. Neiman. **2003**. Ady4p and Spo74p are components of the meiotic spindle pole body that promote growth of the prospore membrane in *Saccharomyces cerevisiae*. *Eukaryot Cell* **2**, 431–445.

5
Dissection of Basal Body and Centriole Function in the Unicellular Green Alga *Chlamydomonas reinhardtii*

Susan K. Dutcher

5.1
Introduction

In 1702, Leeuwenhook recorded his observations of a unicellular green alga. "Their bodies seemed to be composed of particles that represented an oval figure; and therewithal they had two short thin instruments which stuck out a little way from the round contour and wherewith they performed the motions of rolling around and going forward". These two instruments were clearly the two anteriorly placed flagella of a unicellular green alga and these flagella have led many researchers to study the alga, *Chlamydomonas reinhardtii*. At the base of the two flagella are basal bodies, which were not visible with Leeuwenhook's microscopes. Later researchers provided clear evidence that the structures at the base of flagella were similar to the centrioles, structures observed by Boveri near the nucleus and at the poles of spindles [1]. With the advent of electron microscopy, the elaborate structure of basal bodies and centrioles were revealed. Each of the structures is a barrel of nine triplet microtubules with various additional appendages [2]. As a starting point for this chapter, basal bodies refer to the structures that are present at the base of cilia or flagella while centrioles refer to the structures that are present near the nucleus or at the poles of spindles. *Chlamydomonas* has basal bodies during interphase that are converted to centrioles at mitosis. Basal bodies are characterized by the presence of a transition zone and are continuous with the flagellar axoneme. Centrioles lack both the transition zone and flagellar axoneme and instead are associated with the spindle poles.

Centrosomes in Development and Disease. Edited by Erich A. Nigg

ISBN 3-527-30980-2

5.2
Why Study a Green Alga to Learn about Centrioles and Basal Bodies?

Model organisms provide experimentally accessible conditions to study a biological process shared by a larger group of organisms. In much the same way that *Saccharomyces cerevisiae* has provided understanding of the spindle pole body, *Chlamydomonas* has begun to provide information about the role of the basal body/centriole through the analysis of mutations that disrupt basal body duplication and assembly and perturb their cellular functions. *Chlamydomonas* is a unicellular green alga with two anteriorly placed flagella. Its ellipsoidal cell body is about 10 μm in length and the two flagella are about 10–12 μm long. Because it is normally haploid, mutant phenotypes are easily identified. Conversely, recessive lethal mutations can be propagated in diploid organisms. Meiotic progeny can be isolated to allow mapping of mutant phenotypes to one of 17 linkage groups by conventional or physical markers. *Chlamydomonas* is easy to grow in the laboratory in either liquid or solid medium and it has rapid mitotic (8 h) and meiotic (5 days) cycles. Moreover, the molecular genetic techniques of DNA transformation, insertional mutagenesis, and RNA interference (RNAi) are straightforward. A draft of the sequence of the ~ 130-Mb genome has recently become available from DOE/JGI and the genome has about 19000 predicted protein-coding genes. Because *Chlamydomonas* can be grown in large quantities in simple liquid cultures, it is possible to isolate basal bodies in quantities sufficient for biochemical analysis. This combination of properties has made it possible to use *Chlamydomonas reinhardtii* to identify proteins and genes that affect basal body duplication and assembly as well as to begin to understand the roles of basal bodies/centrioles in these cells.

5.3
Structure of the Basal Body and Centriole in *Chlamydomonas*

The basal body in *Chlamydomonas* is an elaborate organelle with a highly complex morphology and many different associated fibers. To understand basal bodies and centrioles, it is easiest to begin with a discussion of their structure. Electron microscopy of *Chlamydomonas* by Ringo [3] and Johnson and Porter [4] provided the first information about the structure and cellular localization of the basal body in *Chlamydomonas*. During interphase, the basal bodies are present at the anterior end of the cell while the cup-shaped chloroplast resides at the posterior end of the cell (Figure 5.1). The basal body and the associated transition zone are about 400 nm in length (Figures 5.2 and 5.3). The morphology of the basal body evolves along the proximal to distal length [5]. At the proximal end is a ring of amorphous electron-dense material, which is less than 40 nm in length. Out of this darkly staining material arise the nine, angled triplet microtubule blades. Each blade has an A, B, and C microtubule or subfiber. The A subfiber is a complete microtubule with 13 protofilaments while the B and C subfibers are incomplete microtubules with only 11 protofilaments. At the proximal end, just above the amorphous material, the nine

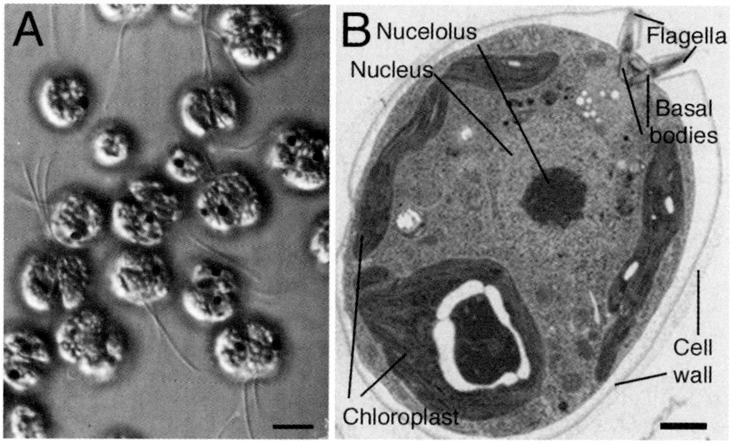

Figure 5.1 Light microscopy (A) and electron microscopy (B) images of *Chlamydomonas reinhardtii.* Panel A show a field of recently mated *Chlamydomonas* cells by differential interference contrast (DIC) optics. The darkly staining spots are the eyespots. Each dikaryon has two eyespots, one contributed by each parent and four flagella, two contributed by each parent. Scale bar, 5 μm. (Reprinted with permission from *Journal of Cell Science* [18]). Panel B shows a thin section of a *Chlamydomonas* cell. The flagella and the associated basal bodies are located at the top of the image. The nucleus and its darkly staining nucleolus are labeled. The chloroplast encompasses the majority of the cell volume. Scale bar, 0.5 μm.

blades are connected by a central pinwheel structure. Although this structure was first observed by Ringo [3], tomographic reconstruction of the basal bodies makes it clear that each spoke of the pinwheel has a dense knob and is connected to the A subfiber of the triplet microtubule [5]. The lumen for the next 200 nm contains lightly staining, nebulous material with no obvious structure. At the distal end of the basal body proper, transitional fibers radiate out from the triplet microtubules with an attachment to all of the subfibers. The transitional fibers are quite dynamic and change rapidly as sections progress up the basal body. The transition fibers change from a triangular striated appearance to slender threads that become Y-shaped. These Y-shaped fibers end on dense knobs that are inserted in the plasma membrane (Figure 5.3). These dense knobs have been hypothesized to be flagellar pores that allow entry of the flagellar axonemal proteins [6]. The basal bodies, as defined by triplet microtubules, end just after the appearance of the transitional fibers. The C subfiber ends and the structures that comprise the transition zone begin. The transition zone has two stellate fiber arrays (Figure 5.3). The first array appears as a nine-pointed star with the vertices centered on the A subfiber and contains a central hub formed of electron-dense triangular points. A central amorphous disk that is only about 10–15 nm thick separates the first stellate array from the second array. This amorphous disk, which can be seen in longitudinal section, cannot be visualized in cross-section by conventional electron microscopy because it is too shallow. The second stellate array also consists of a nine-pointed star, but its hub is much more elaborate. The two arrays

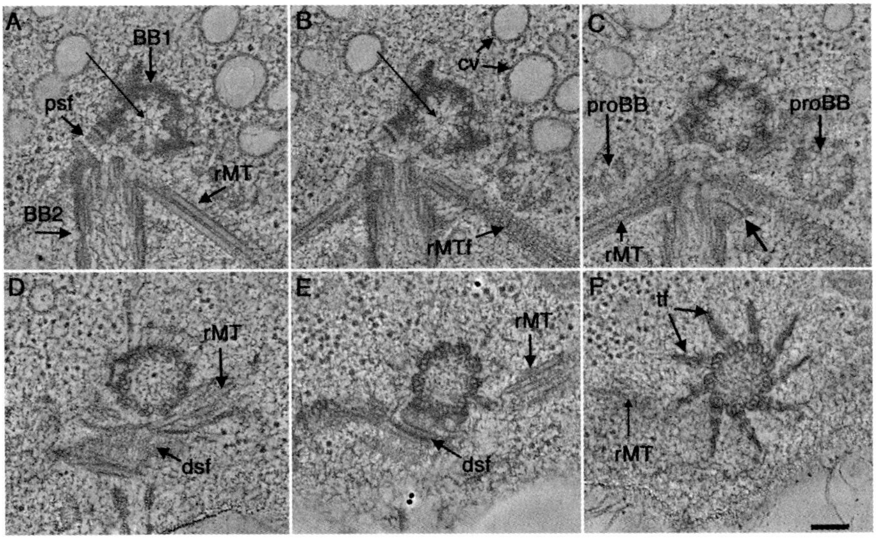

Figure 5.2 Selected tomographic slices from the proximal to the distal region of a wild-type *Chlamydomonas* basal body complex. (A) One of the two basal bodies is shown in cross-section (BB1) and the other basal body is shown in longitudinal view (BB2). The proximal base of BB1 consists of an amorphous, electron-dense ring and there is a nine-fold symmetrical pinwheel structure in its center, indicated by an arrow. A two-membered rootlet microtubule (rMT) is labeled. The proximal striated fiber (psf) connects the two mature basal bodies. (B) The pinwheel structure is formed from three rings (arrow). Contractile vacuoles (CV) are seen in the vicinity of the basal body complex. (C) Two probasal bodies (ProBB) lie adjacent to the mature basal bodies. One of the four-membered rootlet microtubules (rMT) is present in the lower left quadrant. A fiber connecting the BB2 to the rootlet microtubule is present and indicated by an arrow. (D) and (E) The distal striated fiber (dsf) and rootlet microtubules (rMT) are indicated. (F) Transitional fibers (tf) radiate out from the triplets at the distal end of the basal body. (Reprinted with permission of *Molecular Biology of the Cell* [5]).

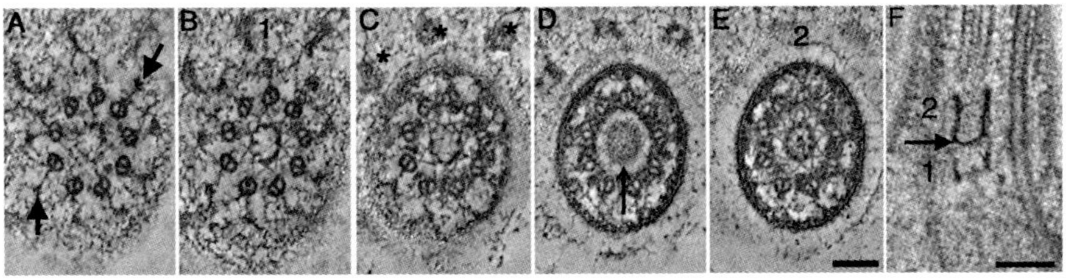

Figure 5.3 Selected tomographic slices from the transition zone of a wild-type *Chlamydomonas* basal body. (A) Proximal region of the transition zone contains doublet microtubules and Y-shaped connectors (arrows). (B) and (C) First stellate fiber array consists of a nine-pointed star that contains a central hub formed from electron-dense triangular points. (D) A central, amorphous disk (arrow) replaces the first stellate fiber array. (E) A second stellate fiber array at the distal end of the transition zone has nine-pointed star with an elaborate center. (F) A longitudinal view of the two stellate fibers appears as an osmophilic H. The crossbar is the amorphous material in panel D. Bar, 100 nm (Reprinted with permission of *Molecular Biology of the Cell* [5]).

are osmophilic and appear H-shaped in longitudinal sections (Figure 5.3). When basal bodies become the centrioles at the poles of the mitotic spindle, the conversion is associated with the loss of the transition zone. The centrioles in *Chlamydomonas* are composed of amorphous material and the pinwheel at the proximal end.

The molecular composition of these various substructures remains largely unknown with a few exceptions. The Y-shaped connectors at the distal tip of the transitional fibers contain p210, which was identified in *Spermatozopsis similes* [7]. Rib43, a coiled-coil protein, is found in isolated preparations of the three protofilaments that are shared between the A and B subfibers and the B and C subfibers [8]. The stellate fibers contain centrin [9], but it is not known what additional proteins are present. The Vfl1 protein is a coiled-coil protein with five leucine-rich repeats. It localizes to the lumen of the transition zone of mature basal bodies and also to newly formed probasal bodies. The protein is localized to the side of the basal body that faces the other basal bodies. Mutations in *vfl1* shows defects in establishment of rotational symmetry and this results in basal body missegregation [10]. Two-dimensional gel electrophoresis of preparations enriched in basal bodies reveal over 150 polypeptides. α- and β-tubulin comprise over 60% of the protein by mass (Figure 5.4). *In vivo* labeling with radioactive phosphate shows that over one-third of the polypeptides are phosphorylated. The MPM-2 antibody, which recognizes a number of phosphorylated epitopes in mammalian cells, stain the poles of mitotic spindles in *Chlamydomonas* [11].

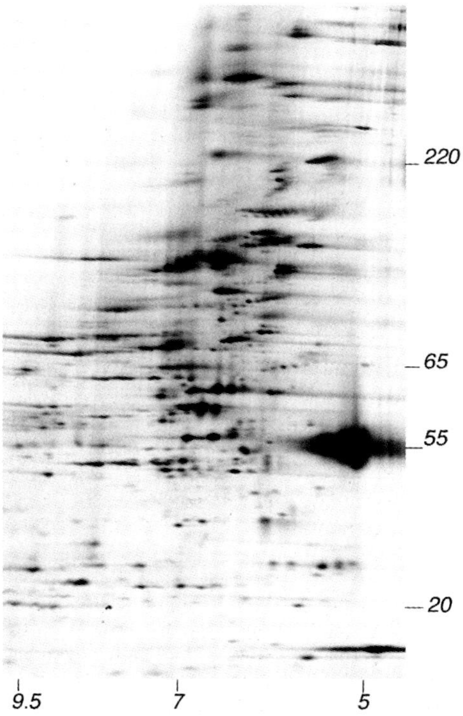

Figure 5.4 Two-dimensional PAGE of isolated basal bodies illustrates the complexity of these organelles. The proteins are separated by molecular weight in the vertical direction. Proteins with relative molecular weights ranging from 220 to 13 kDa are well resolved. In the horizontal dimension, the proteins are separated by charge with pH 9.5 on the right to pH 4.5 on the left. At about 55 kDa and a pI of 5 are α- and β-tubulin.

5.4
Additional Fibers that Connect Basal Bodies and Centrioles

It is also useful to analyze the cytoskeletal elements that hold the two basal bodies together and attach them to other parts of the cell. There are three types of fiber systems (Figure 5.5). Fibers that contain centrin are important to the segregation of the basal bodies. Rootlet microtubules are critically important to cellular polarity and cleavage furrow placement. Finally, non-contractile fibers, the most diverse group, include SF-assemblin and Vfl3.

5.4.1
Contractile Fibers

Centrin is a 20-kDa EF hand-containing protein that is present in the distal and proximal striated fibers as well as in the stellate fibers of the transition zone (Figure 5.5). These fibers connect the two basal bodies to each other. Additional centrin-containing fibers connect the basal bodies to the nucleus [12]. Contraction of centrin fibers is mediated by changes in intracellular calcium [13]. Disruption of centrin function has deleterious consequences for the cell. In RNAi-treated cells or

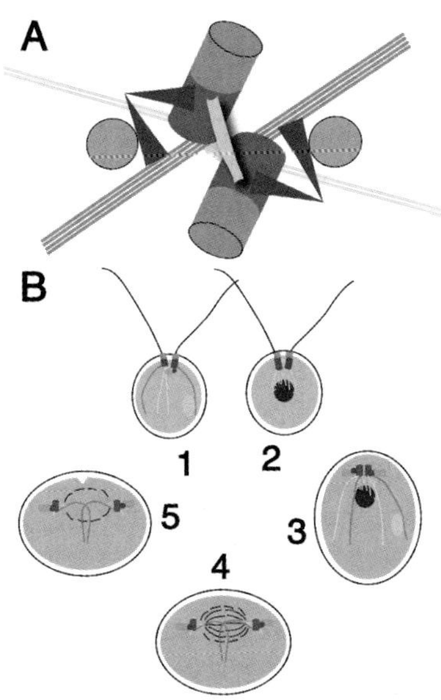

Figure 5.5 The three fiber systems of the basal body complex. (A) The mature basal bodies are shown in red, the transition zones in peach and the probasal bodies are shown in pink. The rootlet microtubules have four microtubules (orange) or two microtubules (yellow) and attach at specific triplet microtubules of the basal body. The distal (solid) and proximal (striped) striated fibers are shown in light blue. They connect the two mature basal bodies at the two ends. The lateral fibers are shown in green. They connect the mature basal body to its daughter probasal body across the rootlet microtubules. (B) Changes in the fiber systems during the cell cycle. 1, During interphase the basal bodies and transition zones are continuous with the flagella. The rootlet microtubules are adjacent to the plasma membrane. One of the four-membered rootlet microtubules lie adjacent to the eyespot (rose). 2, Another view of interphase cells illustrates that the basal bodies are connected to the nucleus and to each other by centrin fibers. 3, At preprophase, the flagella are lost. The probasal bodies elongate. The distal and proximal striated fibers are lost. 4, The two-membered rootlet microtubules shorten. The centrioles (without transition zones) are found at the poles of the spindle. The four-membered rootlet microtubules arc over the spindle. The eyespot is disassembled. 5, Cytokinesis is initiated at one end of the cell. This will be followed by extension of the two-membered rootlet microtubules, the striated fibers, and assembly of new rootlet microtubules and of a new eyespot in association with the new four-membered rootlet microtubules (see Color Plates page XXIV.

cells with a missense mutation, the amount of centrin is reduced and basal bodies missegregate [14, 15]. When the 5′ UTR is deleted, the cells die (K. Mills, E. Trabuco, and S. K. Dutcher, unpublished results). Thus, it is clear that centrin is vital to centriole/basal body function. It is not known if centrin is simply required for segregation of the basal bodies or if it has an essential role in basal body/centriole duplication as is observed for the yeast homolog *CDC31* [16, 17]. The lethal phenotype of the deletion allele suggests centrin has an essential role.

5.4.2 Rootlet Microtubules

Rootlet microtubules consist of four bundles of microtubules arranged in a cross-shaped pattern (Figure 5.5). Two of the bundles have two microtubules in them and these are arranged at 180° from each other. The other two bundles have four microtubules and are also separated by 180°. Together these bundles form a cross-shaped rootlet system. During interphase, the rootlet microtubules are anchored at the basal body and extend about three-quarters of the length of the cell closely apposed to the plasma membrane. Rootlet microtubules are involved in positioning the eyespot, an organelle important for phototactic behavior, which is located on the equator of the cell [18]. During mitosis, the rootlets containing two microtubules shorten considerably. The rootlets with four microtubules arc over the spindle and bend to form a 90° angle [19]. This structure, originally called the metaphase band [4], simply represents the rootlet microtubules that have been recruited for a distinct secondary function during mitosis. During mitosis, these perpendicular rootlet microtubules serve to recruit actin to the cleavage furrow. *Chlamydomonas* cells harboring mutations that result in the loss of B and/or C subfibers have rootlet microtubules that are not anchored properly. This causes the cleavage furrow to be misplaced relative to the two daughter nuclei [5, 20, 21].

5.4.3 Non-contractile Fibers

One set of non-contractile fibers contains SF-assemblin, a 30-kDa coiled-coil protein [22]. These fibers overlay the rootlet microtubules for part of their length (Figure 5.5). They may reinforce the rootlet microtubules. Reduction in the level of message by RNAi constructs for SF-assemblin results in flagellar assembly defects [23].

Vfl3 is a coiled-coil protein of about 65 kDa. It is localized to lateral fibers, which may play a role in attaching the probasal body to the mature basal body (Figure 5.5). Mutations in the *VFL3* gene result in cells with defects in basal body segregation. Homologs of the *VFL3* gene are found in most multicellular organisms [24].

5.5
Overview of the Cell Cycle of *Chlamydomonas*

The events of the mitotic cell cycle in *Chlamydomonas nasuta* were first investigated by Kater in 1929 [25] by light microscopy. The position of *C. natusa* in the *Chlamydomonas* phylogeny based on ribosomal DNA sequence is close to *C. eugametos* (A. Coleman and T. Prehold), but it provides an excellent portrait of cell division in *Chlamydomonas* and serves to underscore the process of conversion of basal bodies to centrioles. When grown in constant light, *Chlamydomonas* spends most of the cell cycle in G1. S-phase, G2, and M occur quickly. Non-mitotic cells are flagellated (Figure 5.6A) and their basal bodies are connected by a darkly staining fiber to the nucleus. This is likely to be the centrin connection between the basal bodies and the nucleus. Preprophase (Figure 5.6B) is characterized by increased granularity in the nucleus, loss of flagella, and a 90° rotation of the protoplast within the cell wall. The rotation is not required, as it can be blocked by bright light and cell division still occurs [18]. The basal bodies begin to move apart (Figure 5.6C) and are subsequently found at the poles of the spindle (Figure 5.6D). During mitosis another darkly-staining fiber spans the spindle in these images. This fiber is likely to be the four-membered rootlet microtubules [19]. At telophase, the beginning of the cleavage furrow is observed (Figure 5.6E) and it is extended from one end of the cell towards the other (Figure 5.6F) until cytokinesis is complete (Figure 5.6G). The new daughter cells re-grow flagella (Figure 5.6H) and emerge from the old cell wall, which is discarded. Remarkably, this early light microscopic characterization has been confirmed by electron [3, 4], immunofluorescence [9], and DIC microscopy [18]. The movement of the centrioles can be visualized in real time using DIC optics (Figure 5.7).

5.6
Duplication of Basal Bodies in *Chlamydomonas*

In mammalian cells, centriole duplication begins at the G1 to S transition [26]. In contrast, *Chlamydomonas* cells exiting mitosis have already assembled probasal bodies [27]. Electron microscopy of *Chlamydomonas* cells in G1, as judged by FACS analysis, shows that they have two probasal bodies adjacent to the two mature basal bodies. These probasal bodies are 40–90 nm in length [5] and are often missed by conventional electron microscopy [3, 4]. This is likely to be due to their short length relative to the normal depth of thin sections used for electron microscopy. The probasal bodies contain the amorphous material that is present in the mature basal bodies, which suggests that this material is important for early events in basal body assembly. The pinwheel structure and the knob-like elaborations are also present. The microtubule blades in these probasal bodies have complete triplet blades. Elongation of the probasal bodies to new, full length basal bodies occurs just prior to prophase (Figure 5.5B). Concurrently, the centrin-containing striated fibers that connect the two mature basal bodies are lost [3] breaking the link be-

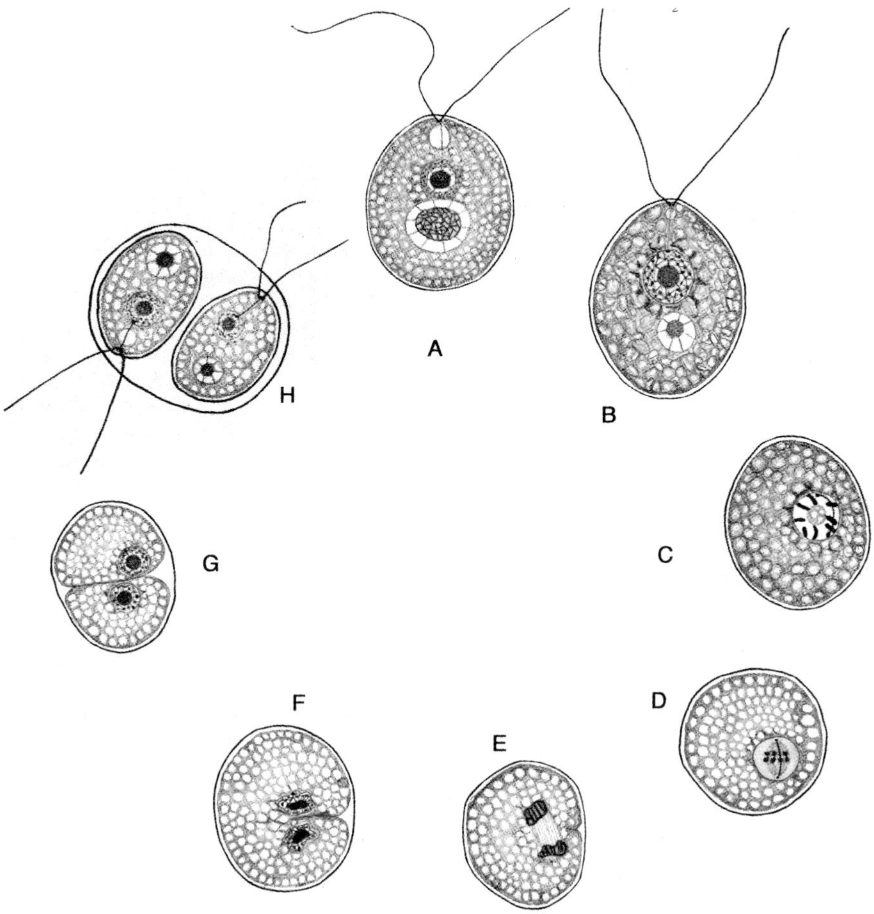

Figure 5.6 Drawings from 1929 showing cell division in *Chlamydomonas nasuta* closely resemble the current view of the cell cycle in *Chlamydomonas reinhardtii.* (A) Interphase cells are flagellated and the basal bodies are connected by a fiber that can be visualized with Schaudinn's fixative. This is likely to be the centrin connection between the basal bodies and the nucleus. (B) Pre-prophase is evidenced by increased granularity in the nucleus. (C) The basal bodies begin to move apart. (D) Basal bodies are found at the poles of the spindle and a more darkly staining fiber connects them. This fiber is likely to be the rootlet microtubules [19]. (E) At telophase, the beginning of the cleavage furrow is observed. (F) The cleavage furrow is extended from one end of the cell towards the other. (G) Cytokinesis is complete. (H) The new daughter cells will re-grow flagella and then emerge from the old cell wall, which is discarded. Reproduced from [25].

tween the two mature basal bodies. This process allows separation of two centriole pairs. Each pair consists of an old basal body/centriole and a new basal body/centriole. Each pair segregates to a pole of the mitotic spindle. The pairs of centrioles are found outside of the perforated nuclear envelope and the spindle microtubules are inserted into the nucleus through these perforations [28].

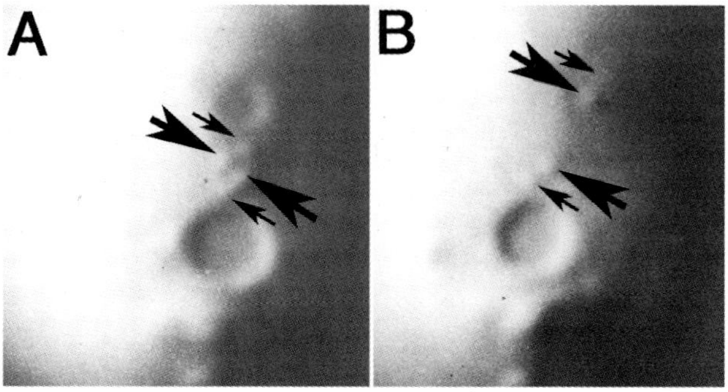

Figure 5.7 Migration of centrioles at mitosis. Differential interference contrast micrograph of *Chlamydomonas* centrioles in preprophase of mitosis. (A) The four centrioles are arranged in a diamond configuration and can be oriented relative to the contractile vacuole that appears as a large crater below the centrioles. Large black arrows point to the mother centrioles and the smaller arrowheads indicate the daughter centrioles ($t = 0$ min.). (B) The mother–daughter pairs of centrioles have separated and are moving towards the spindle poles ($t = 7$ min.). (Reprinted with permission from the *Journal of Cell Science* [18]).

Studies of basal body duplication in *Paramecium* by Dippell [29] elegantly showed intermediates in basal body assembly. A ring of nine singlet microtubules was formed, the B tubule was added, and before all of the B subfiber was added, the C subfiber would begin to be added. Cross-sectional images thus reveal a cylinder of microtubules with varying numbers of B and C subfibers (Figure 5.8A). This type of image has not been observed in *Chlamydomonas*.

Several lines of evidence suggest that the assembly pathway may differ from that proposed by Dippell [29]. The localization of two proteins (p210 and Vfl1) suggests a different pathway. In mature basal bodies, p210 co-localizes with the Y-shaped fibers at the distal end of the transitional fibers in mature basal bodies in *Spermatozopsis* [7] and the Vfl1 protein localizes to the distal end [10]. If basal body duplication were to proceed strictly by the model proposed by Dippell in Figure 5.8A, then one would not expect to see p210 or Vfl1 on probasal bodies. However, in both cases, these proteins are present on probasal bodies and end up at the distal end of the basal bodies. As shown in the diagram in Figure 5.8B, it is suggested that several of these proteins may sit on the plus end of the assembling microtubules and move outward during elongation much like the attachment between kinetochores and microtubules [30]. The phenotype of basal bodies in *bld2; rgn1-1* cells also suggest that disassembly may occur from the minus end of the microtubules of the basal body in these mutant cells. Electron micrographs of these cells show basal bodies with blades missing at the proximal end but intact at the distal end [31].

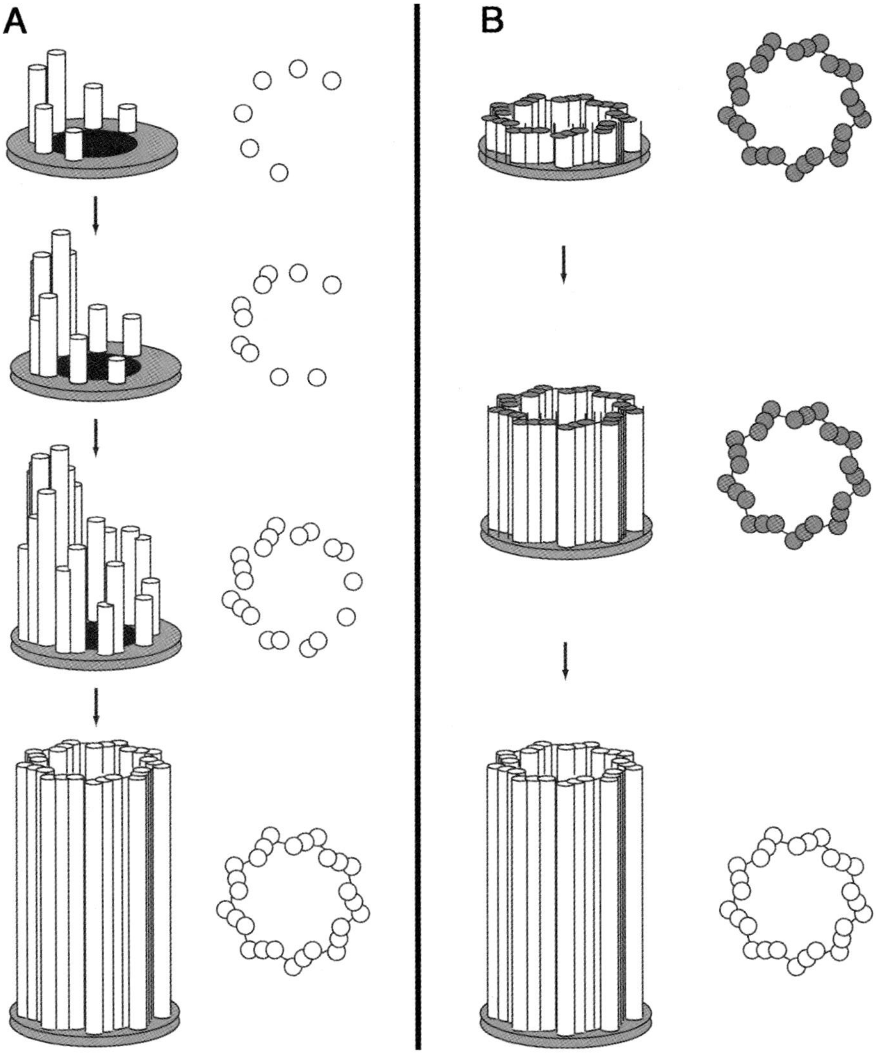

Figure 5.8 (A) Schematic diagram of basal body assembly as suggested by Dippell [29]. The first sign of basal body assembly is the appearance of an amorphous disk. This is followed by singlet microtubules that correspond to the A-tubule. These are followed by B- and, finally, C-tubules. Each stage is not completed simultaneously for all nine blades, so it is possible to observe mixtures of singlet, doublet and triplet microtubules within a given cross-section. (B) Schematic diagram of basal body assembly with a traveling cap model of regulation. Proteins (black) destined for fibers at the distal end (such as p210) travel at the tip of elongating microtubule blades. This complex could also serve as a negative regulator to prevent premature elongation of probasal bodies. These proteins leave the distal end of the mature basal body and move to various fibers once elongation is complete.

5.7 Role of Tubulin Isoforms in Basal Body Duplication

The role of various tubulin isoforms in basal body duplication has been addressed by mutational analysis and gene silencing in *Chlamydomonas*, *Trypanosoma brucei* and *Paramecium tetrauleis*. Gamma (γ)-tubulin plays an important role in all microtubule organizing centers [32]. Its role in basal body duplication has been addressed using gene silencing in *Trypanosoma* [33] and in *Paramecium* [34]. The experiments in *Trypanosoma* also suggest a role for γ-tubulin in assembly of the central pair of microtubules in the flagellar axoneme. This result is consistent with the previous localization of γ-tubulin to the interior of *Chlamydomonas* basal bodies [35]. In *Paramecium*, reduction in γ-tubulin results in defects in basal body duplication. Examination of the proximal end of the triplet microtubules in the tomographic reconstructions of *Chlamydomonas* using the IMOD software, which allows one to rotate the reconstructed image, shows closed microtubule ends [5]. These closed ends have been associated with the presence of γ-tubulin and the gamma-tubulin ring complex proteins [36].

Delta (δ)-tubulin, which is present in most organisms with triplet microtubules in their centrioles, does not appear to play a primary role in basal body duplication [37]. When this gene is deleted in *Chlamydomonas*, basal bodies are able to duplicate but fail to form triplet microtubules along most of the length of the basal body. The majority of the basal body blades contain A and B subfibers. At the distal end of the basal body where the transitional fibers are found, a short stretch of triplet microtubules is observed [5, 21].

Epsilon (ε)-tubulin, which is also present in most organisms with triplet microtubules in their centrioles, is likely to have a more profound role in basal body duplication. This gene is essential in *Chlamydomonas*. When the amount of ε-tubulin is reduced [38] or a truncated form is made ([39] and unpublished data), short incomplete basal bodies result. In *Chlamydomonas*, these incomplete basal bodies are 40–90 nm in length and are primarily composed of singlet microtubules [40]. In *Paramecium*, such basal bodies have singlet, doublet, and triplet microtubules [38]. Similarly, *Xenopus* extracts that have been immunodepleted for ε-tubulin fail to assemble centrin-containing centrioles [41]. It is likely therefore that ε-tubulin plays a key role in basal body duplication.

Recently, eta (η)-tubulin has been found in *Paramecium* [42] as well as in databases for *Ciona*, and *Chlamydomonas*. Two mutations in *Paramecium* suggest that η-tubulin has an important role in basal body duplication. The temperature-sensitive η-tubulin mutation, *sm19*, fails to duplicate basal bodies upon shift to the restrictive temperature. Already assembled basal bodies appear to be nearly wild-type, but new ones fail to assemble. This is consistent with the notion that the assembled η-tubulin remains stable at the restrictive temperature or that η-tubulin provides a scaffolding function only during basal body duplication. Antibodies to η-tubulin in *Paramecium* are not available to discriminate between these two mechanisms.

5.8
Timing of Basal Body/Centriole Duplication in *Chlamydomonas*

The duplication of centrioles and basal bodies are a key step in the cell cycle and yet, little is known about the mechanisms or controls of this event in *Chlamydomonas* or in other organisms. In most organisms, new centrioles form at right angles nearby to pre-existing centrioles, but the reasons are unknown. New centrioles form at specific times in the cell cycle and the molecules that regulate this timing are beginning to be uncovered in other organisms. The direction of assembly of centrioles/basal bodies remains unknown.

Mutations in *Chlamydomonas* have made it possible to ask questions about *de novo* assembly of basal bodies in the absence of existing centrioles or basal bodies. Development in *Xenopus* or *Sciara* requires the contribution of a centriole for proper development [43, 44] while *Spisula* and *Marsilea* are able to generate centrioles *de novo* [45]. To address the reasons why some cells can form centrioles *de novo*, but cells with centrioles only form new ones adjacent to old ones, Marshall and coworkers [46] used a temperature-sensitive mutation in the centrin gene (*vfl2)* to isolate aflagellate cells due to basal body missegregation. Immunofluorescence with antibodies to acetylated α- or γ-tubulin suggested that aflagellate cells lack basal bodies. Pedigree analysis of single cells or examination of bulk cultures showed that these aflagellate cells were able to build basal bodies and flagella. These experiments suggest that basal bodies can arise *de novo* in *Chlamydomonas* with the caveat that probasal bodies (~ 40 nm) may not be easily detected by immunofluorescence.

5.9
Function of Basal Bodies and Centrioles in *Chlamydomonas*

Basal bodies and centrioles in *Chlamydomonas reinhardtii* serve several roles during the cell cycle. During interphase, basal bodies provide a templating function for the two anteriorly placed flagella and act as docking sites for kinesin II and components of intraflagellar transport [47, 48]. In addition, they are required for the organization of the rootlet microtubules [49]. Rootlet microtubules, in turn, are implicated in the positioning of the eyespot, an organelle needed for phototactic behavior [18]. Basal bodies are also required for the correct placement and attachment of centrin fibers as well as the lateral fibers, which connect the basal bodies to the nucleus and to the probasal bodies, respectively. During mitosis, basal bodies convert to centrioles and are found at the poles of the spindle. It is not clear if they are required in the formation of the mitotic spindle. Centrioles continue to have a role in organizing the rootlet microtubules, which are required for the correct placement of the cleavage furrow [20]. The connection of centrioles to the centrin fibers that connect to the nucleus remains important for placement of the nucleus relative to the cleavage furrow [20].

5.10
What Makes a Basal Body Different from a Centriole?

One major difference between basal bodies and centrioles is the presence of the transition zone. In sea urchin blastulae, each cell has a single flagellum that is attached to a basal body and associated transition zone. At the commencement of mitotic preprophase, the flagellum and transition zone are retracted. The basal bodies then migrate to the poles of the spindle [50]. A similar set of events is observed in *Chlamydomonas* [4]. The transition zone is released from the basal body proximal to the stellate fibers to generate a centriole (Figure 5.5B(3)).

5.10.1
Transition Zone and Docking

One role of the transition zone may be in docking the intraflagellar transport proteins and their motors. Kinesin II, which is a heterotrimeric complex of two kinesin proteins and a large associated protein, is found concentrated around the basal bodies in *Chlamydomonas* and is required for transport of some but not all flagellar proteins from the cell body to the tip of the flagella [48, 51]. Two multimeric-complexes that contain at least 16 polypeptides are required for flagellar assembly [48, 52]. They assemble into a complex that is referred to as the intraflagellar transport (IFT) particle [53]. Immunoelectron microscopy suggests that IFT52, one member of the complex, is concentrated at the distal tips of the transitional fibers as they contact the plasma membrane [47]. Mutations in the gene for *IFT52* result in aflagellate cells [54]. The transition zone may be important in anchoring these fibers.

5.10.2
Transition Zone and Autonomy

Another role of the transition zone is in autonomy [55]. *Chlamydomonas* will shed its flagella in response to acid shock, chloral hydrate, dibucaine and other noxious compounds; this is known as autonomy. This process makes sense because most of the cell is covered by a cell wall, while the flagella are surrounded by a plasma membrane. The loss of the flagella decreases the surface area that is exposed to compounds which can cross the exposed plasma membrane, but not the cell wall. The severing of the flagella occurs distal to the stellate fiber arrays (Figure 5.9C) and the plasma membrane re-forms around the basal bodies. The events in deflagellation produce a different organelle than the conversion of basal bodies to centrioles at mitosis. The basal body retains the transition zone following deflagellation, but loses it prior to mitosis (Figure 5.9C).

Two molecules have been proposed to play roles in the deflagellation response to noxious compounds. The first is the centrin located in the stellate arrays (Figure 5.5A), but the evidence for centrin playing a role in the severing process is contradictory [56, 57]. The other molecule is katanin, which is a heterodimer. The p60 subunit of katanin, which is an AAA ATPase, was first identified from

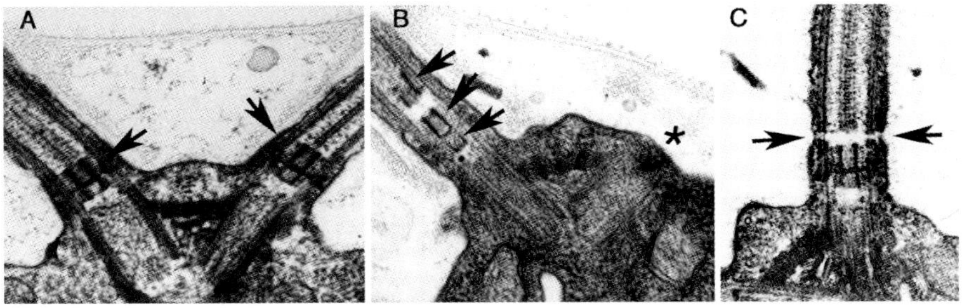

Figure 5.9 Electron micrographs of longitudinal views of flagella and basal bodies. (A) Wild-type flagella and basal bodies with the osmophilic H of the transition zone (arrow) that is present on both basal bodies. (B) Flagella and basal bodies from uni1-1 mutant cells. The older of two basal bodies has material that resembles the osmophilic material of the transition zone in several positions, but the morphology is altered (arrows). The younger of the two basal bodies has no transition zone. This resembles a centriole at the poles of the mitotic spindle (Reprinted with permission form *Cell* [72]). (C) Basal body and flagellum from a cell treated with acid. The flagellum is beginning to detach at the point indicated by the arrow.

sea urchin extracts, which can sever microtubules. The p80 subunit has a WD domain and is thought to target the other subunit to the centrosome [58]. In *Chlamydomonas,* the *PF15* gene encodes a p80 subunit [59]. However, the *pf15* null allele does not have a defect in deflagellation [60]. No mutants are currently available in the p60 subunit. There are several lines of evidence to support a role for katanin in deflagellation. Purified sea urchin katanin will severe flagellar axonemes from *Chlamydomonas.* Antibodies to sea urchin katanin will block calcium-induced *in vitro* deflagellation [61, 62]. Confirmation of the role of katanin with a mutant or by RNAi will confirm its role in deflagellation. It is not known which molecules are needed for the severing of the axonemal doublet microtubules needed for deflagellation.

Mutations that are unable to complete autonomy have been isolated. When subjected to an acid shock, the flagella fail to detach in the *fa1* and *fa2* mutant strains [63]. Fa1 is a novel protein that is localized to the basal body/transition zone, but it has no homologs in other organisms [63, 64]. *FA2* encodes a kinase that is most similar to the Nek kinases in other organisms. In addition to a defect in deflagellation, *fa2-1* cells are slow to transit from G2 to M and are slow to assemble flagella upon exiting from mitosis [65]. The target(s) of Fa2 is unknown, but it is interesting to speculate that it alters a component(s) of the transition zone that is needed for disassembly and then assembly of the transition zone. This would predict that the cells that exit mitosis are slow to assemble transition zones and to build flagella.

5.10.3
Maturation of Basal Bodies

Because centrioles replicate conservatively and segregate semi-conservatively, a cell will have an old or mother centriole and a new or daughter centriole at any one time. New centrioles mature through the cell cycle. In mammalian cells this maturation can be monitored by the appearance of several proteins, which include ninein, ODF2, centriolin, and dynactin [66–69]. Ninein, centriolin, and ODF2 are localized to the subdistal appendages. ε-Tubulin is also present only on the old centriole [70] and immuno-electron microscopy localizes it to the subdistal appendages as well [53]. The role of the subdistal appendages remains unknown, but they may play a role in assembling the primary cilium, which is an internal cilium that is only assembled on old centrioles. The subdistal appendages in mammalian cells and the transition zone in *Chlamydomonas* have similar structures and may play similar roles.

The structure of centrioles has been characterized in isolated centrosomes from lymphoblastoma cells. Subdistal and distal appendages are present on the distal end of mother centrioles but not on daughter centrioles. The subdistal appendages interact with two adjacent microtubule blades of the centriole and are attached at the transition between the triplet and doublet blades [71]. Their fine structure suggests radially assembled fibers that are interrupted by transverse fibers. Subdistal appendages are wider at the base, become thinner as they emanate away from the centriole, and appear flexible. At the very distal end of the centriole are the distal appendages, which are structurally distinct from the subdistal appendages. They appear as a long rod with a swollen tip. In cells with a primary cilium, these fibers help attach the centriole to the plasma membrane [12]. The transition fibers in *Chlamydomomas* also appear at the transition between triplet and doublet microtubules. They are wider at their base and change shape distally along the basal body length until they transform into Y-shaped connectors [5]. The evolution of the transition fibers into the Y-shaped connectors is not obvious in thin-section electron microscopy, but is illuminated by tomographic analysis. It is possible that distal appendages of centrioles are also an extension of the subdistal appendages. The only marker for the Y-shaped connector is a protein in the naked green alga, *Spermatozopsis* [7]. It localizes to probasal bodies and at the distal end of mature basal bodies. Remarkably, ninein and centriolin have a similar behavior during development, localizing to newly forming centrioles and to the subdistal appendages. To date, no markers are currently known for the distal appendages; however, they are less robust in size compared to the subdistal fibers and may have escaped observation. It is not clear whether subdistal appendages and the distal fibers have been well resolved in all studies. The appearance of these immunological markers suggests that the appendages are assembled onto the daughter centrioles in G2 of the cell cycle, while transitional fibers in *Chlamydomonas* are present in G1.

Unlike the pair of mammalian centrioles in a somatic cell, the mother and daughter basal bodies in *Chlamydomonas* must both be mature and competent to assemble flagella. ε-Tubulin in *Chlamydomonas* surrounds both basal bodies but the relationship to the transition zone is not known [39]. As suggested above, the transition zone could be an indication of maturity. In four different *uni1* mutant strains, the two basal bodies differ [72]. One has a transition zone, which is abnormal, and the other has no transition zone (Figure 5.9A and B). The basal body with a transition zone assembles a flagellum while the one without a transition zone lacks a flagellum. The Uni1 gene product may thus play a role in signaling assembly of the transition zone on the new basal body at the end of mitosis. At this time, the Uni1 gene product is unknown.

5.11 Conclusion

At the end of mitosis, *Chlamydomonas* must convert its centrioles back into basal bodies and duplicate two new probasal bodies. The conversion involves assembly of new transition zones on the old centrioles. These basal bodies must assemble flagella and assemble new fiber systems which include centrin fibers, rootlet microtubules, and non-contractile fibers. Formation of new probasal bodies must also occur. These must be attached to the old basal bodies, perhaps via Vfl3, and the probasal bodies must be capped so that further elongation is blocked until the next prophase. Some of the genes involved in these events are known in *Chlamydomonas* (Table 5.1).

In a remarkable paper from 1929, Kater was able to describe many of the key events associated with the interconversion of basal bodies and centrioles using only the light microscope. With the development of genetic and proteomic tools, we are now elucidating the intricate protein and gene networks that regulate this complex and intriguing process.

Acknowledgments

This chapter would not have been possible without the enthusiasm and comments of Dr Naomi Morrissette. Her love of microbes has stimulated many useful discussions and ideas.

This work is supported by a grant from the National Institutes of Health (GM-32843).

Table 5.1 Basal body/centriole proteins in *Chlamydomonas* and their homologs.

Chlamydomonas-protein (gene)	*Localization in Chlamydomonas*	*Function in Chlamydomonas*	*Homologs*	*Localization*
Centrin (*VFL2)*	Distal, proximal striated fibers, stellate arrays	Essential	Present in most organisms	Centriole
δ-Tubulin (*UNI3)*	Basal bodies	Triplet microtubule assembly and flagellar assembly	Present in most organisms with centrioles	Manchette in mice
ε-Tubulin (*BLD2)*	Surrounding basal bodies and centrioles	Essential	Present in most organisms with centrioles	Subdistal appendages of centrioles
Fa1	Transition zone	Flagellar autonomy	Coiled-coil	
Fa2	Unknown	Flagellar autonomy and cell cycle progression	Homolog in most organisms	Unknown
SF-assemblin	Fibers	Basal body segregation	Coiled-coil	
Vfl1	Transition zone	Basal body segregation	Coiled-coil	
Vfl3	Lateral fibers	Basal body and probasal body attachment	Present in most organisms with centrioles	Unknown

References

1. Boveri T. **1901**. Zellen-Studien IV. Uber die Nature der Centrosomen. *Jena Z Med Naturw* **28**:1–220.
2. Fawcett DW, Porter KR. **1954**, A study on the fine structure of ciliated epithelia. *J. Morphol.* **94**:221–282.
3. Ringo DL. **1967**. Flagellar motion and fine structure of the flagellar apparatus in *Chlamydomonas*. *J Cell Biol* **33**:543–571.
4. Johnson UG, Porter KR. **1968**. Fine structure of cell division in *Chlamydomonas reinhardi*. Basal bodies and microtubules. *J Cell Biol* **38**:403–425.
5. O'Toole ET, Giddings TH, McIntosh JR, Dutcher SK. **2003**. Three-dimensional organization of basal bodies from wild-type and delta-tubulin deletion strains of *Chlamydomonas reinhardtii*. *Mol Biol Cell* **14**:2999–3012.
6. Rosenbaum JL, Witman GB. **2002**. Intraflagellar transport. *Nature Rev Mol Cell Biol* **3**:813–825.
7. Lechtreck KF, Teltenkotter A, Grunow A. **1999**. A 210-kDa protein is located in a membrane-microtubule linker at the distal end of mature and nascent basal bodies. *J Cell Sci* **112** (Pt 11): 1633–1644.
8. Norrander JM, deCathelineau AM, Brown JA, Porter ME, Linck RW. **2000**. The Rib43a protein is associated with

forming the specialized protofilament ribbons of flagellar microtubules in Chlamydomonas. *Mol Biol Cell* **11**:201–215.

9. Salisbury JL, Baron AT, Sanders MA. **1988**. The centrin-based cytoskeleton of *Chlamydomonas reinhardtii*: distribution in interphase and mitotic cells. *J Cell Biol* **107**:635–641.
10. Silflow CD, LaVoie M, Tam LW, Tousey S, Sanders M, Wu W, Borodovsky M, Lefebvre PA. **2001**. The Vfl1 Protein in *Chlamydomonas* localizes in a rotationally asymmetric pattern at the distal ends of the basal bodies. *J Cell Biol* **153**:63–74.
11. Harper JD, Rao PN, John PC. **1990**. The mitosis-specific monoclonal antibody MPM-2 recognizes phosphoproteins associated with the nuclear envelope in *Chlamydomonas reinhardtii* cells. *Eur J Cell Biol* **51**:272–278.
12. Baron AT, Greenwood TM, Bazinet CW, Salisbury JL. **1992**. Centrin is a component of the pericentriolar lattice. *Biol Cell* **76**:383–388.
13. Goodenough UW, Shames B, Small L, Saito T, Crain RC, Sanders MA, Salisbury JL. **1993**. The role of calcium in the *Chlamydomonas reinhardtii* mating reaction. *J Cell Biol* **121**:365–374.
14. Wright RL, Adler SA, Spanier JG, Jarvik JW. **1989**. Nucleus–basal body connector in *Chlamydomonas*: evidence for a role in basal body segregation and against essential roles in mitosis or in determining cell polarity. *Cell Motil Cytoskeleton* **14**:516–526.
15. Wright RL, Salisbury J, Jarvik JW. **1985**. A nucleus–basal body connector in *Chlamydomonas reinhardtii* that may function in basal body localization or segregation. *J Cell Biol* **101**:1903–1912.
16. Biggins S, Rose MD. **1994**. Direct interaction between yeast spindle pole body components: Kar1p is required for Cdc31p localization to the spindle pole body. *J Cell Biol* **125**:843–852.
17. Spang A, Courtney I, Grein K, Matzner M, Schiebel E. **1995**. The Cdc31p-binding protein Kar1p is a component of the half bridge of the yeast spindle pole body. *J Cell Biol* **128**:863–877.
18. Holmes JA, Dutcher SK. **1989**. Cellular asymmetry in *Chlamydomonas reinhardtii*. *J Cell Sci* **94** (Pt 2):273–285.
19. Gaffal KP, el-Gammel S. **1990**. Elucidation of the enigma of the metaphase band of *Chlamydomonas reinhardtii*. *Protoplasma* **156**:139–148.
20. Ehler LL, Holmes JA, Dutcher SK. **1995**. Loss of spatial control of the mitotic spindle apparatus in a *Chlamydomonas reinhardtii* mutant strain lacking basal bodies. *Genetics* **141**:945–960.
21. Fromherz S, Giddings TH, Ospina-Gomez N, Dutcher SK **2004**. Mutations in a-tubulin promote basal body maturation and flagellar assembly in the absence of λ-tubulin. *J Cell Sci* **117**:303–314.
22. Lechtreck KF, Melkonian M. **1991**. Striated microtubule-associated fibers: identification of assemblin, a novel 34-kD protein that forms paracrystals of 2-nm filaments *in vitro*. *J Cell Biol* **115**:705–716.
23. Lechtreck KF, Rostmann J, Grunow A. **2002**. Analysis of *Chlamydomonas* SF-assemblin by GFP tagging and expression of antisense constructs. *J Cell Sci* **115**:1511–1522.
24. Silflow CD, Iyadurai KB. **2002**. The *Chlamydomonas VFL3* gene product is required for correct positioning of the basal bodies. *Mol Biol Cell* **13**:326a.
25. Kater JM. **1929**. Morphology and division of *Chlamydomonas* with reference to the phylogeny of the flagellate neuromotor system. *University of California Publication in Zoology* **33**:125–168.
26. Kuriyama R, Dasgupta S, Borisy GG. **1986**. Independence of centriole formation and initiation of DNA synthesis in Chinese hamster ovary cells. *Cell Motil Cytoskeleton* **6**:355–362.
27. Gould RR. **1975**. The basal bodies of *Chlamydomonas reinhardtii*. Formation from probasal bodies, isolation, and partial characterization. *J Cell Biol* **65**:65–74.
28. Coss RA. **1974**. Mitosis in *Chlamydomonas reinhardtii* basal bodies and the mitotic apparatus. *J Cell Biol* **63**:325–329.

29. Dippell RV. **1968**. The development of basal bodies in paramecium. *Proc Natl Acad Sci USA* **61**:461–468.
30. Biggins S, Walczak CE. **2003**. Captivating capture: how microtubules attach to kinetochores. *Curr Biol* **13**:R449–R460.
31. Preble AM, Giddings TH, Jr., Dutcher SK. **2001**. Extragenic bypass suppressors of mutations in the essential gene BLD2 promote assembly of basal bodies with abnormal microtubules in *Chlamydomonas reinhardtii. Genetics* **157**:163–181.
32. Oakley BR. **2000**. An abundance of tubulins. *Trends Cell Biol* **10**:537–542.
33. McKean PG, Baines A, Vaughan S, Gull K. **2003**. Gamma-tubulin functions in the nucleation of a discrete subset of microtubules in the eukaryotic flagellum. *Curr Biol* **13**:598–602.
34. Ruiz F, Beisson J, Rossier J, Dupuis-Williams P. **1999**. Basal body duplication in Paramecium requires gamma-tubulin. *Curr Biol* **9**:43–46.
35. Silflow CD, Liu B, LaVoie M, Richardson EA, Palevitz BA. **1999**. Gamma-tubulin in Chlamydomonas: characterization of the gene and localization of the gene product in cells. *Cell Motil Cytoskeleton* **42**:285–297.
36. Moritz M, Braunfeld MB, Guenebaut V, Heuser J, Agard DA. **2000**. Structure of the gamma-tubulin ring complex: a template for microtubule nucleation. *Nature Cell Biol* **2**:365–370.
37. Dutcher SK, Trabuco EC. **1998**. The UNI3 gene is required for assembly of basal bodies of *Chlamydomonas* and encodes delta-tubulin, a new member of the tubulin superfamily. *Mol Biol Cell* **9**:1293–1308.
38. Dupuis-Williams P, Fleury-Aubusson A, de Loubresse NG, Geoffroy H, Vayssie L, Galvani A, Espigat A, Rossier J. **2002**. Functional role of epsilon-tubulin in the assembly of the centriolar microtubule scaffold. *J Cell Biol* **158**:1183–1193.
39. Dutcher SK, Morrissette NS, Preble AM, Rackley C, Stanga J. **2002**. Epsilon-tubulin is an essential component of the centriole. *Mol Biol Cell* **13**:3859–3869.
40. Goodenough UW, StClair HS. **1975**. BALD-2: a mutation affecting the formation of doublet and triplet sets of microtubules in *Chlamydomonas reinhardtii. J Cell Biol* **66**:480–491.
41. Chang P, Giddings TH, Jr., Winey M, Stearns T. **2003**. Epsilon-tubulin is required for centriole duplication and microtubule organization. *Nature Cell Biol* **5**:71–76.
42. Ruiz F, Krzywicka A, Klotz C, Keller A, Cohen J, Koll F, Balavoine G, Beisson J. **2000**. The SM19 gene, required for duplication of basal bodies in *Paramecium*, encodes a novel tubulin, eta-tubulin. *Curr* Biol **10**:1451–1454.
43. de Saint Phalle B, Sullivan W. **1998**. Spindle assembly and mitosis without centrosomes in parthenogenetic *Sciara* embryos. *J Cell Biol* **141**:1383–1391.
44. Klotz C, Dabauvalle MC, Paintrand M, Weber T, Bornens M, Karsenti E. **1990**. Parthenogenesis in *Xenopus* eggs requires centrosomal integrity. *J Cell Biol* **110**:405–415.
45. Mizukami I, Gall J. **1966**. Centriole replication. II. Sperm formation in the fern, *Marsilea*, and the cycad, *Zamia. J Cell Biol* **29**.97–111.
46. Marshall WF, Vucica Y, Rosenbaum JL. **2001**. Kinetics and regulation of *de novo* centriole assembly. Implications for the mechanism of centriole duplication. *Curr Biol* **11**:308–317.
47. Deane JA, Cole DG, Seeley ES, Diener DR, Rosenbaum JL. **2001**. Localization of intraflagellar transport protein IFT52 identifies basal body transitional fibers as the docking site for IFT particles. *Curr Biol* **11**:1586–1590.
48. Cole DG, Diener DR, Himelblau AL, Beech PL, Fuster JC, Rosenbaum JL. **1998**. *Chlamydomonas* kinesin-II-dependent intraflagellar transport (IFT): IFT particles contain proteins required for ciliary assembly in *Caenorhabditis elegans* sensory neurons. *J Cell Biol* **141**:993–1008.
49. LeDizet M, Piperno G. **1986**. Cytoplasmic microtubules containing acetylated alpha-tubulin in *Chlamydomonas*

reinhardtii: spatial arrangement and properties. *J Cell Biol* **103**:13–22.

50. Masuda M, Sato H. **1984**. Reversible resorption of cilia and the centriole cycle in dividing cells of sea urchin blastulae. *Zoolog Sci* **1**:445–462.

51. Walther Z, Vashishtha M, Hall JL. **1994**, The *Chlamydomonas* FLA10 gene encodes a novel kinesin-homologous protein. *J Cell Biol* **126**:175–188.

52. Piperno G, Mead K. **1997**, Transport of a novel complex in the cytoplasmic matrix of *Chlamydomonas* flagella. *Proc Natl Acad Sci USA* **94**:4457–4462.

53. Cole DG. **2003**. The intraflagellar transport machinery of *Chlamydomonas reinhardtii*. *Traffic* **4**:435–442.

54. Brazelton WJ, Amundsen CD, Silflow CD, Lefebvre PA. **2001**. The bld1 mutation identifies the *Chlamydomonas* osm-6 homolog as a gene required for flagellar assembly. *Curr Biol* **11**:1591–1594.

55. Lewin RA, Lee TH, Fang LS. **1982**. Effects of various agents on flagellar activity, flagellar autotomy and cell viability in four species of *Chlamydomonas* (chlorophyta: volvocales). *Symp Soc Exp Biol* **35**:421–437.

56. Sanders MA, Salisbury JL. **1989**. Centrin-mediated microtubule severing during flagellar excision in *Chlamydomonas reinhardtii*. *J Cell Biol* **108**:1751–1760.

57. Quarmby L. **2000**. Cellular Samurai: katanin and the severing of microtubules. *J Cell Sci* **113** (Pt 16):2821–2827.

58. Hartman JJ, Mahr J, McNally K, Okawa K, Iwamatsu A, Thomas S, Cheesman S, Heuser J, Vale RD, McNally FJ. **1998**. Katanin, a microtubule-severing protein, is a novel AAA ATPase that targets to the centrosome using a WD40-containing subunit. *Cell* **93**:277–287.

59. Smith EF, Lefebvre PA. **1998**. PF15 is required for central microtubules and is homologous to p80 katanin. *Mol Biol Cell* **9**:278a.

60. Adams GM, Huang B, Piperno G, Luck DJ. **1981**. Central-pair microtubular complex of *Chlamydomonas* flagella: polypeptide composition as revealed by analysis of mutants. *J Cell Biol* **91**: 69–76.

61. Lohret TA, McNally FJ, Quarmby LM. **1998**. A role for katanin-mediated axonemal severing during *Chlamydomonas* deflagellation. *Mol Biol Cell* **9**:1195–1207.

62. Lohret TA, Zhao L, Quarmby LM. **1999**. Cloning of *Chlamydomonas* p60 katanin and localization to the site of outer doublet severing during deflagellation. *Cell Motil Cytoskeleton* **43**:221–231.

63. Finst RJ, Kim PJ, Quarmby LM. **1998**. Genetics of the deflagellation pathway in *Chlamydomonas*. *Genetics* **149**:927–936.

64. Finst RJ, Kim PJ, Griffis ER, Quarmby LM. **2000**. Fa1p is a 171 kDa protein essential for axonemal microtubule severing in *Chlamydomonas*. *J Cell Sci* **113** (Pt 11):1963–1971.

65. Mahjoub MR, Montpetit B, Zhao L, Finst RJ, Goh B, Kim AC, Quarmby LM. **2002**. The FA2 gene of *Chlamydomonas* encodes a NIMA family kinase with roles in cell cycle progression and microtubule severing during deflagellation. *J Cell Sci* **115**:1759–1768.

66. Mogensen MM, Malik A, Piel M, Bouckson-Castaing V, Bornens M. **2000**. Microtubule minus-end anchorage at centrosomal and non-centrosomal sites: the role of ninein. *J Cell Sci* **113** (Pt 17):3013–3023.

67. Ou YY, Mack GJ, Zhang M, Rattner JB. **2002**. CEP110 and ninein are located in a specific domain of the centrosome associated with centrosome maturation. *J Cell Sci* **115**:1825–1835.

68. Quintyne NJ, Schroer TA. **2002**. Distinct cell cycle-dependent roles for dynactin and dynein at centrosomes. *J Cell Biol* **159**:245–254.

69. Gromley A, Jurczyk A, Sillibourne J, Halilovic E, Mogensen M, Groisman I, Blomberg M, Doxsey S. **2003**. A novel human protein of the maternal centriole is required for the final stages of cytokinesis and entry into S phase. *J Cell Biol* **161**:535–545.

70. Chang P, Stearns T. **2000**. Delta-tubulin and epsilon-tubulin: two new human centrosomal tubulins reveal new aspects of centrosome structure and function. *Nature Cell Biol* 2:30–35.

71. Paintrand M, Moudjou M, Delacroix H, Bornens M. **1992**. Centrosome organization and centriole architecture: their sensitivity to divalent cations. *J Struct Biol* **108**:107–128.

72. Huang B, Ramanis Z, Dutcher SK, Luck DJ. **1982**. Uniflagellar mutants of *Chlamydomonas*: evidence for the role of basal bodies in transmission of positional information. *Cell* **29**:745–753.

6
The Centrosome in Evolution

Juliette Azimzadeh and Michel Bornens

6.1
Introduction

A major outcome of the systematic sequencing of genomes has been to promote a reappraisal of the traditional phylogenetic trees based on morphology and embryology. This, together with improvements in tree reconstructions, has promoted a deep reorganization of the entire life tree. For example the new molecule-based animal phylogeny proposes a view of the evolution of protostomes and deuterostomes which is clearly more consistent than the previous phylogenies [1].

On the other hand, the considerable progress in the understanding of cell and molecular biology of development has led to a better recognition that evolution is a cellular process. This led Gerhart and Kirschner [2] to suggest that had Theodosius Dobzhansky known that cell function can be explained in terms of gene products, he would have completed his famous comment "Nothing in biology makes sense except in the light of evolution" to "Nothing in evolution makes sense except in the light of cell biology".

In many respects, the centrosome organelle has a specific status in cell organization and evolution. The most compelling feature is that the centrosome is not present in all organisms: plants control their microtubule arrays during interphase as well as during mitosis, and divide without a centrosome. On the other hand, complete parthenogenetic development can take place in frog embryos when the sperm cell is replaced by the centrosome of a somatic cell, whether from the same or from a divergent species. This result emphasizes another compelling feature of this organelle, namely that the core structures of the centrosome from animal cells, the centrioles, can function as basal bodies/kinetosomes to grow cilia or flagella and vice versa.

We will attempt in this brief account to outline the major questions raised by the centrosome organelle and propose tentative answers in the light of evolution.

Centrosomes in Development and Disease. Edited by Erich A. Nigg

ISBN 3-527-30980-2

6.2 The Centriole/Basal Body Structure is a Derived Character[1] of Eukaryotes

Among divergent eukaryotes, the great range of differences in shape and complexity of cellular structures on which microtubules are nucleated, has not arisen as an evolutionary change from the simple to the complex, but rather the other way around: all data from molecular phylogeny indicate that the centriolar structure and the associated (9 + 2) flagellum are very ancient inventions, present at the apparition of the early eukaryotic cells [3]. The structurally simpler centrosome[2] from yeasts or the lack of recognizable centrosomes, for example in seed plants, are derived features. In most cases, a structural association can be observed between centrioles/basal bodies, or other centrosomal structures, and the nucleus.

6.3 The Basal Body/Axoneme is the Ancestral Structure

A general survey of the consensus phylogeny of eukaryotes clearly indicates that the centriole/basal body structure did not appear without an axoneme. Centrioles localized in a central body not directly associated with the plasma membrane are observed in more recent taxa[3], essentially in animals. The basal body/axoneme is maintained in the great majority of taxa as most of them have maintained a flagellate stage in their life cycle (Figure 6.1). As this survey indicates, it would in fact be easier to draw up a list of those taxa in which the flagellum has totally disappeared. Taxa lacking basal bodies/axonemes are found mainly in the green lineage and in-

---

[1] "Derived character" refers to a character state that is present in one or more but not all subclades of a clade (i. e. a monophyletic group of organisms, that share a common ancestor) under consideration. A derived character state is inferred to be a modified version of the primitive condition of that character, and to have arisen later in the evolution of the clade.

[2] We will use the classical term "centrosome" (etymologically "central body") throughout as a generic term to describe any isolatable single-copy organelle such as the SPB of fungi, the NAB of cellular slime molds, the centroplast of heliozoa or the centriole-based centrosome of animal cells, all of which have in common three basic properties: to generally maintain themselves at the cell center due to their microtubule nucleating/anchoring properties, to be physically associated with the nucleus, and to duplicate once during the cell cycle. The acronym MTOC, for Microtubule Organizing Center, usually has limited acceptance as it does not specifically describe an isolable structure. This can cause some difficulties. For example, in myotubes, in which centrioles have been eliminated, the microtubules are nucleated from the nuclear periphery [4], which can thus be described as an MTOC but not as a centrosome. In the pillar cells of the mammalian inner ear, microtubules are nucleated at the centrosome which is located in an apical position and also produces a primary cilium. The microtubules are then transported away from the centrosome and anchored at distant apical sites which, with good reason, are often referred to as MTOCs (see Chapter 15). In other words, whereas a centrosome is necessarily an MTOC, the reverse is not true.

[3] Taxon, taxa (pl.) refers to a group or category of living organisms i. e. the published groups within each of the divisions in the Linnean hierarchy.

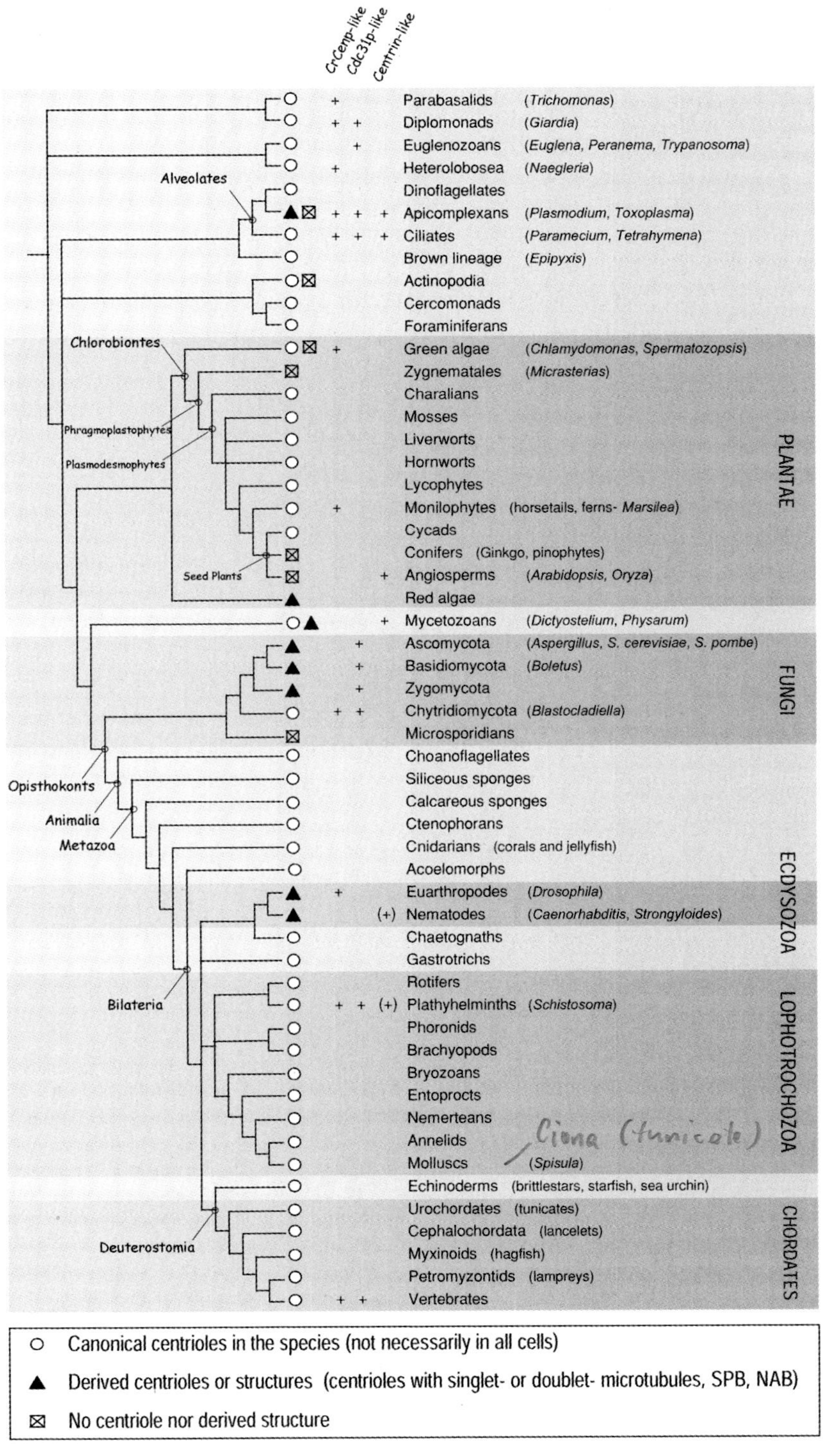

Figure 6.1 Conservation of the centriolar structure in the phylogenetic tree of eukaryotes (adapted from [114]). The data on the phylogenetic distribution of centrin subfamilies presented in Figure 6.2 has also been indicated.

clude the whole taxon of Rhodophytes (red algae) and some of the higher taxa such as the spermatophytes (seed plants) in the chlorobiontes in which angiosperms and conifers may have lost the centriole structure independently, and the zygnematales (unicellular or filamentous green algae formerly known as conjugate algae due to the mirror symmetry of their cell organization).

The remarkable conservation of the canonical centriole/basal body structure, with the nine triplets of microtubules, seems to be necessary for the assembly or the activity of the axoneme. When the axoneme is no longer functional, the basal body structure becomes less constrained. Derived centrioles with doublets or singlets are observed in unicellular organisms, such as some apicomplexans, or in multicellular organisms like arthropods and nematodes from the taxon Ecdysozoa in which cilia are most often derived structures involved in sensory reception. When axonemes are totally absent, basal body/centrioles are also absent. Various centrosomal structures, no longer associated with the plasma membrane, can be observed instead, such as the Spindle Pole Body (SPB) of higher fungi, the Nucleus Associated Body (NAB) of cellular slime molds, and other "simpler" structures in some diatoms or some actinopods. Alternatively, there is no recognizable centrosome organelle as is the case of microsporidia which share a common ancestor with fungi, and have lost the centriole or any type of centrosome, a situation that may be linked to their survival strategy as intracellular parasites.

In metazoan organisms where cell differentiation leads to a great number of cell types proficient in various specific functions, different situations may be encountered. In vertebrates most cells have a centriole pair organizing the centrosome but some cells do not as exemplified by myotubes and muscle cells [4]. Only the male gamete has a motile flagellum and some specialized epithelial cells are characterized by cilia on their apical surfaces which beat with a wave-like motion to mobilize external materials e. g. mucus in the trachea or oocytes in the oviduct. Depending on the type of tissue, post-mitotic differentiated cells may possess a non-motile primary cilium growing from the older centriole of the centrosome, which becomes anchored to the plasma membrane. This transition can be produced reversibly *in vitro* depending on the serum level in the culture medium, illustrating that the same organelle can switch from basal body function in G0-arrested cells to centriole function in proliferating cells.

6.4
Functions Associated with the Flagellar Apparatus

Three major functional modules, all expressing a facet of cell polarity, are associated with the basal body/axoneme: cell locomotion in which it acts as a swimming organelle due to its beating activity, sensory reception in which it acts as a sensory organ due to the concentration of receptors on the ciliary plasma membrane which cannot diffuse beyond a barrier at the base of the cilium where the basal body is anchored, and cell division in which, in agreement with the early views of Boveri [5], the basal body/axoneme acts in coordinating karyokinesis and cytokinesis, due to the duplication process governing basal body/centriole biogenesis and, as we learned after Boveri, to its critical role in the cell-wide organization of the microtubule network.

6.4.1
Cell Locomotion

Although the details of the generic dynein-dependent sliding of doublet microtubules in the axoneme have long been known, the overall control of the flagellar movement is far from being understood. We will only note here that the mechanisms of flagellum-dependent motility can be quite different. For instance, in the euglenid taxon there are examples of the classical beating movement of the flagellum (e. g. *Euglena*) or the movement of an undulating membrane by the flagellum (e. g. trypanosomes). But the flagellum can also produce a very efficient unidirectional gliding motion powered by flagellar surface motility as is the case with *Peranema trichophorum* [6]. The surface motility which can be uncoupled from ciliary beating and may be driven by a retrograde motor along the axoneme seems to move transmembrane mastigonemes. This surface motility serves additional functions in this organism, such as prey capture, a feature which has also been documented in other taxa. The stramenopile *Epipyxis pulchra*, can use its two different flagella as an efficient pair of pliers to catch, engulf, or release, any prey passing at grasping distance [7, 8]. A striking feature of this unicellular organism is that once contact has been made between the different types of prey and the flagellar surface, the organism is able to select the preferred prey, suggesting that a sensory mechanism is also at work in this process. The strong vortex created by the beating of numerous cilia in the *Paramecium* gullet is an alternate and powerful method of capturing prey. These few examples show how the basal body/axoneme and its membrane can be used in various ways to ensure basic functions for cell survival such as polarity, motility, sensory reception, and moreover, can be adapted to various challenges including the capture of prey.

6.4.2
Sensory Reception

In all metazoa cilia have been recruited for sensory reception. This is true in deuterostomes and protostomes and a great range of stimuli seem to be collected in this way. In taxa where the basal body/centriole structure is derived, as in Ecdysozoa (arthropods and nematodes), the only cilia which remain are derived structures involved in sensory reception [9].

In vertebrates there is a growing body of evidence that primary cilia, which are very common in vertebrate tissues, are sensory organelles, the cone and rod cells in the retina being an example of this. Primary cilia might be involved in specific signaling in various tissues, including during early mammalian development where they participate in the left–right patterning of the embryo. With regard to this specific case, a very elaborate use of the cilia is observed in the Hensen node: a population of moving monocilia from centrally-located node cells create a flow and immotile cilia from peripherally-located cells detect the flow in a manner apparently similar to the mechano-reception observed in kidney cells [10].

Genetic and biochemical analysis of flagellar assembly in the unicellular algae *Chlamydomonas reinhardtii* has revealed a strong conservation of the intraflagellar machinery responsible for transporting precursor complexes to the tip of the growing cilia or flagella, a conservation which correlates with the conservation of the axoneme structure among divergent eukaryotes [11]. This study on a unicellular model has a direct impact on human pathology: the severe polycystic kidney disease appears to be due to mutations in genes participating in this highly conserved machinery. This results in either the absence of primary cilia in epithelial cells lining the kidney nephron or in their dysfunction [12, 13]. In both cases, cilia-dependent mechano-reception at the surface of the kidney epithelium is impaired. In addition, kidney cells resume growth, pointing to the long-noted correlation between the presence of a primary cilium and G0-arrest in the cell cycle.

As in the case of locomotion, the use of cilia for sensory reception is thus exhibiting an amazing variation together with a strong conservation of basic principles.

6.4.3
Cell Division

The precise way in which the basal body/axoneme acts as a division organ varies amongst divergent organisms but should not be expected to be limited to karyokinesis. It is beyond the scope of this chapter to analyze the different patterns that have been selected during evolution (see Chapters 5, 9, and 20). Perhaps the euglenid protists provide the best examples of how cell organization is controlled throughout the cell division process by the flagellar apparatus. In trypanosomes for example, the whole division process starts at the duplicating basal body and terminates at the tip of the flagellum [14]. Even the positioning and segregation of the mitochondrial genome is ensured by the duplication of basal bodies through distinct transmembrane structural linkage [15]. The very efficient RNAi inactivation which is now feasible in trypanosomes, has led to the clear demonstration that beyond the temporal order of events, control of the whole process of cell division is dependent upon basal body duplication (see Chapter 20) [16]. A very interesting insight has recently been provided by this approach: when the expression of a protein required for intraflagellar transport and thus necessary for flagellum assembly is knocked out, the length of the flagellum is decreased as expected, but so is the cell size [17]. Indeed, matching the swimming capacity of the cell with the size and weight of the body that has to be moved might have survival value. This remarkable result thus suggests a mechanism by which the flagellum is used as a ruler to sense the size of the cell and is involved in promoting cytokinesis, a possibility which is also supported by the fact that the tip of the flagellum defines the site of cleavage during cytokinesis. Trypanosomes completely devoid of flagella cannot divide.

There are reasons to think that in other systems the microtubule network as a whole could also participate in the coupling of cell growth and cell division. The cortical microtubule network of unicellular organisms shows a unique polarity with regard to the flagellar apparatus, in much the same way as the more internal microtubule network in animal cells does with regard to the centrosome. In both cases, the cellular distribution and to some extent the dynamics of most membrane compartments is dependent upon microtubule organization and dynamics. This represents a powerful means for organizing intracellular compartments. Assembly of new membrane compartments during cell growth would be expected to require more microtubules, and this net increase in microtubule polymer could in some way monitor mass increase (see also below).

A role in cell division is indeed the earliest recognized property of the centrosome. Studying fertilization in *Ascaris* and in sea urchin, Boveri [5] referred to the centrosome as the "cell division organ" without which cell division could not occur. To him, the uniparental inheritance of the centrosome via the male gamete was one way of insuring that the unfertilized egg could not divide, and thus, that parthenogenetic development could not occur. In agreement with this view, unfertilized eggs could be "activated" by pricking them, leading to cytoplasmic reorganization as shown by the displacement of cortical pigments, but would never cleave. This was then challenged by reports suggesting that parthenogenesis could be triggered in amphibian eggs provided the needle used to prick the egg was first dipped into the blood of the frog. The rate of success was very low however, and the postulated blood regulatory factor necessary for parthenogenesis was never characterized. But this claim together with the discovery that plant cells could apparently divide without a centrosome placed Boveri's hypothesis under suspicion. It was not until the 1970s that centriole/basal body-containing neck fractions of sperm cells from *Xenopus* and sea urchin were demonstrated to be the only fractions able to trigger parthenogenetic cleavage [18]. In agreement with Boveri's view, the centrosome was therefore shown to be necessary and sufficient to induce parthenogenesis. In addition, the demonstration that heterologous centrosomes were also competent indicated a functional conservation of centrosome-associated activities among divergent species. Centrosome-induced parthenogenesis in *Xenopus* has since then been extensively used as an approach to analyzing the minimum requirements for centrosome continuity (for a full account of this aspect, see [19]) and *Xenopus* egg extract has been shown to be a powerful tool in analysis of vertebrate centrosome reproduction *in vitro* [20].

Even if we are far from a comprehensive view, a better understanding now exists as to how the centrosome can critically participate in the control of whole cell division processes by organizing the array of microtubules and by acting as a platform where regulatory complexes accumulate. From the initiation of centrosome duplication, which takes place before S phase onset as established by genetic analysis in budding yeast, to mitosis and cytokinesis, the centrosome–microtubule system is directly involved in the physical and regulatory integration of constraints required for the control of each major cell cycle transition (see below and Chapters 8–11).

Another major aspect of the basal body to centriole conversion during fertilization is that the newly formed centrosome is required for the definition of embryo polarity. It is beyond the scope of this chapter to review this aspect, but there is a strong case in several systems for a correlation between the entry point of the sperm and the antero-posterior polarity of the early embryo. *Caenorhabditis elegans* provides a good model system for analyzing this aspect [21] (see Chapter 12), since embryo polarity can be reversed by manipulating the entry point of the sperm [22]. How signaling between the centrosome and the cortex is established is an intriguing question.

6.5
The Conservative Mode of Duplication of the Basal Body/Centriole/SPB: An Essential Clue for Cell Morphogenesis

How cell shape and polarity are conserved over time and faithfully transmitted through cell division are questions critical to our understanding of the evolution of organisms. The duplication mechanism of the centriole/basal body, or of the yeast centrosome, is conservative: a completely new structure forms aside the "old" structure, usually during S phase. The question thus arises as to whether centrosome inheritance has any role in the control of cell morphogenesis over time.

There is growing evidence that there is an age-related centriole/basal body activity that is relevant for cell morphogenesis. An example is provided by unicellular bi-flagellates, such as *C. reinhardtii*, in which the generational asymmetry of the basal bodies and their mode of inheritance provide clues to the maintenance of cortical polarities during cell division such as the location of the eye spot, which is essential for cell survival as an asymmetric organelle allowing the cell to detect the light and orient its swimming towards or away from the source depending on its intensity [23]. There are other cases of defined modes of basal body inheritance which are essential for the reproduction of cell polarities. The mode of inheritance of the two different flagella in heterokonts follows a precise pattern and seems to be controlled by basal body inheritance [24]. A striking example is that of the unicellular flagellate *Pyramimonas octopus*, which has eight flagella that sequentially occupy defined positions before reaching a predetermined location after three cell generations [25].

Yeasts are model systems in which centrosome inheritance has been studied in most detail. As in most animal systems, the yeast SPB is responsible for microtubule organization in interphase cells, duplication before mitosis and organization of the mitotic spindle. Beyond the perfect symmetry of the mitotic spindle necessary for correct karyokinesis, there is an asymmetry characterized by two functionally distinct SPBs, required for exit from cytokinesis. Two conserved GTPase-regulated protein kinase cascades in budding and fission yeasts have been shown to be critical for coordinating completion of mitosis with cytokinesis (for reviews see [26, 27]). In both cases, the regulatory complexes accumulate at the SPB but the two

SPBs of the dividing cell are biochemically distinct with respect to the associated protein kinase cascade. The reason for this is still poorly understood in the fission yeast *Schizosaccharomyces pombe*. In the budding yeast *Saccharomyces cerevisiae* however, it is only the SPB that migrates into the daughter cell that carries the GTPase-regulated cascade complex and this is believed to be the mechanism whereby entry of the nucleus into the bud is monitored [26]. The SPBs are also intrinsically distinct in terms of generation. A defined mode of SPB inheritance has been demonstrated in budding yeast: the "old" SPB always segregates into the bud, but it is the astral microtubule–cortex interaction which seems to control the localization of the GTPase-regulated cascade complex at the SPB [28].

However, recent results suggest that SPB inheritance could also be important. Alignment of the spindle along the division axis and spatial coordination of spindle position with the cleavage apparatus is crucial to ensure proper inheritance of nuclei in the mother and bud-derived cells (see [29]). Partially overlapping mechanisms ensure spindle alignment via the interaction of microtubules with anchoring sites on the cortex (for reviews see [30, 31]). Recent reports indicate that cortical determinants of spindle alignment are not recruited directly to the cortex as previously assumed but are instead delivered to the cortex on the plus ends of polymerizing microtubules [32, 33]. The mechanism controlling the asymmetrical loading of cortical determinants on one SPB is not fully understood, but one possibility is the generational asymmetry of SPBs which might be correlated with structural and biochemical differences.

Recent evidence strongly suggests that not only is the time of assembly important but also the lineage of the SPB: at the end of *S. cerevisiae* meiosis, once the four haploid genomes have been segregated, the outer plates of the four SPBs act as platforms for the assembly of meiotic plaques, a prerequisite for spore formation. Strains heterozygous for components of the meiotic plaque most often produce two spores instead of four, and these spores are always non-sisters [34]. By modulating the stringency of the sporulation medium, asci with three, two or only one spore could be formed (M. Knop, EMBL, personal communication). Remarkably, the spores which are formed are not the result of a random mechanism, but are dependent on the age of the SPB, demonstrating the existence of four different states for SPBs. Similarly, by analyzing the pedigree of dividing cells while studying the asymmetric localization of Fin1 (the homolog of the human Nek2 kinase) on the mitotic SPBs, I. Hagan. Hagan was able to identify four states for the SPB (I. Hagan, Paterson Institute for Cancer Research, personal communication). The biochemical nature of these states of maturation is not known but presumably involves a mitosis-dependent modification of SPBs. These remarkable results led to a reappraisal of the correlation between SPB inheritance and the mitotic asymmetric accumulation of the GTPase-regulated protein kinases controlling cytokinesis in yeasts.

In cells from metazoa, there are also reasons to believe that the four centrioles of a dividing cell are different: established cell lines expressing centrin-GFP often show four intensities of centriole labeling which seem to respect the lineage of the centrioles [35]. It is not yet known whether the asymmetry in centrosome inheritance introduces a constitutive asymmetry into the cell division process. It has recently been shown however, that completion of cytokinesis in cultured human HeLa or murine L929 cells coincides with the migration of the mother but not the daughter centriole towards the cytokinesis site [36]. Such behavior is reminiscent of the situation in yeast and suggests that centrioles and SPBs have a conserved role in the signaling pathway controlling cytokinesis, a possibility which has recently been supported by the discovery of centriolin, a mother centriole-associated protein having similarities with one protein of the protein kinase cascades that control the last steps of division in yeast [37]. Intriguingly, in the majority of cases, the movement of the mother centriole towards the mid-zone occurs only in one of the two daughter cells.

Why would SPB inheritance, instead of random segregation, have any benefit for cell division or cell morphogenesis? SPBs are apparently highly stable organelles with a very low turnover of their core components. The same is true for vertebrate centrioles [38]. This feature could make the SPB/centrosome a stable internal landmark for morphogenesis and polarity through cell generations. Together with other landmarks essential for cell polarity and present in other compartments, for example the cell cortex [39], this would then ensure a faithful process.

That pre-existing "old" structures or patterns could influence the organization of "new" structures or patterns is a finding relevant to the centriole/basal body duplication process itself and, moreover, has been demonstrated for large assemblies of basal bodies in ciliates [40]. Known as "cortical heredity" (because it was mainly observed on cortices where the structural patterns are very precise), or as cytotaxy, this epigenetic process confers a structural memory encompassing several cell generations. For example, a defect introduced into the cortex of a *Paramecium* can be propagated to the progeny of this cell during more than 1000 generations. A similar inheritance process, involving internal long-lived structures such as the centrosome and stable cortical marks, might contribute to the stability of cell shape over time in other eukaryotic cells.

6.6
The Centrosome or Central Body

The centriole-based centrosome of animal cells is derived from the centriole/basal body/flagellum. One could imagine that this has occurred during the apparition of multicellularity from colonial aggregates. At the stage of the colonial aggregates, individual cells may all have been flagellated, with a cortical microtubule network similar to that observed in unicellular flagellates, and would have been capable of locomotion and sensory reception. As the colony became more integrated, specialized cells becoming interdependent have specialized, some of them losing their flagella and thus becoming more proficient in performing other functions. By migrating to the cell center, the pair of basal bodies would have formed the centrosome. Of the three major functions originally associated with the flagellar apparatus (motility, division, sensory reception), the centrosome would have directly conserved a role in coordinating cell division. The two other functions would not have disappeared with the disappearance of the flagellum, but taken on other modalities more adapted to tissue organization. Cell migration, rather than swimming, would have developed together with a conspicuous cortical actin system, while the microtubule network would no longer be sub-cortical as in unicellular flagellates but intracytoplasmic. Sensory reception would have evolved to utilize actin-anchored receptors for cell–cell interactions.

The movement of the centriole pair towards the cell center could have taken place through molecular mechanisms which are still responsible for maintaining the centrosome at the cell centre in contemporary organisms. The interaction of dynein and other proteins at the plus ends of astral microtubules with the cell cortex is known to control nuclear positioning by controlling the positioning of the centrosome with which the nucleus is tightly associated [41].

Strikingly, nuclear positioning seems to be controlled in a highly conserved manner in divergent species. In animal cells, in cellular slime molds [42] and in fungi [31], dynein, dynactin and other plus-end proteins interact with proteins at the cell cortex to generate forces acting on the centrosome where microtubule minus ends are anchored. Nuclear migration in *Aspergillus nidulans* depends upon a conserved gene network which is also involved in nuclear positioning during neuronal migration in humans [43]. A human genetic defect causing lissencephaly (a result of impaired neuronal migration) resides in a homolog of a fungal nuclear migration gene which is required for dynein function. Interestingly, *Arabidopsis,* which lacks centrioles/basal bodies or SPBs, also lacks dynein and dynactin [44]. This correlation suggests that the major function of cytoplasmic dynein is to position the nucleus within the cell via centrosome/SPB-anchored microtubules.

From this remarkable conservation, it can be inferred that the yeast SPB and the NAB of the cellular slime molds are derived from the centriole/basal body/flagellum. The alternative would be that they have arisen through convergent evolution[4]. This is very unlikely with regard to the conservation of the gene network controlling nuclear positioning or centrosome duplication (see below). Moreover, closely related taxa still possess a basal body/flagellum, for example, plasmodial slime molds (*Physarum*) or lower fungi (Chytridiomycetes, see below). In animals the conversion of basal body to centriole, and thus the back and forth movement of these organelles from the cortex to the cell center, can always occur. In fungi or cellular slime molds, the loss of the flagellum would have resulted not only in the loss of the centriolar structure, but also in a permanent and tight association of a derived and structurally simpler organelle with the nucleus at the cell center.

It has already been mentioned that the bud-ward-directed SPB in *S. cerevisiae* acts as a platform delivering spindle alignment determinants to the appropriate place. This identifies a generic property of centrosomes, i. e. their capacity to be used as platforms for assembling regulatory complexes. Several examples have been reported for the regulation of the cell cycle according to very different modalities

[4] Convergent evolution is the evolutionary process in which organisms that are not closely related independently acquire some characteristic(s) in common.

(see Chapters 4, 8, 9, and 10). One such example is the clustering of regulatory complexes which might trigger a specific activity, a possibility recently substantiated in mammalian cells with regard to mitotic onset [45]. Another example of this generic property of centrosomes is that it may be responsible for spatial arrangements, for example, the relative position of the two daughter nuclei with respect to the cleavage furrow (see Section 6.5). This is how many regulatory activities necessary for cytokinesis are first concentrated at the centrosomes before being presented to the cleavage apparatus.

The centrosome can thus be seen as a specific differentiation of the ancestral organelle adapted to a different cellular context to maintain its control over cell division, the reproduction of polarities and cell motion. In contrast, in multicellular organisms sensory reception of external stimuli such as light, sounds, external fluid movements etc., would still be dependent on a ciliary structure present on specialized cells.

6.7 Evolution of Centrosome-associated Gene Products

Centrosomes from divergent species may be very diverse in their architecture, but nevertheless share several molecular components known to be associated with centrosomal properties. Among these is γ-tubulin, a member of the tubulin superfamily required for the nucleation of microtubule assembly, and centrin, a protein closely related to calmodulin and specifically associated with centrosomal structures over a broad range of eukaryotes.

6.7.1 γ-Tubulin

Like α- and β-tubulin, the subunits of microtubules, γ-tubulin is found in all eukaryotes and is necessary for microtubule function. Four other members of the tubulin superfamily have been identified, which appear to have a more restricted distribution among eukaryotes. This is thought to be due to the fact that δ-, ε-, η- and ζ-tubulin are required for the biogenesis of centrioles/basal bodies rather than for microtubule function [46–49] (see Chapter 2). As emphasized by these authors [46–49], genes for these rare tubulins are completely lacking in the genomes of the fly *Drosophila melanogaster* and the nematode *C. elegans*. The phylogeny of η- and ζ-tubulin is unclear due to the great sequence divergence between species, but it seems that δ- and ε-tubulin are monophyletic [47, 50]. This suggests that δ- and ε-tubulin genes have been inherited by a metazoan common ancestor and have been lost secondarily by *Drosophila* and *C. elegans*, possibly by the entire superphylum of Ecdysozoa to which both species belong. These species are known to form "incomplete" centrioles, however: *Drosophila* centrioles are most often composed of doublet microtubules, whereas nematode centrioles are usually composed of nine singlet microtubules as in the case of *C. elegans*, but can also have 10 singlet microtubules [51]. The absence of the canonical centriole triplet structure appears to correlate with the loss of the rare tubulin isoforms [46].

By contrast, γ-tubulin is a highly conserved protein in eukaryotes. Within the subset of organisms shown in Figure 6.2, protein identity is 61–76 % between metazoa and plants, 60–70 % between metazoa and higher fungi, and 56–71 % between metazoa and ciliates (sequences from *C. elegans, S. cerevisiae* and *Candida albicans* not included; see below). γ-Tubulin is not a core component of centrosomes however, and the evolutionary constraints imposed on this gene product are likely to be due to its nucleating properties rather than its association with any particular microtubule-organizing structure. This is illustrated most strikingly by the great sequence conservation of γ-tubulin in flowering plants that completely lack a centrosome. There are noticeable exceptions however: γ-tubulin from *C. elegans* and budding yeast are rather divergent compared to other taxa (comparison with other γ-tubulin proteins shows 24–41 % identity for *C. elegans*, 28–38 % for budding yeast; on average: 37 and 33 %, respectively). This could be due to a high evolution rate in these species, but also possibly to modifications in the number and/or nature of the γ-tubulin interactants. As suggested by McKean et al. [46], one constraint on the γ-tubulin sequence may be the need to interact with some rare tubulins. For instance, γ-tubulin co-sediments with δ-tubulin [52], and a direct interaction is consistent with structural models predicting interactions between tubulin superfamily members [53]. The complete loss of the rare tubulins in organisms such as budding yeast and *C. elegans* could thus contribute to the divergence of their γ-tubulin genes. It is noteworthy that *C. elegans* is the only metazoan with a divergent γ-tubulin and it also constitutes a rare case in which microtubules have 11 or 15 protofilaments instead of the canonical 13 protofilaments [54].

Differences in the composition of γ-tubulin-containing nucleating complexes may also account for sequence divergence in the case of yeast γ-tubulin, Tub4p. Tub4p exists in a stable complex with Spc97p and Spc98p [55, 56]. Spc97p and Spc98p proteins are related to each other and are both capable of binding to γ-tubulin. In animal cells, γ-tubulin is also associated with orthologs of Spc97p and Spc98p within a tetrameric complex, the γ-TuSC [57–60]. All three proteins that constitute the γ-TuSC are ubiquitous in eukaryotes, and the composition and size of γ-TuSCs appear to be conserved between species [56, 60, 61]. In vertebrates and *Drosophila*, however, microtubules are nucleated by a larger complex, the γ-TuRC, which results from multimerization of the γ-TuSC and the addition of at least six other components [57, 58, 60, 62, 63]. Several of these additional components have been identified in human, *Drosophila* and *Xenopus*, and found to be related to Spc97p and Spc98p [63–67]. Three related genes have also been identified in the *Arabidopsis* genome in addition to SPC97 and SPC98 [67, 68]. In *Drosophila*, it has been shown that not only Dgrip84 and Dgrip91, the respective homologs of Spc97p and Spc98p, but also three other γ-TuRC subunits, can interact directly with γ-tubulin [63]. The need for γ-tubulin to interact individually with all the other components of the γ-TuRC may constitute an important constraint on its sequence, as proteins with more interactants are thought to evolve more slowly [69]. The loss of these additional γ-tubulin complex proteins in budding yeast may have

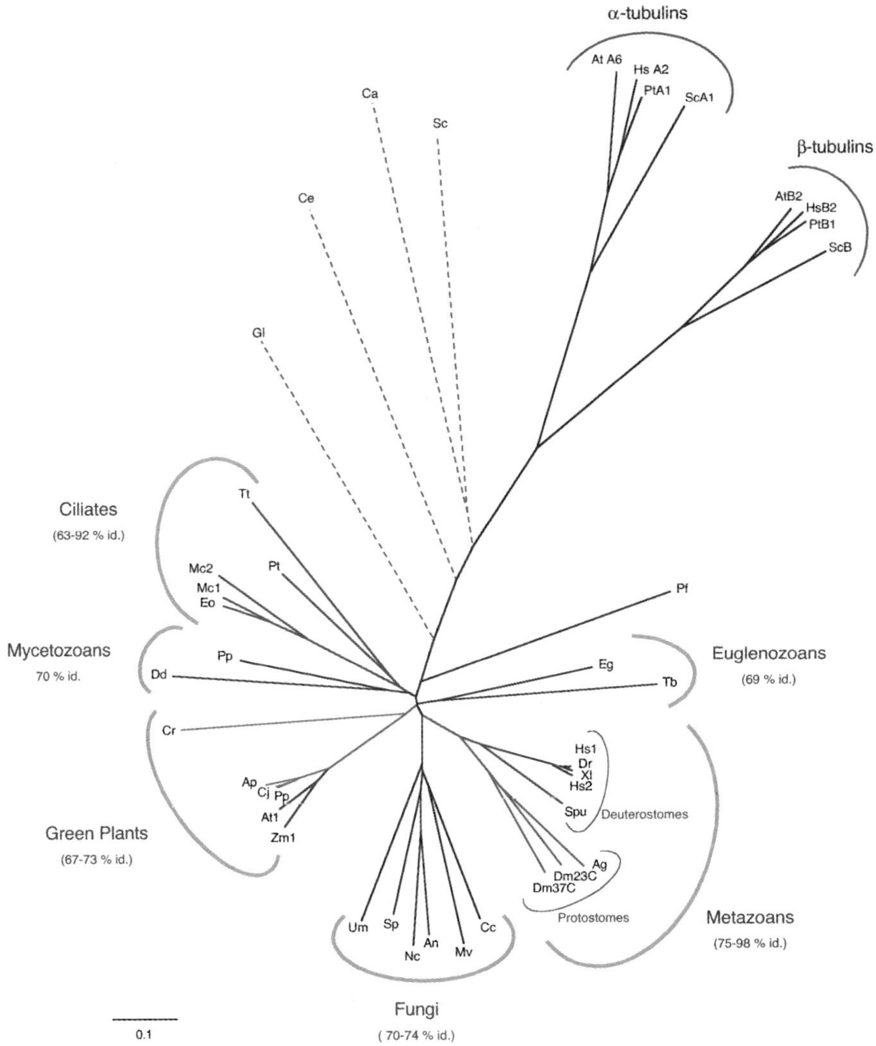

lowered the constraint imposed on the Tub4p sequence and this possibly accounts for its remarkable sequence divergence.

6.7.2 Centrin

Centrin proteins are associated with SPB/centrosomal structures from yeast to vertebrates, but are also expressed in taxa that have lost any type of centrosome such as land plants. Centrins are small proteins of the calcium-binding, EF-hand super-

◀ **Figure 6.2** Phylogenetic tree of the γ-tubulin family constructed with the neighbor-joining (NJ) method [115] using the CLUSTAL W program [116] and drawn using Phylodendron, Version 0.8d (D. G. Gilbert) at www.es.embnet.org. The following γ-tubulin sequences were included: (A) OPISTHOKONTS: **Metazoa**: (i) Chordates: Dr, *Danio rerio* (AAH45486); Hs, *Homo sapiens* TubG1,2p (P23258, Q9NRH3); Xl, *Xenopus laevis* (P23330); (ii) Echinoderms: Spu, *Strongylocentrotus purpuratus* (AAG01846); (iii) Ecdysozoa: Arthropods: Ag, *Anopheles gambiae* (XP_317665); Dm, *Drosophila melanogaster* Tub23C, 37C (CG3157, CG17566); Nematodes: Ce: *Caenorhabditis elegans* (P34475); **Fungi**: (i) Ascomycota: Ca, *Candida albicans* (O93807); An, *Aspergillus nidulans* (P18695); Nc, *Neurospora crassa* (P53377); Sc, *Saccharomyces cerevisiae* (P53378); Sp, *Schizosaccharomyces pombe* (P25295); (ii) Basidiomycota: Cc, *Coprinopsis cinerea* (BAC77342); Mv, *Microbotryum violaceum* (P32348); Um, *Ustilago maydis* (CAD33849); (B) MYCETOZOANS: Dd, *Dictyostelium discoideum* (CAA04130); Ppo, *Physarum polycephalum* (CAA70417); (C) ALVEOLATES: (i) Ciliates: Eo, *Euplotes octocarinatus* (P34786); Pt, *Paramecium tetraurelia* (CAA09991); Mc, *Moneuplotes crassus* TubG1, 2p (P54403, P54404); Tt, *Tetrahymena termophila* (U96076); (ii) Apicomplexans: Pf, *Plasmodium falciparum* (P34787); (D) EUGLENOZOANS: Eg, *Euglena gracilis* (AAK37966); Tb, *Trypanosoma brucei* (CAA68866); (E) DIPLOMONADS: Gl, *Giardia lamblia* (EAA41546); (F) GREEN PLANTS: (i) Ulvophytes: Cr, *Chlamydomonas reinhardtii* (Q39582); (ii) Streptophytes: Mosses: Pp, *Physcomitrella patens* (Q9XFG3); Liverworts: Cj, *Conocephalum japonicum* (AAM44306); Ferns: Ap, *Anemia phyllitidis* (P34785); Angiosperms: Dicots: At, *Arabidopsis thaliana* TubG1p (P38557); Monocots: Zm, *Zea mais* (Q41807). **α-tubulins**: AtA6, *Arabidopsis thaliana* α-tubulin 6 (NP_849388); HsA2, *Homo sapiens* α-tubulin 2 (Q13748); PtA1, *Paramecium tetraurelia* α-tubulin 1 (CAA67847); ScA1, *Saccharomyces cerevisiae* Tub1p(NP_013625); **β-tubulins**: AtB2, *Arabidopsis thaliana* β-tubulin 2 (NM_125664); HsB2, *Homo sapiens* β-tubulin 2 (AAH39175); PtB1, *Paramecium tetraurelia* β-tubulin 1 (S25182); ScB, *Saccharomyces cerevisiae* Tub2p (NP_116616). Species expressing divergent γ-tubulins are represented with dotted lines. Grouping of these unrelated fast evolving sequences with the outgroup, i. e. α- and β-tubulins, is an artefact due to long-branch attraction [117]. The range of protein identity within the different subgroups is indicated.

family of proteins. Like the closely related calmodulin, centrin proteins have been shown to act as regulators in very diverse cellular processes [70–72], but their precise function(s) within centrosomes remains poorly understood [73].

Two distinct roles have been established. First, centrins are involved in SPB/centrosome duplication: mutations in the budding yeast centrin, encoded by the *CDC31* gene, or in the fission yeast homolog, SpCdc31, led to a block in the earliest step of SPB duplication and resulted in mitotic arrest with a monopolar spindle [74–76]. Similarly, it was recently demonstrated that knock-down of human centrin 2 (HsCEN2) by RNAi prevents centriole duplication in cultured cells [77]. Second, centrins are constituents of calcium-sensitive fibers that connect the basal bodies to one another and to the nucleus in diverse flagellate green algae [78]. These fibers contract in response to elevated calcium concentrations. They are implicated in basal body localization and segregation, as shown by mutations or knock-down of the *Chlamydomonas reinhardtii* centrin (CrCEN) gene [79].

Phylogenetic analysis of centrins from a wide variety of eukaryotes (Figure 6.3) shows that the two founding members of the centrin family, i. e. CrCenp (*C. reinhardtii*) and Cdc31p (*S. cerevisiae*), belong to two different, well-defined subfamilies to which many of the known centrin genes belong. The first subfamily contains centrins from green algae and lower land plants, insects, and vertebrate homologs

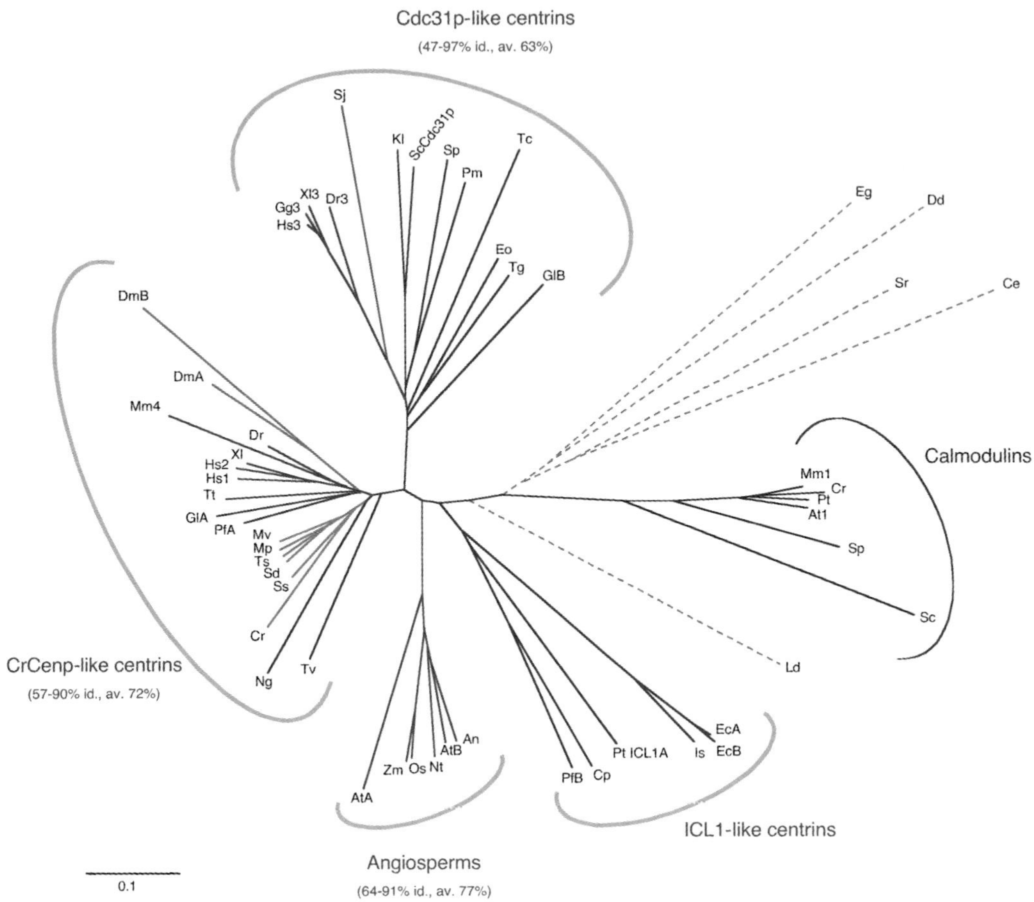

of human centrin 2, an ubiquitously expressed isoform, and centrins 1 and 4, two isoforms specific for ciliated epithelia [80–85]. The second main subfamily is formed by fungal Cdc31-like centrin proteins and vertebrate homologs of human centrin 3, the other ubiquitously expressed human isoform [86]. These two subfamilies not only differ in their sequence, but also appear to be functionally distinct. This has been illustrated by expression studies of human centrins in yeast and *Xenopus* two cell-stage embryos [76, 87, 88], which showed that HsCen3p, but not HsCen2p, has a dominant-negative effect on SPB or centrosome duplication. This has led to the proposal that Cen3p, like yeast Cdc31p, may be preferentially implicated in centrosome duplication, whereas Cen2p might participate in other cell division events such as centrosome segregation or cytokinesis [88] This proposal is no longer tenable however, as one of its predictions, namely that *C. reinhardtii* should also express a Cdc31p-like isoform, has not been fulfilled. In addition, the

◀ **Figure 6.3** Phylogenetic tree of the centrin family based on the neighbor-joining method (see Figure 6.1). The NJ tree was constructed from alignment of protein sequences covering the four EF-hand domains (residues 29–162 for *Chlamydomonas reinhardtii* centrin). The following centrin sequences were used: (A) OPISTHOKONTS: **Metazoa**: (i) Chordates: Dr, *Danio rerio* Cenp (CF269323, BQ450470), Cen3p (BM141295); Gg, *Gallus gallus* (CD216537); Hs, *Homo sapiens* Cen1, 2, 3p (Q12798, P41208, O15182); Mm, *Mus musculus* Cen4p (NP_665824); Xl, *Xenopus laevis* Cenp (AAA79194), Cen3p (AAG30507); (ii) Lophotrochozoans: Platyhelmynthes: Eg, *Ecchinococcus granulosus* (BI244113); Sj, *Schistosoma japonicum* (BU780242); (iii) Ecdysozoa: Arthropods: Dm, *Drosophila melanogaster* CenA, Bp (RE19335, AT22559); Nematodes: Sr, *Strongyloides ratti* (BI0876); Ce, *Caenorhabditis elegans* (NM_066585); **Fungi**: (i) Ascomycota: Kl, *Kluyveromyces lactis* (AL427669); Sc, *Saccharomyces cerevisiae* Cdc31p (S47549); Sp, *Schizosaccharomyces pombe* SpCdc31p (T41061); (ii) Basidiomycota: Pm, *Pisolithus microcarpus* (CB011055); (B) MYCETOZOANS: Dd, *Dictyostelium discoideum* Crp (C84755); (C) ALVEOLATES: (i) Ciliates: Ec, *Entodinium caudatum* CenA, Bp (AF065248, AF065247); Eo, *Euplotes octocarinatus* (Y18899); Is, *Isotricha* sp. BBF-2003 (CF181956); Pt, *Paramecium tetraurelia* ICL1A (Q27177); Tt, *Tetrahymena termophila* (AAF66602); (ii) Apicomplexans: Cp, *Cryptosporidium parvum* (B88508); Pf, *Plasmodium falciparum* CenA, Bp (NP_703273, NP_702332); Tg, *Toxoplasma gondii* (CB755050); (D) EUGLENOZOANS: Ld, *Leishmania donovanii* (AF406767); Tc, *Trypanosoma cruzi* (AW621079); (E) DIPLOMONADS: Gl, *Giardia lamblia* CenA, Bp (U59300, U42428); (F) HETEROLOBOSEA: Ng, *Naegleria gruberi* (U21725); (G) PARABASALIDS: Tv, *Trichomonas vaginalis* (CAB55607); (H) GREEN PLANTS: (i) Ulvophytes: Cr, *Chlamydomonas reinhardtii* (P05434); Mp, *Micromonas pusilla* (CAA58718); Sd, *Scherffelia dubia* (Q06827); Ss, *Spermatozopsis similis* (P43645); Ts, *Tetraselmis striata* (P43646); (ii) Streptophytes: Ferns: Mv, *Marsilea vestita* (U92973); Angiosperms: Dicots: An, *Atriplex nummularia* (P41210); At, *Arabidopsis thaliana* CenA, Bp (CAB16762, T45582); Nt, *Nicotiana tabacum* (AAF07221); Monocots: Os, *Oryza sativa* (BAC79876); Zm, *Zea mais* (CF384542). **Calmodulins**: AtCaM1 (P25854); CrCaM (P04352); MmCaM1 (AAH54805); PtCaM (P07463); ScCaM (P06787); SpCaM (P05933). Dotted lines indicate artefactual positioning due to long-branch attraction [117]. The range of protein identity within the different subgroups is indicated.

Drosophila genome does not contain any sequence encoding a Cen3p isoform. On the other hand, inactivation of CrCenp in *C. reinhardtii* causes defects in basal body replication [89]. Similarly, depletion of HsCen2p by RNAi blocks centrosome duplication [77], a result which indicates that in vertebrate cells, Cen2p and Cen3p may have not only very similar if not identical localizations [86, 87], but might also cooperate in the same functions.

The widespread but not universal distribution of Cdc31p and CrCenp subfamilies indicates that the duplication of their common ancestor is an early event in eukaryotic evolution and that the absence of one or other isoform in some taxa is a derived character. Loss of one isoform would be a secondary event that may have occurred several times (Figure 6.1). The duplication of the common ancestor probably occurred before the separation of bikonts, which gave rise to green plants and major protozoan groups, and opisthokonts, which are the common ancestors of fungi, animals, and choanozoans, since the two isoforms are expressed in the protist *Giardia intestinalis*. During later evolution, one of the two isoforms was lost early in some taxa, leaving only the CrCenp-like isoform as in green algae or the Cdc31p-like isoform as in higher fungi. In other taxa, for example vertebrates, the two isoforms have both been conserved. Additional divergence seems to have taken place among the chlorobiontes in flowering plants, and among the bilateria in Ecdysozoa, where loss of the Cdc31p-like isoform is observed in the fly and in nematodes, together with an important sequence divergence of the CrCenp-like centrin in worms (Figure 6.3). Finally, organisms such as alveolates, i.e. ciliates and apicomplexans, express an additional subfamily of centrins related to the *Paramecium* ICL1 (InfraCiliary Lattice 1) centrins. *Paramecium* infraciliary lattice is a contractile cytoskeletal network that subtends the whole cellular surface, and that contains ~ 20 isoforms of ICL centrins [90, 91]. Silencing of ICL1 genes leads to complete disorganization of the ICL, but has no apparent effect on cell growth or morphology [91, 92], indicating that proper basal body duplication and ciliogenesis do not require this subfamily of centrins. These functions are likely to involve other *bona fide* centrin genes recently identified in the *Paramecium* genome, that are related to either CrCenp- or Cdc31p-like centrins (F. Koll, CGM/CNRS, personal communication).

The distribution of CrCenp subfamily members suggests that these genes could be specifically required for basal body/axoneme-related functions. First, centrin genes that are specifically expressed in flagellated or ciliated cells are of the CrCenp type. This is exemplified for instance, by *Naegleria* centrin, which is expressed only during differentiation from amoebae to flagellates, concomitant with the *de novo* formation of basal bodies and flagella [93], and by mammalian centrin 1 [82–84] and centrin 4 [85] which are expressed in ciliated epithelial cells. Second, centrins of the CrCenp subfamily are present in the transition zone linking the basal body to the axoneme, and along the axoneme of diverse organisms [84, 94, 95]. These centrins have been shown to be light chains associated with inner dynein arms in *Chlamydomonas* and in the ciliate *Tetrahymena* [95, 96], and to participate in the control of the ciliary beat in *Tetrahymena* and in human ciliated epithelial cells [84, 95].

This view is also supported by the evolution of CrCenp-like centrins in eukaryotes. In the green lineage, basal bodies and axonemes are found in green algae, as well as in motile gametes of lower land plants such as mosses, ferns, and ancient seed plants like cycads and Ginkgo. The centrin gene of the water fern *Marsilea vestita*, which is expressed during spermiogenesis concomitant with the formation of the motile apparatus [97], is closely related to algal centrins (Figure 6.3). In contrast, the flowering plants which have also evolved from Chlorophytes, never assemble basal bodies. Interestingly, this correlates with a significant divergence of their centrin genes which form a fourth well-defined subfamily. This suggests that in angiosperms, centrin genes have been released from the evolutionary constraints imposed on other CrCenp-like centrins by the conservation of a motile apparatus, and have probably been maintained for their involvement in other cellular functions.

In higher fungi, the CrCenp subfamily is completely missing: it is not found in the fully sequenced genomes of the yeasts *S. cerevisiae* and *S. pombe*, and sequences identified in other higher fungi all correspond to a Cdc31p homolog. As these organisms are expected to have inherited both types of centrin from an opisthokont common ancestor, they must have lost the missing centrin gene secondarily. This event, like divergence of centrins in angiosperms, seems to correlate with the loss of the motile apparatus. To support this view, it would be of great interest to study a group of fungi situated at the base of the fungal tree in the class of chytridiomycetes which form flagellated gametes. Remarkably, a very recent identification of centrin genes in the chytridiomycete *Blastocladiella emersonii* has demonstrated that it possesses both CrCenp- and Cdc31p-like isoforms (K. F. Ribichich and S. L. Gomes, University of Sao Paulo, personal communication). These primitive fungi could prove to be very useful for studying centrin function, and possibly many other aspects of basal body/centriole biology.

6.7.3 Centrin-binding Proteins

Characterization of the recently identified yeast centrin-binding protein Sfi1p and related proteins in other species [98] will certainly lead to a greater understanding of the mechanisms involving centrins in centrosome function, especially those based on centrin-based fiber contraction. In budding yeast, both Sfi1p and Cdc31p localize to the half-bridge, a structure on which the assembly of the new SPB is initiated, and which is essential for SPB duplication. Sfi1p contains multiple internal repeats, each individual repeat being able to bind Cdc31p, and is proposed to form centrin-associated calcium-dependent contractile fibers. In contrast to other SPB components interacting with Cdc31 identified previously, Sfi1p has a vertebrate homolog. hSfi1p also contains multiple centrin-binding repeats, and localizes to the centrioles in human cells [98]. Although the function of hSfi1p remains to be determined, these results suggest that it could also play a role in centrosome duplication. Interestingly, hSfi1p binds both Cen2p and Cen3p *in vitro*. This could account for the fact that both centrins seem to be involved in mammalian centrosome duplication [76, 77, 88].

So far, potential homologs of *SFI1* are found in vertebrates (human, mouse and ESTs from zebrafish and *Xenopus*), fungi (*S. cerevisiae, S. pombe, Candida albicans* and *Aspergillus*), and the protozoa *Giardia intestinalis* (EAA38810; Sfi1p repeats were identified by ScanProsite [99] as described in [98]). Interestingly, homologs are found in green algae as well, notably in *C. reinhardtii* (K. F. Lechtreck, University of Cologne, personal communication) and in *Spermatozopsis similis*. In the latter case, the potential *SFI1* homolog has been shown to bind centrin specifically [100]. Whether these proteins are constituents of the different fiber systems in green algae has still to be determined, but it is a very tempting hypothesis. This would suggest that during evolution both centrin and an ancient protein complex have been recruited to different centrosomes, and possibly for different functions associated with these organelles. Determination of the precise localization of vertebrate and algal Sfi1p homologs within the centrosome/flagellar apparatus will be very interesting with regard to the hypothesis that some structural aspects of duplication may have been conserved, particularly the possibility proposed by Adams and Kilmartin [101], that a structure functionally equivalent to the yeast half-bridge exists in vertebrate centrosomes.

Interestingly, we were not able to identify any *SFI1* homolog in *Drosophila* and *C. elegans* genomes. The worm expresses a very divergent centrin gene however. *C. elegans* centrin is related to CrCenp subfamily members at its C-terminal half but greatly diverges at its N-terminus, lacking the two N-terminal EF-hand domains. It is expressed only in a subset of neural cells, and knock-down of its expression by RNAi indicates that it is a non-essential gene in *C. elegans* (see www.wormbase.org: Gene summary for R08D7.5). This is further supported by comparison with another nematode, *Strongyloides ratii*, which also expresses a divergent centrin gene (Figure 6.3). Although related, the two worm centrins share only 30% protein identity, indicating that they have been released from the evolutionary constraints imposed on other centrin genes. Assuming that vertebrate Sfi1p is functionally similar to yeast Sfi1p and that the mechanisms underlying duplication of fungal SPBs and animal centriole-based centrosomes share common principles, this would seem to imply that nematodes have evolved different mechanisms to initiate centrosome duplication. This would be consistent with accumulating data on *C. elegans*, showing that most of the worm centrosomal components have no obvious counterparts (as judged by currently available bioinformatics) in other systems [102–105]. Moreover, divergence of the centrin gene in the worm correlates with the absence of motile cilia in the worm body as well as in sperm, which have an amoeboid movement. The absence of an *SFI1* homolog is more intriguing for *Drosophila*, were two CrCenp-like centrins have been identified and localized (J. Azimzadeh and M. Bornens, in preparation). This may suggest that in flies, as in nematodes, initiation of centrosome duplication does not involve quite the same molecular mechanisms as in other opisthokonts. Conservation of CrCenp-like centrins in *Drosophila* could thus merely reflect their requirement for a motile apparatus-related, Sfi1p-independent function, as *Drosophila*, in contrast to *C. elegans*, forms flagellated male gametes.

An interesting and perhaps disappointing outcome of this survey is that classical model organisms for studying development, notably *Drosophila* or *C. elegans*, may have intrinsic limitations with regard to the unraveling of the centriole/basal body assembly pathway. It is likely therefore, that understanding centriole/basal body assembly in animal systems will benefit from the use of unicellular model systems such as *Paramecium, Tetrahymena, Chlamydomonas* or *Trypanosoma*, all systems in which gene inactivation is now possible. It may be significant, however, that in *C. elegans*, the most conserved components so far are those associated with intraflagellar transport [11]. The recent discovery of a basal body component, which seems to be mutated in the Bardet–Biedl syndrome (characterized by kidney malformation among other defects) and which is conserved in nematodes [106] may contribute an important piece to the puzzle.

An experimental method to estimate the global conservation of the centrosome in divergent species is the parthenogenetic assay in the *Xenopus* egg (see above). It has been convincingly demonstrated that centrosomes from deuterostomes, sea urchin [18] and various vertebrates [107] as well as from lophotrochozoa (protostomes) such as the clam *Spisula* (R. Palazzo and M. Bornens, unpublished data), were able to induce parthenogenetic cleavage of *Xenopus* eggs. However, centrosomes from *Drosophila*, either from syncitial embryos or from cultured cells, although they assemble a robust aster of microtubules from egg tubulin, were unable to induce parthenogenesis [108]. Perhaps this is related to the absence of *SFI1* in the genome of this species.

6.8
Conclusion: The Centrosome – A Cell Individuation[5] Organ?

The conservation of the centriole/basal body from the early eukaryotes illustrates the considerable success of this cytoskeletal structure. The flagellar apparatus and the centrosome appear as two versions of the same organelle ensuring similar functions in different cell contexts. The conversion of basal body/kinetosome to centriole, back and forth, which occurs at each generation in metazoans seems to recapitulate the long evolution of this organelle and of its connections with either the plasma membrane or the nuclear periphery.

6.8.1
Survival Value of Coupling Basic Functional Modules on the Same Organ

One may wonder why cell locomotion, sensory reception and cell division are intimately associated with the flagellar apparatus and whether these functional modules were progressively recruited by the ancestral apparatus or were all associated from the start. Compartmentalized eukaryotic cells are much larger than prokaryotic cells, and the cytoskeleton is believed to have allowed cells to enlarge (although bacteria are now known to have an elaborate cytoskeleton). A cytoskeleton evolving in an hypothetical primitive eukaryotic cell which has managed to feed efficiently on the surrounding medium, would have to participate in the division process. Since the volume of the cell would increase faster than the surface of exchange, cell division could be seen as the more urgent function to which the cytoskeleton would be recruited. Sub-cortical distribution of the cytoskeleton may become necessary to sense cell growth and to organize cell division accordingly (see diagram in Figure 6.4). However the benefit would have been limited as the dividing cells would feed on the same medium. If the dividing apparatus is also a motile system, polarizing the sub-cortical cytoskeleton, this would be an important benefit as the

[5] Individuation is a process or an act whereas individualization is the result of an individuation process or human decision. The principle of individuation has been debated from the beginning. One school of thought placed the emphasis on what distinguishes one individual from another, thus the principle was matter or substance. The other school of thought placed the emphasis on what gives a being its practical existence; in this case the principle involved time and space or shape. Applied to cells, the former school of thought would probably see the principle of cell individuation in the transcriptome/proteome, whereas the latter school would be concerned with what gives cells their dynamic stability and robustness. In the second school of thought the most appropriate stage for analysis is when a new cell separates from the mother cell.

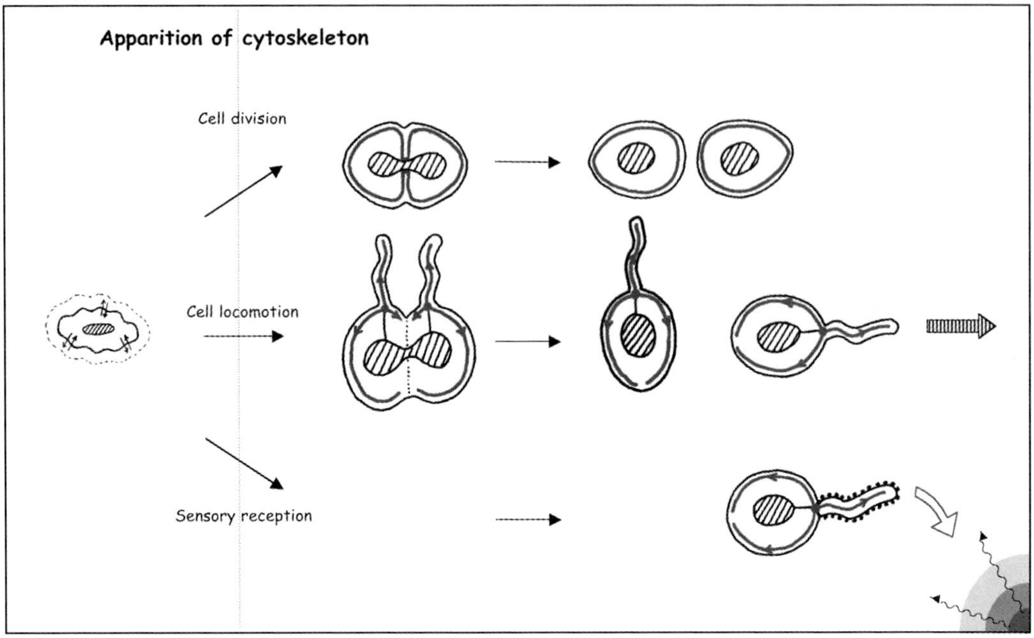

Figure 6.4 Diagram naively illustrating the decisive survival value for a large cell to associate modules for cell division, cell locomotion and sensory reception on the same polarized cytoskeleton apparatus (represented by a gray line parallel to the cell contour). Only the cell contour, the nucleus (hatched) and its link to the cytoskeleton have been represented.

daughter cells would then be able to explore new territories. If the motile system is also a sensory device however, there is a decisive benefit as the dividing cells "know" where to go to find food. Coupling modules for sensory reception and for locomotion on the same organ might thus have been a critical requirement for large cells. The association of the three basic modules with the same cytoskeleton apparatus – the centriole/basal body axoneme and the polarized cortical microtubule array – would have given optimal survival value to primitive eukaryotic cells.

6.8.2
Co-Evolution of the Centrosome and the Cleavage Apparatus

The centriole/basal body-associated functional modules are essential to cell polarity and require regulatory integration at the cell level. We have noted that the centrosome acts as a regulatory platform, in particular for the organization and the activation of the mitotic exit pathway coordinating karyokinesis with cytokinesis by fission. It is noteworthy that the complete loss of centrioles or derived structures in the green lineage is correlated with major changes in cell economy. The most conspicuous difference for cell division is that it is not achieved by fission as in fungi or animal cells. A long evolution among the phragmoplastophytes has led to the progressive loss of the centriole correlated with the loss of a contractile cleavage furrow which has been replaced by the phragmoplast growing from the center to the cell periphery. Intermediary situations can be observed in plasmodesmophytes. The complete loss of the centriole or a derived structure is observed in higher spermatophytes. Strikingly, in this case cytokinesis is not completed and the plasma membrane is continuous around sister cells through plasmodesmata; this leads to a synplasm rather than a tissue formed of independent cells. Another example is the Rhodophyta, which represent an early loss of the centriole/basal body in the green lineage: here too a different cleavage mechanism has evolved and division is not completed. This clearly indicates a co-evolution of the centrosome and the actin-dependent constriction of the plasma membrane. Complete cell division seems to be possible only when a centrosome-based aster of microtubules is dynamically interacting with cortical actin to control the formation of a ring-shaped cleavage furrow, suggesting that all these components constitute a generic integrated module for cell division.

6.8.3 The Biological Significance of Having a Cell Center

The centralized organization of the microtubule array appears critical for cell individuation, possibly because the microtubule-nucleating structure acts as a platform where regulatory complexes are assembled or activated, and which can then regulate peripheral events through microtubule-dependent transport, and in turn can be regulated through feedback loops. Another feature is probably critical for cell individuation, and that is the ability to control the positioning of the nucleus within cells. This is possible through the association of the nucleus with the centrosome. All these features are required for cell migration which is essential during metazoan development. In contrast, plant cells are immotile and multicellularity has been acquired by completely different mechanisms. As mentioned above, *Arabidopsis* has lost dynein and dynactin together with the centriole, thus losing the ability to move the nucleus within the cell in a microtubule-dependent manner [109].

Cleavage of the unfertilized *Xenopus* egg provides a direct illustration of the fact that centralized organization of the microtubule array is critical for cell individuation. Cleavage can be triggered by injection of a somatic centrosome whereas activation of the egg by pricking it does not lead to the individuation of two blastomeres in spite of an equivalent complement of microtubules in the egg. Another example is when starfish eggs are enucleated after fertilization: due to the sequential duplication of the sperm centrosome, individuation of 2^9 blastomeres lacking nuclei is observed [110, 111]. Similarly, when supernumerary centrosomes are artificially induced in the cellular slime mold *D. discoideum*, each of those centrosomes triggers the individuation of an enucleated cytoplast at the time of cytokinesis [112].

Individuation is a critical issue in living systems[6]. Many organisms have complex life cycles in which unicellular and multicellular phases, or multinucleated plasmodes, are observed sequentially. A survey of the centrosome throughout evolution, as well as during the development and differentiation of animal embryos in which multinucleation occurs by fusion of monucleated precursors or by interrupted cytokinesis, strongly suggests that the centrosome is required for cell individuation. This could explain the conservation of this organelle throughout evolution. This may also imply that deciphering the properties of the centrosome will be critical for a comprehensive understanding of growth and form in living systems.

Acknowledgments

We wish to thank Drs Anne Fleury, Karl-F. Lechtreck and Hervé Philippe for very helpful discussions, and Drs Iain Hagan, France Koll, Michael Knop, and Suely L. Gomes for sharing unpublished results.

[6] In living systems individuation processes can involve very different substrates. One of many examples that have occurred during evolution is the individuation of a new species or a multicellular organism from a colony of unicellular protists. During development of an embryo, formation of tissues is another example of individuation processes. In the human species individuation processes include psychosomatic development. Indeed the separation–individuation process leading to adulthood [113] and the separation–individuation of cells during cytokinesis are metaphorically identical, leading to the 'mother and daughter cell' terminology.

References

1. Adoutte, A., G. Balavoine, N. Lartillot, O. Lespinet, B. Prud'homme, and R. de Rosa. **2000**. The new animal phylogeny: reliability and implications. *Proc Natl Acad Sci USA* 97:4453–4456.
2. Gerhart, J., and M. Kirschner. **1997**. *Cells, Embryos, and Evolution. Towards a Cellular and Developmental Understanding of Phenotypic Variation and Evolutionary Adaptability.* Blackwell Science. Preface pp. IX–XI.
3. Baroin, A., R. Perasso, L. Qu, G. Brugerolle, J. Bachellerie, and A. Adoutte. **1988**. Partial phylogeny of the unicellular eukaryotes based on rapid sequencing of a portion of 28S ribosomal RNA. *Proc Natl Acad Sci USA* 85:3474–3478.
4. Tassin, A. M., B. Maro, and M. Bornens. **1985**. Fate of microtubule-organizing centers during myogenesis *in vitro*. *J Cell Biol* 100:35–46.
5. Boveri, T. **1901**. Über die Natur der Centrosomen. *Jena Z Med Naturw* 28: 1–220.
6. Saito, A., Y. Suetomo, M. Arikawa, G. Omura, S. Khan, S. Kakuta, E. Suzaki, K. Kataoka, and T. Suzaki. **2003**. Gliding movement in *Peranema trichophorum* is powered by flagellar surface motility. *Cell Motil Cytoskeleton* 55:244–253.
7. Wetherbee, R., and R. A. Andersen. **1988**. Flagella of a chrysophycean alga play an active role in prey capture and selection. *Protoplasma* 166:1–7.

8. Andersen, R. A., and R. Wetherbee. **1992**. Microtubules of the flagellar apparatus are active during prey capture in the chrysophycean alga *Epipyxis pulchra*. *Protoplasma* 166:8-20.

9. Eberl, D., R. Hardy, and M. Kernan. **2000**. Genetically similar transduction mechanisms for touch and hearing in Drosophila. *J Neurosci*. 20:5981–5988.

10. McGrath, J., S. Somlo, S. Makova, X. Tian, and M. Brueckner. **2003**. Two populations of node monocilia initiate left–right asymmetry in the mouse. *Cell* 114:61–73.

11. Rosenbaum, J., and G. Witman. **2002**. Intraflagellar transport. *Nature Rev Mol Cell Biol* 3:813–825.

12. Pazour, G., and G. Witman. **2003**. The vertebrate primary cilium is a sensory organelle. *Curr Opin Cell Biol* 15:105–110.

13. Nauli, S., F. Alenghat, Y. Luo, E. Williams, P. Vassilev, X. Li, A. Elia, W. Lu, E. Brown, S. Quinn, D. Ingber, and J. Zhou. **2003**. Polycystins 1 and 2 mediate mechanosensation in the primary cilium of kidney cells. *Nature Genet* 33:129–137.

14. Bastin, P., T. MacRae, S. Francis, K. Matthews, and K. Gull. **1999**. Flagellar morphogenesis: protein targeting and assembly in the paraflagellar rod of trypanosomes. *Mol Cell Biol* 19:191–200.

15. Ogbadoyi, E., D. Robinson, and K. Gull. **2003**. A high-order trans-membrane structural linkage is responsible for mitochondrial genome positioning and segregation by flagellar basal bodies in trypanosomes. *Mol Biol Cell* 14:1769–1779.

16. Gull, K. **2003**. Host–parasite interactions and trypanosome morphogenesis: a flagellar pocketful of goodies. *Curr Opin Microbiol* 6:365–370.

17. Kohl, L., D. Robinson, and P. Bastin. **2003**. Novel roles for the flagellum in cell morphogenesis and cytokinesis of trypanosomes. *EMBO J* 22:5336–5346.

18. Maller, J., D. Poccia, D. Nishioka, P. Kido, J. Gerhart, and H. Hartman. **1976**. Spindle formation and cleavage in *Xenopus* eggs injected with centriole-containing fractions from sperm. *Exp. Cell Res* 99:285–294.

19. Tournier, F., and M. Bornens. **2001**. Centrosomes and parthenogenesis. *Methods Cell Biol* 67:213–224.

20. Piel, M., and M. Bornens. **2001**. Centrosome reproduction *in vitro*: mammalian centrosomes in *Xenopus* lysates. *Methods Cell Biol* 67:289–304.

21. O'Connell, K. **2000**. The centrosome of the early *C. elegans* embryo: inheritance, assembly, replication and developmental roles. In *The Centrosome in Cell Replication and Early Development*, Palazzo R. E. and Schatten, G. P. (Eds). Academic Press, 365–384.

22. Goldstein, B. and Hird, S. N. **1996**. Specification of the anteroposterior axis in *Caenorhabditis elegans*. *Development* 122: 1467–1474.

23. Holmes, J., and S. Dutcher. **1989**. Cellular asymmetry in *Chlamydomonas reinhardtii*. *J Cell Sci* 94:273–285.

24. Beech, P., K. Heimann, and M. Melkonian. **1991**. Development of the flagellar apparatus during the cell cycle in unicellular algae. *Protoplasma* 164:23–37

25. Moestrup, O., and T. Hori. **1989**. Ultrastructure of the flagellar apparatus in *Pyramimonas octopus* (Prasinophyceae). II. Flagellar roots, connecting fibers and numbering of individual flagella in green algae. *Protoplasma* 148:41–56.

26. McCollum, D., and K. L. Gould. **2001**. Timing is everything: regulation of mitotic exit and cytokinesis by the MEN and SIN. *Trends Cell Biol* 11:89–95.

27. Bardin, A., and A. Amon. **2001**. Men and sin: what's the difference? *Nature Rev Mol Cell Biol* 2:815–826.

28. Pereira, G., T. Tanaka, K. Nasmyth, and E. Schiebel. **2001**. Modes of spindle pole body inheritance and segregation of the Bfa1p-Bub2p checkpoint protein complex. *EMBO J* 20:6359–6370.

29. Kusch, J., D. Liakopoulos, and Y. Barral. **2003**. Spindle asymmetry: a compass for the cell. *Trends Cell Biol* 13:562–569.

30. Segal, M., and K. Bloom. **2001**. Control of spindle polarity and orientation in

Saccharomyces cerevisiae. *Trends Cell Biol* 11:160–166.
31. Schuyler, S., and D. Pellman. **2001**. Search, capture and signal: games microtubules and centrosomes play. *J Cell Sci* 114:274–255.
32. Liakopoulos, D., J. Kusch, S. Grava, J. Vogel, and Y. Barral. **2003**. Asymmetric loading of Kar9 onto spindle poles and microtubules ensures proper spindle alignment. *Cell* 112:561–574.
33. Maekawa, H., T. Usui, M. Knop, and E. Schiebel. **2003**. Yeast Cdk1 translocates to the plus end of cytoplasmic microtubules to regulate bud cortex interactions. *EMBO J* 22:438–449.
34. Wesp, A., S. Prinz, and G. Fink. **2001**. Conservative duplication of spindle poles during meiosis in *Saccharomyces cerevisiae*. *J Bacteriol* 183:2372–2375.
35. Piel, M., P. Meyer, A. Khodjakov, C. L. Rieder, and M. Bornens. **2000**. The respective contributions of the mother and daughter centrioles to centrosome activity and behavior in vertebrate cells. *J Cell Biol* 149:317–329.
36. Piel, M., J. Nordberg, U. Euteneuer, and M. Bornens. **2001**. Centrosome-dependent exit of cytokinesis in animal cells. *Science* 291:1550–1553.
37. Gromley, A., A. Jurczyk, J. Sillibourne, E. Halilovic, M. Mogensen, I. Groisman, M. Blomberg, and S. Doxsey. **2003**. A novel human protein of the maternal centriole is required for the final stages of cytokinesis and entry into S phase. *J Cell Biol* 161:535–545.
38. Kochanski, R. S., and G. G. Borisy. **1990**. Mode of centriole duplication and distribution. *J Cell Biol* 110:1599–605.
39. Chen, T., T. Hiroko, A. Chaudhuri, F. Inose, M. Lord, S. Tanaka, J. Chant, and A. Fujita. **2000**. Multigenerational cortical inheritance of the Rax2 protein in orienting polarity and division in yeast. *Science* 290:1975–1978.
40. Beisson, J., and T. Sonnenborn. **1965**. Cytoplasmic inheritance of the organization of the cell cortex in *Paramecium aurelia*. *Proc Natl Acad Sci USA* 53:275–282.
41. Grill, S., P. Gonczy, E. Stelzer, and A. Hyman. **2001**. Polarity controls forces governing asymmetric spindle positioning in the *Caenorhabditis elegans* embryo. *Nature* 409:630–633.
42. Koonce, M., J. Kohler, R. Neujahr, J. Schwartz, I. Tikhonenko, and G. Gerisch. **1999**. Dynein motor regulation stabilizes interphase microtubule arrays and determines centrosome position. *EMBO J* 18:6786–6792.
43. Morris, N. **2003**. Nuclear positioning: the means is at the ends. *Curr Opin Cell Biol* 15:54–59.
44. Lawrence, C., N. Morris, R. Meagher, and R. Dawe. **2001**. Dyneins have run their course in plant lineage. *Traffic* 2:362–363.
45. Jackman, M., C. Lindon, E. Nigg, and J. Pines. **2003**. Active cyclin B1-Cdk1 first appears on centrosomes in prophase. *Nature Cell Biol* 5:143–148.
46. McKean, P. G., S. Vaughan, and K. Gull. **2001**. The extended tubulin superfamily. *J Cell Sci* 114:2723–2733.
47. Dutcher, S. K. **2001**. The tubulin fraternity: alpha to eta. *Curr Opin Cell Biol* 13:49–54.
48. Dupuis-Williams, P., A. Fleury-Aubusson, N. de Loubresse, H. Geoffroy, L. Vayssie, A. Galvani, A. Espigat, and J. Rossier. **2002**. Functional role of epsilon-tubulin in the assembly of the centriolar microtubule scaffold. *J Cell Biol* 158:1183–1193.
49. Beisson, J., and M. Wright. **2003**. Basal body/centriole assembly and continuity. *Curr Opin Cell Biol* 15:96–104.
50. Ruiz, F., A. Krzywicka, C. Klotz, A. Keller, J. Cohen, F. Koll, G. Balavoine, and J. Beisson. **2000**. The *SM19* gene, required for duplication of basal bodies in *Paramecium*, encodes a novel tubulin, eta-tubulin. *Curr Biol* 10:1451–1454.
51. Mansir, A., and J. Justine. **1995**. Centriole with ten singlets in spermatozoa of the parasitic nematode *Heligmosomoides polygyrus*. In *Advances In Spermatozoal Phylogeny and Taxonomy*, Vol. 166. B. Jamieson, Ausio, J. & Justine, JL., (Eds). Mém. Mus. natn. Hist. nat: Paris, 119–128.
52. Smrzka, O., N. Delgehyr, and M. Bornens. **2000**. Tissue-specific expression and subcellular localisation of mam-

malian delta-tubulin. *Curr Biol* 10:413–416.

53. Inclan, Y. F., and E. Nogales. **2001**. Structural models for the self-assembly and microtubule interactions of gamma-, delta- and epsilon-tubulin. *J Cell Sci* 114:413–422.

54. Chalfie, M., and J. Thomson. **1982**. Structural and functional diversity in the neuronal microtubules of *Caenorhabditis elegans*. *J Cell Biol* 93:15–23.

55. Knop, M., G. Pereira, S. Geissler, K. Grein, and E. Schiebel. **1997**. The spindle pole body component Spc97p interacts with the gamma-tubulin of *Saccharomyces cerevisiae* and functions in microtubule organization and spindle pole body duplication. *EMBO J* 16:1550–1564.

56. Knop, M., and E. Schiebel. **1997**. Spc98p and Spc97p of the yeast gamma-tubulin complex mediate binding to the spindle pole body via their interaction with Spc110p. *EMBO J* 16:6985–6995.

57. Martin, O. C., R. N. Gunawardane, A. Iwamatsu, and Y. X. Zheng. **1998**. Xgrip109: A gamma tubulin-associated protein with an essential role in gamma tubulin ring complex (gammaTuRC) assembly and centrosome function. *J Cell Biol* 141:675–687.

58. Murphy, S. M., L. Urbani, and T. Stearns. **1998**. The mammalian gamma-tubulin complex contains homologues of the yeast spindle pole body components Spc97p and Spc98p. *J Cell Biol* 141:663–674.

59. Tassin, A. M., C. Celati, M. Moudjou, and M. Bornens. **1998**. Characterization of the human homologue of the yeast Spc98p and its association with gamma-tubulin. *J Cell Biol* 141:689–701.

60. Oegema, K., C. Wiese, O. C. Martin, R. A. Milligan, A. Iwamatsu, T. J. Mitchison, and Y. Zheng. **1999**. Characterization of two related drosophila gamma-tubulin complexes that differ in their ability to nucleate microtubules. *J Cell Biol* 144:721–733.

61. Vinh, D. B.N., J. W. Kern, W. O. Hancock, J. Howard, and T. N. Davis. **2002**. Reconstitution and characterization of budding yeast gamma-tubulin complex. *Mol Biol Cell* 13:1144–1157.

62. Zheng, Y. X., M. L. Wong, B. Alberts, and T. Mitchison. **1995**. Nucleation of microtubule assembly by a gamma-tubulin-containing ring complex. *Nature* 378:578–583.

63. Gunawardane, R., O. Martin, and Y. Zheng. **2003**. Characterization of a new gammaTuRC subunit with WD repeats. *Mol Biol Cell* 14:1017–1026.

64. Fava, F., B. RaynaudMessina, J. LeungTack, L. Mazzolini, M. Li, J. C. Guillemot, D. Cachot, Y. Tollon, P. Ferrara, and M. Wright. **1999**. Human 76p: a new member of the gamma-tubulin-associated protein family. *J Cell Biol* 147:857–868.

65. Gunawardane, R., O. Martin, K. Cao, L. Zhang, K. Dej, A. Iwamatsu, and Y. Zheng. **2000**. Characterization and reconstitution of Drosophila gamma-tubulin ring complex subunits. *J Cell Biol* 151:1513–1524.

66. Zhang, L., T. Keating, A. Wilde, G. Borisy, and Y. Zheng. **2000**. The role of Xgrip210 in gamma-tubulin ring complex assembly and centrosome recruitment. *J Cell Biol* 151:1525–1536.

67. Murphy, S. M., A. M. Preble, U. K. Patel, K. L. O'Connell, D. P. Dias, M. Moritz, D. Agard, J. T. Stults, and T. Stearns. **2001**. GCP5 and GCP6: two new members of the human gamma-tubulin complex. *Mol Biol Cell* 12:3340–3352.

68. Erhardt, M., V. StoppinMellet, S. Campagne, J. Canaday, J. Mutterer, T. Fabian, M. Sauter, T. Muller, C. Peter, A. M. Lambert, and A. C. Schmit. **2002**. The plant Spc98p homologue colocalizes with gamma-tubulin at microtubule nucleation sites and is required for microtubule nucleation. *J Cell Sci* 115:2423–2431.

69. Fraser, H., A. Hirsh, L. Steinmetz, C. Scharfe, and M. Feldman. **2002**. Evolutionary rate in the protein interaction network. *Science* 296:750–752.

70. Sullivan, D. S., S. Biggins, and M. D. Rose. **1998**. The yeast centrin, Cdc31p, and the interacting protein kinase, Kic1p, are required for cell integrity. *J Cell Biol* 143:751–765.

71. Araki, M., C. Masutani, M. Takemura, A. Uchida, K. Sugasawa, J. Kondoh, Y. Ohkuma, and F. Hanaoka. **2001**. Centrosome protein centrin 2/caltractin 1 is part of the xeroderma pigmentosum group C complex that initiates global genome nucleotide excision repair. *J Biol Chem* 276:18665–18672.
72. Pulvermuller, A., A. Giessl, M. Heck, R. Wottrich, A. Schmitt, O. Ernst, H. Choe, K. Hofmann, and U. Wolfrum. **2002**. Calcium-dependent assembly of centrin–G-protein complex in photoreceptor cells. *Mol Cell Biol* 22:2194–2203.
73. Levy, Y. Y., E. Y. Lai, S. P. Remillard, M. B. Heintzelman, C. Fulton. **1996**. Centrin is a conserved protein that forms diverse associations with centrioles and MTOCs in *Naegleria* and other organisms. *Cell Motil Cytoskeleton* 33:298–323.
74. Huang, B., A. Mengersen, and V. D. Lee. **1988**. Molecular cloning of cDNA for caltractin, a basal body-associated Ca^{2+}-binding protein: homology in its protein sequence with calmodulin and the yeast CDC31 gene product. *J Cell Biol* 107:133–140.
75. Spang, A., I. Courtney, K. Grein, M. Matzner, and E. Schiebel. **1995**. The Cdc31p-binding protein Kar1p is a component of the half bridge of the yeast spindle pole body. *J Cell Biol* 128:863–877.
76. Paoletti, A., N. Bordes, R. Haddad, C. Schwartz, F. Chang, and M. Bornens. **2003**. Fission yeast cdc31p is a component of the half-bridge and controls SPB duplication. *Mol Biol Cell* 14:2793–2808.
77. Salisbury, J. L., K. M. Suino, R. Busby, and M. Springett. **2002**. Centrin-2 is required for centriole duplication in mammalian cells. *Curr Biol* 12:1287–1292.
78. Salisbury, J. L., A. T. Baron, and M. A. Sanders. **1988**. The centrin-based cytoskeleton of *Chlamydomonas reinhardtii*: distribution in interphase and mitotic cells. *J Cell Biol* 107:635–41.
79. Taillon, B. E., S. A. Adler, J. P. Suhan, and J. W. Jarvik. **1992**. Mutational analysis of centrin: an EF-hand protein associated with three distinct contractile fibers in the basal body apparatus of *Chlamydomonas*. *J Cell Biol* 119:1613–1624.
80. Lee, V. D., and B. Huang. **1993**. Molecular cloning and centrosomal localization of human caltractin. *Proc Natl Acad Sci USA* 90:11039–11043.
81. Errabolu, R., M. A. Sanders, and J. L. Salisbury. **1994**. Cloning of a cDNA encoding human centrin, an EF-hand protein of centrosomes and mitotic spindle poles. *J Cell Sci* 107:9–16.
82. Wolfrum, U., and J. L. Salisbury. **1998**. Expression of centrin isoforms in the mammalian retina. *Exp Cell Res* 242:10–17.
83. Hart, P. E., J. N. Glantz, J. D. Orth, G. M. Poynter, and J. L. Salisbury. **1999**. Testis-specific murine centrin, Cetn1: genomic characterization and evidence for retroposition of a gene encoding a centrosome protein. *Genomics* 60:111–120.
84. Laoukili, J., E. Perret, S. Middendorp, O. Houcine, C. Guennou, F. Marano, M. Bornens, and F. Tournier. **2000**. Differential expression and cellular distribution of centrin isoforms during human ciliated cell differentiation *in vitro*. *J Cell Sci* 113:1355–1364.
85. Gavet, O., C. Alvarez, P. Gaspar, and M. Bornens. **2003**. Centrin4p, a novel mammalian centrin specifically expressed in ciliated cells. *Mol Biol Cell* 14:1818–1834.
86. Middendorp, S., A. Paoletti, E. Schiebel, and M. Bornens. **1997**. Identification of a new mammalian centrin gene, more closely related to *Saccharomyces cerevisiae CDC31* gene. *Proc Natl Acad Sci USA* 94:9141–9146.
87. Paoletti, A., M. Moudjou, M. Paintrand, J. L. Salisbury, and M. Bornens. **1996**. Most of centrin in animal cells is not centrosome-associated and centrosomal centrin is confined to the distal lumen of centrioles. *J Cell Sci* 109:3089–3102.
88. Middendorp, S., T. Kuntziger, Y. Abraham, S. Holmes, N. Bordes, M. Paintrand, A. Paoletti, and M. Bornens.

2000. A role for centrin 3 in centrosome reproduction. *J Cell Biol* 148:405–416.

89. Koblenz, B., J. Schoppmeier, A. Grunow, and K. Lechtreck. **2003**. Centrin deficiency in *Chlamydomonas* causes defects in basal body replication, segregation and maturation. *J Cell Sci* 116:2635–2646.

90. Madeddu, L., C. Klotz, J. Le Caer, and J. Beisson. **1996**. Characterization of centrin genes in *Paramecium*. *Eur J Biochem* 238:121–128.

91. Ruiz, F., L. Vayssie, C. Klotz, L. Sperling, and L. Madeddu. **1998**. Homology-dependent gene silencing in *Paramecium*. *Mol Biol Cell* 9:931–943.

92. Beisson, J., J. Clerot, A. Fleury-Aubusson, N. Garreau de Loubresse, F. Ruiz, and C. Klotz. **2001**. Basal body-associated nucleation center for the centrin-based cortical cytoskeletal network in Paramecium. *Protist* 152:339–354.

93. Levy, Y. Y., E. Y. Lai, S. P. Remillard, and C. Fulton. **1998**. Centrin is synthesized and assembled into basal bodies during *Naegleria* differentiation. *Cell Motil Cytoskeleton* 40:249–260.

94. Sanders, M. A., and J. L. Salisbury. **1994**. Centrin plays an essential role in microtubule severing during flagellar excision in *Chlamydomonas reinhardtii*. *J. Cell Biol* 124:795–805.

95. Guerra, C., Y. Wada, V. Leick, A. Bell, and P. Satir. **2003**. Cloning, localization, and axonemal function of Tetrahymena centrin. *Mol Biol Cell* 14:251–261.

96. LeDizet, M., and G. Piperno. **1995**. The light chain p28 associates with a subset of inner dynein arm heavy chains in *Chlamydomonas* axonemes. *Mol Biol Cell* 6:697–711.

97. Hart, P. E., and S. M. Wolniak. **1998**. Spermiogenesis in *Marsilea vestita*: a temporal correlation between centrin expression and blepharoplast differentiation. *Cell Motil Cytoskeleton* 41:39–48.

98. Kilmartin, J. **2003**. Sfi1p has conserved centrin-binding sites and an essential function in budding yeast spindle pole body duplication. *J Cell Biol.* 162:1211–1221.

99. Gattiker, A., E. Gasteiger, and A. Bairoch. **2002**. ScanProsite: a reference implementation of a PROSITE scanning tool. *Appl Bioinform* 1:107–108.

100. Steinkötter, J. **1997**. Centrin und centrinbindende Proteine in grünalgen. PhD thesis, Mathematisch-Naturwissenschaftliche Fakultät, Cologne.

101. Adams, I. R., and J. V. Kilmartin. **2000**. Spindle pole body duplication: a model for centrosome duplication? *Trends Cell Biol* 10:329–335.

102. O'Connell, K., C. Caron, K. Kopish, D. Hurd, K. Kemphues, Y. Li, and J. White. **2001**. The *C. elegans* zyg-1 gene encodes a regulator of centrosome duplication with distinct maternal and paternal roles in the embryo. *Cell Motil Cytoskeleton* 105:547–558.

103. Hamill, D., A. Severson, J. Carter, and B. Bowerman. **2002**. Centrosome maturation and mitotic spindle assembly in *C. elegans* require SPD-5, a protein with multiple coiled-coil domains. *Dev Cell* 3:673–684.

104. Kirkham, M., T. Muller-Reichert, K. Oegema, S. Grill, and A. Hyman. **2003**. SAS-4 is a *C. elegans* centriolar protein that controls centrosome size. *Cell* 112:575–587.

105. Leidel, S., and P. Gonczy. **2003**. SAS-4 is essential for centrosome duplication in *C. elegans* and is recruited to daughter centrioles once per cell cycle. *Dev Cell* 4:431–439.

106. Ansley, S., J. Badano, O. Blacque, J. Hill, B. Hoskins, C. Leitch, J. Kim, A. Ross, E. Eichers, T. Teslovich, A. Mah, R. Johnsen, J. Cavender, R. Lewis, M. Leroux, P. Beales, and N. Katsanis. **2003**. Basal body dysfunction is a likely cause of pleiotropic Bardet–Biedl syndrome. *Nature* 425:628–633.

107. Tournier, F., M. Cyrklaff, E. Karsenti, and M. Bornens. **1991**. Centrosomes competent for parthenogenesis in *Xenopus* eggs support procentriole budding in cell-free extracts. *Proc Natl Acad Sci USA* 88:9929–9933.

108. Tournier, F., Y. Bobinnec, A. Debec, P. Santamaria, and M. Bornens. **1999**. *Drosophila* centrosomes are unable to trigger parthenogenetic development of *Xenopus* eggs. *Biol Cell* 91:99–108.

109. Ketelaar, T., C. Faivre-Moskalenko, J. Esseling, N. de Ruijter, C. Grierson, M. Dogterom, and A. Emons. **2002**. Positioning of nuclei in *Arabidopsis* root hairs: an actin-regulated process of tip growth. *Plant Cell* 14:2941–2955.

110. Nagano, H., S. Hirai, K. Okano, and S. Ikegami. **1981**. Achromosomal cleavage of fertilized starfish eggs in the presence of aphidicolin. *Dev Biol* 85:409–415.

111. Picard, A., J. C. Labbe, and M. Doree. **1988**. The cell cycle can occur in starfish oocytes and embryos without the production of transferable MPF (maturation-promoting factor). *Dev Biol* 128:129–135.

112. Gräf, R., C. Daunderer, and M. Schliwa. **2000**. Dictyostelium DdCP224 is a microtubule-associated protein and a permanent centrosomal resident involved in centrosome duplication. *J Cell Sci* 113:1747–1758.

113. Gnaulati, E., and B. J. Heine **2001**. Separation–individuation in late adolescence: an investigation of gender and ethnic differences. *J Pathol* 135:59–70.

114. Pennisi, E. **2003**. "Tree of Life". *Science* 300:1694–1695.

115. Saitou, N., and M. Nei. **1987**. The neighbor-joining method: a new method for reconstructing phylogenetic trees. *Mol Biol Evol* 4:406–425.

116. Thompson, J. D., D. G. Higgins, and T. J. Gibson. **1994**. CLUSTAL W: improving the sensitivity of progressive multiple sequence alignment through sequence weighting, positions-specific gap penalties and weight matrix choice. *Nucleic Acids Res* 22:4673–4680.

117. Gribaldo, S., and H. Philippe. **2002**. Ancient phylogenetic relationships. *Theor Popul Biol* 61:391–408.

Part II
The Integration of Centrosome and Chromosome Cycles

Centrosomes in Development and Disease. Edited by Erich A. Nigg

ISBN 3-527-30980-2

7
A Proteomic Approach to the Inventory of the Human Centrosome

Christopher J. Wilkinson, Jens S. Andersen, Matthias Mann and Erich A. Nigg

7.1
Introduction

Over the past decade, our ability to determine the protein composition of complex biological samples has improved greatly, thanks to the advent of high resolution, high sensitivity mass spectrometers on the one hand, and the computing power and databases needed to interpret the results obtained with such instruments on the other hand. As a consequence, proteomic approaches have met with great success in recent years. Both organelles, for example nucleoli and mitochondria, and organisms, notably yeast and *Plasmodium falciparum* – the parasite responsible for malaria – have been studied in this way [1–5].

The centrosome is another obvious, albeit challenging target for a proteomic analysis. Despite more than 100 years of study and the impact of molecular biology, a complete description of its component parts still constitutes an important goal for the future. It also seems clear that traditional approaches, driven mainly by functional screens, may not discover all components. Hence, mass spectrometry-based approaches hold great promise for a thorough analysis of this complex organelle. The first important application of mass spectrometric proteomics to a microtubule-organizing center was the analysis of the yeast spindle pole body by Kilmartin, Mann and co-workers several years ago [6]. This yeast equivalent of the centrosome yielded to mass spectrometric analysis after a heroic purification effort, emphasizing the difficulty of purifying low abundance organelles. For a comprehensive analysis of the centrosome, the difficulties in a proteomic approach were expected to be similar. Although the centrosome is by no means the smallest amongst the cellular organelles, its purification is made difficult by its low copy number (Figure 7.1) and its interactions with both the cytoskeleton and other subcellular organelles, notably the Golgi apparatus and the nucleus. Lastly, there is no delineating membrane. So, the more extensively the centrosome is purified, the greater the danger of losing components.

Centrosomes in Development and Disease. Edited by Erich A. Nigg

ISBN 3-527-30980-2

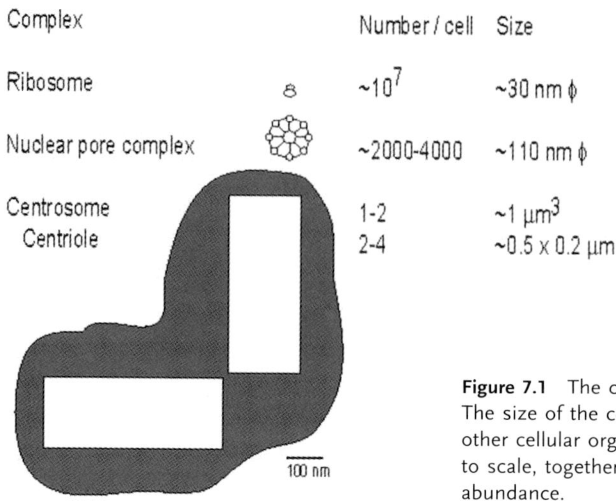

Figure 7.1 The centrosome in perspective. The size of the centrosome compared to other cellular organelles in the cell is shown to scale, together with estimates of their abundance.

7.2 What is a Centrosome Component?

The absence of a clearly visible (membranous) boundary raises the general question of how to define a centrosomal component. With our limited current knowledge of centrosome structure it is in fact difficult to determine exactly where this organelle ends. Boundaries are further blurred by the fact that many components, although clearly enriched at the centrosome, also occur in large cytoplasmic pools. Prominent examples for proteins existing both at the centrosome and in the cytoplasm include γ-tubulin and centrin [7, 8]. Nevertheless, these proteins can readily be used as markers in fluorescence microscopy, because their enrichment at the centrosome results in a high signal density. Many microtubule (MT)-associated proteins are also expected to be present in relatively high concentrations at the centrosome. To distinguish such proteins from the genuine (core) components, cells are usually exposed to either cold or nocodazole treatment, both of which depolymerize MTs. Those proteins whose centrosome localization resists cold or nocadozole treatment are operationally defined as core centrosome components.

Of particular importance is the highly dynamic structure of the centrosome which displays major structural alterations and compositional changes during the cell cycle [9, 10] (see also Chapter 9). Several proteins have been described whose centrosomal localization is severely diminished at the onset of mitosis e. g. PCM-1, C-Nap1, and Nlp [11–13] while others, notably the γ-tubulin ring complex, are recruited in greater quantity as the centrosome undergoes so-called maturation at the G2/M transition [14]. Yet other proteins localize to the spindle poles only during mitosis, some of which, like NuMA [15], are nuclear proteins during interphase. In these latter cases, electron microscopy may be required to determine the exact location of such proteins in the area of the poles. As exemplified by plants, spindle poles can clearly exist in the absence of centrosomes. Thus, although the terms "spindle pole" and "(mitotic) centrosome" are often used interchangeably, they do not necessarily describe the exact same structure. Thus, some proteins may accumulate in the vicinity of the poles without being components of the centrosome proper.

From a technical perspective, there are a number of issues that are becoming increasingly important the more widely the centrosome is being studied. In particular, it is important to bear in mind that animal sera occasionally contain antibodies with anti-centrosome activity even prior to immunization [16]. To make sure that a newly raised antibody does not erroneously report on centrosome association of a protein under study, it is critical, therefore, to perform control experiments with pre-immune rather than non-immune serum, i. e. serum collected from the same animal that was subsequently used for immunization rather than another individual of the same species. Furthermore, it is generally useful to confirm the immunocytochemical localization of a novel endogenous protein by showing that an epitope-tagged product of the corresponding cDNA also localizes to the centrosome. However, such experiments do not necessarily constitute an infallible method either. In addition to the pitfall that overexpression might produce artefactual associations, we have occasionally (albeit rarely) observed that different tags yielded different results (e. g. [17], C. J. Wilkinson and E. A. Nigg, unpublished results). This may reflect misfolding or masking of targeting signals caused by particular tags. As with all (immuno-)cytochemical approaches, complementary biochemical data are thus desirable. This is true in spite of the fact that none of the available purification methods affords centrosomes without significant contamination by cellular proteins, so that biochemical approaches provide corroborating rather than unequivocal evidence.

7.3 Composition of the Human Centrosome: A Proteomic Approach

The exact number of centrosomal proteins is difficult to establish with confidence. Nevertheless, a survey of the literature suggests that nearly a 100 proteins associate with centrosomes at some stage of the cell cycle in different species. Of these, about 60 are expected to be present on the interphase centrosome in humans. Many centrosomal proteins have been discovered through genetic analyses or through the production of antibodies to purified centrosomes (or MT-binding proteins). The search for interaction partners of centrosomal components by yeast two-hybrid screens and immunoprecipitation experiments has further expanded the catalog. Most recently, mass spectrometry has been used to determine the composition of the γ-tubulin ring complex [18]. Of great promise is the application of this technology to whole centrosomes which is now becoming feasible [19–21], as described in more detail below.

A priori, it might seem desirable to improve current protocols for centrosome purification, prior to subjecting such samples to a proteomic analysis. However, as the centrosome lacks a clear boundary, harsher purification methods inevitably entail the danger of losing peripheral components. In a recent study [21], we therefore decided to analyze a partially pure sample of centrosomes, taking advantage of both the increased sensitivity of contemporary mass-spectrometry instruments and advanced data analysis algorithms. In brief, two methods were used to generate peptides for mass spectrometric analysis. In one protocol, whole centrosome preparations were digested with trypsin, in the other, samples were first subjected to one-dimensional gel electrophoresis, followed by in-gel trypsin digestion of 15 slices. Peptides were then separated by nano-liquid chromatography before injection into a quadrupole time-of-flight mass spectrometer by an electrospray source.

This analysis identified over 2000 peptides corresponding to 500 proteins [21]. Since the centrosomes had been purified from asynchronously growing cells (with about 70 % in G1, the remainder in S, G2 and M), we expected to find primarily, proteins associated with the interphase centrosome. Of the approximately 60 described components of the interphase centrosome, 47 were indeed found (Table 7.1). These included both prominent structural proteins (e. g. pericentrin,

Table 7.1 Known centrosome components detected by mass-spectrometric analysis of purified centrosomes*.

Structural components	**Regulatory molecules**
Alpha tubulin	Cell division control protein 2 (Cdc2), Cdk1
Beta tubulin	cAMP-dependent protein kinase type II – alpha regulatory chain
Gamma tubulin	cAMP-dependent protein kinase – alpha catalytic subunit
Gamma-tubulin complex component 2	Serine/threonine-protein kinase Plk1
Gamma-tubulin complex component 3	Serine/threonine-protein kinase Nek2
Gamma-tubulin complex component 4, h76p	Serine/threonine protein kinase Sak
Gamma-tubulin complex component 5	Casein kinase I, delta and epsilon isoforms
Gamma-tubulin complex component 6	Protein phosphatase 2A
Centrin 2	Protein phosphatase 1 alpha isoform
Centrin 3	14-3-3 proteins, epsilon and gamma isoforms
AKAP450 (AKAP350, GC-Nap)	
Pericentrin/Kendrin (alternatively spliced proteins)	**Motor and related proteins**
Ninein	Dynein heavy chain
Pericentriolar material 1 (PCM-1)	Dynein intermediate chain
ch-TOG protein	Dynein light chain
C-Nap1, Cep250, Cep2	Dynactin 1, p150 Glued
Centriole associated protein CEP110, Cep1, centriolin	Dynactin 2, p50
Centrosomal P4.1-associated protein (CPAP)	Dynactin 3
CLIP-associating proteins CLASP1 and CLASP2	
ODF2, cenexin	**Heat shock proteins**
Lis1	Heat shock protein Hsp90
Nudel	TCP subunits
EB1	Heat shock protein Hsp73
Centractin	
Myomegalin	

*Centrosome proteins were identified as described in [21].

ninein, C-Nap1) as well as components of the γ-tubulin ring complex and regulatory proteins (e.g. the protein kinase Nek2). Of the other proteins identified by mass spectrometry, about 90 had not previously been studied in any detail and about 350 had previously been characterized in contexts that bear no obvious relation to centrosomes.

Considering that the purification protocol used in the above experiments was known to yield preparations in which centrosomes are enriched but far from pure, we were not surprised to find that many of the proteins identified by mass spectrometry appeared to be contaminants, at least by sequence criteria. Such proteins included ribosome subunits, translation initiation factors, actin binding proteins or nuclear components, i.e. proteins that have not previously been observed at the centrosome, although they have been studied extensively. As described elsewhere in this volume (Chapter 8), it is increasingly recognized that centrosomes may function not only as microtubule-organizing centers, but may also provide solid-state platforms for facilitating or enhancing intracellular signaling processes. Thus, it would be premature to conclude that all of the "non-centrosomal" factors described above merely represent contaminants, although this presently constitutes the most parsimonious interpretation. Clearly, as the sensitivity of mass spectrometry improves, the more contaminants one expects to identify. This then raises the important issue of how to distinguish bona fide centrosomal components from contaminants.

When all identified proteins were sorted by number of peptide hits, which is a reflection of both the abundance and the size of the parental protein, the top proteins on the list turned out to be mostly centrosomal or novel. It was these latter proteins we therefore focused on for further testing, using two complementary approaches. First, about 30 novel proteins were selected on the basis of their ranking and the availability of cDNA clones from public collections [22, 23] and tested for centrosomal localization. As the production of antibodies to such a large number of proteins was not practical, we sought to verify their centrosome association by expressing them as tagged proteins in cultured human cells, and then observing their cellular localization by (immuno)fluorescence microscopy. To minimize the possibility of overexpression artefacts and avoid the problems associated with tagging as discussed above, all proteins were tagged in three different ways: GFP at the N- or C-termini, or myc at the N-terminus. Fortunately, our results agreed in all but one case, where only one of the tags reported on a centrosome association, so that a positive localization remains tentative. For all other proteins that scored positive for centrosome localization (19 in total), staining of two dots, co-localizing with γ-tubulin, was generally observed at low expression levels. At higher levels of expression, several proteins in addition formed large, globular aggregates in the cell. Not surprisingly, these proteins all displayed extensive regions of predicted coiled-coil structure. The other centrosomally localized proteins additionally showed a diffuse cytoplasmic staining. A representative selection of these localization data is shown in Figure 7.2.

The second method we used for identifying bona fide centrosomal proteins was termed protein correlation profiling (PCP) [21]. This method is based on an algo-

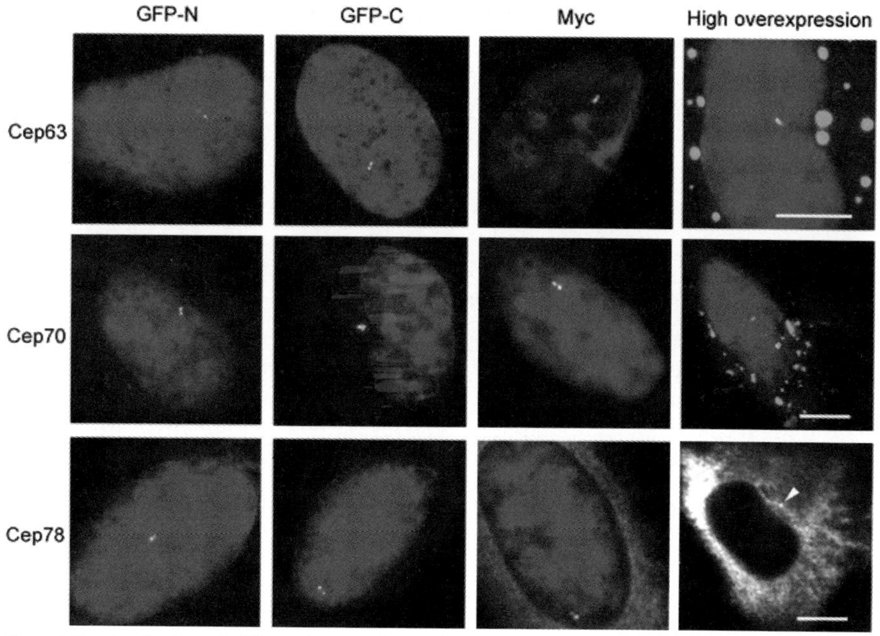

Figure 7.2 A selection of differently tagged, novel centrosome proteins. Rows from top to bottom show Cep63, Cep70 and Cep78. Columns from left to right show N-terminal GFP, C-terminal GFP and N-terminal myc-tagged proteins, respectively. The most right-hand column shows the results of very high overexpression of these proteins (tagged at the N-terminus with GFP), generating large aggregates or a high cytoplasmic background. Green, ectopically expressed centrosomal proteins; red, γ-tubulin; blue, DNA (DAPI). The arrowhead points to the position of the centrosome. Scale bars, 10 μm; panels in the three left columns are to the same scale as the top right panel (see Color Plates page XXV).

rithm that allowed us to track the relative abundance of peptides through the different sucrose fractions that were obtained at the final stage of centrosome purification. In brief, while it is difficult to use mass spectrometry for absolute quantification, the relative signal intensity of the same peptide in different samples analyzed consecutively is a fairly reliable measure of its relative abundance. Using this approach, peptide elution profiles could be established for hundreds of peptides and then compared to the profiles obtained for known centrosomal proteins. Based on co-elution profiles, a "goodness of fit" could be established that allowed the classification of proteins as likely centrosomal components or likely contaminants. Although definitive validation of candidate centrosomal proteins predicted by PCP will have to await more direct localization studies, the invaluable advantage of the PCP approach is that it provides an opportunity to extract information on hundreds of proteins from mass spectrometry data, thereby obviating the need to raise antibodies against hundreds of proteins merely for explorative purposes. In general, the PCP method should prove useful in the analysis of any multiprotein complex that can be enriched by fractionation but is difficult to obtain in pure form.

In summary, through our follow-up on candidate centrosomal proteins identified by mass spectrometry, we identified 19 novel centrosome components through epitope tagging. In addition, four of the unknown proteins under study were independently identified by other laboratories as bona fide centrosome components, anticipating our validation by tagging (Table 7.2). All these localization data were inde-

Table 7.2 Proteins newly identified and shown to localize to human centrosomes.

Protein name	*Clone*	*Motifs, comments*	*Accession number*
Cep27	FLJ10460	One coiled-coil domain	NP_060567
Cep41	FLJ22445	One coiled-coil domain, Rhodanese-like	NP_061188
Cep57	KIAA0092	Two coiled-coil domains	TREMBL:Q9BVF9
Cep63	FLJ13386	Six coiled-coil domains	TREMBL:Q9H8N0
Cep68	KIAA0582		NP_055962
Cep70	FLJ13036	Two coiled-coil domains, tetratricopeptide (TPR)	NP_077817
Cep72	KIAA1519	Leucine Rich Repeat (LRR)	TREMBL:Q9P209
Cep76	FLJ12542		NP_079175
Cep78	FLJ12643		TREMBL:Q9H9N3
Cep131	KIAA1118	Four coiled-coil domains, troponin	TREMBL:Q9UPN4
Cep135	KIAA0635	13 coiled-coil domains, (suggested in [54])	NP_055460
Cep152	KIAA0912	Eight coiled-coil domains	TREMBL:O94986
Cep164	KIAA1052	Seven coiled-coil domains, WW (two conserved tryptophans)	NP_055771
Cep192	KIAA1569 FLJ00145		NP_115518
Cep215	KIAA1633	Nine coiled-coil domains, homology to rat myomegalin	NP_060719
Cep290	KIAA0373	Nine coiled-coil domains, homology to human CTCL tumor antigen	SWISS-PROT:O15078
ALMS1 (463 kDa)	KIAA0328	Alstrom syndrome, disease linked	NP_055935
OFD-1 (117 kDa)		Four coiled-coil domains, LisH (validated in [31])	SWISS-PROT:O75665
NA-14 (13 kDa)		One coiled-coil domain, (validated in [55])	SWISS-PROT:O43805
CCCAP (74 kDa)		Five coiled-coil domains (validated in [56])	TREMBL:O60527
CP110	KIAA0419	Two coiled-coil domains, (validated in [57])	SWISS-PROT:O43303

Table 7.2 (continued)

Protein name	*Clone*	*Motifs, comments*	*Accession number*
Rootletin (230 kDa)	KIAA0445	14 coiled-coil domains, (independently validated in [58])	NP_055490
FOP (43 kDa)		LisH, Chromosomal translocation, disease linked	NP_008976

References are given for proteins independently verified elsewhere in the course of this screen. Novel factors are termed Cep where Cep stands for centrosomal protein and is followed by the molecular weight calculated from the full length sequence. Adapted from [21].

pendently confirmed also by PCP, attesting to the predictive power of this novel method. Furthermore, PCP identified another 41 proteins (Table 7.3) whose peptide profiling scores indicate that they are likely associated with centrosomes [21].

Remarkably, almost all recently described new centrosome proteins have also been identified in this survey, raising the question of how near to completion the present inventory is. We believe that the vast majority of the core components has probably been identified, although, clearly, the inventory is not yet quite complete. A few of the proteins previously reported to be associated with interphase centrosomes were not observed. In some cases, this might call for a re-examination of the published evidence. In other cases, we presume that low abundance may explain our inability to detect particular proteins. This latter explanation may apply to Nlp [12] and to δ- and ε-tubulin [24]. More difficult to explain is the absence of TACC in our inventory, a fairly abundant protein we would thus have expected to be present [25]. Perhaps, this protein was lost during purification. Finally, considering that the preparation analyzed in this study consisted primarily of interphase centrosomes, the absence of prominent spindle pole-associated components, such as NuMA [26], was not surprising. This strongly predicts, however, that proteins associating preferentially with mitotic spindle poles remain to be discovered.

7.4 Inspection of Novel Centrosome Proteins by Sequence Analysis

Most previously known components of the centrosome appear to play primarily structural roles. In line with this conclusion, a bias for structural proteins is apparent also in the novel components identified through proteomics [21]. Known functional domains or motifs were relatively rare, and, as a consequence, sequence analysis provides few clues as to the possible functions of the novel proteins. Strikingly, however, about half of the novel proteins are predicted to contain extensive coiled-coil regions. Proteins with a propensity to adopt coiled-coil structures were previously known to be common amongst components of both the centrosome and the spindle pole body (see Chapter 4 by Winey), but why the centrosome requires so many distinct structural proteins is yet another mystery of this organelle.

Table 7.3 Candidate centrosome components predicted by PCP (adapted from [21]).

Accession number	*Protein name*
NP_055625	CAP350
TREMBL:Q9C0D2	KIAA1731
NP_055730	KIAA1074
NP_055627	KARP-1-binding protein
NP_005886	Golgin-160
TREMBL:Q8WU14	KIAA0542
NP_663622	FLJ31872
TREMBL:Q9H7P7	FLJ00020
TREMBL:Q96DK7	KIAA1764
SWISS-PROT:P22314	Ubiquitin-activating enzyme E1
TREMBL:Q96SE1	NGAP-like protein
TREMBL:Q13025	Autoantigen
TREMBL:Q8N137	Lyst-interacting protein LIP8
NP_653319	AY099107
TREMBL:Q8NDE8	FLJ38327
TREMBL:Q9HCJ8	FLJ12902
TREMBL:Q8WXW3	Progesterone-induced blocking factor 1
TREMBL:Q96NL6	FLJ30655
TREMBL:Q8TC05	Mdm1
SWISS-PROT:O00139	Kinesin-like protein KIF2
NP_659436	MGC20806
TREMBL:O94927	KIAA0841
TREMBL:Q8NA30	NEDD1
TREMBL:Q96RI5	Unconventional myosin 1G methonine form
TREMBL:O43606	IT1
NP_066300	FEZ1
TREMBL:Q8NA72	FLJ35779
NP_116205	FLJ14640
TREMBL:Q8N3K0	DKFZp761A078
NP_758440	TUWD12
TREMBL:Q9Y6R9	BC282485_1
NP_079280	FLJ13215
SWISS-PROT:Q9P2S5	WD-repeat protein 8
NP_060610	FLJ10565
TREMBL:Q8NCB8	FLJ90366
NP_056241	FLJ90808
TREMBL:Q9NS50	FLJ32194
NP_803546	C14orf61
SWISS-PROT:Q9Y5B8	Nucleoside diphosphate kinase 7
NP_060285	FLJ22363
NP_078824	FLJ23047

One emerging idea is that the centrosome functions not only as a MT-organizing center but also as a solid-state platform for regulatory molecules and a junction for signaling processes [9, 27–29] (see Chapter 8). From this perspective, it would seem plausible that a complex structural scaffold may be required to allow for the specific and orderly binding of numerous factors. The many coiled-coil proteins may therefore act as a core structure to which weakly interacting proteins can bind transiently.

Another possible explanation for the multitude of centrosomal coiled-coil proteins is that interactions between such proteins may be ideally suited to confer the striking flexibility that recent experiments have brought to light [30]. Live cell imaging performed on interphase centrosomes has in fact revealed that the two centrioles within a given centrosome display a much more dynamic behavior than had previously been appreciated. This plasticity may require a dynamic architecure that a rigid matrix, composed of only few components, would be less likely to provide. Similarly, it is possible that numerous structural components are required to generate a sufficiently malleable and responsive structure to allow for cell cycle- or differentiation-dependent alterations in centrosome structure and composition.

Two proteins identified in the survey, termed ALMS1 and OFD1, have previously been identified genetically as being linked to human diseases. In particular, the C-terminal half of ALMS1 was localized to the centrosome by tagging [21] and, independently, antibodies to OFD1 have been shown to decorate the centrosome [31]. Both proteins need further investigation to confirm their association with the centrosome but their identification as candidate centrosome components is intriguing. The diseases caused by defects in these genes are relatively rare and poorly understood. Patients with Alstrom syndrome (ALMS1) display a complex set of symptoms. Childhood obesity starts at the early age of 6 months and many patients develop type 2 diabetes. The disease is also associated with neurosensory defects and subsets of patients show dilated cardiomyopathy, hepatic dysfunction, hypothyroidism, male hypogonadism, short stature and mild developmental delay [32, 33]. The symptoms of oral–facial–digital syndrome type I (OFD1) are more straightforward. Typically, the patients have malformations of the digits and face or oral cavity such as a cleft palate. In addition, two-fifths of patients have defects in the central nervous system [34]. Thus, in both of these diseases, a considerable variety of tissues appears to be affected. With the identification of the mutated proteins as putative centrosome components, it is tempting to speculate that the cause of these diseases is disruption of centrosome (or basal body) function during development. This adds to emerging evidence suggesting a critical role of the centrosome/basal body in forming the architecture of particular tissues [35, 36]. Perhaps we are about to witness the dawn of “centrosomopathies”.

7.5 Cell Cycle Changes in Centrosome Composition

The structure and behavior of the centrosome changes markedly throughout the cell cycle. This is reflected by the appearance or disappearance of individual centrosomal proteins at particular cell cycle stages, notably at the G2/M transition when the centrosome undergoes maturation in preparation for spindle formation [37, 38]. Other potentially important changes concern the establishment and dissolution of different types of linkages that are thought to connect the two parental centrioles to each other, and each parental centriole (mother centriole) to its growing pro-centriole (daughter centriole), respectively (Figure 7.3A). Critical changes undoubtedly occur also during duplication of the centrioles in S phase, and it has long been known that the two parental centrioles can be distinguished from each other by the cell cycle-regulated appearance of appendages at the distal and sub-distal end of only the older ("mature") centriole [39]. The functions of these appendages remain to be fully understood, but roles in microtubule anchoring and the formation of the primary cilium have been documented [39].

Cell cycle-dependent changes in centrosome structure have been extensively described at the electron microscopic level, but their detection by light microscopy remains difficult. This is a reflection of both the lower resolution of light microscopy and the internal organization of the centrosome. Antibodies against most PCM components (including γ-tubulin) will stain material associated with both centrioles and, therefore, give rise to two closely spaced dots under the microscope. Depending on the plane of focus and the spatial orientation of the organelle, however, the two dots will occasionally coalesce into one. As cells progress through the cycle, the two-dot staining pattern will not visibly change, even though the two centrioles duplicate (Figure 7.3B). This is because parental centrioles and their closely apposed pro-centrioles are not easily visualized as separate entities by antibodies directed against PCM components.

A further complication arises from the fact that in some G1 phase cells the two parental centrioles will be very close to each other (thus difficult to recognize as two distinct dots), whereas in others they split over distances of several microns (hence clearly producing two dots) (e. g. [40, 41]). Unfortunately, these split single centrosomes are frequently – but erroneously – considered as two already duplicated centrosomes (Figure 7.3C). Because of the small size of centrioles, the scarcity of centriolar markers and the geometric considerations described above, a rigorous and reliable quantitative analysis of centrosome duplication by light microscopy remains a difficult task.

Using cell lines stably expressing GFP-centrin and high resolution microscopy, Bornens and colleagues have been able to visualize both centrioles and growing pro-centrioles [30, 42]. By carefully studying centrosome duplication in S phase cells, these workers were able to observe two faint dots (representing pro-centrioles) appearing next to the two more intensely labeled dots (representing the parental centrioles). Subsequently, the brightness of the pro-centriolar signals increased during S phase progression, so that by G2, four dots could be seen (repre-

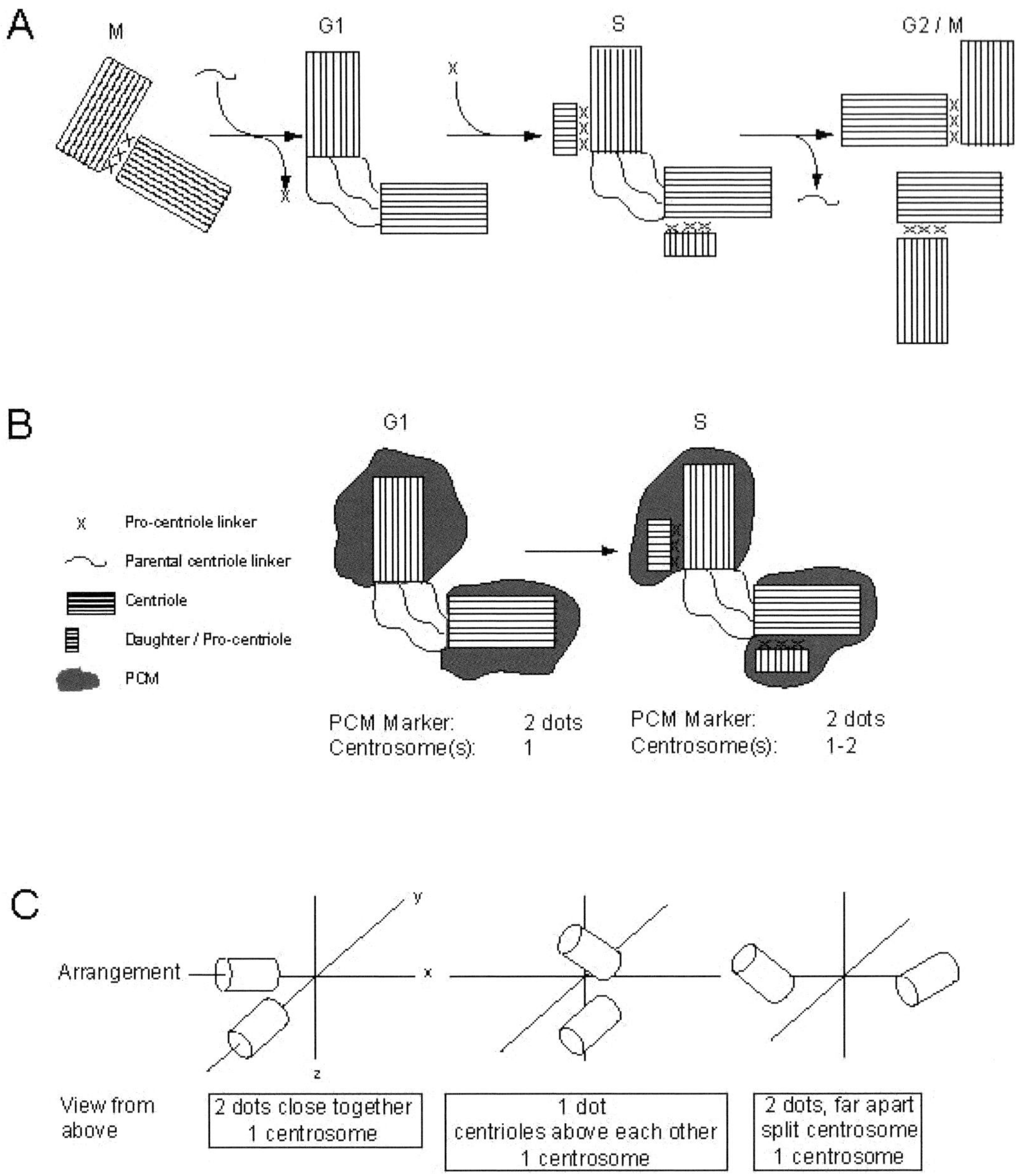

senting two doublets of duplicated centrioles). Thus, by studying centriolar markers under optimal conditions, it is possible to monitor centrosome duplication by observing the increase in the number of fluorescently labeled dots from two to four. However, as signals frequently coalesce, depending on geometry and plane of focus, the unequivocal counting of these signals is far from trivial.

◀ **Figure 7.3** Cell cycle dynamics of the centrosome. This figure summarizes, in schematic form, a few considerations that are relevant to the study of centrosome dynamics by light microscopy. (A) Two different centriolar linkers probably exist in the centrosome, one linking the emerging or newly formed (daughter) centriole to the parental centriole, the other linking two parental centrioles that will themselves give rise to pro-centrioles during S phase. The latter connection must be broken during G2 to allow the two duplicated centrosomes to separate for spindle formation. The former link must be dissolved at the end of mitosis to allow the two centrioles to move separately and, most likely, to allow subsequent duplication. (B) The limits of resolution of light microscopy and the broad distribution of many pericentriolar matrix proteins make it difficult to monitor centrosome duplication by fluorescence microscopy. (C) The different possible orientations of multiple centrioles relative to each other and the observer and the ability of the centrosome to split (because of centrioles separating from each other) constitute further potential sources of confusion when counting centrosomes.

7.6
The Impact of MS on Centrosome Analysis during Cell Cycle and Development

We are confident that the ability to analyse the centrosome by mass spectrometry will greatly help in monitoring changes occurring at this organelle as cells proliferate and differentiate. In particular, it should be possible to purify centrosomes from cells synchronized at different stages of the cell cycle. With increasing sensitivity of mass spectrometry and a concomitant decrease in the amount of material needed for analysis, it will also become attractive to apply mass spectrometry to study changes in centrosome composition during the development of different tissues. A comparison of the components in such preparations should theoretically reveal proteins that are added or lost during the cell cycle and/or during differentiation. An important advance favoring this type of study is the emerging ability to use non-radioactive heavy isotopes for protein labeling [43–46]. Since the masses of peptides derived from isotope-labeled cells will be shifted relative to those from unlabeled cells, it is possible to carry out quantitative comparisons of protein levels between two different cell populations by mass spectrometry. Particularly powerful are experimental protocols in which labeled and unlabeled samples are mixed prior to organelle purification and mass spectrometric analysis, eliminating errors due to variations in sample processing.

Another, wide open field for future investigation concerns the role of posttranslational modifications, notably phosphorylation, in the control of centrosome structure and activity [47]. Clearly, the increasing sensitivity of mass spectrometry and the continued development of peptide fractionation and isolation procedures [48–51] hold great promise for elucidating these regulatory events. In the case of the centrosome, these approaches appear particularly important and attractive, as conventional biochemical approaches for studying posttranslational modifications are severely limited by the low amounts of centrosomal proteins that can be prepared for study.

7.7 Expanding Proteomic Information into Knowledge about Function

The described proteomic approach has provided an invaluable source of information on the component parts of the human centrosome [21], and we expect that mass spectrometry will continue to provide insight into cell-cycle and developmental changes in centrosome composition. Clearly, though, the persisting key task is to use this information to derive models of how the whole centrosome works. In the past, centrosome proteins have often been discovered through genetic screens or biochemical searches for interacting partners of already characterized proteins. And in at least some of these cases, clues to the functions of newly discovered proteins were thus available. For proteins discovered through proteomics, with no obvious functional links or insight from sequence information, knowing where to start to decipher a function is less obvious. *A priori*, overexpression of a novel protein may lead to observations (e. g. γ-tubulin recruitment) that suggest a specific function (e. g. [12]). Conversely, depletion of a protein, notably by siRNA technology, may yield information on a loss-of-function or hypomorphic phenotype [52, 53]. However, by siRNA it may be difficult to deplete some centrosomal proteins sufficiently to produce a clear-cut phenotype. In particular, in cases where large cytoplasmic pools exist and turnover at the centrosome occurs, very extensive depletion may be required before the function of the centrosomal pool is impaired. Thus, in spite of the undisputed power of siRNA approaches, it will ultimately be important to assess the function of selected centrosomal proteins by gene knock-out strategies. Finally, another persisting challenge is to develop novel assays to probe various aspects of centrosome function. These assays should focus not only on the roles of centrosomes in microtubule organization, cell polarity and motility, but also in cell cycle progression and development.

7.8 Conclusion and Prospects

The use of proteomic approaches to investigate the centrosome has been remarkably successful. It thus seems legitimate to hope for a comprehensive description of centrosome composition in a not-too-distant future. Furthermore, mass spectrometry holds great promise for monitoring changes in centrosome function. This is true regardless of whether changes in centrosome behavior during cell cycle progression or differentiation involve changes in protein composition or in the activity of particular components in response to posttranslational modifications. As in many other fields to which mass spectrometry has been applied as an analytical technique, it is thus likely to develop into a core tool for centrosome research.

Acknowledgments

We thank Thibault Mayor for his centrosome purifications in the early stages of the project, Peter Mortensen for his important contribution to the development of PCP, Monika Matzner for expert assistance with the tagging of candidate proteins, Roman Körner for critical reading of the manuscript and many laboratory members for helpful discussions. The work in the authors' laboratory was generously supported by the Max-Planck-Society (CJW and EAN) and the Danish National Research Foundation (JSA and MM).

References

1. Andersen, J. S., C. E. Lyon, A. H. Fox, A. K. Leung, Y. W. Lam, H. Steen, M. Mann, and A. I. Lamond. **2002**. Directed proteomic analysis of the human nucleolus. *Curr Biol* 12:1–11.
2. Lasonder, E., Y.Ishihama, J. S.Andersen, A. M. Vermunt, A. Pain, R. W. Sauerwein, W. M. Eling, N. Hall, A. P. Waters, H. G. Stunnenberg, and M.Mann. **2002**. Analysis of the Plasmodium falciparum proteome by high-accuracy mass spectrometry. *Nature* 419:537–542.
3. Taylor, S. W., E. Fahy, B. Zhang, G. M. Glenn, D. E. Warnock, S. Wiley, A. N. Murphy, S. P. Gaucher, R. A. Capaldi, B. W. Gibson, and S. S. Ghosh. **2003**. Characterization of the human heart mitochondrial proteome. *Nature Biotechnol* 21:281–286.
4. Washburn, M. P., D. Wolters, and J. R. Yates, III. **2001**. Large-scale analysis of the yeast proteome by multidimensional protein identification technology. *Nature Biotechnol* 19:242–247.
5. Florens, L., M. P. Washburn, J. D. Raine, R. M. Anthony, M. Grainger, J. D. Haynes, J. K. Moch, N. Muster, J. B. Sacci, D. L. Tabb, A. A. Witney, D. Wolters, Y. Wu, M. J. Gardner, A. A. Holder, R. E. Sinden, J. R. Yates, and D. J.Carucci. **2002**. A proteomic view of the Plasmodium falciparum life cycle. *Nature* 419:520–526.
6. Wigge, P. A., O. N. Jensen, S. Holmes, S. Soues, M. Mann, and J. V. Kilmartin. **1998**. Analysis of the Saccharomyces spindle pole by matrix-assisted laser desorption/ionization (MALDI) mass spectrometry. *J Cell Biol* 141:967–977.
7. Moudjou, M., N. Bordes, M. Paintrand, and M.Bornens. **1996**. gamma-Tubulin in mammalian cells: the centrosomal and the cytosolic forms. *J Cell Sci* 109:875–887.
8. Paoletti, A., M. Moudjou, M. Paintrand, J. L. Salisbury, and M. Bornens. **1996**. Most of centrin in animal cells is not centrosome-associated and centrosomal centrin is confined to the distal lumen of centrioles. *J Cell Sci* 109:3089–3102.
9. Doxsey, S. **2001**. Re-evaluating centrosome function. *Nature Rev Mol Cell Biol* 2:688–698.
10. Meraldi, P. and E. A. Nigg. **2002**. The centrosome cycle. *FEBS Lett* 521:9–13.
11. Balczon, R., L. Bao, and W. E. Zimmer. **1994**. PCM-1, A 228-kD centrosome autoantigen with a distinct cell cycle distribution. *J Cell Biol* 124:783–793.
12. Casenghi, M., P. Meraldi, U. Weinhart, P. I. Duncan, R. Korner, and E. A. Nigg. **2003**. Polo-like kinase 1 regulates Nlp, a centrosome protein involved in microtubule nucleation. *Dev Cell* 5:113–125.
13. Mayor, T., Y. D. Stierhof, K. Tanaka, A. M. Fry, and E. A. Nigg. **2000**. The centrosomal protein C-Nap1 is required for cell cycle-regulated centrosome cohesion. *J Cell Biol* 151:837–846.

14. Zheng, Y., M. K. Jung, and B. R. Oakley. **1991**. Gamma-tubulin is present in *Drosophila melanogaster* and *Homo sapiens* and is associated with the centrosome. *Cell* 65:817–823.
15. Price, C. M. and D. E. Pettijohn. **1986**. Redistribution of the nuclear mitotic apparatus protein (NuMA) during mitosis and nuclear assembly. Properties of purified NuMA protein. *Exp Cell Res* 166:295–311.
16. Nigg, E. A. **1988**. Nuclear function and organization: the potential of immunochemical approaches. *Int Rev Cytol* 110:27–92.
17. Stucke, V. M., H. H. Sillje, L. Arnaud, and E. A. Nigg. **2002**. Human Mps1 kinase is required for the spindle assembly checkpoint but not for centrosome duplication. *EMBO J* 21:1723–1732.
18. Murphy, S. M., A. M. Preble, U. K. Patel, K. L. O'Connell, D. P. Dias, M. Moritz, D. Agard, J. T. Stults, and T. Stearns. **2001**. GCP5 and GCP6: two new members of the human gamma-tubulin complex. *Mol Biol Cell* 12:3340–3352.
19. Mollinari, C., B. Lange, and C. Gonzalez. **2002**. Miranda, a protein involved in neuroblast asymmetric division, is associated with embryonic centrosomes of *Drosophila melanogaster*. *Biol Cell* 94:1–13.
20. Lange, B. M., A. Bachi, M. Wilm, and C. Gonzalez. **2000**. Hsp90 is a core centrosomal component and is required at different stages of the centrosome cycle in Drosophila and vertebrates. *EMBO J* 19:1252–1262.
21. Andersen, J. S., C. J. Wilkinson, T. Mayor, P. Mortensen, E. A. Nigg, and M. Mann. **2003**. Proteomic characterization of the human centrosome using protein correlation profiling. *Nature* (in press).
22. Strausberg, R. L., E. A. Feingold, R. D. Klausner, and F. S. Collins. **1999**. The mammalian gene collection. *Science* 286:455–457.
23. Ohara, O., T. Nagase, K. Ishikawa, D. Nakajima, M. Ohira, N. Seki, and N. Nomura. **1997**. Construction and characterization of human brain cDNA libraries suitable for analysis of cDNA clones encoding relatively large proteins. *DNA Res* 4:53–59.
24. Chang, P. and T. Stearns. **2000**. Delta-tubulin and epsilon-tubulin: two new human centrosomal tubulins reveal new aspects of centrosome structure and function. *Nature Cell Biol* 2:30–35.
25. Gergely, F., C. Karlsson, I. Still, J. Cowell, J. Kilmartin, and J. W. Raff. **2000**. The TACC domain identifies a family of centrosomal proteins that can interact with microtubules. *Proc Natl Acad Sci USA* 97:14352–14357.
26. Tang, T. K., C. J. Tang, Y. J. Chao, and C. W. Wu. **1994**. Nuclear mitotic apparatus protein (NuMA): spindle association, nuclear targeting and differential subcellular localization of various NuMA isoforms. *J Cell Sci* 107:1389–1402.
27. Lange, B. M. **2002**. Integration of the centrosome in cell cycle control, stress response and signal transduction pathways. *Curr Opin Cell Biol* 14:35–43.
28. Rieder, C. L., S. Faruki, and A. Khodjakov. **2001**. The centrosome in vertebrates: more than a microtubule-organizing center. *Trends Cell Biol* 11:413–419.
29. Jackman, M., C. Lindon, E. A. Nigg, and J. Pines. **2003**. Active cyclin B1-Cdk1 first appears on centrosomes in prophase. *Nature Cell Biol* 5:143–148.
30. Piel, M., P. Meyer, A. Khodjakov, C. L. Rieder, and M. Bornens. **2000**. The respective contributions of the mother and daughter centrioles to centrosome activity and behavior in vertebrate cells. *J Cell Biol* 149:317–330.
31. Romio, L., V. Wright, K. Price, P. J. Winyard, D. Donnai, M. E. Porteous, B. Franco, G. Giorgio, S. Malcolm, A. S. Woolf, and S. A. Feather. **2003**. OFD1, the gene mutated in oral–facial–digital syndrome type 1, is expressed in the metanephros and in human embryonic renal mesenchymal cells. *J Am Soc Nephrol* 14:680–689.
32. Collin, G. B., J. D. Marshall, A. Ikeda, W. V. So, I. Russell-Eggitt, P. Maffei,

S. Beck, C. F. Boerkoel, N. Sicolo, M. Martin, P. M. Nishina, and J. K. Naggert. **2002**. Mutations in ALMS1 cause obesity, type 2 diabetes and neurosensory degeneration in Alstrom syndrome. *Nature Genet* 31:74–78.

33. Hearn, T., G. L. Renforth, C. Spalluto, N. A. Hanley, K. Piper, S. Brickwood, C. White, V. Connolly, J. F. Taylor, I. Russell-Eggitt, D. Bonneau, M. Walker, and D. I. Wilson. **2002**. Mutation of ALMS1, a large gene with a tandem repeat encoding 47 amino acids, causes Alstrom syndrome. *Nature Genet* 31:79–83.

34. Ferrante, M. I., G. Giorgio, S. A. Feather, A. Bulfone, V. Wright, M. Ghiani, A. Selicorni, L. Gammaro, F. Scolari, A. S. Woolf, O. Sylvie, L. Bernard, S.Malcolm, R.Winter, A.Ballabio, and B.Franco. **2001**. Identification of the gene for oral–facial–digital type I syndrome. *Am J Hum Genet* 68:569–576.

35. Ansley, S. J., J. L. Badano, O. E. Blacque, J. Hill, B. E. Hoskins, C. C. Leitch, J. C. Kim, A. J. Ross, E. R. Eichers, T. M. Teslovich, A. K. Mah, R. C. Johnsen, J. C. Cavender, R. A. Lewis, M. R. Leroux, P. L. Beales, and N. Katsanis. **2003**. Basal body dysfunction is a likely cause of pleiotropic Bardet–Biedl syndrome. *Nature* 425:628–633.

36. Ibanez-Tallon, I., N. Heintz, and H. Omran. **2003**. To beat or not to beat: roles of cilia in development and disease. *Hum Mol Genet* 12:R27–R35.

37. Khodjakov, A. and C. L. Rieder. **1999**. The sudden recruitment of gamma-tubulin to the centrosome at the onset of mitosis and its dynamic exchange throughout the cell cycle, do not require microtubules. *J Cell Biol* 146:585–596.

38. Palazzo, R. E., J. M. Vogel, B. J. Schnackenberg, D. R. Hull, and X. Wu. **2000**. Centrosome maturation. *Curr Top Dev Biol* 49:449–470.

39. Bornens, M. **2002**. Centrosome composition and microtubule anchoring mechanisms. *Curr Opin Cell Biol* 14:25–34.

40. Fry, A. M., P. Meraldi, and E. A. Nigg. **1998**. A centrosomal function for the human Nek2 protein kinase, a member of the NIMA family of cell cycle regulators. *EMBO J* 17:470–481.

41. Faragher, A. J. and A. M. Fry. **2003**. Nek2A kinase stimulates centrosome disjunction and is required for formation of bipolar mitotic spindles. *Mol Biol Cell* 14:2876–2889.

42. Keryer, G., O. Witczak, A. Delouvee, W. A. Kemmner, D. Rouillard, K. Tasken, and M. Bornens. **2003**. Dissociating the centrosomal matrix protein AKAP450 from centrioles impairs centriole duplication and cell cycle progression. *Mol Biol Cell* 14:2436–2446.

43. Blagoev, B., I. Kratchmarova, S. E. Ong, M. Nielsen, L. J. Foster, and M. Mann. **2003**. A proteomics strategy to elucidate functional protein–protein interactions applied to EGF signaling. *Nature Biotechnol* 21:315–318.

44. Ranish, J. A., E. C. Yi, D. M. Leslie, S. O. Purvine, D. R. Goodlett, J. Eng, and R. Aebersold. **2003**. The study of macromolecular complexes by quantitative proteomics. *Nature Genet* 33:349–355.

45. Gygi, S. P., B. Rist, S. A. Gerber, F. Turecek, M. H. Gelb, and R. Aebersold. **1999**. Quantitative analysis of complex protein mixtures using isotope-coded affinity tags. *Nature Biotechnol* 17:994–999.

46. Ong, S. E., B. Blagoev, I. Kratchmarova, D. B. Kristensen, H. Steen, A. Pandey, and M. Mann. **2002**. Stable isotope labeling by amino acids in cell culture, SILAC, as a simple and accurate approach to expression proteomics. *Mol Cell Proteomics* 1:376–386.

47. Mayor, T., P. Meraldi, Y. D. Stierhof, E. A. Nigg, and A. M. Fry. **1999**. Protein kinases in control of the centrosome cycle. *FEBS Lett* 452:92–95.

48. Ficarro, S. B., M. L. McCleland, P. T. Stukenberg, D. J. Burke, M. M. Ross, J. Shabanowitz, D. F. Hunt, and F. M. White. **2002**. Phosphoproteome analysis by mass spectrometry and its application to *Saccharomyces cerevisiae*. *Nature Biotechnol* 20:301–305.

49. Zhou, H., J. D. Watts, and R. Aebersold. **2001**. A systematic approach to the analysis of protein phosphorylation. *Nature Biotechnol* 19:375–378.

50. Oda, Y., T. Nagasu, and B. T. Chait. **2001**. Enrichment analysis of phosphorylated proteins as a tool for probing the phosphoproteome. *Nature Biotechnol* 19:379–382.

51. Mann, M., S. E. Ong, M. Gronborg, H. Steen, O. N. Jensen, and A. Pandey. **2002**. Analysis of protein phosphorylation using mass spectrometry: deciphering the phosphoproteome. *Trends Biotechnol* 20:261–268.

52. Kirkham, M., T. Muller-Reichert, K. Oegema, S. Grill, and A. A. Hyman. **2003**. SAS-4 is a *C. elegans* centriolar protein that controls centrosome size. *Cell* 112:575–587.

53. Gonczy, P., C. Echeverri, K. Oegema, A. Coulson, S. J. Jones, R. R. Copley, J. Duperon, J. Oegema, M. Brehm, E. Cassin, E. Hannak, M. Kirkham, S. Pichler, K. Flohrs, A. Goessen, S. Leidel, A. M. Alleaume, C. Martin, N. Ozlu, P. Bork, and A. A. Hyman. **2000**. Functional genomic analysis of cell division in *C. elegans* using RNAi of genes on chromosome III. *Nature* 408:331–336.

54. Ohta, T., R. Essner, J. H. Ryu, R. E. Palazzo, Y. Uetake, and R. Kuriyama. **2002**. Characterization of Cep135, a novel coiled-coil centrosomal protein involved in microtubule organization in mammalian cells. *J Cell Biol* 156:87–99.

55. Pfannenschmid, F., V. C. Wimmer, R. M. Rios, S. Geimer, U. Krockel, A. Leiherer, K. Haller, Y. Nemcova, W. Magcs. **2003**. Chlamydomonas DIP13 and human NA14: a new class of proteins associated with microtubule structures is involved in cell division. *J Cell Sci* 116:1449–62.

56. Kenedy, A. A., K. J. Cohen, D. A. Loveys, G. J. Kato, C. V. Dang. **2003**. Identification and characterization of the novel centrosome-associated protein CCCAP. *Gene* 303:35–46.

57. Chen, Z., V. B. Indjeian, M. McManus, L. Wang, B. D. Dynlacht. **2002**. CP110, a cell cycle-dependent CDK substrate, regulates centrosome duplication in human cells. *Dev Cell* 3:339–50.

58. Yang, J., X. Liu, G. Yue, M. Adamian, O. Bulgakov, T. Li. **2002**. Rootletin, a novel coiled-coil protein, is a structural component of the ciliary rootlet. *J Cell Biol* 159:431–40.

8
The Role of the Centrosome in Cell Cycle Progression

Andrew M. Fry and Rebecca S. Hames

8.1
Introduction

The centrosome is a tiny subcellular organelle present in only one or two copies per cell. Yet, through its role as the primary site of microtubule nucleation and organization, it contributes to numerous cellular processes including anteriograde and retrograde transport, positioning of organelles and chromosome segregation. If this was not enough, there is now a growing body of evidence that the centrosome plays additional important roles in orchestrating many of the key transition events that occur during cell cycle progression including mitotic entry, anaphase onset, cytokinesis and S-phase entry. It seems to do this by acting as a solid-phase signaling platform providing docking surfaces on which key enzymes can be brought into contact with their substrates and upstream regulators. In this review, we will consider the evidence that has implicated the centrosome in regulating specific cell cycle transitions and discuss the consequences that this new and exciting information has on our understanding of cell cycle control.

Progression through the eukaryotic cell cycle requires the precise coordination and integration of many critical biochemical events. Frequently, these involve protein phosphorylation and dephosphorylation reactions, as well as the activation of targeted proteolysis. However, the cell is not a test-tube and in reality these enzymatic processes are carefully regulated in a spatial as well as temporal fashion [1]. At the simplest level, enzymes and substrates may be physically separated into different subcellular compartments until the appropriate time in the cell cycle. However, on a more sophisticated level, activators or inhibitors may be brought into close physical proximity to their targets through immobilization on particular subcellular structures.

The animal cell centrosome is a discrete non-membranous organelle that sits in the cytoplasm close to the nucleus [2, 3]. Its fungal counterpart, the spindle pole body (SPB), may also be found just outside the nuclear envelope or, in some species, within the nuclear envelope [4, 5]. The core structural components of the higher eukaryotic centrosome include two barrel-shaped centrioles, composed pre-

Centrosomes in Development and Disease. Edited by Erich A. Nigg

ISBN 3-527-30980-2

dominantly of nine highly stable microtubule triplets, and the surrounding pericentriolar material (PCM). The PCM contains proteins required for microtubule nucleation and anchoring that are held within a fibrous lattice somehow attached to the walls of the centrioles (see Chapters 3, 5, and 15 for a detailed description of centrosome structure and microtubule nucleation). Because the centrosome lacks a surrounding lipid bilayer, its three-dimensional architecture is maintained through specific protein–protein interactions. Not surprisingly, then, most centrosomal proteins contain protein interaction domains, with coiled-coil motifs featuring prominently. These protein interaction domains were originally assumed to contribute solely to the maintenance of the centrosome structure itself and to the recruitment of proteins involved in regulating microtubule nucleation or centriole duplication. However, the growing list of diverse enzymes detected at the centrosome (see [6, 7]) has challenged this assumption and raised the possibility that the centrosome acts as a docking platform for a wide range of regulatory molecules that do not necessarily have a function directly related to centrosome biology itself. In this scenario, the centrosome behaves as a command center integrating signals from different pathways and ensuring the correct output. This activity may be entirely independent of microtubules or it may utilize microtubules to facilitate movement of signaling molecules to and from the centrosome.

Evidence for a centrosomal role in externally-regulated signal transduction pathways remains rather sketchy at the present time. In contrast, there is now abundant and persuasive evidence that the centrosome acts as a scaffold for coordinating intrinsic cell cycle events [8, 9]. How the centrosome plays a key role in regulating mitotic entry, the metaphase–anaphase transition, execution of cytokinesis and S-phase entry will form the central debate of this chapter.

8.2
Cell Cycle Dynamics of Centrosome Structure

An excellent description of the morphological events and biochemical regulation of the centriole duplication cycle is included elsewhere in this volume. In the context of this chapter it is nevertheless worth briefly reviewing the gross structural changes that take place in the centrosome during cell cycle progression. As cells progress from G1 into S phase, the process of centriole duplication begins. This continues during S and G2 with the elongation of pro-centrioles. By late G2, centriole duplication is mostly complete and the cell possesses two centrosomes that each contains two closely juxtaposed centrioles. Throughout interphase, the two centrosomes are held in close proximity as the result of some component part of the PCM that acts as a linker connecting the two fully-formed centrioles [7]. Prior to mitotic entry this tether is disassembled in a process referred to as centrosome disjunction [10]. But centrosome disjunction can also be thought of as just one part of a more global change known as centrosome maturation that takes place at the G2/M transition [11]. During centrosome maturation, many centrosomal proteins, e. g. Plk1 [12] or NuMA [13], are recruited for the first time in the cell cycle while others, e. g.

C-Nap1 [14] or Nlp [15], are discarded. These changes in protein composition may trigger centrosome disjunction at the same time as increasing the rate of microtubule nucleation. Generally, there is an overall increase in centrosome bulk at this time in line with the increased requirement for microtubule nucleation. As a result of centrosome maturation, it is fair to say that the composition of an interphase centrosome is quite different from that of a mitotic spindle pole. Late in mitosis, there is a similar, albeit opposite, reversion of the spindle pole to an interphase centrosome. As this happens, the two centrioles within a single pole lose their close apposition and start to behave in a remarkably independent manner [16]. The ensuing movements of the centrioles at this stage in the cell cycle might contribute to the fidelity of cytokinesis, as described later in this chapter, and elsewhere in this volume. Finally, as cells re-enter G1 they inherit a single centrosome with two centrioles ready to begin the centriole duplication cycle once again.

8.3
Old and New Functions of the Centrosome

From a functional perspective, the centrosome, and its homologous structures in other species, is first and foremost the primary site of microtubule nucleation within the cell. It therefore contributes enormously to cellular properties that are dependent upon the microtubule network, including segregation of duplicated chromosomes on a microtubule-based spindle (reviewed in [17]). This function was established many years ago and disruption of centrosome number or architecture has now been intimately linked with chromosome segregation defects, aneuploidy, chromosome instability and loss of cell polarity, all of which are classic hallmarks of malignant tumor cells [18–20]. However, more recent research into centrosome biology, partly fuelled by the cancer connection, has thrown up some intriguing and unexpected results with respect to the exact role of the centrosome in animal cells. Astral arrays of microtubules and bipolar spindles can form both *in vivo* and *in vitro* in the absence of centrosomes, challenging the view that the centrosome is essential for spindle formation and chromosome segregation (e. g. [21, 22]). Indeed, early embryonic divisions in many rodents occur in the absence of centrosomes and acentrosomal cell lines that are viable have been isolated from *Drosophila* [23]. Flies can even develop to maturity in the presence of centrosomes deficient in certain core centrosomal proteins [24]. Critically, though, parthenogenesis (the complete development to adulthood of an egg without fertilization) of frogs and mammals can only occur if a centrosome is present demonstrating that centrosomes do provide an essential function in the development of vertebrates [25]. Although this function may relate to its central role in microtubule nucleation or organization, it may equally reflect a non-microtubule related activity. Indeed, work from a number of laboratories has now demonstrated that, in addition to its microtubule-related function, the centrosome plays key roles in regulating specific cell cycle transitions, and these might explain the essential nature of the centrosome. Furthermore, failure of these non-microtubule-related functions may also contribute to the centrosome-associated phenotypes of cancer cells. In the following sections, we will examine some of the possible mechanisms by which the centrosome may contribute to different cell cycle transitions.

8.4
The Centrosome in G2/M Control

Early research into how the G2/M transition is regulated laid the foundations for our entire understanding of eukaryotic cell cycle control. Genetic studies on fission and budding yeast cell division and biochemical studies into the process of oocyte maturation led to the conclusion that cell cycle transitions are universally regulated by cyclin-dependent kinases in conjunction with their regulatory subunits, cyclins [26, 27]. In essence, it was demonstrated that the G2/M transition requires activation of Cdk1/cyclin B and, as we now know, Cdk1/cyclin A. Much, however, remained at that time to be understood about how Cdk/cyclin complexes are regulated in time and space.

Cdk1 is now recognized to require more than cyclin binding for its activation. It is through a complex modulation of its phosphorylation status that Cdk1 becomes turned on at the G2/M transition [28, 29]. It must be phosphorylated on an activating threonine (Thr-161 in human Cdk1) within the T-loop of the catalytic domain and dephosphorylated on Thr-14 and Tyr-15 residues in the ATP-binding pocket. The activating threonine is phosphorylated by the Cdk-activating kinase, CAK. The inhibitory phosphorylation sites in the ATP-binding cleft are phosphorylated, in higher eukaryotes, by Myt1 (Thr-14) and Wee1 (Tyr-15), and both are dephosphorylated by the dual-specificity phosphatase, Cdc25. Cdk1 can phosphorylate and activate Cdc25, thereby creating a positive feedback loop once a small fraction of Cdk1 has become active [30, 31]. Cdc25 is also subject to phosphorylation and activation by the Polo-like kinase, Plk1/Plx1 [32]. It has not been proven, but seems likely, that Wee1 is simultaneously inhibited by Plk1 [33], thereby further promoting the activation of Cdk1. However, the observation that mitotic entry can occur in the absence of Polo kinase activity in certain systems suggests that the primary role of Polo at this stage of the cell cycle is to influence the rate of mitotic commitment rather than mitotic commitment *per se* [34].

The questions that we wish to address are to what extent Cdk1 activation takes place at the centrosome and whether the centrosome is required for Cdk1 activation. Cdk1 has long been known to localize to centrosomes from the onset of mitosis [35, 36]. As major changes in microtubule nucleation capacity occur at this time and Cdk1 is capable of regulating microtubule dynamics [37], a reasonable proposition was that a fraction of Cdk1 is required to localize to the centrosome in order to initiate these changes. In other words, Cdk1 is targeted to the centrosome simply to regulate its microtubule nucleation capacity. This view however has been challenged by the localization of upstream regulators of Cdk1/cyclin B to mitotic centrosomes. In particular, Polo kinases have been shown in a wide variety of organisms to localize to mitotic centrosomes and SPBs from very early in mitosis to around the time of anaphase [12, 38–42]. Hence, Cdk1 activation may occur first at the centrosome through localized activation of Cdc25 by Polo kinase. It is probably not that simple though, as association of Polo kinase with the SPB in fission yeast does require some Cdk1 and Cdc25 activity [41]. Importantly, though, the efficient amplification of the positive feedback loop for mitotic commitment may be specifically promoted at the centrosome through bringing partially active Cdk1 into close physical proximity with Cdc25 and Plk1 (Figure 8.1). In support of this, a report suggested that inhibitory Tyr-15 phosphorylation is lost from centrosomal (and cytoplasmic) Cdk1 before it disappears from nuclear Cdk1 [43].

In a careful study performed in human cells, the timing of Cdk1/cyclin B1 activation was followed using phosphospecific antibodies [44]. The activation of the Cdk1/cyclin B1 complex is usually associated with phosphorylation on particular sites within the cytoplasmic retention sequence of cyclin B1, including Ser-126 and Ser-133. Ser-126 is an autophosphorylation site and therefore directly reflects Cdk1 activation. Using antibodies that recognize these phosphorylated sites, Jackman and coworkers clearly demonstrated that Cdk1/cyclin B1 activation is detected on centrosomes in prophase as well as weakly in the cytoplasm, prior to its appear-

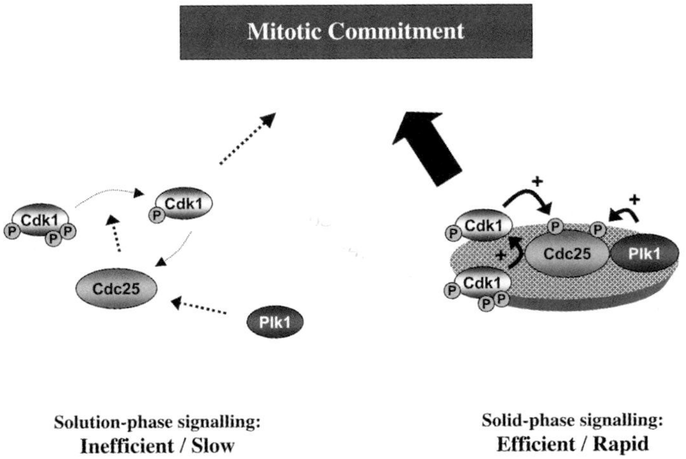

Figure 8.1 The centrosome acts as a solid-phase platform for cell cycle-related signaling events. Through recruitment and concentration of signaling pathway components (right-hand figure), the centrosome (or SPB in fungi) can act as a scaffold or platform dramatically enhancing the efficiency of signal transduction events as compared to that seen in solution (left-hand figure). This is illustrated here using components that promote mitotic commitment as an example. Recruitment to the centrosome of Cdk1, together with its upstream regulators Cdc25 and Plk1, triggers rapid and efficient generation of active Cdk1, a pre-requisite for the sudden passage of cells from G2 to M. This is achieved through Plk1 phosphorylating and activating Cdc25 which, in turn, stimulates the removal of inhibitory phosphorylation sites on Cdk1. Concentration at the centrosome also facilitates positive feedback loops, with Cdk1 being able to further activate Cdc25 by phosphorylation. Localization at the centrosome is likely to be transient and so, once activated, Cdk1 can migrate to other sites in the cell for substrate phosphorylation. The one caveat to this model is that hard evidence for Cdc25 localization to the centrosome at the G2/M transition is still lacking.

ance in the nucleus [44]. Ser-133 may be phosphorylated slightly before Ser-126, and was found to be a Plk1 phosphorylation site. Hence, co-localization at the centrosome also promotes phosphorylation of cyclin B1 by Plk1. Contrary to a previous report [45] Jackman and colleagues also found that Plk1 did not directly promote the translocation of cyclin B1 into the nucleus. Thus, it appears that translocation of the complex to the nucleus is not the trigger event that activates Cdk1 as had been previously proposed, but that Cdk1 is first activated perhaps weakly in the cytoplasm and then, through concentration of positive feedback components, strongly on centrosomes.

In the filamentous fungus *Aspergillus nidulans*, another serine/threonine protein kinase, NIMA, is absolutely required for the G2/M transition [46]. The mechanistic reason for this is not clear although it may also relate to the local concentration of mitotic regulators, as *nimA* mutants fail to correctly localize Cdk1/cyclin B to either the nucleus or SPB [47]. NIMA itself localizes to the SPB, as does the fission yeast equivalent Fin1 [47, 48]. Although not an essential gene, temperature sensitive *fin1* mutants exhibit spindle defects and *fin1* null alleles require an intact spindle checkpoint for viability, suggesting an intrinsic role in SPB spindle function [49]. However, there is interesting circumstantial evidence that Fin1, like NIMA, also contributes to mitotic commitment. A dominant mutant of a fission yeast SPB component, Cut12, is able to initiate mitotic entry in the absence of Cdc25 function [50]. In this *cut12/stf1* mutant, it was found that the Polo kinase, Plo1, localized to the SPB throughout the cell cycle, rather than only at mitosis as happens in wild-type cells [41]. This premature recruitment is abolished in a *fin1*^{ts} mutant, while artificially upregulating Fin1 expression promoted the early appearance of Plo1 on interphase SPBs [49]. Thus, Fin1, and perhaps NIMA, may accelerate mitotic entry through recruitment of mitotic kinases such as Polo to the fungal equivalent of the centrosome. The closest structural homolog to NIMA and Fin1 in vertebrates is Nek2 [51]. Nek2 strongly localizes to the centrosome where it is thought to regulate centrosome structure at G2/M and thus contribute to spindle formation [52, 53]. Whether Nek2, or one of the 10 other NIMA-related kinases present in mammals, has a direct role in mitotic commitment remains to be seen.

The definitive proof that centrosomes are required for G2/M entry seemed to have come when centrosomes were microsurgically removed from cells leading to an apparent arrest in G2 [54]. However, by combining this same approach with live cell imaging, it became apparent that the karyoplasts lacking centrosomes primarily arrested in G1 after progressing through mitosis ([55] and discussed in more detail later). So are centrosomes dispensable for the G2/M transition after all? Available evidence suggests that centrosomes may not be strictly required for mitosis in animal cells (as they are not required in higher plants and female meiosis in many animal species), but that they are likely to contribute to the efficiency and/or accuracy of mitosis. Experimentally, acentrosomal cultured cells can enter mitosis, but perhaps in a delayed and less coordinated fashion. Centrosomes certainly accelerate mitotic entry and activation of Cdk1/cyclin B in frog eggs [56] and, following microinjection, can release starfish oocytes from a G2 arrest [57]. In summary then, it seems reasonable to propose that centrosomes promote a high precision mitotic entry by facilitating the positive feedback activation loop of Cdk1 by Plk1, Cdc25 and perhaps NIMA. Once activated, Cdk1 may leave the centrosome and trigger further activation of Cdk1 elsewhere in the cell, thereby leading to the sudden and dramatic entry of cells into mitosis.

8.5 Initiation of Cyclin B Destruction at the Centrosome

Following mitotic entry, the next critical transition point in the cell cycle is the metaphase to anaphase transition. This is under the control of the spindle assembly checkpoint, which prevents anaphase onset until all chromosomes have achieved attachment to opposite spindle poles (reviewed in [58]). In an elegant approach that made use of cells containing two spindles, it was shown that one spindle could initiate anaphase despite the presence of mono-orientated chromosomes on the second spindle [59]. This implied for the first time that the molecular components of the spindle assembly checkpoint are physically restricted to the spindle structure itself. The obvious location for components of the checkpoint is the kinetochore/centromere region where attachment of microtubules and tension generated by bipolar attachment can be monitored. In support of this model, multiple spindle checkpoint proteins including Mad1, Mad2, BubR1 (Mad3), Bub1, Bub3, Mps1, Aurora B, Rod and Zw10, have all been localized to the kinetochore [58].

The main target of the spindle assembly checkpoint is the multi-subunit ubiquitin ligase known as the anaphase promoting complex or cyclosome (APC/C) [60]. Amongst other substrates, the APC/C polyubiquitylates securin and cyclin B targeting them for proteasome-mediated degradation. Destruction of securin initiates anaphase by releasing separase which in turn cleaves the centromeric cohesin molecules that tether the sister chromatids [61]. Destruction of cyclin B promotes mitotic exit. Recognition of substrates by the APC/C requires an additional adaptor subunit that at the time of the metaphase/anaphase transition is the Cdc20 (Fizzy/Fzy) protein, and in late mitosis/G1 is the Cdh1 (Fizzy-related/Fzr) protein [62]. The current view on how the spindle assembly checkpoint prevents anaphase onset is that checkpoint proteins, notably Mad2, Bub3 and BubR1, form a mitotic checkpoint complex (MCC) with Cdc20 preventing it from interacting with and activating the APC/C [63–66]. Once full bipolar attachment has been achieved, Cdc20 is no longer assembled into checkpoint complexes thereby allowing APC/C-Cdc20 complexes to form initiating the polyubiquitylation of substrates.

The question we are interested in here is where the APC/C-Cdc20 is first activated: is it primarily at the kinetochore or could it rather occur at the centrosome/spindle pole? Circumstantial evidence has come from the localization of APC/C components, notably Cdc16 and Cdc27, as well as Cdc20 to the centrosome, first, in fixed cells [67, 68] and, more recently, in live cells [70, 71, 126]. In reality, though, the localization of these proteins is both complex and dynamic with the APC/C and Cdc20 being found at a number of locations in mitosis including kinetochores and the cytoplasm. Furthermore, in *Drosophila* embryos, Cdc16 and Cdc27 do not show entirely overlapping patterns of localization or complex size raising the possibility that there may be multiple, distinct versions of the APC/C operating at different sites [69].

Although present at many sites throughout the cell, there is good evidence that the APC/C is first activated towards cyclin B at the spindle poles. Live cell imaging of GFP-tagged cyclin B destruction in HeLa cells [72], cellularized *Drosophila* embryos [126] and yeast cells [73] reveals a loss of protein that starts at the spindle pole before moving in a wave along the rest of the spindle and only then onto the cytoplasm (or nucleus in yeast). Perhaps even more exciting is the demonstration that, in a particular *Drosophila* mutant (*cfo*), centrosomes detach from their spindles and whilst this does not prevent destruction of cyclin B at the spindle poles, there is no destruction on spindles and embryos arrest in anaphase [74]. This experiment provides powerful evidence that destruction of cyclin B begins on spindle poles and requires an intact physical connection to the spindle to propagate the wave of destruction. It is also possible that a checkpoint is activated in response to centrosome detachment preventing further destruction of cyclin B. Intriguingly, in syncytial (early stage) *Drosophila* embryos, cyclin B is only destroyed on the spindle, whereas it remains present in the cytoplasm [71]. This spatially restricted pattern of destruction suggests that it cannot be the global activation of the APC/C itself that controls the timing of cyclin B destruction, since the APC/C is not spatially restricted. Instead, it is possible that spatial restriction of Cdc20 could be critical. In fact, Cdc20 does appear to be restricted to the spindle during the syncytial stage of insect cell development leading Raff and colleagues to propose that this is why destruction of cyclin B is limited to the spindle in these embryos [71]. Furthermore, they hypothesize that destruction of cyclin B throughout the rest of the cell might depend upon Cdh1 which is only expressed after cellularization, although contrary to the proposed model, Cdh1 is highly concentrated on centrosomes throughout the cell cycle. It remains to be tested whether such spatial restriction of Cdc20 and Cdh1 can explain the temporal pattern of cyclin B destruction in adult vertebrate cells where Cdc20 binds to the APC/C before Cdh1.

Another possibility is that Cdc20 is released from checkpoint complexes in the vicinity of the spindle pole thereby making this the first place that APC/C-Cdc20 can form (Figure 8.2). Following microtubule attachment at the kinetochore

Figure 8.2 Spatial regulation of cyclin B1 destruction at the metaphase/anaphase transition. ▶ In early metaphase, cyclin B1 is stable and localized to the spindle pole and spindle fibres. Live cell studies in *Drosophila* embryos indicate that the APC/C ubiquitin ligase is present at spindle poles at this time but is inactive due to the absence of the essential adaptor subunit Cdc20. Cdc20 is thought to be assembled into inhibitory complexes with spindle checkpoint proteins including Mad2, BubR1 and Bub3 at unattached kinetochores. Following microtubule attachment, these complexes are transported along spindle fibers towards spindle poles by minus end-directed motors such as cytoplasmic dynein. Once at the poles, the checkpoint complexes are somehow disassembled allowing Cdc20 to bind and activate the APC/C. For correct operation of the checkpoint, release of Cdc20 and activation of the APC/C depends upon complete attachment of all kinetochores to the spindle. It is not clear how or where this is controlled, but time-lapse imaging in human cells and *Drosophila* embryos reveals that destruction of cyclin B1 is first observed at spindle poles before spreading outwards along spindle fibres.

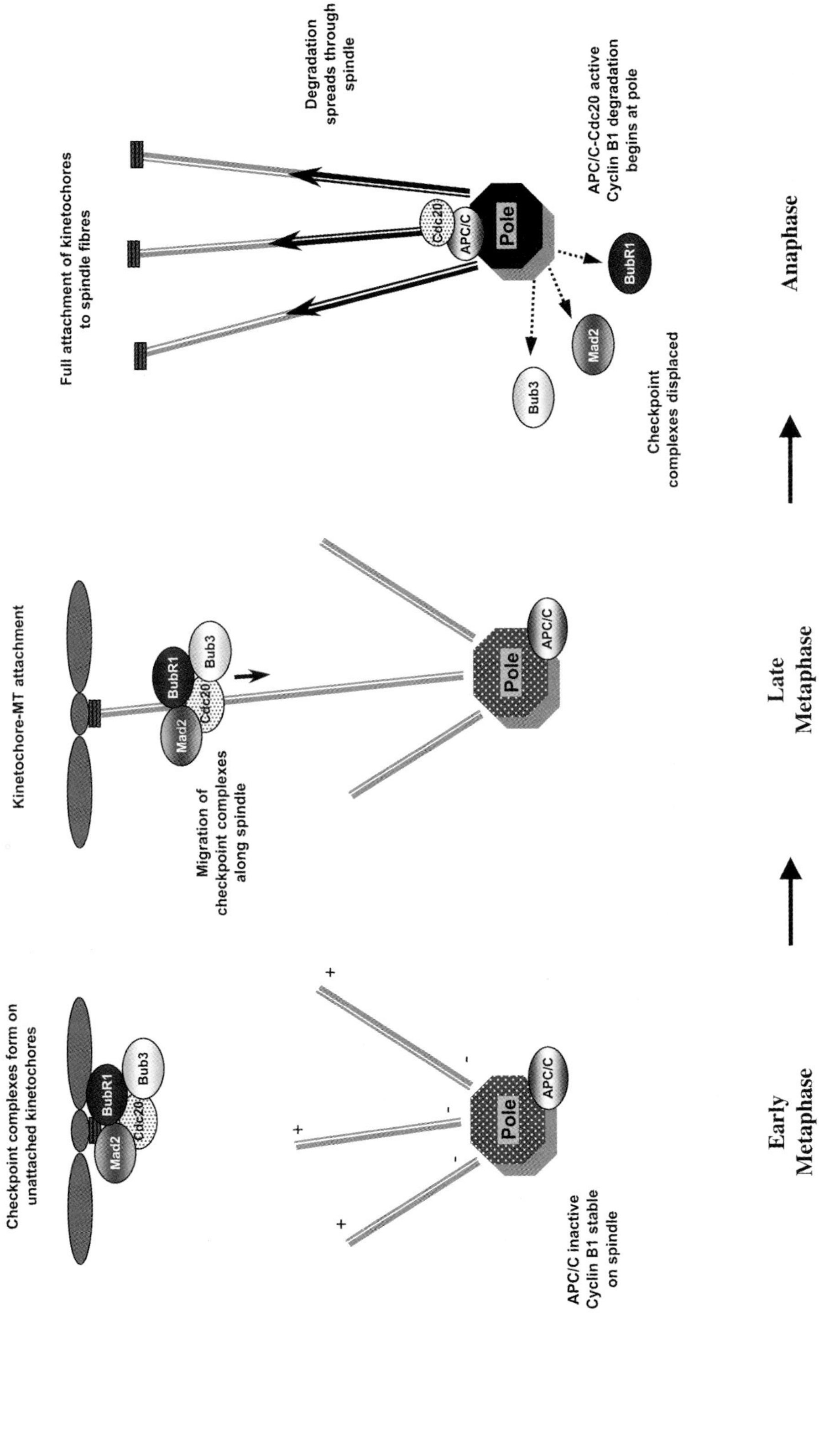
Checkpoint complexes form on unattached kinetochores
Mad2
Cdc20
BubR1
Bub3
+
-
Pole
APC/C
APC/C inactive Cyclin B1 stable on spindle
Early Metaphase
Kinetochore-MT attachment
Migration of checkpoint complexes along spindle
Late Metaphase
Full attachment of kinetochores to spindle fibres
Degradation spreads through spindle
APC/C-Cdc20 active Cyclin B1 degradation begins at pole
Checkpoint complexes displaced
Anaphase

many checkpoint proteins including Mad2, BubR1, CENP-E, Rod and Zw10, exhibit unidirectional migration from the kinetochores to the spindle poles along spindle fibers [75–77]. Based on the rate of these movements, it is believed that they are mostly driven by the minus end-directed microtubule motor, cytoplasmic dynein [77, 78]. This may be a mechanism to disseminate the checkpoint complexes throughout the spindle or to turn off the checkpoint following microtubule attachment. The next question though is how the Cdc20-checkpoint protein complexes are disassembled. Cdc20 is phosphorylated in early mitosis by Cdk1 and MAPK to promote its association with spindle checkpoint proteins and prevent it from binding the APC/C [79]. If the dephosphorylation of Cdc20 that followed checkpoint inactivation occurred primarily at the spindle pole this would lead to disassembly of the checkpoint complexes and restricted formation of APC/C–Cdc20 at this site. Currently, this is pure speculation but could perhaps be addressed with Cdc20 phosphosite-specific antibodies in a similar approach to that described above for showing that Cdk1–cyclin B1 activation occurs first at the centrosome.

Finally, as well as binding of Cdc20, activation of the APC/C requires phosphorylation of APC/C subunits by Cdk1 and Plk1, and dephosphorylation of sites phosphorylated by PKA [80–82]. The localization of Cdk1 and Plk1 to centrosomes during mitosis has already been discussed, so what about PKA? A fraction of PKA clearly localizes to interphase centrosomes [83, 84] as a result of binding to A-kinase anchoring proteins including AKAP450 (also known as AKAP350 or CG-NAP) and pericentrin that are concentrated at the centrosome [85]. Displacement of specific pools of PKA from mitotic spindle poles may involve a shift in binding preference from centrosomal to non-centrosomal AKAPs [86]. Equally important is the localization of the phosphatase, possibly PP1, which removes the phosphates added by PKA. So, although we argued above that APC/C proteins are not spatially restricted, they could still be locally activated by changes in their phosphorylation state.

Clearly, there are many experiments that still need to be done to prove whether or not the APC/C is activated first at spindle poles and, if so, to determine the mechanism for this and whether it relates to Cdc20 localization, activation or possibly phosphorylation of the APC/C. The above discussion has mostly focused on the destruction of cyclin B. Yet the destruction of other substrates may depend upon activation of the APC/C at other sites. Securin is localized throughout the cell as well as on the spindle in mitosis and, temporally, destruction of securin is coincident with that of cyclin B, at least within the constraints of current time-lapse imaging [87]. The localization pattern therefore does not preclude the possibility that securin destruction is initiated at spindle poles, but equally there is no strong evidence to say that it is. Despite this current gap in our understanding, there is growing acceptance that mitotic protein destruction is spatially regulated and that the spindle poles have an important role to play at least in initiating the destruction of cyclin B.

8.6
The Contribution of Centrosomes to Cytokinesis

Following the separation of chromosomes, cytokinesis, or division of cytoplasm, must occur to ensure an equal distribution of genetic material to the two daughter cells (reviewed in [88]). The spatial cues for cytokinesis are coordinated with chromosome segregation as the orientation of cell division is determined by the position of the mitotic spindle [89]. The first visible sign of cytokinesis is the formation of an acto-myosin based contractile ring, which forms perpendicular to the central spindle in late anaphase. As this begins to constrict, the plasma membrane invaginates and the cleavage furrow appears, a process that requires synthesis of new plasma membrane. As the furrow further constricts, the microtubule bundles of the central spindle become confined to the ill-defined structure known as the midbody that connects the dividing cells. The final step of mitosis is abscission, when the last remnants of cytoplasmic connections are broken to produce two identical daughter cells, signaling the end of cell division. The molecular processes of cytokinesis and abscission are complex and still far from understood. Intriguingly, though, there is now a wealth of evidence that implicates the centrosome in a number of distinct events that ultimately lead to cytokinesis (summarized in Figure 8.3A).

Using different technologies, three groups recently asked whether cells lacking centrosomes can complete cytokinesis. Firstly, Khodjakov and Rieder used highly focused lasers to selectively obliterate centrosomes (see Chapter 10). Surprisingly, this did not prevent formation of a bipolar spindle [90], but it did interfere with spindle orientation presumably due to loss of astral microtubules [91]. The consequence of having spindles that lacked cortical attachment was incomplete chromosome separation and the formation of thin chromatin bridges connecting the daughter nuclei, a feature known to inhibit cytokinesis [92]. Indeed, 30–50 % of cells with laser-ablated centrosomes failed to complete cytokinesis. Secondly, Hinchcliffe and Sluder used needle microsurgery to remove centrosomes together with a portion of cytoplasm from BSC-1 cells. Again, a significant fraction of the acentrosomal karyoplasts were delayed in mitosis and failed to complete cytokinesis [51]. Thirdly, Piel and Bornens showed that an acentrosomal *Drosophila* cell line, 1182-4, frequently exhibited incomplete cytokinesis leading to the accumulation of two or more connected interphase cells [93]. Taken together, these independent experimental approaches clearly indicate that centrosomes are essential for a robust separation of chromosomes, which in turn is needed for subsequent progression through cytokinesis. However, alone they do not necessarily reveal an intrinsic role for the centrosome in the biochemical pathways leading to cytokinesis.

In the context of this chapter, we are particularly interested in whether the centrosome acts as a signaling platform to direct events leading to cytokinesis, beyond simply determining the extent of chromosome separation or the plane of cell division. A more direct role for the centrosome in coordinating the timing of cell abscission is suggested by the behavior of individual centrioles during late

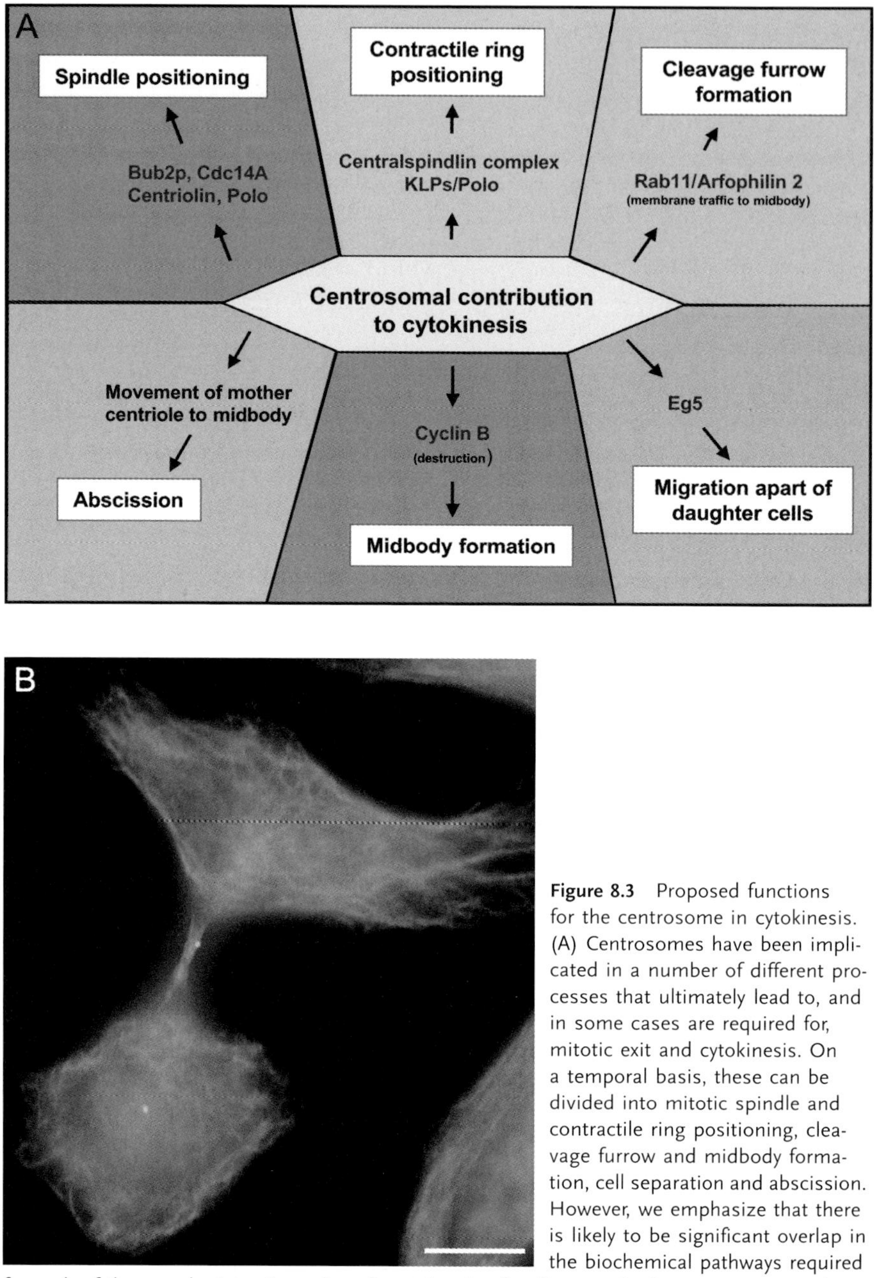

Figure 8.3 Proposed functions for the centrosome in cytokinesis. (A) Centrosomes have been implicated in a number of different processes that ultimately lead to, and in some cases are required for, mitotic exit and cytokinesis. On a temporal basis, these can be divided into mitotic spindle and contractile ring positioning, cleavage furrow and midbody formation, cell separation and abscission. However, we emphasize that there is likely to be significant overlap in the biochemical pathways required for each of these endpoints. Examples of proteins that localize to mitotic centrosomes and are implicated in these pathways are indicated in dark blue. (B) One of the most intriguing questions relating to the role of the centrosome in cytokinesis is why the mother centriole migrates towards the midbody prior to cell abscission. HeLa cells are shown following methanol fixation and staining with antibodies against α-tubulin (green) and γ-tubulin (red). DNA is stained with Hoechst 33258 (blue). Scale bar, 10 μm (see Color Plates page XXVI).

mitosis. Careful imaging of fixed and live cells revealed unexpectedly that just prior to cytokinesis there is a dramatic splitting of the two centrioles in each spindle pole [93, 94]. The mother centriole moves into the midbody, whilst the daughter centriole remains stationary in the center of its respective cytoplasm (see Figure 8.3B). As the mother centriole moves back into the cell center, cytokinesis is completed. The regulation of the movement of the mother centriole is not fully understood but the protein kinase p160^{ROCK} (Rho-associated kinase) may be required since a small molecule inhibitor of this kinase can trigger premature migration of the mother centriole to the midbody and early mitotic exit [95]. These movements also depend on remodeling of the post-anaphase microtubule network as addition of nocodazole causes the mother centriole to remain at the midbody, inhibiting abscission, while removal of nocodazole causes immediate abscission [93] Thus, in certain cell types, mother centriole movements correlate very closely with the timing of cell abscission, although whether they are a necessary prerequisite still remains to be proven.

What could be the purpose of these dramatic centriolar migrations? One possibility is that as the mother centriole migrates away from the midbody, the interphase cytoskeleton is established allowing the generation of opposite forces which propel migrating cells apart [96]. However, during tissue morphogenesis in the animal, cell division and cytokinesis do not usually require cell migration. An alternative hypothesis is that the mother centriole actively transports signaling factors necessary for abscission into close proximity with the midbody. In fact a number of important regulatory proteins localize to the centrosome during early mitosis and then move to the midbody during late mitosis. These include protein kinases, such as polo, and motor proteins, such as the kinesins Eg5 or MKlps [12, 97–99]. MKlp1 can bind MgcRacGAP (CYK-4 in *C. elegans*), a Rho family GTPase-activating protein, to make the centralspindlin complex. This complex is thought to mediate the microtubule bundling that occurs in the central region of dividing cells [100]. It has been proposed that, in *Drosophila,* the centralspindlin complex of Pav-KLP (the homolog of MKlp1) and RacGAP50C interacts with Pebble, a Rho1-GEF (guanine nucleotide exchange factor), and together this trimeric complex somehow positions the contractile ring and coordinates cytoskeletal remodelling during cytokinesis [101].

The MKlps also physically associate with and are phosphorylated by polo kinases and these processes may be promoted by centrosomal recruitment. Mutations in *pavarotti*, the *Drosophila* gene encoding Pav-KLP, or depletion of MKlp2 in mammalian cells leads not only to mislocalization of polo, but also to failure of cytokinesis [97, 102]. Likewise, disruption of polo activity either through genetic mutation or the use of siRNA oligonucleotides results in a failure to complete cytokinesis [103, 104]. The mechanism by which polo regulates cytokinesis is not fully understood although evidence suggests that polo-dependent phosphorylation of NudC (nuclear distribution gene C) is required [105]. Polo kinases also have an important role to play in activating the mitotic exit network in yeast (see below). Thus, a major function of the kinesin motors may be to transport polo, and other regulators, from the centrosome to the midbody. However, as this transport can occur along microtubules, it still does not explain why the mother centriole itself should need to visit the site of abscission.

Apart from the centrosome, other cellular organelles including the Golgi complex and endoplasmic reticulum contribute to the regulation of late mitotic events. In particular, it seems likely that endosomal pathways are required to deliver membrane to the site of cleavage furrow formation. However, even in this process, the centrosome may have a role to play. Arfophilin-2, an ADP ribosylation factor binding protein, is implicated in cytokinesis due to sequence homology with *Drosophila* nuclear fallout, a centrosomal protein implicated in cellularization and cytokinesis [106]. Arfophilin-2 binds Rab11, another protein implicated in regulating traffic through the recycling endosome compartment. Importantly, RNA-mediated interference of Rab11 in *C. elegans* leads to specific regression of the cleavage furrow at the final stage of abscission [107]. Both arfophilin-2 and Rab11 have been localized to the perinuclear region in the vicinity of the centrosome implying that centrosomes may contribute to cytokinesis through integrating distinct signals in the endosomal recycling pathway [108].

The completion of cytokinesis in fungal cells is absolutely dependent upon a checkpoint mechanism called the mitotic exit network (MEN) in budding yeast and septum initiation network (SIN) in fission yeast (reviewed in [109, 110]). These checkpoints operate though GTPase-regulated protein kinase cascades, many components of which are associated with the SPBs. Hence, it is entirely plausible that the SPB is acting as a solid phase platform to promote these signaling events in much the same way as described earlier for control of the G2/M transition. Importantly, the mitotic exit checkpoint in budding yeast is dependent on the cellular position of the SPB ensuring that mitotic exit and cytokinesis only occur after migration of the nucleus into the bud. The GTPase Tem1p binds to the spindle pole that migrates into the bud via the Bfa1p–Bub2p GAP complex and is kept inactive until late anaphase. Both Tem1p and Bub1p are associated with the daughter SPB. At this point Lte1p, a putative GEF, is released from the cortex of the bud and activates Tem1p, which binds cdc15, Dbf2 and Mob1 triggering the release of the phosphatase Cdc14p from the nucleolus. This in turn dephosphorylates Cdk substrates, promoting Cdk inactivation and allowing mitotic exit. Defects in cytoplasmic microtubule interactions with the cell cortex and misalignment of the spindle delays Tem1p activation and mitotic exit, thus coordinating cell cycle progression with spindle positioning [111] (see also Chapter 4).

In animal cells, misaligned spindles also delay mitotic progression, raising the possibility that there is conservation of these processes between SPBs and centrosomes, and possibly a conserved spindle positioning checkpoint [112]. Evidence for this comes from the existence of mammalian homologs of some of the MEN components such as Cdc14p, Bub2p and Mob1p [113, 114]. Human Cdc14A phosphatase localizes to the centrosome and its overexpression causes chromosome segregation defects and cytokinesis failure [115, 116]. Centriolin, a novel protein that localizes to the mother centriole as well as the midbody, shares a limited region of homology with the budding yeast MEN component Nud1p, which anchors the MEN complex to the SPB through direct interactions with Bub2p [117]. Depletion of centriolin by siRNAs causes cytokinesis failure and, ultimately, G1 arrest with chains of cells remaining interconnected by long intercellular bridges. These data support the idea that mammalian cells may possess a regulatory pathway similar to the MEN/SIN that coordinates the final stages of cell division. It is intriguing to speculate that the dependency of fungal cytokinesis on SPB positioning in some way reflects the way that abscission in mammalian cells may be dependent on the repositioning of the mother centriole. Hence, the mother centriole could anchor a regulatory pathway that controls the final stages of mitosis and promotes cytokinesis.

Clearly, the centrosome is intimately involved in late mitotic events. The challenge now is to understand how the centrosome contributes to cytokinesis at the molecular level. We discussed earlier how cyclin B1 destruction is initiated at the centrosome, and it has long been known that failure to degrade cyclin B1 prevents midbody formation and ultimately cytokinesis [118]. However, it is highly unlikely that the sole purpose of the centrosome with respect to cytokinesis is to degrade cyclin B1. The impressive migration of the mother centriole alone suggests a much more direct role for this organelle perhaps in transporting proteins to their site of action at the midbody. These proteins may form complexes with other components of the same signaling pathway whilst still at the centrosome or else after they arrive at the midbody. Either way, this would provide a mechanism whereby active complexes only exist when the inactive constituents come together at a specific localization within the cell, thus regulating the spatial and temporal aspects of the signaling cascade and checkpoints involved in transit through cytokinesis.

8.7
A Role for Centrosomes in G1/S Progression?

Perhaps the most startling finding to arise from recent experiments on centrosome function is the apparent dependency on centrosomes of the G1/S transition. The same techniques used to define a cytokinesis function have also revealed a centrosome requirement for S-phase entry [55, 91]. As has already been discussed, the removal of the centrosome by microsurgery or laser ablation leads to acentrosomal cells which exhibit a prolonged mitosis and frequent failure of cytokinesis. However, those cells which do make it through cytokinesis never progress into the subsequent S-phase in the absence of centrosomes. This is most elegantly demonstrated if only one centrosome is destroyed by laser microsurgery during prophase. In this case, the acentrosomal daughter cell becomes arrested in G1 phase, whereas the centrosome-containing offspring progresses to the next mitosis [91]. Acentrosomal G1 cells are capable of assembling a microtubule organizing center containing γ-tubulin and pericentrin, but not centrioles [5]. This raises two intriguing possibilities: (i) the presence of a checkpoint that monitors the existence of core centrosomal structures such as centrioles, or (ii) a dependency on the centrosome for G1/S promoting activity.

A less attractive possibility, but one which must be considered, is that cytokinesis was never properly completed in these experiments despite the appearance of post-mitotic cells. In this case, failure to completely divide may result in thin chromosomal or cytoplasmic bridges that are sufficient to activate a checkpoint. Certainly, a p53-dependent post-mitotic checkpoint has been well characterized in response to tetraploidization and it is possible that this is also activated in cells that are still partially connected [119]. So the critical question is whether the lack of a centrosome prevents S-phase entry in cells that have completed mitotic division events without error. In the experiments using needle microsurgery to remove the centrosome, cells were not synchronized and so one might expect some of these to have been in G1 when their centrosome was removed. A small fraction of these cells never did progress to the subsequent mitosis leaving open the possibility that they were arrested prior to G1/S [55]. Likewise, laser ablation of centrosomes has been performed in G1 cells but it was not reported whether these cells entered S phase or not [120]. Centriole disassembly has also been induced by microinjection of polyglutamylated tubulin antibodies and this did not appear to cause a G1/S arrest, but whether the centrioles were completely disrupted is hard to tell [121]. Experiments in one specialized cell system, fertilized mouse oocytes, do support a requirement for the centrosome for S-phase progression. In these cells, assembly of the zygotic centrosome depends upon maternally-derived centrosomal material coalescing into a functional microtubule organizing center. Antibody inhibition of the PCM-1 protein prevents the assembly of this maternal centrosome and, concomitantly, leads to an interphase arrest [122].

The idea of a G1 checkpoint that monitors centriole number or integrity is an attractive one but there is currently little experimental data to support it. It is already clear that the centriole duplication cycle is carefully integrated with the chromosomal replication cycle with both events depending upon activation of Cdk2 and inactivation of Rb [10]. It is possible then that G1 checkpoint proteins such as Rb, or indeed p53, may respond to loss of centrioles. The important question this raises is what molecular components of the centriole is the checkpoint monitoring? A number of proteins have been identified which are mother centriole-specific, including ε-tubulin, cenexin, centriolin and ninein. On the basis that cells should always possess a mother centriole, any one of these proteins could signal the presence of centrioles if its localization to the mother centriole led to its activation or stabilization.

The alternative to the checkpoint hypothesis is that centrosomes are required in a positive fashion to promote pathways required for G1/S progression. This may occur through recruitment or concentration of molecules that are essential for the initiation of DNA synthesis in a similar manner to that already described for G2/M entry. Cyclin E, which activates Cdk2 at the G1/S transition, is concentrated in the region of the centrosome in *Xenopus* embryos [123] and, presumably Cdk2 must come into contact with the G1 centrosome to phosphorylate substrates such as nucleophosmin [124]. The importance of the phosphorylation status of Cdk2 for its activation is not as well studied as that for Cdk1, but it is possible that members of the Cdc25 family and Plk family have a role to play. The fact that embryonic and adult cell extracts as well the *Drosophila* acentrosomal cell line can cycle from G1 into S phase implies that there is no absolute requirement for a centrosome function in this cell cycle transition. However, as for the G2/M transition, one can still speculate that the centrosome facilitates efficient activation of the G1/S transition via a scaffolding function.

8.8
In Conclusion

The emerging theme that we have tried to emphasize in this chapter is that the centrosome is more than just a microtubule organizing center, playing a vital role in controlling cell cycle transitions in both mitosis and interphase. We have attempted to show how the centrosome might facilitate cell cycle transitions by acting as a multivalent signaling platform that ensures switches required to trigger the next phase of the cell cycle are flipped in an efficient and irreversible manner. In essence, we are proposing that the centrosome performs a scaffolding function for integrating, regulating and amplifying signaling pathways that control cell cycle transitions. At the molecular level, this means that the primary function of some, and maybe many, of the large coiled-coil proteins that inhabit the centrosome, is to provide binding surfaces for regulatory enzymes. Indeed, scaffold proteins have taken on great importance in understanding how signaling pathways such as the MAPK pathways integrate and respond to different extracellular cues [125]. The detection of signaling molecules such as PKA, Ca^{2+}/CaM-dependent protein kinase, PI-3-K, fyn, PKC-θ and casein kinase I raise the possibility that the centrosome also plays a part in the regulation of externally-derived signal transduction events, perhaps to promote G1/S progression. Unraveling the significance of the centrosome in receptor-based cell signaling may yet be a rich vein for future centrosome research. Certainly, it would seem that the days when biochemical events leading to cell signaling and cell cycle transitions were thought to take place in a cytosolic soup, are numbered.

Acknowledgments

We would like to thank members of our laboratory for useful discussion and comments on the manuscript. The authors' laboratory is funded by grants from The Wellcome Trust, Cancer Research UK and the AICR; A. M.F. is a Lister Institute Research Fellow.

References

1. Pines, J. **1999**. Four-dimensional control of the cell cycle. *Nature Cell Biol* **1,** E73–E79.
2. Bornens, M. **2002**. Centrosome composition and microtubule anchoring mechanisms. *Curr Opin Cell Biol* **14,** 25–34.
3. Doxsey, S. **2001**. Re-evaluating centrosome function. *Nature Revs: Mol Cell Biol* **2,** 688–698.
4. Francis, S. E., and Davis, T. N. **2000**. The spindle pole body of Saccharomyces cerevisiae: architecture and assembly of the core components. *Curr Top Dev Biol* **49,** 105–132.
5. Hagan, I. M., and Petersen, J. **2000**. The microtubule organizing centers of *Schizosaccharomyces pombe*. *Curr Top Dev Biol* **49,** 133–159.
6. Brinkley, B. R., and Goepfert, T. M. **1998**. Supernumerary centrosomes and cancer: Boveri's hypothesis resurrected. *Cell Motil Cytoskeleton* **41,** 281–288.
7. Mayor, T., Meraldi, P., Stierhof, Y.-D., Nigg, E. A., and Fry, A. M. **1999**. Protein kinases in control of the centrosome cycle. *FEBS Lett* **452,** 92–95.
8. Doxsey, S. J. **2001**. Centrosomes as command centres for cellular control. *Nature Cell Biol* **3,** E105–E108.
9. Rieder, C. L., Faruki, S., and Khodjakov, A. **2001**. The centrosome in vertebrates: more than a microtubule organizing centre. *Trends Cell Biol* **11,** 413–419.
10. Hinchcliffe, E. H., and Sluder, G. (**2001**). "It takes two to tango": understanding how centrosome duplication is regulated throughout the cell cycle. *Genes Dev* **15,** 1167–1181.
11. Blagden, S. P., and Glover, D. M. **2003**. Polar expeditions – provisioning the centrosome for mitosis. *Nature Cell Biol* **5,** 505–511.
12. Golsteyn, R. M., Mundt, K. E., Fry, A. M., and Nigg, E. A. **1995**. Cell cycle regulation of the activity and subcellular localization of PLK1, a human protein kinase implicated in mitotic spindle function. *J Cell Biol* **129,** 1617–1628.
13. Merdes, A., Ramyar, K., Vechio, J. D., and Cleveland, D. W. **1996**. A complex of NuMA and cytoplasmic dynein is essential for mitotic spindle assembly. *Cell* **87,** 447–458.
14. Fry, A. M., Mayor, T., Meraldi, P., Stierhof, Y.-D., Tanaka, K., and Nigg, E. A. **1998**. C-Nap1, a novel centrosomal coiled-coil protein and candidate substrate of the cell cycle-regulated protein kinase Nek2. *J Cell Biol* **141,** 1563–1574.
15. Casenghi, M., Meraldi, P., Weinhart, U., Duncan, P. I., Korner, R., and Nigg, E. A. **2003**. Polo-like kinase 1 regulates Nlp, a centrosome protein involved in microtubule nucleation. *Dev Cell* **5,** 113–125.
16. Piel, M., Meyer, P., Khodjakov, A., Rieder, C. L., and Bornens, M. **2000**. The respective contributions of the mother and daughter centrioles to centrosome activity and behaviour in vertebrate cells. *J Cell Biol* **149,** 317–329.
17. McIntosh, J. R., Grishchuk, E. L., and West, R. R. **2002**. Chromosome-microtubule interactions during mitosis. *Ann Rev Cell Dev Biol* **18,** 193–219.
18. Brinkley, B. R. **2001**. Managing the centrosome numbers game: from chaos to stability in cancer cell division. *Trends Cell Biol* **11,** 18–21.
19. Nigg, E. A. **2002**. Centrosome aberrations: cause or consequence of cancer progression? *Nature Rev Cancer* **2,** 815–825.
20. Pihan, G. A., and Doxsey, S. J. **1999**. The mitotic machinery as a source of genetic instability in cancer. *Semin Cancer Biol* **9,** 289–302.
21. Heald, R., Tournebize, R., Blank, T., Sandaltzopoulos, R., Becker, P., Hyman, A., and Karsenti, E. **1996**. Self-organization of microtubules into bipolar spindle around artifical chromosomes in *Xenopus* egg extracts. *Nature* **382,** 420–425.

22. Rodionov, V. I., and Borisy, G. G. **1997**. Self-centering activity of cytoplasm. *Nature* **386**, 170–173.

23. Debec, A., Detraves, C., Montmory, C., Geraud, G., and Wright, M. **1995**. Polar organization of gamma-tubulin in acentriolar mitotic spindles of *Drosophila melanogaster* cells. *J Cell Sci* **108**, 2645–2653.

24. Megraw, T. L., Kao, L.-R., and Kaufman, T. **2001**. Zygotic development without functional mitotic centrosomes. *Curr Biol* **11**, 116–120.

25. Kuntziger, T., and Bornens, M. **2000**. The centrosome and parthenogenesis. *Curr Top Dev Biol* **49**, 1–25.

26. Maller, J. M. **1995**. Maturation-promoting factor in the early days. *Trends Biochem Sci* **20**, 524–528.

27. Nurse, P. **2000**. A long twentieth century of the cell cycle and beyond. *Cell* **100**, 71–78.

28. O'Farrell, P. H. **2001**. Triggering the all-or-nothing switch into mitosis. *Trends Cell Biol* **11**, 512–519.

29. Ohi, R., and Gould, K. L. **1999** Regulating the onset of mitosis. *Curr Opin Cell Biol* **11**, 267–273.

30. Hoffmann, I., Clarke, P. R., Marcote, M. J., Karsenti, E., and Draetta, G. **1993**. Phosphorylation and activation of human Cdc25C by Cdc2-cyclin B and its involvement in the self-amplification of MPF at mitosis. *EMBO J* **12**, 53–63.

31. Izumi, T., and Maller, J. L. **1993**. Elimination of cdc2 phosphorylation sites in the cdc25 phosphatase blocks initiation of M-phase. *Mol Biol Cell* **4**, 1337–1350.

32. Kumagai, A., and Dunphy, W. G. **1996**. Purification and molecular cloning of Plx1, a cdc25-regulatory kinase from *Xenopus* egg extracts. *Science* **273**, 1377–1380.

33. Bartholomew, C. R., Woo, S. H., Chung, Y. S., Jones, C., and Hardy, C. F. **2001**. Cdc5 interacts with the Wee1 kinase in budding yeast. *Mol Cell Biol* **21**, 4949–4959.

34. Abrieu, A., Brassac, T., Galas, S., Fisher, D., Labbe, J. C., and Doree, M. **1998**. The Polo-like kinase Plx1 is a component of the MPF amplification loop at the G_2/M-phase transition of the cell cycle in *Xenopus* eggs. *J Cell Sci* **111**, 1751–1757.

35. Bailly, E., Doree, M., Nurse, P., and Bornens, M. **1989**. p34cdc2 is located in both nucleus and cytoplasm; part is centrosomally associated at G2/M and enters vesicles at anaphase. *EMBO J* **8**, 3985–3995.

36. Pockwinse, S. M., Krockmalnic, G., Doxsey, S. J., Nickerson, J., Lian, J. B., van Wijnen, A. J., Stein, J. L., Stein, G. S., and Penman, S. **1997**. Cell cycle independent interaction of CDC with the centrosome, which is associated with the nuclear matrix-intermediate filament scaffold. *Proc Natl Acad Sci USA* **94**, 3022–3027.

37. Verde, F., Labbe, J.-C., Doree, M., and Karsenti, E. **1990**. Regulation of microtubule dynamics by cdc2 protein kinase in cell-free extracts of *Xenopus* eggs. *Nature* **343**, 233–238.

38. Arnaud, L., Pines, J., and Nigg, E. A. **1998**. GFP-tagging reveals human Polo-like kinase 1 at the kinetochore/centromere region of mitotic chromosomes. *Chromosoma* **107**, 424–429.

39. Lee, K., Yuan, Y.-L., Kuriyama, R., and Erikson, R. **1995**. Plk is an M phase-specific protein kinase and interacts with a kinesin-like protein, CHO1/MKLP-1. *Mol Cell Biol* **15**, 7143–7151.

40. Logarinho, E., and Sunkel, C. E. **1998**. The *Drosophila* POLO kinase localizaes to multiple compartments of the mitotic apparatus and is required for the phosphorylation of MPM-2 reactive epitopes. *J Cell Sci* **111**, 2897–2909.

41. Mulvihill, D. P., Petersen, J., Ohkura, H., Glover, D. M., and Hagan, I. M. **1999**. Plo1 kinase recruitment to the spindle pole body and its role in cell division in *Schizosaccharomyces pombe*. *Mol Biol Cell* **10**, 2771–2785.

42. Qian, Y.-W., Erikson, E., Li, C., and Maller, J. **1998**. Activated polo-like kinase Plx1 is required at multiple points during mitosis in *Xenopus laevis*. *Mol Cell Biol* **18**, 4262–4271.

43. De Souza, C. P. C., Ellem, K. A. O., and Gabrielli, B. G. **2000**. Centrosomal and cytoplasmic Cdc2/cyclin B1 activation precedes nulcear mitotic events. *Exp Cell Res* **257**, 11–21.

44. Jackman, M., Lindon, C., Nigg, E. A., and J., P. **2003**. Active cyclin B1-Cdk1 first appears on centrosomes in prophase. *Nature Cell Biol* **5**, 143–148.

45. Toyoshima-Morimoto, F., Taniguchi, E., Shinya, N., Iwamatsu, A., and Nishida, E. **2001**. Plo-like kinase 1 phosphorylates cyclin B1 and targets it to the nucleus during prophase. *Nature* **410**, 215–220.

46. O'Connell, M. J., Krien, M. J. E., and Hunter, T. **2003**. Never say never. The NIMA-related protein kinases in mitotic control. *Trends Cell Biol* **13**, 221–228.

47. Wu, L., Osmani, S. A., and Mirabito, P. M. **1998**. A role for NIMA in the nuclear localization of cyclin B in *Aspergillus nidulans*. *J Cell Biol* **141**, 1575–1587.

48. Krien, M. J., West, R. R., John, U. P., Koniaras, K., McIntosh, J. R., and O'Connell, M. J. **2002**. The fission yeast NIMA kinase Fin1p is required for spindle function and nuclear envelope integrity. *EMBO J* **21**, 1713–1722.

49. Grallert, A., and Hagan, I. M. **2002**. *S. pombe* NIMA related kinase, Fin1, regulates spindle formation, and an affinity of Polo for the SPB. *EMBO J* **21**, 3096–3107.

50. Bridge, A. J., Morphew, M., Bartlett, R., and Hagan, I. M. **1998**. The fission yeast SPB component Cut12 links bipolar spindle formation to mitotic control. *Genes Dev* **12**, 927–942.

51. Fry, A. M. **2002**. The Nek2 protein kinase: a novel regulator of centrosome structure. *Oncogene* **21**, 6184–6194.

52. Faragher, A. J., and Fry, A. M. **2003**. Nek2A kinase stimulates centrosome disjunction and is required for formation of bipolar mitotic spindles. *Mol Biol Cell* **14**, 2876–2889.

53. Fry, A. M., Meraldi, P., and Nigg, E. A. **1998**. A centrosomal function for the human Nek2 protein kinase, a member of the NIMA-family of cell cycle regulators. *EMBO J* **17**, 470–481.

54. Maniotis, A., and Schliwa, M. **1991**. Microsurgical removal of centrosomes blocks cell reproduction and centriole regeneration in BSC-1 cells. *Cell* **67**, 495–504.

55. Hinchcliffe, E. H., Miller, F. J., Cham, M., Khodjakov, A., and Sluder, G. **2001**. Requirement of a centrosomal activity for cell cycle progression through G1 into S-phase. *Science* **291**, 1547–1550.

56. Perez-Mongiovi, D., Beckhelling, C., Chang, P., Ford, C. C., and Houliston, E. **2000**. Nuclei and microtubule asters stimulate maturation/M phase promoting factor (MPF) activation in Xenopus eggs and egg cytoplasmic extracts. *J Cell Biol* **150**, 963–974.

57. Picard, A., Karsenti, E., Dabauvalle, M. C., and Doree, M. **1987**. Release of mature starfish oocytes from interphase arrest by microinjection of human centrosomes. *Nature* **327**, 170–172.

58. Musacchio, A., and Hardwick, K. G. **2002**. The spindle checkpoint: structural insights into dynamic signalling. *Nature Rev Mol Cell Biol* **3**, 731–741.

59. Rieder, C. L., Khodjakov, A., Paliulis, L. V., Fortier, T. M., Cole, R. W., and Sluder, G. **1997**. Mitosis in vertebrate somatic cells with two spindles: implications for the metaphase/anaphase transition checkpoint and cleavage. *Proc Natl Acad Sci USA* **94**, 5107–5112.

60. Page, A. M., and Hieter, P. **1999**. The anaphase-promoting complex: new subunits and regulators. *Ann Rev Biochem* **68**, 583–609.

61. Nasmyth, K. **2002**. Segregating sister genomes: the molecular biology of chromosome separation. *Science* **297**, 559–565.

62. Morgan, D. O. **1999**. Regulation of the APC and the exit from mitosis. *Nature Cell Biol* **1**, E47–E53.

63. Hwang, L. H., Lau, L. F., Smith, D. L., Mistrot, C. A., Hardwick, K. G., Hwang, E. S., Amon, A., and A. W., M. **1998**. Budding yeast Cdc20: a target of the spindle checkpoint. *Science* **279**, 1041–1044.

64. Kim, S. H., Lin, D. P., Matsumoto, S., Kitazono, A., and Matsumoto, T. (1998). Fission yeast Slp1: an effector of the Mad2-dependent spindle checkpoint. *Science* **279**, 1045–1047.

65. Sudakin, V., Chan, G., and Yen, T. J. **2001**. Checkpoint inhibition of the APC/C in HeLa cells is mediated by a complex of BubR1, Bub3, Cdc20, and Mad2. *J Cell Biol* **154**, 925–936.

66. Tang, Z., Bharadwaj, R., Li, B., and Yu, H. **2001**. Mad2-independent inhibition of APC-Cdc20 by the mitotic checkpoint protein BubR1. *Dev Cell* **1**, 227–237.

67. Kallio, M. J., Weinstein, J., Daum, J. R., Burke, D. J., and Gorbsky, G. J. **1998**. Mammalian p55CDC mediates association of the spindle checkpoint protein Mad2 with the cyclosome/anaphase promoting complex and is involved in regulating anaphase onset and late mitotic events. *J Cell Biol* **141**, 1393–1406.

68. Tugendreich, S., Tomkiel, J., Earnshaw, W., and Hieter, P. **1995**. CDC27Hs colocalizes with CDC16Hs to the centrosome amd mitotic spindle amd is essential for the metaphase to anaphase transition. *Cell* **81**, 261–268.

69. Huang, J.-Y., and Raff, J. W. **2002**. The dynamic localization of the *Drosophila* APC/C: evidence for the existence of multiple complexes that perform distinct functions and are differentially localized. *J Cell Sci* **115**, 2847–2856.

70. Kallio, M. J., Beardmore, V. A., Weinstein, J., and Gorbsky, G. J. **2002**. Rapid microtubule-independent dymanics of Cdc20 at kinetochores and centrosomes in mammalian cells. *J Cell Biol* **158**, 841–847.

71. Raff, J. W., Jeffers, K., and Huang, J.-Y. **2002**. The roles of Fzy/Cdc20 and Fzr/Cdh1 in regulating the destruction of cyclin B in space and time. *J Cell Biol* **157**, 1139–1149.

72. Clute, P., and Pines, J. **1999**. Temporal and spatial control of cyclin B1 destruction in metaphase. *Nature Cell Biol* **1**, 82–87.

73. Yanagida, M., Yamashita, Y. M., Tatebe, H., Ishii, K., Kumada, K., and Nakaseko, Y. **1999**. Control of metaphase–anaphase progression by proteolysis: cyclosome function regulated by the protein kinase A pathway, ubiquitination and localization. *Phil Trans R Soc Lond B Biol Sci* **354**, 1559–1570.

74. Wakefield, J., Huang, J.-Y., and Raff, J. W. **2000**. Centrosomes have a role in regulating the destruction of cyclin B in early *Drosophila* embryos. *Curr Biol* **10**, 1367–1370.

75. Basto, R., Gomes, R., and Karess, R. E. **2000**. Rough deal and Zw10 are required for the metaphase checkpoint in *Drosophila*. *Nature Cell Biol* **2**, 939–943.

76. Chan, G. K., Jablonski, S. A., Starr, D. A., Goldberg, M. L., and Yen, T. J. **2000**. Human Zw10 and ROD are mitotic checkpoint proteins that bind to kinetochores. *Nature Cell Biol* **2**, 944–947.

77. Howell, B. J., McEwan, B. F., Canman, J. C., Hoffman, D. B., Farrar, E. M., Rieder, C. L., and Salmon, E. D. **2001**. Cytoplasmic dynein/dynactin drives kinetochore protein transport to the spindle poles and has a role in mitotic spindle checkpoint inactivation. *J Cell Biol* **155**, 1159–1172.

78. Wojcik, E., Basto, R., Serr, F., Scaerou, F., Karess, R. E., and Hays, T. **2001**. Kinetochore dynein: its dynamics and role in the transport of the rough deal checkpoint protein. *Nature Cell Biol* **3**, 1001–1007.

79. Chung, E., and Chen, R.-H. **2003**. Phosphorylation of Cdc20 is required for its inhibition by the spindle checkpoint. *Nature Cell Biol* **5**, 748–753.

80. Descombes, P., and Nigg, E. A. **1998**. The polo-like kinase Plx1 is required for M phase exit and destruction of mitotic regulators in *Xenopus* egg extracts. *EMBO J* **17**, 1328–1335.

81. Kotani, S., Tugendreich, S., Fujii, M., Jorgensen, P.-M., Watanabe, N., Hoog, C., Hieter, P., and Todokoro, K. **1998**. PKA and MPF-activated polo-like kinase regulate anaphase-promoting complex activity and mitosis progression. *Mol Cell* **1**, 371–380.

82. Shirayama, M., Zachariae, W., Ciosk, R., and Nasmyth, K. **1998**. The polo-like kinase Cdc5p and the WD-repeat protein Cdc20p/fizzy are regulators and substrates of the anaphase promoting complex in *Saccharomyces cerevisiae*. *EMBO J* **17**, 1336–1349.

83. De Camilli, P., Moretti, M., Donini, S. D., Walter, U., and Lohmann, S. M. **1986**. Heterogeneous distribution of the cAMP receptor protein RII in the nervous system: evidence for its intracellular accumulation on microtubules, microtubule-organizing centers, and in the area of the Golgi complex. *J Cell Biol* **103**, 189–203.
84. Nigg, E. A., Schäfer, G., Hilz, H., and Eppenberger, H. M. **1985**. Cyclic-AMP-dependent protein kinase type II is associated with the Golgi complex and with centrosomes. *Cell* **41**, 1039–1051.
85. Diviani, D., and Scott, J. D. **2001**. AKAP signalling complexes at the cytoskeleton. *J Cell Sci* **114**, 1431–1437.
86. Carlson, C. R., Witczak, O., Vossebein, L., Labbe, J.-C., Skalhegg, B. S., Keryer, G., Herberg, F. W., Collas, P., and Tasken, K. **2001**. CDK1-mediated phosphorylation of the RIIa regulatory subunit of PKA works as a molecular switch that promotes dissociation of RIIa from centrosomes at mitosis. *J Cell Sci* **114**, 3243–3254.
87. Hagting, A., den Elzen, N., Vodermaier, H. C., Waizenegger, I. C., Peters, J.-M., and Pines, J. **2002**. Human securin proteolysis is controlled by the spindle checkpoint and reveals when the APC/C switches from activation by Cdc20 to Cdh1. *J Cell Biol* **157**, 1125–1137.
88. Glotzer, M. **2001**. Animal cell cytokinesis. *Annu Rev Cell Dev Biol* **17**, 351–386.
89. Rappaport, R. **1996**. *Cytokinesis in Animal Cells*. Cambridge University Press: Cambridge.
90. Khodjakov, A., Cole, R. W., Oakley, B. R., and Rieder, C. L. **2000**. Centrosome-independent mitotic spindle formation in vertebrates. *Curr Biol* **10**, 59–67.
91. Khodjakov, A., and Rieder, C. L. **2001**. Centrosomes enhance the fidelity of cytokinesis in vertebrates and are required for cell cycle progression. *J Cell Biol* **153**, 237–242.
92. Mullins, J. M., and Biesele, J. J. **1977**. Terminal phase of cytokinesis in D-98s cells. *J Cell Biol* **73**, 672–684.
93. Piel, M., Nordberg, J., Euteneuer, U., and Bornens, M. **2001**. Centrosome-dependent exit of cytokinesis in animal cells. *Science* **291**, 1550–1553.
94. Mack, G., and Rattner, J. B. **1993**. Centrosome repositioning immediately following karyokinesis and prior to cytokinesis. *Cell Motil Cytoskeleton* **26**, 239–247.
95. Chevrier, V., Piel, M., Collomb, N., Saoudi, Y., Frank, R., Paintrand, M., Narumiya, S., Bornens, M., and Job, D. **2002**. The Rho-associated protein kinase p160ROCK is required for centrosome positioning. *J Cell Biol* **157**, 807–817.
96. Ou, Y. Y., and Rattner, J. B. **2002**. Post-karyokinesis centrosome movement leaves a trail of unanswered questions. *Cell Motil Cytoskeleton* **51**, 123–132.
97. Adams, R. R., Tavares, A. A., Salzberg, A., Bellen, H. J., and Glover, D. M. **1998**. pavarotti encodes a kinesin-like protein required to organize the central spindle and contractile ring for cytokinesis. *Genes Dev* **12**, 1483–1494.
98. Sawin, K., and Mitchison, T. **1995**. Mutations in the kinesin-like protein Eg5 disrupting localization to the mitotic spindle. *Proc Natl Acad Sci USA* **92**, 4289–4293.
99. Uzbekov, R., Prigent, C., and Arlot-Bonnemains, Y. **1999**. Cell cycle analysis and synchronization of the *Xenopus laevis* XL2 cell line: study of the kinesin related protein XlEg5. *Microsc Res Tech* **45**, 31–42.
100. Mishima, M., Kaitna, S., and Glotzer, M. **2002**. Central spindle assembly and cytokinesis require a kinesin-like protein/RhoGAP complex with microtubule bundling activity. *Dev Cell* **2**, 41–54.
101. Somers, W. G., and Saint, R. **2003**. A RhoGEF and Rho family GTPase-activating protein complex links the contractile ring to cortical microtubules at the onset of cytokinesis. *Dev Cell* **4**, 29–39.
102. Neef, R., Preisinger, C., Sutcliffe, J., Kopajtich, R., Nigg, E. A., Mayer, T. U., and Barr, F. A. **2003**. Phosphorylation of mitotic kinesin-like protein 2 by

polo-like kinase 1 is required for cytokinesis. *J Cell Biol* **162**, 863–875.

103. Carmena, M., Riparbelli, M. G., Minestrini, G., Tavares, A. M., Adams, R., Callaini, G., and Glover, D. M. **1998**. *Drosophila* polo kinase is required for cytokinesis. *J Cell Biol* **143**, 659–671.

104. Liu, X., and Erikson, R. L. **2002**. Activation of Cdc2/cyclin B and inhibition of centrosome amplification in cells depleted of Plk1 by siRNA. *Proc Natl Acad Sci USA* **99**, 8672–8676.

105. Zhou, T., Aumais, J. P., Liu, X., Yu-Lee, L. Y., and Erikson, R. L. **2003**. A role for Plk1 phosphorylation of NudC in cytokinesis. *Dev Cell* **5**, 127–138.

106. Rothwell, W. F., Fogarty, P., Field, C. M., and Sullivan, W. **1998**. Nuclear-fallout, a *Drosophila* protein that cycles from the cytoplasm to the centrosomes, regulates cortical microfilament organisation. *Development* **125**, 1295–1303.

107. Skop, A. R., Bergmann, D., Mohler, W. A., and White, J. G. **2001**. Completion of cytokinesis in *C. elegans* requires a brefeldin A-sensitive membrane accumulation at the cleavage furrow apex. *Curr Biol* **11**, 735–746.

108. Hickson, G. R., Matheson, J., Riggs, B., Maier, V. H., Fielding, A. B., Prekeris, R., Sullivan, W., Barr, F. A., and Gould, G. W. **2003**. Arfophilins are dual Arf/Rab 11 binding proteins that regulate recycling endosome distribution and are related to *Drosophila* nuclear fallout. *Mol Biol Cell* **14**, 2908–2920.

109. McCollum, D., and Gould, K. L. **2001**. Timing is everything: regulation of the mitotic exit network and cytokinesis by the MEN and SIN. *Trends Cell Biol* **11**, 89–95.

110. Pereira, G., and Schiebel, E. **2001**. The role of the yeast spindle pole body and the mammalian centrosome in regulating late mitotic events. *Curr Opin Cell Biol* **13**, 762–769.

111. Pereira, G., Hofken, T., Grindlay, J., Manson, C., and Schiebel, E. **2000**. The Bub2p spindle checkpoint links nuclear migration with mitotic exit. *Mol Cell* **6**, 1–10.

112. O'Connell, C. B., and Wang, Y. L. **2000**. Mammalian spindle orientation and position respond to changes in cell shape in a dynein-dependent fashion. *Mol Biol Cell* **11**, 1765–1774.

113. Richardson, P., and Zon, L. **1995**. Molecular cloning of a cDNA with a novel domain present in the tre-2 oncogene and the yeast cell cycle regulators BUB2 and cdc16. *Oncogene* **11**, 1139–1148.

114. Wang, Y., Hu, F., and Elledge, S. **2000**. The Bfa1/Bub2 GAP complex comprises a universal checkpoint required to prevent mitotic exit. *Curr Biol* **10**, 1379–1382.

115. Kaiser, B. K., Zimmerman, Z. A., Charbonneau, H., and Jackson, P. K. **2002**. Disruption of centrosome structure, chromosome segregation, and cytokinesis by misexpression of human Cdc14A phosphatase. *Mol Biol Cell* **13**, 2289–2300.

116. Mailand, N., Lukas, C., Kaiser, B. K., Jackson, P. K., Bartek, J., and Lukas, J. **2002**. Deregulated human Cdc14A phosphatase disrupts centrosome separation and chromosome segregation. *Nature Cell Biol* **4**, 317–322.

117. Gromley, A., Jurczyk, A., Sillibourne, J., Halilovic, E., Mogensen, M., Groisman, I., Blomberg, M., and Doxsey, S. **2003**. A novel human protein of the maternal centriole is required for the final stages of cytokinesis and entry into S phase. *J Cell Biol* **161**, 535–545.

118. Wheatley, S. P., Hinchcliffe, E. H., Glotzer, M., Hyman, A. A., Sluder, G., and Wang, Y. **1997**. CDK1 inactivation regulates anaphase spindle dynamics and cytokinesis *in vivo*. *J Cell Biol* **138**, 385–393.

119. Andreassen, P. R., Lohez, O. D., Lacroix, F. B., and Margolis, R. L. **2001**. Tetraploid state induces p53-dependent arrest of nontransformed mammalian cells in G1. *Mol Biol Cell* **12**, 1315–1328.

120. Khodjakov, A., Rieder, C. L., Sluder, G., Cassels, G., Sibon, O., and Wang, C. L. **2002**. *De novo* formation of centrosomes in vertebrate cells arrested during S phase. *J Cell Biol* **158**, 1171–1181.

121. Bobinnec, Y., Moudjou, M., Fouquet, J. P., Desbruyeres, E., Edde, B., and Bornens, M. **1998**. Glutamylation of centriole and cytoplasmic tubulin in proliferating non-neuronal cells. *Cell Motil Cytoskeleton* **39**, 223–232.

122. Balczon, R., Simerly, C., Takahashi, D., and Schatten, G. **2002**. Arrest of cell cycle progression during first interphase in murine zygotes microinjected with anti-PCM-1 antibodies. *Cell Motil Cytoskeleton* **52**, 183–192.

123. Hinchcliffe, E. H., Li, C., Thompson, E. A., Maller, J. L., and Sluder, G. **1999**. Requirement of Cdk2-cyclin E activity for repeated centrosome reproduction in *Xenopus* egg extracts. *Science* **283**, 851–854.

124. Okuda, M., Horn, H. F., Tarapore, P., Tokuyama, Y., Smulian, A. G., Chan, P.-K., Knudsen, E. S., Hofmann, I. A., Snyder, J. D., Bove, K. E., and Fukasawa, K. **2000**. Nucleophosmin/B23 is a target of CDK2/cyclin E in centrosome duplication. *Cell* **103**, 127–140.

125. Morrison, D. K., and Davis, R. J. **2003**. Regulation of MAP kinase signalling modules by scaffold proteins in mammals. *Ann Rev Cell Dev Biol* **19**, 91–118.

126. Huang, J.-Y., and Raff, J. W. **1999**. The disappearance of cyclin B at the end of mitosis is regulated spatially in *Drosophila* cells. *EMBO J* **18**, 2184–2195.

9
Centrosome Duplication and its Regulation in the Higher Animal Cell

Greenfield Sluder

9.1
Introduction

As the primary microtubule-organizing center of the higher animal cell, the centrosome has a profound influence on all microtubule-dependent processes. Before the cell enters mitosis the single interphase centrosome duplicates exactly once, and after nuclear envelope breakdown, the two sister centrosomes nucleate the astral arrays that contribute most of the microtubules to the formation of the spindle. Centrosomes, through these astral microtubules, determine spindle polarity, spindle position/orientation in the cell, and the plane of cleavage. The presence of more than two centrosomes at the onset of mitosis, a condition called centrosome amplification, greatly increases the chances that the cell will assemble a multipolar spindle and distribute chromosome unequally (for examples see [1–3]). The penalties for mistakes in chromosome distribution can be severe for the organism; genomic instability due to whole chromosome losses or gains can lead to loss of normal alleles for tumor suppressor genes and other genetic imbalances that can promote unregulated growth characteristics and a diminished apoptotic response to cellular damage (reviewed in [4–6]) Genomic instability due to unequal chromosome distribution at mitosis is thought to be a major driving force in multi-step carcinogenesis [7–10]. Centrosome amplification is an intractable problem for the cell because extra centrosomes are not eliminated and there is no checkpoint that aborts mitosis when spindle polarity is abnormal [3]. Thus, it is of the greatest importance for the cell to have two and only two centrosomes when it enters mitosis.

Centrosome reproduction, or duplication, is the process whereby the single interphase centrosome exactly doubles before mitosis. The cell must ensure four seemingly simple things: the centrosome must duplicate; the two sister centrosomes must separate; duplication must be limited so that one centrosome becomes only two; and the centrosome must duplicate at the right time in relation to nuclear events in the cell cycle. Although all of this is elementary in concept, making sure that all four conditions are met with no mistakes, cell cycle after cell cycle is not a simple proposition. The regulation of centrosome duplication involves

Centrosomes in Development and Disease. Edited by Erich A. Nigg

ISBN 3-527-30980-2

multiple mechanisms, some intrinsic to the centrosome and others based in the activities of kinases that control nuclear events in the cell cycle. None of these controls have been fully explored and perhaps there are more to be identified.

9.2 The Events of Centrosome Reproduction

On the basis of morphological events, centrosome reproduction has been broken down into four events: centriole disorientation, centriole duplication, centrosome disjunction and sister centrosome separation (Figure 9.1). These events are defined by what can be seen chronologically by light and electron microscopy and by what can be inferred from the behavior of centrosomal components under normal and experimental conditions.

9.2.1 Centriole Disorientation

Mother and daughter centrioles are thought to be tethered to each other. This is inferred from observations that centrosome isolation produces centrosomes with paired centrioles and centrioles are in close proximity in an orthogonal arrangement during G1 in some cell types. In other cells, such as HeLa, the centrioles separate widely in G1 [11–13], but return closer together during S, G2, and M. For cells in which the centrioles remain in close proximity, the slight separation and

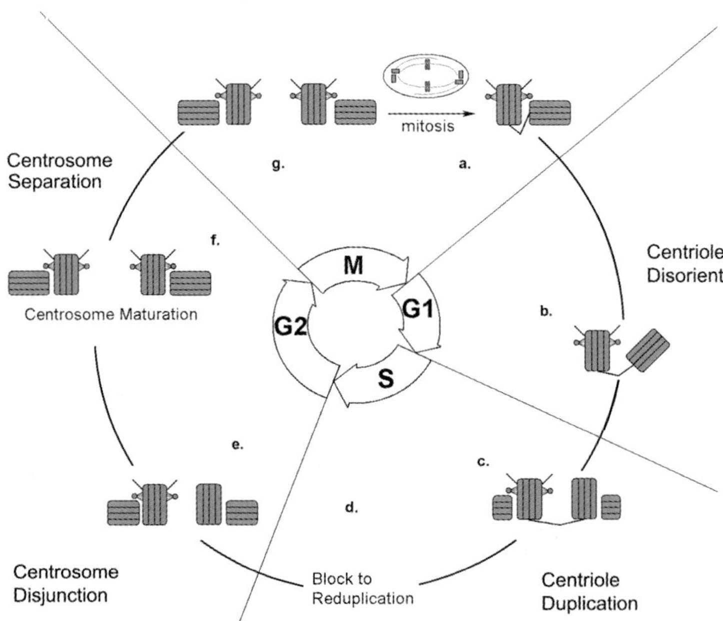

loss of orthogonal relationship between the centriole pair, observed during late G_1, is commonly said to be the leading event in centrosome reproduction (Figure 9.1b) [14–16]. Although this morphological change has been interpreted to mean a disjunction event, a proposition supported by the behavior of centrioles in *Xenopus* egg extracts [16], it is not rigorously known if this represents the functional separation of the mother and daughter centrioles or rather a relaxation in their spatial association without loss of physical connection in somatic cells. Although the centrioles separate slightly at this time, the centrosome as a whole does not split until later in interphase.

9.2.2 Centriole Duplication

Centriole duplication is first seen at the beginning of S phase or during S phase by the appearance of short daughter centrioles, or pro-centrioles, at right angles to and separated slightly from the two parental centrioles at their proximal ends (Figure 9.1c) [14, 15, 17–19]. These pro-centrioles elongate during S and G_2, reaching mature length in mitosis or the following G_1 [14, 20]. Daughter centrioles (the elongating pro-centrioles) do not acquire distal and subdistal appendages until they fully mature in the subsequent cell cycle [21–23]. Although we can observe when assembled pro-centrioles first appear, pro-centriole assembly could in principle, be the physical manifestation of initial reproductive processes that began in G_1, mitosis, or even in the previous cell cycle.

◀ **Figure 9.1** Schematic representation of the centrosome duplication cycle. Centrioles are shown as shaded barrels; the appendages on the older centrioles represent sub-distal and distal appendages. The pericentriolar material is not shown; in some cell types it surrounds both centrioles and in other types it is associated primarily with the older centriole. **a.** At the end of mitosis each daughter cell receives a single centrosome containing a pair of centrioles in close proximity. The centrioles are shown to be connected by a fibrous link. **b.** Centriole disorientation is seen as a relaxation of the tight orthogonal relationship between the older and younger centrioles in late G1. Although this is said to be the leading event of centrosome duplication, it is not known if this represents a disjunction event or the relaxation of a persistent connection. **c.** The start of centriole duplication is seen by the assembly of short pro-centrioles at right angles to the proximal ends of the two parental centrioles. The pro-centrioles elongate throughout the rest of interphase, reaching their mature length in mitosis or the following G1. **d.** In normal cells a centrosome intrinsic mechanism blocks re-duplication of centrioles. The two centrosomes, each containing a parent centriole and a pro-centriole, remain tethered to each other. **e.** The two centrosomes disjoin, or lose their connection, at a variable time in G2 due to a change in the balance of Nek2 and PP1 activities as well as Cdc14 phosphatase activity. **f.** The severing of this connection allows the sister centrosomes to spatially separate around the nucleus. During G2 the centrosomes mature and both parental centrioles have distal and sub-distal appendages. Centriole duplication is said to be conservative because the pro-centriole is assembled from subunits in the cytoplasm, not from components of the mother centriole. Centriole distribution to sister centrosomes is said to be semi-conservative because parental centrioles are distributed to both centrosomes. **g.** At mitosis centriole pairs are located at each spindle pole. The mother and daughter centrioles may or may not be linked to each other. Figure adapted from Hinchcliffe and Sluder [127] and Nigg [6].

9.2.3 Centrosome Disjunction

During and after the duplication of the centrioles, the two sister centrosomes are thought to be physically linked by a tether [21] that will be degraded or severed at a variable time in G_2 when the two sister centrosomes undergo centrosome disjunction with a mother–pro-centriole pair in each sister centrosome (Figure 9.1e) [24, 25]. Centrosome disjunction is the unseen event that cuts the physical link between duplicated centrosomes and is distinct from the actual separation of the sister centrosomes that indicates that disjunction has occurred.

9.2.4 Centrosome Separation

Centrosome separation is the spatial separation of centrosomes around the nucleus during prophase (Figure 9.1f), driven by a combination of plus and minus end-directed microtubule motor proteins. The extent to which aster separation occurs before the onset of mitosis can vary between cells in the same population. In some cases the two centrosomes remain close together until nuclear envelope breakdown, while in others both asters are well separated around the nucleus before the end of prophase [26].

The disjunction of the sister centrosomes, as seen by their spatial separation, is attributed to the activity of centriole-associated Nek2 kinase when its antagonist – protein phosphatase 1α – is inactivated at the onset of mitosis [27] (reviewed in [28]). In G2 Nek2A phosphorylates C-Nap1, a protein located at the proximal ends of the two parental centrioles, but not pro-centrioles (Figure 9.2). This phosphorylation leads to the eventual loss of C-Nap1 from the centrioles and the loss of connection between the sister centrosomes thereby allowing them to separate. It is not presently clear if the connecting fiber between the centrioles is composed of C-Nap1 or if this protein serves as the interface that anchors other connecting fiber proteins to the centrioles as suggested by immunoelectron microscopy. Late in mitosis or early G1, C-Nap1 is again observed at the centrosomes. The extent to which centrioles are linked during mitosis bears further investigation. On the one hand, the mother and daughter centrioles are arranged in an orthogonal arrangement and in close proximity to each other during mitosis, suggesting that they are physically linked (see [12, 13, 29]). On the other hand, the centrioles in each centrosome will separate to establish independent spindle poles when mitosis is prolonged [30–33]. This either reflects a lack of attachment or is due to changes in kinase/phosphatase equilibria during prolonged mitosis that lead to a loss of centriole cohesion.

Lest we think that this is the complete story for centrosome cohesion/disjunction, recent studies raise the possibility that there are more players to be discovered. Quantitative fluorescence work with inducible Nek2A wild-type and kinase-dead constructs suggests that Nek2A activation alone may not be sufficient to fully displace C-Nap1 from centrosomes and the loss of C-Nap1 from centrosomes

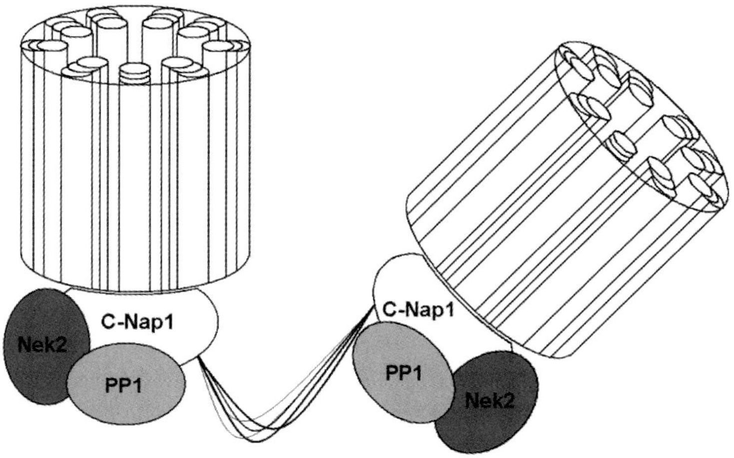

Figure 9.2 Model for the spatial localization of proteins involved in regulating the connection between centrioles. The centrioles are connected by an extensible link of unknown composition shown here as a group of fibers. The C-Nap1 protein, concentrated at the proximal ends of the centrioles, anchors the linking fibers to the centrioles. C-Nap1 exists in a ternary complex with the kinase Nek2 and the phosphatase PP1. During G2 the phosphorylation of C-Nap1 by Nek2 leads to the release of the connecting fiber from the centrioles. The extent of C-Nap1 phosphorylation, however, is determined by a balance in the activities of Nek2 and PP1. Figure adapted from Fry [28].

is not sufficient for centrosome disjunction [34]. Also, Mailand and co-workers [35] have reported that Cdc14A phosphatase activity is involved in centrosome disjunction. This centrosome-localized phosphatase must act at a different level than protein phosphatase 1α because overexpression of Cdc14A leads to precocious separation of the sister centrosomes and downregulation of its abundance leads to a failure in centrosome separation. Finally, phosphorylation of the centrosomal protein centrin during G_2/prophase correlates with centrosome disjunction [36].

The mechanism that determines that only one daughter centriole will be assembled at right angles to and seemingly slightly separated from its mother is not understood. Our current thinking is that the parent centriole provides a site or docking location for proteins that initiate the self assembly of a template for the assembly of triplet microtubules and core structures of the pro-centrioles. This notion is supported by the report of specific precursor structures, seen as an annular ring or a looped fiber containing nine densely staining foci that later elaborate into triplet microtubules, next to the mother basal body in ciliate and *Chlamydomonas* basal body duplication [37, 38] (reviewed in [39]). However, it appears that the mother centriole is capable of providing more than one site for pro-centriole formation, at least in *Drosophila*. At restrictive temperatures somatic cells with a temperature-sensitive mutation in Cyclin-dependent kinase 1 (Cdk1) show cycles of DNA endoreduplication and the formation of more than one daughter centriole in close association with the mother centriole [40].

In higher animal cells, exceptions to the spatial specificity of pro-centriole formation at the mother centriole are found in the *de novo* formation of centrioles after parthenogenetic activation of sea urchin eggs [41, 42], the formation of multiple basal bodies from specialized generative structures during differentiation of ciliated epithelia [43, 44], and centrosome assembly in somatic cells with compromised p53 function after the resident centrosome is laser ablated [45]. In addition, centriole formation is under developmental control in the mouse zygote. Centrioles are not seen in the early mitotic divisions but are found later in development [46]. It is not known if this represents the *de novo* formation of centrioles or the propagation of cryptic templating structures for centrioles (see [47, 48] for the establishment of this cryptic template paradigm in the ciliate *Naeglaria* and the surf clam *Spisula* oocyte, respectively). The existence of cryptic centriole determinants is favored by the fact that centrioles arise in proper copy number in each mouse embryo cell, in sharp contrast to the multiple centrioles that are assembled in variable number during *de novo* centriole formation during parthenogenesis in zygotes and in somatic cells from which the centrosome is laser ablated.

The assembly of just one daughter centriole at a slight distance from its mother during centriole duplication has recurrently brought to mind DNA replication, the modern paradigm for a templated reproductive process in which information and copy number are under rigid control. Almost 40 years ago workers started considering the notion that centrioles and basal bodies could be semi-autonomous organelles with their own DNA, much like mitochondria. A variant on this theme was the hypothesis that centrioles, like ribosomes, contain RNA that serves a structural role during their assembly. These possibilities inspired numerous studies, the vast majority of which concentrated on trying to demonstrate the existence of centriole/basal body specific DNA or RNA under the assumption that presence implies function. Since all of this work was fraught with serious technical problems and ultimately produced inconclusive observations, we will not review this field but rather refer the curious to two reviews [49, 50]. Suffice it to say that presently there is no credible evidence for the existence of DNA in centrioles/basal bodies or for the direct involvement of RNA in centrosome duplication.

9.2.5 Some Proteins Needed for Centrosome Reproduction

Recent studies have started to identify a number of centrosomal proteins that are required for centrosome duplication. Since these may be structural proteins and there is no evidence that their availability is limiting under normal circumstances, we will not treat them as participants in the normal control pathways for centrosome duplication, at least for the moment. Although we will only enumerate some of these studies, we note that the further investigation of these proteins will become increasingly important as we seek to discover and understand the molecular interactions involved in the assembly of daughter centrioles. Parenthetically, we add that such investigations might provide insight into the developmental regulation of basal body/centriole assembly from putative cryptic determinants in *Naeglaria* and early mouse embryos.

Gamma tubulin is a key component of the gamma tubulin ring complexes in the pericentriolar material that nucleates microtubules [51]. RNA interference of gamma tubulin expression in *Paramecium* [52] and its downregulation in *Tetrahymena* [53] has revealed that this protein is required for basal body duplication. Also, other tubulin isoforms, such as *epsilon-tubulin* and *delta-tubulin*, are required for the complete assembly of basal bodies [54–58] (reviewed in [39, 59, 60]; see also Chapters 2 and 5). Beyond the tubulin family, various isoforms of *centrin*, EF-hand proteins found concentrated in the lumen of the centriole [61] and in a more dispersed form throughout the pericentriolar material, are required for centriole duplication in HeLa cells [62] and *Xenopus* embryos [36]. Lastly, two groups identified the SAS-4 gene in *C. elegans* whose protein product is localized to the centriole or closely associated structures throughout the cell cycle [63, 64]. When the expression of this protein is diminished with RNAi, the centrosome as a whole does not duplicate. It remains to be determined if this is due to a failure of centriole duplication or separation (see also Chapter 12).

9.3 Control of Centrosome Duplication

Control of centrosome duplication is exercised by limits that are intrinsic to the centrosome itself and by extrinsic controls imposed by changing cytoplasmic conditions during cell cycle progression. Limits intrinsic to the centrosome determine the number of new centrosomes that arise from the original centrosome; cytoplasmic controls determine when the centrosome duplicates in relation to the progression of nuclear events such as DNA synthesis and mitosis.

9.3.1 Control of Centrosome Number: Intrinsic Mechanisms

There is a “counting” mechanism that ensures that each round of duplication will produce only one new centrosome. This numerical control is determined by the cycle of centriole disjunction and duplication. Since centrioles act to localize the pericentriolar material, the number of centriole pairs determines the number of spindle poles. This limit to the number of centrosomes is of critical importance for all cells and particularly so for the early cleavage divisions during development. Zygotes contain at fertilization enough centrosomal subunits to assemble many complete centrosomes [65, 66], yet they normally assemble only one new centrosome at each cell cycle.

The evidence that centrioles are the counting mechanism originated with the remarkable finding of Mazia and co-workers [67] that it is possible to experimentally manipulate the reproductive capacity of centrosomes in sea urchin zygotes. When mitosis is prolonged by any of several independent methods, the two spindle poles split during mitosis to yield four functional poles that will not further subdivide even when mitosis is prolonged to 20 times its normal duration (Figure 9.3a–c) [30, 32]. Ultrastructural analysis of such tetrapolar spindles reveals that each pole contains only one centriole, confirming that the centrosomes have split, not duplicated [31, 32]. After the cell divides into four, these half centrosomes each assemble a daughter centriole, thus becoming complete normal centrosomes with full reproductive capacity (Figure 9.3d). However, they do not undergo centriole separation or centrosome disjunction, and each cell assembles a monopolar spindle at the next mitosis (Figure 9.3e). In some cases two of the four spindle poles at first mitosis do not fully separate and, as a consequence, the zygote divides into three with one blastomere inheriting two spindle poles (Figure 9.3f, lower daughter cell). At second mitosis, this daughter assembles a functional bipolar spindle and divides in a normal fashion (Figure 9.3g). At subsequent cell cycles, the centrosomes duplicate normally. This indicates that the centrosome of monopolar spindles is functionally normal.

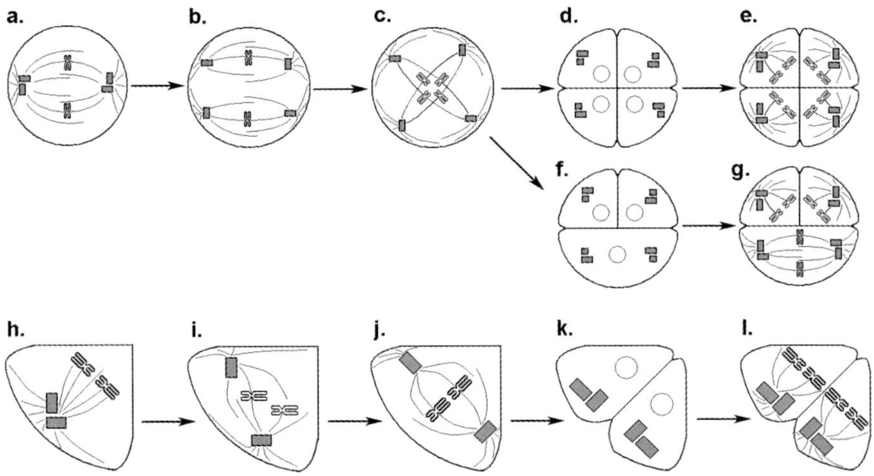

Figure 9.3 Diagrammatic representation of the experimental manipulation of the reproductive capacity of spindle poles in sea urchin zygotes. **a.** The first division spindle has a pair of centrioles at each spindle pole. **b–c.** When prometaphase is prolonged, the centriole pairs with associated pericentriolar material split without duplicating. The four spindle poles, each containing a single centriole, separate and a tetrapolar spindle is formed. **d.** In telophase, as the cell divides into four, the singlet centrioles duplicate. At the next mitosis monopolar spindles are assembled, and each centrosome has the normal complement of two centrioles. **f–g.** If two centrosomes of the tetrapolar spindle fail to completely separate (lower cell), a single nucleus is re-formed in one blastomere. At mitosis a bipolar spindle is assembled; its poles have a normal complement of centrioles and consequently reproduce in a normal fashion in subsequent cell cycles. **h–l.** Mitosis in blastomeres containing a monopolar spindle (continuation of stage shown in diagram **e**). **h.** Blastomere containing a monopolar spindle – enlarged diagram. **i–j.** Prometaphase is often substantially longer than normal due to unattached kinetochores. The spindle pole splits and the spindle re-organizes into a bipolar configuration with one centriole in each centrosome. **k.** In telophase each daughter cell inherits a single centriole which later duplicates. **l.** At mitosis monopolar spindles are again assembled. During prolonged prometaphase the one spindle pole splits yielding a bipolar spindle with a single centriole in each centrosome (not shown).

If a daughter cell with a monopolar spindle remains in mitosis longer than normal, as often happens due to unattached kinetochores, the centrosome of the monopolar spindle will split to give two functional spindle poles with one centriole apiece (Figure 9.3h–j). These poles undergo centriole duplication without disjunction during interphase, and monopolar spindles are once again formed at the following mitosis (Figure 9.3k–l). The importance of centrioles in the control of centrosome number was further substantiated by the finding that zygotes lacking centrioles form a single microtubule organizing center that does not double between mitoses [68]. These observations are not peculiar to embryonic cells; when mitosis is prolonged in mammalian cells cultured by transfection with a non-degradable cyclin B construct, the spindle poles often double from two to four [33].

9.3.2
Block to Re-replication

If mitosis is to be normal, it is of obvious importance for the cell to ensure that the centrosome duplicates only once during each cell cycle. How the cell accomplishes this has been a matter of some confusion because a number of cultured cell types and many zygotes exhibit multiple rounds of centrosome duplication when they are arrested in S phase [32, 69–71] (discussed in [72]). These findings raised the problematical question of why centrosomes normally do not re-duplicate during S phase, especially when DNA synthesis is slowed by environmental perturbations. Observations that the period of centrosome re-duplication during S phase arrest is on average more than twice as long as the entire cell cycle in sea urchin zygotes and CHO cells [32, 71] suggested that, under normal circumstances, S phase does not last long enough for centrosomes to re-duplicate.

This rationalization was never particularly strong for mammalian somatic cells, because those with an intact p53 pathway duplicate their centrosome only once even though the cell cycle is arrested in S phase [73, 74] (G. Sluder, unpublished data). Indeed, a recent study reveals that the normal cell does not take chances on the duration of S phase for an event as important as centrosome duplication [74]. These workers fused cells in different phases of interphase to determine if an already duplicated centrosome would re-duplicate in S-phase cytoplasm, conditions that are supportive of centrosome duplication. The results of this study are most clearly illustrated by the behavior of centrosomes in cells resulting from the experimental fusion of cells in G1 and G2. As reported earlier (reviewed in [75]) the G1 nucleus undergoes DNA synthesis but the G2 nucleus does not and later both nuclei enter mitosis synchronously. These fused cells each start with an unduplicated centrosome from the G1 cell and a duplicated centrosome (seen as a pair of centrosomes) originating from the G2 cell. The revealing finding of the study by Wong and Stearns was that the G1 centrosome duplicated while the G2 centrosomes did not, even though both were exposed to S-phase conditions. These results indicate that there is a block to re-duplication; it is not based in a putative inhibitor present in G2 cytoplasm; and it is intrinsic to the centrosome itself.

Additional support for the existence of this block to re-duplication comes from findings that the constitutive overexpression of cyclin E in rat and human cells leads to the precocious duplication of the centrosome before S phase but not to a high incidence of centrosome re-duplication [76]. As we will discuss later, the late G1 rise in the activity of Cyclin-dependent kinase 2 (Cdk2) complexed with cyclin E or cyclin A drives initiation of centrosome duplication.

9.3.3
Time of Centrosome Duplication: Extrinsic Controls

The cell must ensure that the events of centrosome reproduction are properly coordinated with nuclear events in the cell cycle if it is to have just two centrosomes at the onset of mitosis.

In principle, a logical way for the cell to coordinate nuclear and centrosomal events would be for nuclear activities, such as the synthesis of DNA, to determine when the centrosome can reproduce. A direct test of these possibilities using enucleated sea urchin zygotes revealed that neither the presence of the nucleus nor its activities are required for repeated cycles of centrosome reproduction [77]. The finding that all the centrosomes duplicated synchronously within each zygote suggests that the temporal control of centrosome reproduction is under cytoplasmic control.

9.3.4
Cyclin-dependent Kinases in the Control of Centrosome Reproduction

The apparent cytoplasmic control over the time of centrosome duplication suggested the possibility that the activities of the Cdks that control cell cycle progression might also control centrosome duplication. This would provide a logical way for the cell to coordinate centrosome duplication with nuclear events in the cell cycle. The involvement of Cdk1 (also known as $p34^{cdc2}$) in centrosome duplication was first examined. The fact that centrosomes repeatedly reproduce when the zygote cell cycle is arrested in interphase by complete inhibition of protein synthesis from the time of fertilization [65, 66] indicates that centrosome duplication cannot be driven in any simple way by the cyclic rise and fall of Cdk1–cyclin A or B activity. These cyclins are proteolytically degraded at the end of mitosis/meiosis and Cdk1 activity requires their synthesis anew at each cell cycle, something that does not happen in the presence of inhibitors to protein synthesis. Nevertheless, it was possible that the absolute value of Cdk1-B or Cdk1-A activity during interphase might provide conditions that gate the ability of centrosomes to duplicate in a fashion analogous to the Cdk-dependent block to re-duplication of DNA [78]. However, tests of this possibility revealed that centrosomes repeatedly reproduce in zygotes arrested during S phase regardless of whether Cdk1 activity is high or low [32]. That centrosome reproduction is independent of the absolute value of Cdk1 activity at the cell cycle stage that normally supports centrosome duplication, argues against its involvement in the control of this process.

Attention then turned to Cdk2 because the activation of Cdk2–cyclin E and Cdk2–cyclin A are normally required for the G_1/S transition and the maintenance of S-phase progression [79, 80], times in the cell cycle when centriole duplication occurs. In the late 1990s a number of laboratories investigated the role of Cdk2–cyclin E activation in centrosome reproduction, and a series of papers (here categorized by experimental system) appeared in rapid succession.

9.3.4.1 Zygote Systems

Hinchcliffe and co-workers [81] developed an S phase-arrested *Xenopus* egg extract that supports multiple rounds of sperm centrosome duplication *in vitro*. When Cdk2–cyclin E activity was inhibited by recombinant Δ34Xic-1, an NH_3-terminal truncated form of *Xenopus* Cdk inhibitor Xic-1^{p27}, the asters doubled once but repeated doubling of the asters did not occur. The basis for the one-time doubling of the asters is not known but may have represented the splitting and separation of the two sperm centrioles each of which organizes an aster (see [31]). Multiple rounds of aster duplication were restored when an excess of purified Cdk2–cyclin E was added to the Δ34Xic-1 treated extracts. At the concentration used Δ34Xic-1 specifically inhibits the activity of Cdk2–cyclin E but not Cdk1–cyclin A or Cdk1–cyclin B [82]. Cdk2–cyclin A activity was not a factor in these experiments, because Cdk2 does not complex with cyclin A until after the mid-blastula transition in *Xenopus* [83]. Since the majority of S phase-promoting activity is provided by Cdk1–cyclin A activity [80], which is not inhibited by Δ34Xic-1 at the concentrations used, the inhibition of Cdk2–cyclin E should not have driven the cell cycle out of S phase.

Lacey and co-workers [16] developed and used a different *Xenopus* egg extract assay system in which isolated mammalian centrosomes were used to examine centriole disjunction as a function of Cdk2–cyclin E activity. Centriole separation was used as a measure of centrosome duplication because centriole disorientation was thought to be the leading event in centrosome reproduction [14]. In control extracts the mother–daughter centriole pairs disjoined while extracts containing the Cdk2 kinase inhibitors p21 or p27 [84, 85] did not show centriole disjunction. These results were confirmed by microinjecting the Cdk2 inhibitors p21 or p27 into *Xenopus* embryos arrested in interphase with protein synthesis inhibitors, which allow repeated centrosome duplication [66]. The centrosomes in the injected blastomeres did not repeatedly duplicate, while those in the uninjected blastomeres of the same zygote re-duplicated.

9.3.4.2 Mammalian Somatic Cells

An early indication that Cdk2–cyclin E activity determines when the centrosome in mammalian somatic cells duplicates came from the report of Mantel and co-workers [86] that inhibiting $p21^{cip1/waf\text{-}1}$ in human hematopoetic cells increased Cdk2 activity and caused the cells to accumulate multiple centrosomes.

When Chinese hamster ovary (CHO) cells are arrested in S phase with hydroxyurea for prolonged periods of time, the centrosome duplicates multiple times

[71]. Matsumoto et al. [87] used this experimental system to demonstrate that centrosome re-duplication is inhibited when the activity of Cdk2 is blocked by drugs or the overexpression of $p21^{cip1/waf-1}$. However, an independent study provided evidence that Cdk2–cyclin A is more effective than Cdk2–cyclin E in restoring multiple rounds of centrosome duplication in CHO cells arrested at the G_1/S transition by transfection with a mutant form of Rb that lacks Cdk phosphorylation sites [88] (also see [89]). While co-expression of cyclin A restored significant levels of repeated centrosome duplication, overexpression of cyclin E did not. These findings also suggest that the action of Cdk2–cyclin A in centrosome duplication occurs downstream of its role in the Rb phosphorylation pathway needed to drive the G_1/S transition.

Although these observations, taken together, suggest that somatic cells and early cleavage stage zygotes may use different Cdk2–cyclin complexes to regulate centrosome reproduction, these cell types may not use fundamentally different control strategies. Perhaps centrosomes are responsive to both Cdk2–cyclin A and Cdk2–cyclin E. Since Cdk2 does not complex with cyclin A until the mid-blastula transition in *Xenopus* zygotes, Cdk2–cyclin E may be the only kinase complex available. In somatic cells that contain both Cdk2–cyclin A and Cdk2–cyclin E kinase complexes, Cdk2–cyclin A activity may also play an important role in promoting centrosome duplication. Even though these interesting details bear further investigation, the theme that arose from these studies is that the cell's entry into S phase and centrosome duplication are linked through a rise in Cdk2 activity.

However, more recent studies indicate that this notion is overly simple; there must be functional redundancy for kinases that promote centrosome duplication. Berthet and co-workers [90] and Ortega and co-workers [91] independently generated Cdk2 knockout mice that were viable, albeit with defects in germ cell development. Immunoprecipitates of cyclin E1 complexes from Cdk2−/− animal tissue extracts showed no kinase activity, at least towards histone H1, while cyclin A2 immunoprecipitates were active. In addition, Geng and co-workers [92] generated cyclin E1 and cyclin E2 knockout mice that developed normally and were viable, although the cyclin E2−/− males were sterile. When these mice were crossed, the E1/E2 double knockout was an embryonic lethal due to problems with placental development. Nevertheless, these embryos survived until the 10th day of gestation indicating that cell proliferation occurs in the absence of cyclin E. These observations make a strong argument that cell proliferation and presumably centrosome duplication do not absolutely require Cdk2 or cyclin E. Perhaps cyclin A with a yet to be identified kinase partner or other kinases are able to compensate under these extraordinary experimental circumstances. Although these new observations rule out the simple notion that Cdk2–cyclin E and Cdk2–cyclin A activities are uniquely required for centrosome duplication, particularly in mammalian somatic cells, they do not eliminate a role for these kinase complexes in the initiation of centrosome duplication in normal cells. Cdk2–cyclin E/A activity, although not essential, may nevertheless be important for the timeliness and fidelity of centrosome duplication. This notion is consistent with findings that Cdk2 and cyclin E are not required for DNA synthesis but without them this process is not normal. Cdk2−/− primary fibroblasts are delayed in entry into S phase [90] and cyclin E-deficient cells fail to incorporate MCM proteins into DNA replication origins [92].

9.3.5
Targets of Cdk2–Cyclin E Kinase

When present, Cdk2–cyclin E or A appear to directly phosphorylate proteins of the centrosome and influence pathways that act upon the centrosome. For isolated centrosomes *in vitro* Okuda and co-workers [93] found that Cdk2–cyclin E phosphorylated only nucleophosmin or NO38/B23, a previously identified phosphoprotein component of the nucleolus implicated in ribosome biogenesis [94]. Immunofluorescence analysis indicated that nucleophosmin appeared on centrosomes during mitosis and during the ensuing interphase remained localized to the interphase centrosome until the centrosome duplicated, at which time the nucleophosmin immunoreactivity was lost. Later, when the cells returned to mitosis, nucleophosmin was again observed at the centrosomes. Functional evidence for nucleophosmin involvement in centrosome reproduction came from the finding that centriole/centrosome reproduction was inhibited by the expression of non-phosphorylatable mutant nucleophosmins that remained at the centrosome. Microinjections of antibodies to nucleophosmin, that may sterically block its phosphorylation by Cdk2–cyclin E, also blocked centrosome doubling. Together, these data led the authors to propose that Cdk2–cyclin E-specific phosphorylation of nucleophosmin on threonine 199 causes it to come off of the centrosome thereby allowing duplication to begin [93, 95]. What specific event of centrosome duplication is inhibited by the presence of nucleophosmin is not certain, but the observation of orthogonally arranged, unseparated centrioles in inhibited cells suggests that it is an early event. However, it is not clear that nucleophosmin, in any simple way, limits centriole disjunction because centrioles can split apart during prolonged mitosis, a cell cycle stage when nucleophosmin should be associated with the centrosome [31–33]. Another possible centrosomal target of Cdk2–cyclin E is CP110, a protein localized to the centrioles or their immediate vicinity [96]. This protein is phosphorylated by Cdk2–cyclin E, Cdk2–cyclin A, and Cdk1–cyclin B at sites that are phosphorylated *in vivo*. RNAi-mediated reductions in CP110 abundance blocks centrosome re-duplication during S phase arrest in Saos2 cells which otherwise exhibit such re-duplication.

Cdk2–cyclin E may also participate indirectly in centrosome duplication by stabilizing the cellular levels of Mps1p kinase [97], originally identified as essential for the duplication of the spindle pole body in budding yeast [98]. Endogenous mouse Mps1p kinase is localized to centrosomes throughout the cell cycle and was reported at centrosomes in living cells stably expressing mMps1p-GFP. Functional evidence that mMps1p kinase activity is required for centrosome duplication came from the finding that kinase-dead mMps1p localizes to the centrosomes *in vivo* and diminishes centrosome duplication during the cell cycle and centrosome re-duplication in S-phase arrested cells that would otherwise show centrosome re-duplication during prolonged S phase. Also, when mMps1p is overexpressed during S-phase arrest, correlative light and serial section electron microscopy showed that the centrosomes/centrioles re-duplicated in a cell line that does not show centrosome re-duplication during S-phase arrest. Cdk2–cyclin E activity appears to stabilize mMps1 protein levels; when Cdk2–cyclin E activity is blocked by drug treatments or by overexpression of p21 or p27, the cellular level of mMps1p dramatically drops and its localization to the centrosome is lost.

However, the importance of Mps1p in centrosome reproduction for human cells was brought into question by Stucke and co-workers, [99]. For U2OS cells, a human osteosarcoma-derived cell line, these workers did not find Mps1p at the centrosomes by immunofluorescence with monoclonal antibodies and failed to find functional evidence for an involvement of this kinase in centrosome duplication by antibody injections, expression of kinase dead constructs, or siRNA-mediated reduction of mMps1p protein levels. These surprising differences between the two studies have been re-investigated by Fisk and co-workers [100], who conducted a similar battery of experiments with a previously described polyclonal antibody to hMps1p [101] and several human cell lines, including U2OS cells. The results support their previous conclusions that Mps1 is present at centrosomes and its activity is required for centrosome duplication. Since a detailed comparison of the experiments published by these two groups is beyond the scope of this chapter the reader is referred to the original works.

9.3.6
Other Kinases Involved in Centrosome Duplication

Calcium/calmodulin dependent kinase II (CaMKII) activity is required for centrosome duplication, at least in *Xenopus* embryo extracts [102]. This investigation was prompted by reports that calcium and calmodulin are required for the G1–S transition [103], that periodic calcium oscillations during the *Xenopus* egg cell cycle correlate with this cell cycle transition [104], and that CaMKII localized to centrosomes phosphorylates centrosomal proteins *in vitro* [105, 106]. Matsumoto and Maller, using an S phase arrested extract that supports multiple rounds of centrosome duplication [102], found that centrosome duplication was blocked by chelating calcium with BAPTA and inhibiting the inositol 1,4,5-triphosphate (IP3) receptor with heparin, both of which block calcium transients in egg extracts [107]. In addition, direct inhibition of CaMKII activity with a specific pseudosubstrate peptide led to an immediate cessation of centrosome duplication. The addition of extra CaMKII plus calmodulin to such inhibited extracts rescued centrosome duplication. An interesting issue raised by this study is the difference in the way inhibition of Cdk2–cyclin E and CaMKII effect centrosome duplication. Inhibition of Cdk2–cyclin E activity consistently allows one doubling of the asters initially assembled around the sperm centrosomes, which may represent either splitting or duplication of each centrosome. In contrast, inhibition of CaMKII activity blocked even this initial doubling of the asters. This finding raises the possibility that CaMKII activity is required for an early event in centrosome duplication.

The ZYG1 gene product, a putative kinase, is required for centrosome duplication in *C. elegans*. First identified in a study of embryonic lethal mutants [108], ZYG-1 activity is important at all stages of development, even post-embryonic development [109]. The demonstration of its role in centrosome duplication has come from an analysis of the early cleavage divisions of zygotes resulting from reciprocal crosses in which one parent was effectively null for ZYG-1 [110]. Since paternal ZYG-1 is required for the assembly of a second centriole during spermiogenesis, normal eggs fertilized with mutant sperm receive only a single centriole. During the first cell cycle this centriole duplicates in response to the activity of maternal ZYG-1 activity, and at first mitosis the zygote assembles a monopolar spindle containing just a pair of centrioles. During the second cell cycle this centrosome duplicates normally, and the subsequent division is bipolar. Alternatively, mutant eggs fertilized with wild-type sperm receive a pair of centrioles. During the first cell cycle, these centrioles split apart, but fail to duplicate due to the lack of maternal ZYG-1 activity. This gives rise to a bipolar spindle, with each pole containing a single centriole. In the second cell cycle these single centrioles do not reproduce, and the subsequent mitotic spindles are both monopolar (reviewed in [111]). Immunofluorescence microscopy revealed that ZYG-1 localizes to a small spot at the center of the centrosome primarily in anaphase–telophase but is absent from centrosomes during interphase. Since S phase and centrosome duplication in early embryos begins in telophase [112], ZYG-1 may be at the right place at the right time to promote centrosome duplication. It is too early to know what ZYG-1 is doing; it is an orphan kinase with no obvious sequence similarity to members of established kinase subfamilies. Nevertheless, it will be interesting to determine if there are functional homologs to ZYG-1 in other organisms.

9.3.7
Ubiquitin-mediated Proteolysis in the Control of Centrosome Duplication

Ubiquitin-mediated proteolysis of cell-cycle regulatory proteins, including cyclins, proteins involved in maintaining chromatid cohesion, and Cdk inhibitors, is of fundamental importance for the execution and irreversibility of a number of transition points in the cell cycle – including the G1–S transition when centrioles duplicate [113]. In the same vein, proteolysis of centrosomal proteins might eliminate proteins that limit the start of centrosome duplication, participate with Nek2A in severing the connection between duplicated centrosomes in G2, and block re-duplication by degrading proteins on the centrosome necessary for duplication. Consistent with this notion, a variety of components of the SCF proteolysis pathways as well as the 26-S proteasome have been localized to centrosomes throughout the cell cycle in human cells [114–117]. The investigation of the role of proteolysis in the control of centrosome duplication is in its infancy and this body of work has been reviewed elsewhere; thus, we will cover only its outlines here (see [118, 119]).

Specific proteins are targeted for degradation by the covalent attachment of a chain of ubiquitin proteins by multicomponent ubiquitin ligase complexes or

E3s (reviewed in [120]). Such ubiquitinated proteins are recognized by the 26-S proteasome and completely degraded while the ubiquitin is recycled. Various E3 complexes have different substrate specificities that determine which particular proteins are targeted for proteolysis at any given time [118, 120].

Functional evidence that SCF-mediated E3 activity is involved in initiating centrosome reproduction comes from the finding that antibody inhibition of either Skp1 or Cul1 greatly diminished the percentage of mother–daughter centriole pairs that split apart in a *Xenopus* egg extract and that this was independent of the SCF-mediated degradation of p27 required for the activation of Cdk2–cyclin E [115]. The importance of proteolysis was confirmed by the observation that inhibition of the 26-S proteasome with high concentrations of a proteasome inhibitor blocked the splitting of the mother–daughter centriole pairs in *Xenopus* egg extracts and the re-duplication of centrosomes in early *Xenopus* embryos treated with cycloheximide.

This story has been complicated by reports suggesting that SCF-mediated proteolysis is also required to prevent centrosomes from accumulating to abnormally high numbers. Nakayama and co-workers [121], produced a mouse knockout for Skp2, an F-box protein involved in targeting the degradation of – amongst other proteins – the Cdk inhibitor $p27^{kip1}$ [122]. The fact that $Skp2^{-/-}$ mice are viable indicates that centrosome function and duplication are not grossly abnormal. However, their finding that 38 % of $Skp2^{-/-}$mouse embryonic fibroblasts contained 3–12 centrosomes per cell suggested that defects in substrate-specific proteolysis led to centrosome re-duplication. However, many $Skp2^{-/-}$ cells showed evidence of problems with spindle assembly and/or function which raises the possibility that centrosomes might accumulate through cleavage failure or other mitotic dysfunctions (see [123]). Also of interest is the report of Wojcik and co-workers [124] that 66 % of the neuroblasts of crd *Drosophila* embryos contained 3–17 centrosomes per cell. Crd is a mutant allele of supernumerary limbs (Slimb), a previously identified F-box protein [125]. Also, a recent study has provided compelling evidence that centrosome amplification due to null mutations in skpA, the most abundant *Drosophila* skp1-related gene product, is not due to the over-accumulation of cyclin E [126].

It is puzzling, at first glance, that SCF-mediated proteolysis is required to initiate centrosome reproduction but also appears to be important for preventing centrosome re-duplication. The resolution of this apparent paradox may lie in the substrate specificities of different E3 complexes. Perhaps Skp1- or Cul1-containing SCF complexes are needed to target certain centrosomal proteins for degradation thereby allowing duplication, whereas Skp2- and/or Slimb-containing SCF complexes are required to degrade other proteins in order to block the reproductive capacity of duplicated centrosomes (see [74]).

9.4 Closing Remarks

It has become evident that the controls for centrosome duplication are more complex than we expected. Perhaps this is not surprising because centrosome duplication involves the highly regulated assembly of complex, multifunctional organelles such as the centrioles and the pericentriolar material, which in turn must interface with the pathways that regulate nuclear events in the cell cycle. Since the penalties to the organism for having cells with abnormal centrosome content can be severe, one could expect that it is not a simple proposition to organize a control strategy for centrosome duplication that ensures no mistakes, cell cycle after cell cycle.

Acknowledgments

This work was supported by grants to GS from the NIH (GM 30758). I thank Drs. Anna Krzywicka and particularly Erich Nigg for constructive comments on this manuscript. I also thank Mr. Chris English for his patience and expert help in preparing the figures.

I respectfully dedicate this chapter to the memory of Berth Woodward, the laboratory manager for the Shinya Inoue group who took such wonderful care of us for many years.

References

1. Heneen, W. K. **1970**. *In situ* analysis of normal and abnormal patterns of the mitotic apparatus in cultured rat-kangaroo cells. *Chromosoma*. 29:88–117.
2. Heneen, W. K. **1975**. Kinetochores and microtubules in multipolar mitosis and chromosome orientation. *Exp Cell Res* 91:57–62.
3. Sluder, G., E. A. Thompson, F. J. Miller, J. Hayes, and C. L. Rieder. **1997**. The checkpoint control for anaphase onset does not monitor excess numbers of spindle poles or bipolar spindle symmetry. *J Cell Sci* 110:421–429.
4. Orr-Weaver, T. L. and R. A. Weinberg. **1998**. A checkpoint on the road to cancer. *Nature* 392:223–224.
5. Brinkley, B. R. **2001**. Managing the centrosome numbers game: from chaos to stability in cancer cell division. *Trends Cell Biol* 11:18–21.
6. Nigg, E. A. **2002**. Centrosome aberrations: cause or consequence of cancer progression? *Nature Rev Cancer* 2:815–825.
7. Lengauer, C., K. W. Kinzler, and B. Vogelstein. **1998**. Genetic instabilities in human cancers. *Nature* 396:643–649.
8. D'Assoro, A. B., W. L. Lingle, and J. L. Salisbury. **2002**. Centrosome amplification and the development of cancer. *Oncogene* 21:6146–6153.
9. Goepfert, T. M., Y. E. Adigun, L. Zhong, J. Gay, D. Medina, and W. R. Brinkley. **2002**. Centrosome amplification and overexpression of aurora A are early events in rat mammary carcinogenesis. *Cancer Res* 62:4115–4122.
10. Pihan, G. A., J. Wallace, Y. Zhou, and S. J. Doxsey. **2003**. Centrosome ab-

normalities and chromosome instability occur together in pre-invasive carcinomas. *Cancer Res* 63:1398–1404.

11. Mack, G. and J. B. Rattner. **1993**. Centrosome repositioning immediately following karyokinesis and prior to cytokinesis. *Cell Motil Cytoskeleton* 26:239–247.
12. Piel, M., P. Meyer, A. Khodjakov, C. L. Rieder, and M. Bornens. **2000**. The respective contributions of the mother and daughter centrioles to centrosome activity and behavior in vertebrate cells. *J Cell Biol* 149:317–330.
13. Piel, M., J. Nordberg, U. Euteneuer, and M. Bornens. **2001**. Centrosome-dependent exit of cytokinesis in animal cells. *Science* 291:1550–1553.
14. Kuriyama, R. and G. G. Borisy. **1981**. Centriole cycle in Chinese hamster ovary cells as determined by whole-mount electron microscopy. *J Cell Biol* 91:814–821.
15. Wheatley, D. N. **1982**. *The Centriole: A Central Enigma of Cell Biology.* Elsevier Biomedical Press: Amsterdam, 232.
16. Lacey, K. R., P. K. Jackson, and T. Stearns. **1999**. Cyclin-dependent kinase control of centrosome duplication. *Proc Natl Acad Sci USA* 96:2817–2822.
17. Robbins, E. L., G. Jentzsch, and A. Micali. **1968**. The centriole cycle in synchronized HeLa cells. *J Cell Biol* 36:329–339.
18. Vorobjev, I. A. and Y. S. Chentsov. **1982**. Centrioles in the cell cycle I epithelial cells. *J Cell Biol* 93:938–949.
19. Alvey, P. L. **1985**. An investigation of the centriole cycle using 3T3 and CHO cells. *J Cell Sci* 78:147–162.
20. Lange, B. M., A. J. Faragher, P. March, and K. Gull. **2000**. Centriole duplication and maturation in animal cells. *Curr Top Dev Biol* 49:235–249.
21. Paintrand, M., M. Moudjou, H. Delacroix, and M. Bornens. **1992**. Centrosome organization and centriole architecture: their sensitivity to divalent cations. *J Struct Biol* 108:107–128.
22. Marshall, W. F. **2001**. Centrioles take center stage. *Curr Biol* 11:R487–R496.
23. Bornens, M. **2002**. Centrosome composition and microtubule anchoring mechanisms. *Curr Opin Cell Biol* 14:25–34.
24. Aubin, J. E., M. Osborn, and K. Weber. **1980**. Variations in the distribution and migration of centriole duplexes in mitotic PtK_2 cells studied by immunofluorescence microscopy. *J Cell Sci* 43:177–194.
25. Kochanski, R. S. and G. G. Borisy. **1990**. Mode of centriole duplication and distribution. *J Cell Biol* 110:1599–1605.
26. Rieder, C. L. **1990**. Formation of the astral mitotic spindle: ultrastructural basis for the centrosome-kinetochore interaction. *Electron Microsc Rev* 3:269–300.
27. Meraldi, P. and E. A. Nigg. **2001**. Centrosome cohesion is regulated by a balance of kinase and phosphatase activities. *J Cell Sci* 114:3749–3757.
28. Fry, A. M. **2002**. The Nek2 protein kinase: a novel regulator of centrosome structure. *Oncogene* 21:6184–6194.
29. Rieder, C. L. and G. G. Borisy. **1981**. The attachment of kinetichores to the pro-metaphase spindle in PtK_1 cells. *Chromosoma (Berl).* 82:693–716.
30. Sluder, G. and D. A. Begg. **1985**. Experimental analysis of the reproduction of spindle poles. *J Cell Sci* 76:35–51.
31. Sluder, G. and C. L. Rieder. **1985**. Centriole number and the reproductive capacity of spindle poles. *J Cell Biol* 100:887–896.
32. Hinchcliffe, E. H., G. O. Cassels, C. L. Rieder, and G. Sluder. **1998**. The coordination of centrosome reproduction with nuclear events during the cell cycle in the sea urchin zygote. *J Cell Biol* 140:1417–1426.
33. Gallant, P. and E. A. Nigg. **1992**. Cyclin B2 undergos cell cycle-depentant nuclear translocation and, when expressed as a non-destructable mutant, causes mitotic arrest in HeLa cells. *J Cell Biol* 117:213–224.
34. Faragher, A. J. and A. M. Fry. **2003**. Nek2A kinase stimulates centrosome disjunction and is required for formation of bipolar mitotic spindles. *Mol Biol Cell* 14:2876–2889.

35. Mailand, N., C. Lukas, B. K. Kaiser, P. K. Jackson, J. Bartek, and J. Lukas. **2002**. Deregulated human Cdc14A phosphatase disrupts centrosome separation and chromosome segregation. *Nature Cell Biol* 4:317–322.
36. Lutz, W., W. L. Lingle, D. McCormick, T. M. Greenwood, and J. L. Salisbury. **2001**. Phosphorylation of centrin during the cell cycle and its role in centriole separation preceding centrosome duplication. *J Biol Chem* 276:20774–20780.
37. Dippell, R. V. **1968**. The development of basal bodies in paramecium. *Proc Natl Acad Sci USA* 61:461–468 available.
38. Gould, R. R. **1975**. The basal bodies of *Chlamydomonas reinhardtii*: formation from probasal bodies, isolation, and partial characterization. *J Cell Biol* 65:65–74.
39. Beisson, J. and M. Wright. **2003**. Basal body/centriole assembly and continuity. *Curr Opin Cell Biol* 15:96–104.
40. Vidwans, S. J., M. L. Wong, and P. H. O'Farrell. **2003**. Anomalous centriole configurations are detected in Drosophila wing disc cells upon Cdk1 inactivation. *J Cell Sci* 116:137–143.
41. Von Ledebur-Villiger, M. **1972**. Cytology and nucleic acid synthesis of parthenogenetically activated sea urchin eggs. *Exp Cell Res* 72:285–308.
42. Kallenbach, R. J. and D. Mazia. **1982**. Origin and maturation of centrioles in association with the nuclear envelope in hypertonic-stressed sea urchin eggs. *Eur J Cell Biol* 28:68–76.
43. Sorokin, S. P. **1968**. Reconstructions of centriole formation and ciliogenesis in mammalian lungs. *J Cell Sci* 3:207–230.
44. Dirksen, E. R. **1991**. Centriole and basal body formation during ciliogenesis revisited. *Biol Cell* 72:31–38.
45. Khodjakov, A., C. L. Rieder, G. Sluder, G. Cassels, O. Sibon, and C. L. Wang. **2002**. De novo formation of centrosomes in vertebrate cells arrested during S phase. *J Cell Biol* 158:1171–1181.
46. Szollosi, D., P. Calarco, and P. Donahue. **1972**. Absence of centrioles in the first and second meiotic spindles of mouse oocytes. *J Cell Sci* 11:521–541.
47. Fulton, C. and A. D. Dingle. **1971**. Basal bodies, but not centrioles, in *Naegleria*. *J Cell Biol* 51:826–836.
48. Palazzo, R. E., E. Vaisberg, R. W. Cole, and C. L. Rieder. **1992**. Centriole dulication in lysates of *Spisula solidissima* oocytes. *Science* 256:219–221.
49. Fulton, C. **1971**. Centrioles. In *Origin and Continuity of Cell Organelles*, C. Beerman, et al. (eds). Springer-Verlag; New York, 170–213.
50. Marshall, W. F. and J. L. Rosenbaum. **2000**. Are there nucleic acids in the centrosome? *Curr Top Dev Biol* 49:187–205.
51. Wiese, C. and Y. Zheng. **1999**. Gamma-tubulin complexes and their interaction with microtubule-organizing centers. *Curr Opin Struct Biol* 9:250–259.
52. Ruiz, F., J. Beisson, J. Rossier, and P. Dupuis-Williams. **1999**. Basal body duplication in Paramecium requires gamma-tubulin. *Curr Biol* 9:43–46.
53. Shang, Y., B. Li, and M. A. Gorovsky. **2002**. *Tetrahymena thermophila* contains a conventional gamma-tubulin that is differentially required for the maintenance of different microtubule-organizing centers. *J Cell Biol* 158:1195–1206.
54. Dutcher, S. K. and E. C. Trabuco. **1998**. The UNI3 gene is required for assembly of basal bodies of Chlamydomonas and encodes delta-tubulin, a new member of the tubulin superfamily. *Mol Biol Cell* 9:1293–1308.
55. Chang, P. and T. Stearns. **2000**. Delta-tubulin and epsilon-tubulin: two new human centrosomal tubulins reveal new aspects of centrosome structure and function (comment). *Nature Cell Biol* 2:30–35.
56. Ruiz, F., A. Krzywicka, C. Klotz, A. Keller, J. Cohen, F. Koll, G. Balavoine, and J. Beisson. **2000**. The SM19 gene, required for duplication of basal bodies in *Paramecium*, encodes a novel tubulin, eta-tubulin. *Curr Biol* 10:1451–1454.
57. Dutcher, S. K., N. S. Morrissette, A. M. Preble, C. Rackley, and J. Stanga. **2002**. Epsilon-tubulin is an essential compo-

nent of the centriole. *Mol Biol Cell* 13:3859–3869.

58. Chang, P., T. H. Giddings, Jr, M. Winey, and T. Stearns. **2003**. Epsilon-tubulin is required for centriole duplication and microtubule organization. *Nature Cell Biol* 5:71–76.

59. Dutcher, S. K. **2001**. The tubulin fraternity: alpha to eta. *Curr Opin Cell Biol* 13:49–54.

60. Dutcher, S. K. **2001**. Motile organelles: the importance of specific tubulin isoforms. *Curr Biol* 11:R419–R422.

61. Paoletti, A., M. Moudjou, M. Paintrand, J. L. Salisbury, and M. Bornens. **1996**. Most of centrin in animal cells is not centrosome-associated and centrosomal centrin is confined to the distal lumen of centrioles. *J Cell Sci* 109:3089–3102.

62. Salisbury, J. L., K. M. Suino, R. Busby, and M. Springett. **2002**. Centrin-2 is required for centriole duplication in mammalian cells. *Curr Biol* 12:1287–1292.

63. Kirkham, M., T. Muller-Reichert, K. Oegema, S. Grill, and A. A. Hyman. **2003**. SAS-4 is a *C. elegans* centriolar protein that controls centrosome size. *Cell* 112:575–587.

64. Leidel, S. and P. Gonczy. **2003**. SAS-4 is essential for centrosome duplication in *C. elegans* and is recruited to daughter centrioles once per cell cycle. *Dev Cell* 4:431–439.

65. Sluder, G., F. J. Miller, R. Cole, and C. L. Rieder. **1990**. Protein synthesis and the cell cycle: centrosome reproduction in sea urchin eggs is not under translational control. *J Cell Biol* 110:2025–2032.

66. Gard, D. L., S. Hafezi, T. Zhang, and S. J. Doxsey. **1990**. Centrosome duplication continues in cycloheximide-treated *Xenopus* blastulae in the absence of a detectable cell cycle. *J. Cell Biol* 110:2033–2042.

67. Mazia, D., P. Harris, and T. Bibring. **1960**. The multiplicity of the mitotic centers and the time-course of their duplication and separation. *Biophys Biochem Cytol* 7:1–20.

68. Sluder, G., F. J. Miller, and C. L. Rieder. **1989**. Reproductive capacity of sea urchin centrosomes without centrioles. *Cell Motil Cytoskeleton*. 13:264–273.

69. Sluder, G. and K. Lewis. **1987**. Relationship between nuclear DNA synthesis and centrosome reproduction in sea urchin eggs. *J Exp Zool* 244:89–100.

70. Raff, J. W. and D. M. Glover. **1988**. Nuclear and cyctoplasmic mitotic cycles continue in *Drosophila* embryos in which DNA synthesis is inhibited by aphidicolin. *J Cell Biol* 107:2009–2019.

71. Balczon, R., L. Bao, W. E. Zimmer, K. Brown, R. P. Zinkowski, and B. R. Brinkley. **1995**. Dissociation of centrosome replication events from cycles of DNA Synthesis and mitotic division in hydroxyurea-arrested Chinese hamster ovary cells. *J Cell Biol* 130:105–115.

72. Winey, M. **1999**. Cell cycle: driving the centrosome cycle. *Curr Biol* 9:R449–R452.

73. Borel, F., O. D. Lohez, F. B. Lacroix, and R. L. Margolis. **2002**. Multiple centrosomes arise from tetraploidy checkpoint failure and mitotic centrosome clusters in p53 and RB pocket protein-compromised cells. *Proc Natl Acad Sci USA* 99:9819–9824.

74. Wong, C. and T. Stearns. **2003**. Centrosome number is controlled by a centrosome-intrinsic block to re-duplication. *Nature Cell Biol* 5:539–544.

75. Johnson, R. and P. Rao. **1971**. Neuclocytoplasmic interactions in the achievement of nuclear synchrony in DNA synthesis and mitosis in multinucleate cells. *Biol Rev* 46:97–155.

76. Mussman, J. G., H. F. Horn, P. E. Carroll, M. Okuda, P. Tarapore, L. A. Donehower, and K. Fukasawa. **2000**. Synergistic induction of centrosome hyperamplification by loss of p53 and cyclin E overexpression. *Oncogene* 19:1635–1646.

77. Sluder, G., F. J. Miller, and C. L. Rieder. **1986**. The Reproduction of Centrosomes: Nuclear versus Cytoplasmic Controls. *J Cell Biol* 103:1873–1881.

78. Nguyen, V. Q., C. Co, and J. J. Li. **2001**. Cyclin-dependent kinases prevent DNA re-replication through multiple mechanisms. *Nature* 411:1068–1073.

79. Dulic, V., E. Lees, and S. I. Reed. **1992**. Association of human cyclin E with a periodic G1-S phase protein kinase. *Science* 257:1958–1961.
80. Strausfeld, U. P., M. Howell, P. Descombes, S. Chevalier, R. Rempel, J. Adamczewski, J. L. Maller, T. Hunt, and J. J. Blow. **1996**. Both cyclin A and cyclin E have S-phase promoting (SPF) activity in *Xenopus* egg extracts. *J Cell Sci* 109:1555–1563.
81. Hinchcliffe, E. H., C. Li, E. A. Thompson, J. L. Maller, and G. Sluder. **1999**. Requirement of Cdk2-cyclin E activity for repeated centrosome reproduction in *Xenopus* egg extracts. *Science* 283:851–854.
82. Su, J.-Y., R. E. Rempel, E. Erikson, and J. L. Maller. **1995**. Cloning and characterization of the *Xenopus* cyclin-dependant kinase inhibitor $p27^{XIC1}$. *Proc Natl Acad Sci USA* 92:10187–10191.
83. Rempel, R. E., S. B. Sleight, and J. L. Maller. **1995**. Maternal *Xenopus* Cdk2–Cyclin E complexs function during meiotic and early embryonic cell cycles that lack a G_1 phase. *J Biol Chem* 270:6843–6855.
84. Elledge, S. J. and J. W. Harper. **1994**. Cdk inhibitors: on the threshold of checkpoints and development. *Curr Opin Cell Biol* 6:847–852.
85. Jackson, P. K., S. Chevalier, M. Philippe, and M. W. Kirschner. **1995**. Early events in DNA replication require cyclin E and are blocked by p21CIP1. *J ell Biol* 130:755–769.
86. Mantel, C., S. E. Braun, S. Reid, O. Henegariu, L. Liu, G. Hangoc, and H. E. Broxmeyer. **1999**. p21(cip-1/waf-1) deficiency causes deformed nuclear architecture, centriole overduplication, polyploidy, and relaxed microtubule damage checkpoints in human hematopoietic cells. *Blood* 93:1390–1398.
87. Matsumoto, Y., K. Hayashi, and E. Nishida. **1999**. Cyclin-dependent kinase 2 (Cdk2) is required for centrosome duplication in mammalian cells. *Curr Biol* 9:429–432.
88. Meraldi, P., J. Lukas, A. M. Fry, J. Bartek, and E. A. Nigg. **1999**. Centrosome duplication in mammalian somatic cells requires E2F and Cdk2–cyclin A. *Nature Cell Biol* 1:88–93.
89. Balczon, R. C. **2001**. Overexpression of cyclin A in human HeLa cells induces detachment of kinetochores and spindle pole/centrosome overproduction. *Chromosoma* 110:381–392.
90. Berthet, C., E. Aleem, V. Coppola, L. Tessarollo, and P. Kaldis. **2003**. Cdk2 knockout mice are viable. *Curr Biol* 13:1775–1785.
91. Ortega, S., I. Prieto, J. Odajima, A. Martin, P. Dubus, R. Sotillo, J. L. Barbero, M. Malumbres, and M. Barbacid. **2003**. Cyclin-dependent kinase 2 is essential for meiosis but not for mitotic cell division in mice. *Nature Genet* 35:25–31.
92. Geng, Y., Q. Yu, E. Sicinska, M. Das, J. E. Schneider, S. Bhattacharya, W. M. Rideout, R. T. Bronson, H. Gardner, and P. Sicinski. **2003**. Cyclin E ablation in the mouse (comment). *Cell* 114:431–443.
93. Okuda, M., H. F. Horn, P. Tarapore, Y. Tokuyama, A. G. Smulian, P. K. Chan, E. S. Knudsen, I. A. Hofmann, J. D. Snyder, K. E. Bove, and K. Fukasawa. **2000**. Nucleophosmin/B23 is a target of CDK2/cyclin E in centrosome duplication. *Cell* 103:127–140.
94. Schmidt-Zachmann, M. S., B. Hugle-Dorr, and W. W. Franke. **1987**. A constitutive nucleolar protein identified as a member of the nucleoplasmin family. *EMBO J* 6:1881–1890.
95. Tokuyama, Y., H. F. Horn, K. Kawamura, P. Tarapore, and K. Fukasawa. **2001**. Specific phosphorylation of nucleophosmin on Thr(199) by cyclin-dependent kinase 2-cyclin E and its role in centrosome duplication. *J Biol Chem* 276:21529–21537.
96. Chen, Z., V. B. Indjeian, M. McManus, L. Wang, and B. D. Dynlacht. **2002**. CP110, a cell cycle-dependent CDK substrate, regulates centrosome duplication in human cells. *Dev Cell* 3:339–350.
97. Fisk, H. A. and M. Winey. **2001**. The mouse Mps1p-like kinase regulates centrosome duplication. *Cell* 106:95–104.

98. Winey, M., L. Goetsch, P. Baum, and B. Byers. **1991**. MPS1 and MPS2: novel yeast genes defining distinct steps of spindle pole body duplication. *J Cell Biol* 114:745–754.

99. Stucke, V. M., H. H. Sillje, L. Arnaud, and E. A. Nigg. **2002**. Human Mps1 kinase is required for the spindle assembly checkpoint but not for centrosome duplication. *EMBO J* 21:1723–1732.

100. Fisk, H., C. Mattison, and M. Winey. **2003**. The human Mps1 protein kinase is required for centrosome duplication and normal mitotic progression. *Proc Natl Acad Sci USA* (in press).

101. Liu, S. T., G. K. Chan, J. C. Hittle, G. Fujii, E. Lees, and T. J. Yen. **2003**. Human MPS1 kinase is required for mitotic arrest induced by the loss of CENP-E from kinetochores. *Mol Biol Cell* 14:1638–1651.

102. Matsumoto, Y. and J. L. Maller. **2002**. Calcium, calmodulin, and CaMKII requirement for initiation of centrosome duplication in *Xenopus* egg extracts (comment). *Science* 295:499–502.

103. Means, A. R. **1994**. Calcium, calmodulin and cell cycle regulation. *FEBS Lett* 347:1–4.

104. Swanson, C. A., A. P. Arkin, and J. Ross. **1997**. An endogenous calcium oscillator may control early embryonic division. *Proc Natl Acad Sci USA* 94:1194–1199.

105. Ohta, Y., T. Ohba, and E. Miyamoto. **1990**. Ca^{2+}/calmodulin-dependent protein kinase II: localization in the interphase nucleus and the mitotic apparatus of mammalian cells. *Proc Natl Acad Sci USA* 87:5341–5345.

106. Pietromonaco, S. F., G. A. Seluja, and L. Elias. **1995**. Identification of enzymatically active Ca^{2+}/calmodulin-dependent protein kinase in centrosomes of hemopoietic cells. *Blood Cells, Mol Dis* 21:34–41.

107. Tokmakov, A. A., K. I. Sato, and Y. Fukami. **2001**. Calcium oscillations in *Xenopus* egg cycling extracts. *J Cell Biochem* 82:89–97.

108. Wood, W. B., R. Hecht, S. Carr, R. Vanderslice, N. Wolf, and D. Hirsh. **1980**. Parental effects and phenotypic characterization of mutations that affect early development in *Caenorhabditis elegans*. *Dev Biol* 74:446–469.

109. O'Connell, K. F., C. M. Leys, and J. G. White. **1998**. A genetic screen for temperature-sensitive cell-division mutants of *Caenorhabditis elegans*. *Genetics* 149:1303–1321.

110. O'Connell, K. F., C. Caron, K. R. Kopish, D. D. Hurd, K. J. Kemphues, Y. Li, and J. G. White. **2001**. The *C. elegans* zyg-1 gene encodes a regulator of centrosome duplication with distinct maternal and paternal roles in the embryo. *Cell* 105:547–558.

111. O'Connell, K. F. **2002**. The ZYG-1 kinase, a mitotic and meiotic regulator of centriole replication. *Oncogene* 21:6201–6208.

112. Edgar, L. G. and J. D. McGhee. **1988**. DNA synthesis and the control of embryonic gene expression in *C. elegans*. *Cell* 53:589–599.

113. King, R. W., R. J. Deshaies, J.-M. Peters, and M. W. Kirschner. **1996**. How proteolysis drives the cell cycle. *Science* 274:1652–1659.

114. Tugendreich, S., J. Tomkiel, W. Earnshaw, and P. Hieter. **1995**. CDC27Hs colocalizes with CDC16Hs to the centrosome and mitotic spindle and is essential for the metaphase to anaphase transition. *Cell* 81:261–268.

115. Freed, E., K. R. Lacey, P. Huie, S. A. Lyapina, R. J. Deshaies, T. Stearns, and P. K. Jackson. **1999**. Components of an SCF ubiquitin ligase localize to the centrosome and regulate the centrosome duplication cycle. *Genes Dev* 13:2242–2257.

116. Gstaiger, M., A. Marti, and W. Krek. **1999**. Association of human SCFSKP2 subunit p19 (SKP1) with interphase centrosomes and mitotic spindle poles. *Exp Cell Res* 247:554–562.

117. Wigley, W. C., R. P. Fabunmi, M. G. Lee, C. R. Marino, S. Muallem, G. N. DeMartino, and P. J. Thomas. **1999**. Dynamic association of proteasomal machinery with the centrosome. *J Cell Biol* 145:481–490.

118. Hansen, D. V., J. Y. Hsu, B. K. Kaiser, P. K. Jackson, and A. G. Eldridge. **2002**. Control of the centriole and centro-

some cycles by ubiquitination enzymes. *Oncogene* 21:6209–6221.

119. Hinchcliffe, E. H. and G. Sluder. **2002**. Two for two: Cdk2 and its role in centrosome doubling. *Oncogene* 21:6154–6160.

120. Jackson, P. K., A. G. Eldridge, E. Freed, L. Furstenthal, J. Y. Hsu, B. K. Kaiser, and J. D. Reimann. **2000**. The lore of the RINGs: substrate recognition and catalysis by ubiquitin ligases. *Trends Cell Biol* 10:429–439.

121. Nakayama, K., H. Nagahama, Y. A. Minamishima, M. Matsumoto, I. Nakamichi, K. Kitagawa, M. Shirane, R. Tsunematsu, T. Tsukiyama, N. Ishida, M. Kitagawa, and S. Hatakeyama. **2000**. Targeted disruption of Skp2 results in accumulation of cyclin E and p27(Kip1), polyploidy and centrosome overduplication. *EMBO J* 19:2069–2081.

122. Carrano, A. C., E. Eytan, A. Hershko, and M. Pagano. **1999**. SKP2 is required for ubiquitin-mediated degradation of the CDK inhibitor p27. *Nature Cell Biol* 1:193–199.

123. Meraldi, P., R. Honda, and E. A. Nigg. **2002**. Aurora-A overexpression reveals tetraploidization as a major route to centrosome amplification in p53−/− cells. *EMBO J* 21:483–492.

124. Wojcik, E. J., D. M. Glover, and T. S. Hays. **2000**. The SCF ubiquitin ligase protein slimb regulates centrosome duplication in *Drosophila*. *Curr Biol* 10:1131–1134.

125. Jiang, J. and G. Struhl. **1998**. Regulation of the Hedgehog and Wingless signalling pathways by the F-box/WD40-repeat protein Slimb. *Nature* 391:493–496.

126. Murphy, T. **2003**. *Drosophila skpA*, a component of SCF ubiquitin ligases, regulates centrosome duplication independently of cyclin E accumulation. *J Cell Sci* 116:2321–2332.

127. Hinchcliffe, E. H. and G. Sluder. **2001**. "It takes two to tango": understanding how centrosome duplication is regulated throughout the cell cycle. *Genes Dev* 15:1167–1181.

10
A Synergy of Technologies: Using Green Fluorescent Protein Tagging and Laser Microsurgery to Study Centrosome Function and Duplication in Vertebrates

Alexey Khodjakov and Conly L. Rieder

10.1
Introduction

Although Flemming first described the centrosome in the eggs of the freshwater mussel *Anodonta* in 1875, it was not understood to be a permanent cell organelle until the work of Boveri (who named it in 1888) and van Beneden (see [1]). The aura of mystery that continues to surround this beautiful structure has been fueled by the fact that the centrosome possesses a number of properties that make it unlike any other organelle. For example, the centrosome is the only cytoplasmic structure in a diploid cell that is present in a single copy. However, at the structural level, each centrosome appears to be doubled, in that it consists of two similar-looking cylindrical structures (termed the centrioles) that may (or may not) function together as a single unit. Also, like DNA, but unlike other organelles, the centrosome is replicated only once per cell cycle, during the S period. Moreover, during mitosis the centrosome's constituent centrioles become distributed between the daughter cells in a semi-conservative fashion (that is, each cell inherits one "old" and one "new" centriole).

Another intriguing property of the centrosome is that it appears to be an essential organelle in those animal cells which normally possess it, whereas most plant cells thrive without it. Indeed, the fact that there are no viable mutants that lack centrosomes implies that it performs one or more vital functions in animals. Despite more than 125 years of work on the topic, it is still easier to define what this organelle does not do rather than what it does do. For example, in the 1970s, the conspicuous and ubiquitous ability of the centrosome to nucleate microtubules led to the consensus that its essential roles revolve around this function. However, it is now clear that most of the activities of the centrosome that relate to its role as a microtubule organizing center can be achieved via parallel, centrosome-independent pathways, even the formation of the mitotic spindle [2–6].

The various roles that the centrosome plays during the life of a cell have been difficult to define. Perhaps the major difficulty is that the function of the organelle,

Centrosomes in Development and Disease. Edited by Erich A. Nigg

ISBN 3-527-30980-2

as a structural entity, may mask the roles played by its constituents elsewhere in the cell. Individual centrosomal components are studied using genetics and antibody-blocking approaches. These studies have revealed that elimination of certain centrosomal proteins from the cell, such as γ-tubulin, is lethal to the cell and to the organism [7, 8]. By contrast, elimination of those components responsible for binding γ-tubulin to the centrosome has little effect on the viability and behavior of a cell. This is because only a few copies of those proteins that give the centrosome its form and functions are associated with the organelle at any one time. The rest are scattered throughout the cytoplasm [9, 10], and the centrosome-bound and cytoplasmic fractions are often in dynamic exchange [11]. As a result, a centrosomal protein can function independently in the cytoplasm even when its ability to do its job at the centrosome is inhibited. A striking demonstration of this has recently been reported by Megraw and coworkers [12]. They found that, even though centrosomes in *Drosophila* centrosomin mutants (*cnn*$^-$) fail to nucleate microtubules during mitosis, development still occurs at a normal rate, to produce a mature but sterile fly. Mutations in one centrosomal component may affect only a subset of centrosomal functions, leaving other, more essential functions intact. In the case of the *cnn*$^-$ mutants, the cells still possess centrioles and a functionally normal centrosome during interphase, even though the centrosome's capacity to nucleate microtubules is absent during mitosis. Thus, the fact that a particular centrosomal function, such as nucleation of microtubules during mitosis, is not required for the viability of a cell does not mean that the centrosome as an organelle is dispensable. This being the case, the only way to elucidate the role(s) of the centrosome is to selectively remove it from the cell. In this way, those individual components of the centrosome (e. g. γ-tubulin) which are essential for viability are still present.

There are only two ways to eliminate the centrosome from a cell. The first is to physically remove it (along with part of the cytoplasm), and the second is to destroy it within the cell. Through the years, the first approach has been used by a number of investigators. For example, when synchronized cytochalasin-treated L929 cells (transformed mouse fibroblasts) are enucleated by centrifugation through a Ficoll gradient, centrosomes remain in the cytoplasmic fragments (i. e. cytoplasts) while karyoplasts (i. e. nucleus-containing cell fragments) are formed that lack a centrosome [13]. Similarly, when fertilized sea urchin eggs are forced through a nylon screen, some of the egg fragments produced contain a nucleus but not a centrosome [14]. Studies on karyoplasts have led to the idea that the centriole defines the centrosome, or at least its reproductive capacity. The most successful study in this area was that of Maniotis and Schliwa [15], who used conventional needle-microsurgery to separate the centrosome-containing part of the cytoplasm from BSC-1 (African green monkey kidney) cells. The resulting karyoplasts remained viable for several days, grew, and re-organized a focused array of microtubules, but never regenerated a centrosome. Perhaps the most unexpected finding of this study was that the cells also never entered mitosis. This implies that the centrosome is essential for cell cycle progression in vertebrates. Although the evidence at the time suggested that cells lacking centrosomes arrest in G2, a recent re-evaluation using the same methods and cell line reveals that this block actually occurs during G1 [16].

While relatively simple, removal of the centrosome with a glass micro-needle is limited, in that it can be applied to very few types of cultured cells (the only success has been in the BSC-1 cell line), and then only during interphase and not mitosis [17]. Furthermore, this approach also requires removal of a significant portion of the cytoplasm and the organelles that it contains, along with the centrosome. These include the Golgi apparatus, which is usually positioned near the centrosome. As an alternative to microsurgery, the centrosome can be selectively destroyed *in situ* by focused pulses of high-energy light (usually generated by a laser). As we will demonstrate throughout this chapter, this "ablative photo-decomposition" approach has several unique features that make it the method of choice for eliminating the centrosome (or other organelles) from cultured cells. Among its advantages are that it allows the centrosome to be reproducibly and selectively destroyed within a few seconds in a wide variety of cell types, with minimal collateral damage to other cell components. Furthermore, the operation can be conducted during any period of the cell cycle, including mitosis.

In the past, the utility of laser microsurgery for ablating the centrosome was limited by the fact that, in most cells, this organelle is not resolvable by phase-contrast or DIC microscopy (see [18, 19]). This limitation has, however, been overcome through the use of Green Fluorescent Protein (GFP) fusions to make the centrosome visible, as well as to delineate its boundaries in living cells. Combining GFP labeling and laser microsurgery also allows the extent of centrosome damage to be determined in near-real time, through simply monitoring the destruction of the GFP signal. Furthermore, the use of GFP has improved the precision of the laser microsurgery approach to the extent that it is now possible to selectively destroy only a part of the centrosome, for example, only one of the two centrioles.

In this chapter we review the development of laser microsurgery as a tool for studying the centrosome, and we describe how we have used the approach to discover some new and unexpected properties of this organelle.

10.2 Laser Microsurgery

10.2.1 A Brief History of Development

The earliest attempt to selectively destroy part of a cell with light can be traced back to the work of Tchakhotine, who, in 1912, reduced the image of a UV light source to microscopic dimensions using refracting lenses (see [20]). His approach formed the basis of a technique that was used over the next 25 years to study the effects of irradiating small portions of large cells, mostly marine oocytes and embryos. A problem associated with this technique was one of aiming the light source accurately, and this however, limited its utility for studying smaller cells in culture. These problems were overcome in the early 1950s by Zirkle and colleagues. They routed a micro-beam of UV (or proton) radiation, produced by passing a macroscopic beam through a small aperture, into a phase contrast microscope equipped with a reflecting objective [21, 22]. The resolution and precision of their approach was sufficient to enable them to destroy individual kinetochores and part of the primary constriction on newt chromosomes [23]. These early "partial cell irradiation" studies confirmed that directed chromosome motion requires the presence of the kinetochore, and they also suggested that kinetochores play an important role in preventing a premature anaphase [23].

Most cellular components can be easily destroyed by near-UV (200–308 nm) light because they absorb intensely in this part of the spectrum. As a result a popular approach for selectively destroying components or regions of cells, especially cells in mitosis, has been to pass the filtered output of a microscope lamp through an aperture or slit, to generate a UV microbeam (e. g. [24–28]). However, the same characteristics that make UV an effective destroyer of cellular components also produce side-effects that make data interpretation difficult. This problem arises because different cell components have different action spectra for UV absorption, and the intensity of light needed to destroy a specific target may also produce significant collateral damage to more sensitive out-of-focus organelles [29]. Also, since the spectral composition of UV light varies depending on the system, the same procedure can produce different results in different laboratories [30].

UV microbeam systems can also be constructed around lasers (e.g. [31, 32]). Compared to those based on microscope lamps, laser systems produce a coherent monochromatic beam that can be finely tuned. Such systems have become increasingly popular for dissecting various biological processes, especially those in large and light-tolerant embryos (e. g. *Caenorhabditis elegans* [32, 33]). Although the intensity of the UV radiation is greatest within the focal plane of the objective lens, the rest of the cell still receives a considerable dose because it is also experiences a cone of UV light above and below this plane. Therefore, although laser-based UV systems can improve the reproducibility of an experiment, the extreme sensitivity of many cellular systems to UV light often produces unavoidable nonspecific side-effects.

During the development of laser microsurgery as a biological tool, it became evident that any component in the cell can also be destroyed by pulses of light in the *visible spectrum*, as long as the energy is sufficient (reviewed in [34]). There are two important features of this phenomenon that currently remain poorly understood [see, e. g. discussions in [35, 36]). The first is that the distribution of energy across the beam face can be tuned so that the destruction occurs only within the diffraction-limited spot (the Airy disk), where the focused laser light is the most intense (reviewed in [20, 36]). The second is that this type of laser microsurgery does not require that the cell be pre-loaded with exogenous chromophores; that is, the structure of interest need not be selectively "sensitized" to the wavelength of laser light with a vital dye [37]. It is also clear that not all structures are equally sensitive to the same cumulative energy dose: chromosomes are easily cut at energy doses of pulsed green (532 nm) laser light that have no apparent effects on spindle microtubules [36, 38]. However, light at the same wavelength and frequency, at higher energy levels, can be used destroy any intracellular component including centrioles [34], which are among the most stable structures in the cell [39].

In our work we use 7-ns pulses of green (532-nm) light, obtained by doubling the frequency of the 1064-nm output of a Nd : YAG laser, to ablate or cut cell structures in cultured vertebrate cells [38]. As a rule we run our system at 10 pulses s^{-1}, and for most experiments each pulse contains $\sim$ 400 nJ at the focal plane. Correlative LM (light microscopy) and EM (electron microscopy) studies reveal that just one pulse from this system is sufficient to sever a microtubule or destroy a mitochondrion in a living cell, and only 2–3 pulses are required to destroy a kinetochore [40]. For a 60-100X 1.4 NA lens, the area of damage caused by each pulse corresponds to an ellipsoid-shaped volume of $\sim 0.4 \times 0.4 \times 0.6$ μm [34, 36, 38, 40, 41].

In order to achieve the highest resolution and detectability in the transmitted-light mode, we have centered our system around an inverted microscope equipped with de Senarmont compensation DIC optics. At one time, the major technical problem in combining high-quality fluorescence and laser microsurgery was that both the laser beam and the epi-fluorescence excitation light needed to be steered toward the lens through the same epi-port. In order to switch between the laser-cutting and GFP-observation modes, the dichroic mirrors used for the laser beam and for the excitation light had to be moved. However, since microscopes are now available that are equipped with two independent epi-ports (e. g. Nikon TE-2000E), this problem has been alleviated. With the advent of such microscopes, assembling a laser-microsurgery system becomes a feasible and affordable task for an average-sized cell-biology laboratory. This advance will certainly make laser microsurgery an increasingly popular method, particularly in the centrosome field.

10.2.2
Utility for Removing the Centrosome

The first attempts to remove the centrosome by laser light were those of Berns and his colleagues [18, 19, 42] (see also [31]). These pioneers used the blue (473-nm) or green (514-nm) wavelength from an argon ion laser, to destroy the centrosome in cultured rat-kangaroo kidney cells (PtK_2) that had been pre-treated with acridine orange to make the organelle more light-sensitive. Acridine orange binds to nucleic acids, which at that time were thought to be a centrosomal component. It was clear from these studies, most of which were conducted on cells entering mitosis, that the approach worked. However, it was also evident that there were two problems that impeded any reliable interpretation of the data. First, the centrosome was barely resolvable with the imaging methods available in the late 1970s, which predated video-enhanced microscopy. As a result, the position of the centrosome against a background of similarly sized particles was difficult to define with any certainty. Second, because the centrosome lacks a limiting membrane, its boundary within the living cell was too vague and could not be accurately determined, even under the best imaging conditions of the time. As a result, the failure or success of an experiment (that is, whether the centrosome was destroyed or damaged or left unharmed after the operation) could not be determined immediately. Instead, the extent of damage could only be assayed after the study, by fixing the irradiated cell for a subsequent serial-section EM analysis. The intensive labor required for these types of analyses necessarily limited the numbers of cells that could be followed in the studies.

The advent and widespread use of GFP technology in the mid-1990s prompted us to re-explore the potential of using laser microsurgery to selectively destroy the centrosome in tissue-culture cells. We reasoned that the position of this organelle and also its boundary could be clearly defined under low-light level imaging conditions in live cells expressing a GFP-tagged construct that targets the centrosome. We chose γ-tubulin [11], a protein which is required for microtubule nucleation and which is distributed throughout the pericentriolar material. In 1997, using high-energy (300–400 nJ) pulses of 532-nm (green) laser light, we proved that this "synergy of technologies" approach could be used to reproducibly, selectively, and completely to destroy the centrosome within 1–2 s (10–20 pulses) in cultured

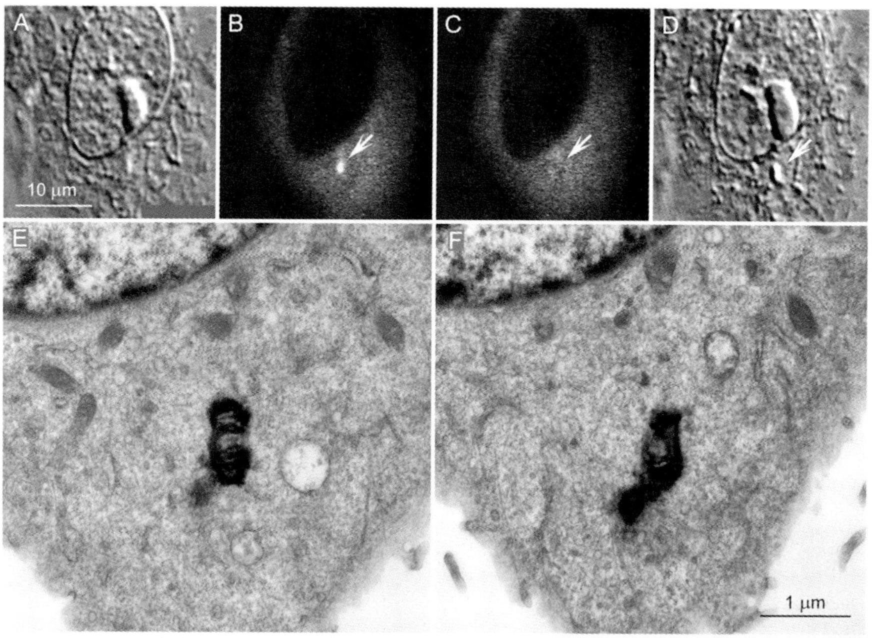

Figure 10.1 Laser-induced ablative photo-decomposition initially converts the centrosome into a dense aggregate of denatured material that is destroyed over time. In this example, the centrosome in an interphase PtK$_1$ cell expressing GFP/γ-tubulin (B, arrow) was destroyed by 15 pulses of laser light (C). The operation produces the formation of a scar, which is clearly visible by DIC microscopy (arrow in D). This cell was then fixed 2 min after the operation and processed for EM. An analysis of serial semi-thick (0.25-μm) sections (E and F) revealed that the scar, which was formed from the contents of the centrosome, is composed of an amorphous, electron-opaque material. Video–LM studies reveal that, over time, the scar will wander randomly within the cell until it finally disappears.

cells expressing GFP-γ-tubulin [34]. It was apparent from same-cell correlative LM and EM studies that the mother and daughter centrioles within an irradiated centrosome, as well as the surrounding pericentriolar material, were converted by the operation into a dense, irregular, and electron-opaque coagulation of denatured protein (Figure 10.1). This "scar" (which we termed a "singlet", see [38] and Figure 10.1E and D) to the material produced during the operation eventually degraded, so we were not able to detect it > 2 h post-irradiation within the cell. Our studies also revealed that the pulses of high energy laser light did not even photobleach the GFP fluorescence immediately adjacent to the irradiated area (e. g. Figure 10.2B); as a result, the extent of the destruction could be easily monitored in live cells during the operation. Subsequent immunological assays revealed that, once the γ-tubulin/GFP signal associated with the centrosome was completely

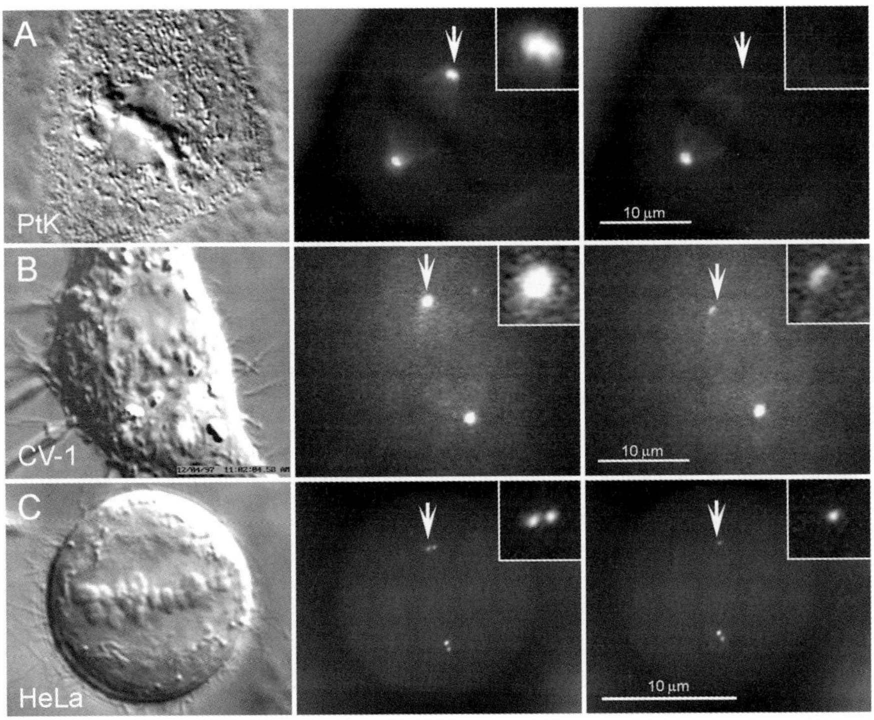

Figure 10.2 Laser microsurgery can be used to selectively destroy the centrosome, or even a single centriole, in various cultured vertebrate cells expressing fluorescent constructs that target the pericentriolar material (A and B) or the centriole (C). In these three examples, the cell is pictured by DIC (left-hand panel) and low light level fluorescent (center panel) microscopy, just prior to laser microsurgery. The same cell is then shown after the operation (right-hand panel), which only takes a few seconds. The inserts depict a higher magnification (3 ×) of the experimental centrosome. Note that the damage inflicted by the laser beam is confined to the Airy disk so that the part of the centrosome immediately adjacent to the irradiated region is not even photobleached (B). The precision of the method allows it to be used to destroy only one of the centrioles (arrow in C). The cell types represented include PtK_1 (A), CV-1 (B), both expressing γ-tubulin/GFP; and HeLa (C), expressing centrin I.

ablated, the centrosome also became unrecognizable to antibodies against many different centrosomal proteins [6]. This fact clearly shows that laser ablation destroys all proteins (and likely non-protein components as well) within the irradiated area and, in contrast to chromophore-assisted light inactivation (CALI), not just merely the protein that is labeled with GFP.

As expected, laser ablation completely eliminates the centrosome's ability to nucleate microtubules. This became evident when we destroyed the centrosome in interphase cells, and then cooled them to 4 °C to disassemble all of the pre-existing microtubules. When we then re-warmed the cells to 37 °C, and examined them 2 min later by α-tubulin immunofluorescence, we found large numbers of short microtubules scattered randomly throughout the cytoplasm (Figure 10.3A). The total amount of tubulin polymer did not dramatically differ between the acentrosomal cells and cells with normal centrosomal microtubule arrays (Figure 10.3B). Over time (~ 1 h), microtubules in the acentrosomal cells became arranged into a cytoplasmic microtubule complex similar to that found in adjacent cells containing a centrosome (see Figures 2 and 4 in [43]). Clearly, removal of the centrosome

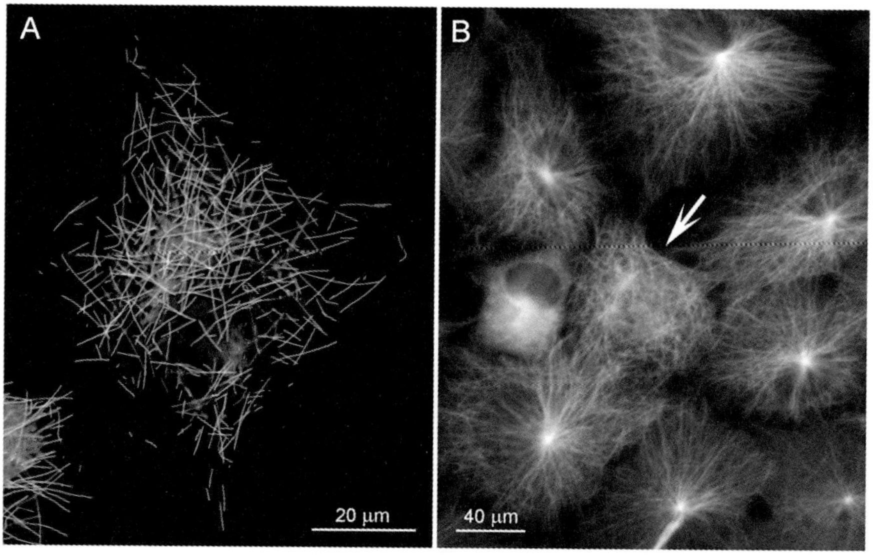

Figure 10.3 Cells lacking a centrosome form a normal-looking interphase array of cytoplasmic microtubules. After ablation of the centrosome by laser microsurgery, CV-1 cells were chilled to 4 °C for 30 min to disassemble their microtubules, and then re-warmed to 37 °C. The cell in A was fixed for a 3-D deconvolution immunofluorescence analysis of its microtubule content and distribution 2 min after warming. The cells in B were fixed 5 min after re-warming, and were analyzed by lower magnification conventional wide-field fluorescence LM. When microtubules are allowed to reassemble in the presence of a centrosome, they are nucleated by the centrosome (not shown). However, in the absence of a centrosome, they are nucleated spontaneously throughout the cytoplasm (A). Over time, they become arranged into a relatively normal-looking cytoplasmic microtubule complex (B). In B, the experimental cell is easy to differentiate (white arrow) because it is surrounded by other cells containing centrosomes that act as prominent foci for microtubules.

from cultured vertebrate cells in interphase does not prevent them from organizing and maintaining a relatively normal-looking array of cytoplasmic microtubules.

Thus far, these "proof-of-concept" studies revealed that laser microsurgery can be used to destroy the centrosome as an entity, without eliminating any of its individual protein components from the cytoplasmic pool. However, the incredible precision of laser microsurgery provides additional capabilities that were not obvious when we initiated our work.

There are now many cell lines that constitutively express a variety of centrosomal proteins, some of which (like centrin), reside only in the centrioles [44]. Through correlative LM/EM and microtubule depolymerization/re-polymerization assays, we found the size of the irradiated zone to be sufficiently small for a single centriole to be selectively destroyed. This is true even when it resides within a common "diplosome", for example during mitosis (Figure 10.2C). Further, because centrin preferentially accumulates within the mother centriole, we can clearly differentiate between the mother and daughter centrioles [44, 45] and can target just one of them for destruction.

The ability to selectively destroy part or all of the centrosome in a living cell, at any stage in its life cycle, provides a powerful approach for answering a number of longstanding questions regarding the function of this organelle, and how its replication is controlled.

10.3 Roles of the Centrosome during Cell Division

10.3.1 Role of the Centrosome during Spindle Assembly

Our original research goal in developing the laser microsurgery approach was to answer the long-standing question of whether the centrosome is required for spindle assembly during mitosis in vertebrates. It has been known for decades that centrosomes define the spindle poles in somatic cells, and that cells with supernumerary centrosomes produce multipolar spindles (see [46, 47]). However, in plants and in some animal systems (e. g. rodent embryos, and oocytes in many species), bipolar spindles are assembled in the absence of centrosomes. One idea, detailed by Hyman and Karsenti [5], is that, when centrosomes are present, they dominate in the process of organizing the microtubules for spindle assembly, because they provide a kinetic advantage for microtubule nucleation. A prediction of this hypothesis is that, in the absence of centrosomes, vertebrate cells will still form a spindle via an acentrosomal self-assembly route, similar to that in *Xenopus* egg extracts. Under this circumstance formation of the bipolar spindle would be achieved as molecular motors sort and organize microtubules, nucleated randomly in the vicinity of the chromatin (reviewed in [48]). The only way to test this idea was to remove one or both centrosomes from a cell just before it enters mitosis.

Our initial series of experiments were conducted in two cell lines: one derived from CV-1 fibroblasts (green monkey kidney) and one derived from an epithelial PtK_1 cell line [6]. We found that, when both centrosomes were destroyed during prophase, the cells formed bipolar spindles during prometaphase, and subsequently completed mitosis. This result showed that a pathway for spindle assembly does exist independent of the centrosome in vertebrate somatic cells. Another important outcome of this study was its finding that the presence of centrosomes is not necessary for normal progression into or through mitosis in vertebrates [6]. Since publishing these results, we have extended our conclusions to several other cell types, including CHO (Chinese hamster ovary) and HeLa (human) cells. Thus, these findings likely reflect a general phenomenon that occurs in all somatic cells.

However, as our attempts to determine how centrosomal and centrosome-independent mechanisms contribute to spindle assembly progressed, we concluded that the relationships among these mechanisms are more complex than was originally predicted. If the two centrosomes fail to separate after nuclear envelope breakdown, a single polar region is formed around which the centrosomes become grouped; that is, a monopolar spindle is formed (e. g. [49, 50]). However, when we ablated just one centrosome during late G2/prophase, the cells subsequently, and reproducibly, formed functional bipolar spindles. These spindles differed from those in control cells only in that the acentrosomal poles consistently lacked astral microtubules (Figure 10.4A; also see [6]). Thus, the spindle was bipolar but contained only one aster, resembling the monoastral mitotic spindles described in the *Drosophila "urchin"* mutants [51].

In order for a bipolar spindle to be formed in the presence of just one centrosome, either the existing centrosome's microtubule nucleation potential must be suppressed, or else the nucleation of free microtubules in the cytoplasm must be promoted [5]. One possibility that is consistent with our results is that when we ablate one centrosome, it generates a signal that downregulates the activity of the remaining centrosome. This would then increase the relative contribution made by the motor-based spindle-formation pathway, and would rescue spindles in cells that have damaged centrosomes. The fact that centrosomal activity can be downregulated during mitosis in response to damaging factors has been previously demonstrated in *Drosophila* embryos [52, 53]. To test this possibility, we ablated one or more centrosomes in tetraploid cells, which enter mitosis with four centrosomes. Usually, these cells form a tetrapolar spindle [54]. However, when we ablated one of the four centrosomes, the cells consistently formed tripolar spindles (Figure 10.4B). When the number of centrosomes was decreased to two, the cells formed bipolar spindles (Figure 10.4C). These observations demonstrate that the activity of the remaining centrosomes is not downregulated after destruction of one centrosome with the laser. It also becomes clear that spindle formation in cells with multiple centrosomes is not governed by the same rules as is spindle formation in cells with just one remaining centrosome. How can this difference be explained?

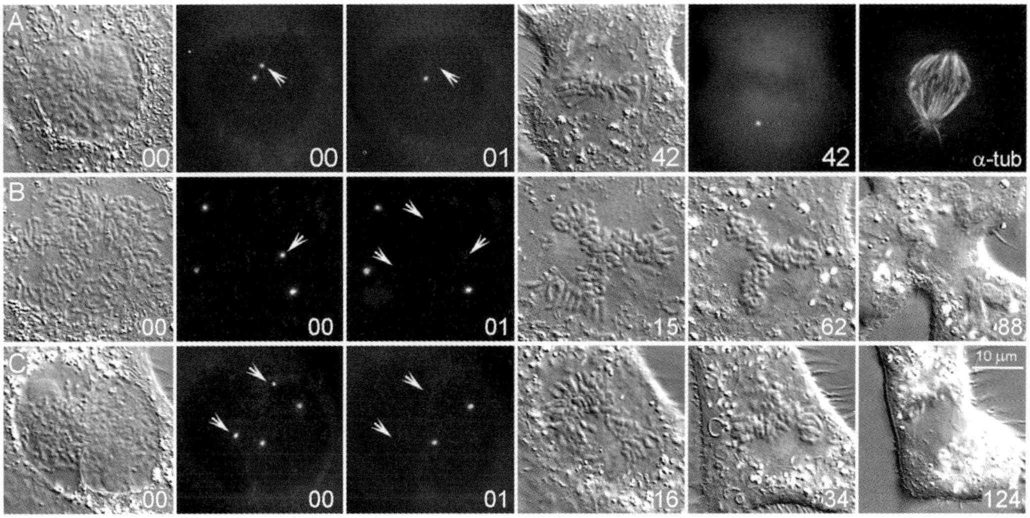

Figure 10.4 Mitotic spindle formation in cells with different number of centrosomes. When one of the two centrosomes was ablated in a diploid cell during prophase the cell formed a bipolar spindle that contained one centrosomal and one acentrosomal pole. A subsequent immunofluorescence analysis revealed that the acentrosomal pole lacked an array of associated astral microtubules (A). However, ablation of supernumerary centrosomes in polyploid cells decreases the number of spindle poles (B and C). In B, one centrosome was destroyed (arrow in fluorescent images) near the time of nuclear envelope breakdown, while in C two centrosomes were destroyed in binucleated CV-1 cells as they enter mitosis with four centrosomes. As a result of the operation the cells, which would have formed tetrapolar spindles, ended up forming tripolar (B) and bipolar (C) spindles. DIC and fluorescence images. Time is in minutes.

It is now evident that not all spindle microtubules are nucleated by the centrosome during the G2/M transition. A significant number persist or form in the cytoplasm, and are then incorporated into the spindle via interactions with, and transport along, the astral microtubules [55, 56]. This mechanism ensures that free microtubule minus ends are promptly sorted toward the centrosomes. Thus, as the spindle is built, it becomes properly oriented within the cell. The efficiency of this "search-and-capture" mechanism for spindle construction obviously depends on the robustness of the astral microtubule arrays associated with the centrosomes. Here we postulate that a single centrosome does not generate an aster that is sufficiently efficient to sweep the cytoplasm free of randomly oriented microtubules. Because the density of astral microtubules is low in the presence of just one centrosome, many of the free microtubules (which would normally be incorporated into the spindle) are left in the cytoplasm, where they self-organize to form the acentrosomal half of a bipolar spindle. An important feature of this hypothesis is that it readily explains why cells with two unseparated centrosomes (which produce two adjacent astral arrays) form monopolar spindles, whereas cells with a single centrosome form bipolar spindles.

10.3.2
Role of the Centrosome during Cytokinesis

If centrosomes are not required for spindle formation, then why does the cell use them for this process – especially considering the deleterious consequences of entering mitosis with extra centrosomes? The answer to this question became evident from our observations on how acentrosomal spindles behave once they have formed: we found that acentrosomal spindles were incapable of changing their orientation in response to changes in cell shape.

Normally, as a cell progresses through mitosis (in tissue culture), it becomes less firmly attached to the substrate. This "rounding" occurs gradually, and is due to a drastic re-organization of the cytoskeleton. As a result, the orientation of the long axis in elongate or ovoid cells may change such that the interpolar spindle axis becomes positioned "across" the cell (perpendicular to the long axis of the cell). Under normal circumstances, such improper spindle orientation would delay exit from mitosis until corrected [57]. The correction process involves a compensatory rotation of the spindle, until it becomes once again oriented parallel to the long axis of the cell. This occurs as cytoplasmic dynein, anchored in the cell cortex, acts on the astral microtubules emanating from the centrosome [57, 58]. Since spindles that lack centrosomes also consistently lack astral microtubules (Figures 10.4A and 10.5; also see [6, 43]), it was expected that they would not be able to re-orient in response to changes in cell shape. Indeed, we found that it was not uncommon for anaphase to start when the long axis of the spindle was perpendicular to the long axis of the cell. In such cases, the spindle failed to properly elongate during anaphase B. As a result, during telophase, the two daughter nuclei were often connected by one or more long chromosome arms (each of which was longer than a half-spindle, see [43]). Thus, even though a normal cleavage furrow formed in the expected position in these cells, perpendicular to the pole-to-pole spindle axis, it ultimately regressed, because cells are unable to cleave through a chromosomal bridge [59].

In addition, anaphase in the presence of an improperly oriented spindle frequently resulted in the formation of furrows that failed to propagate normally. This was especially true whenever a cell with an excessively long axis attempted to assemble a contractile ring around its perimeter. Finally, some cells lacking centrosomes formed multiple furrows in various parts of the cell; these furrows then wandered randomly through the cytoplasm before regressing. All of these phenotypes are consistent with the idea that astral microtubules are necessary in order to position the furrow properly during cytokinesis [60].

It is noteworthy that ~ 50 % of cells lacking centrosomes exhibited one or more of the above phenotypes, and failed to complete cytokinesis. In the other 50 %, cytokinesis was normal, and two independent daughter cells without centrosomes were produced. Clearly, the centrosome is not essential for the completion of cytokinesis. Rather, the phenotypes that we observed were due to a randomization of spindle positioning: when the orientations of the spindle and cell axes happened to be parallel to one another at the onset of anaphase, the terminal stages of cyto-

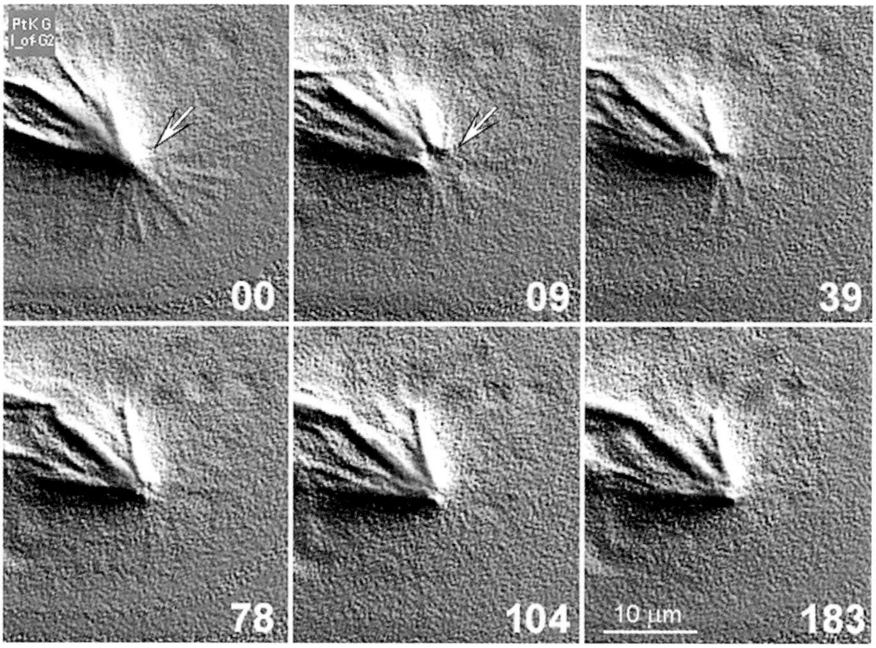

Figure 10.5 Laser ablation of a centrosome during mitosis leads to the rapid disappearance of its associated astral microtubules. Time-lapse sequence of astral microtubules in a prometaphase PtK_1 cell expressing GFP/α-tubulin prior to (arrow in 00), and after (arrow in 09) one of its centrosomes was destroyed. Each frame represents a single fluorescent image, presented in pseudo-DIC to accentuate individual astral microtubules. Note that the damage from the laser is confined to a well-defined small area (arrow in 09), which is rapidly re-populated with microtubules (cf. 09 and 78). However, all astral microtubules associated with the pole rapidly disappear. In the absence of a centrosome, these microtubules never reappear. Time in seconds.

kinesis proceeded normally. Thus, contrary to the interpretation put forward in some recent studies [45, 61], centrosomes are not required for cytokinesis in animal cells. Centrosomes are necessary, however, for ensuring the fidelity of cytokinesis by properly positioning the spindle before anaphase via the asters that they generate.

10.4 The Centrosome in the Cell Cycle

10.4.1 Role of the Centrosome in Progression through the Cell Cycle

The first evidence that the centrosome plays a role in cell cycle progression was obtained through conventional microsurgery. In some cells, like the BSC-1 line, the position of the centrosome is evident, because the organelles are normally arrayed in radial fashion around it. Using a fine glass micropipette, Maniotis and Schliwa [15] surgically removed the centrosome, along with a portion of the cytoplasm, from interphase BSC-1 cells. Over time, the resultant karyoplasts re-established a microtubule organizing center near the nucleus, and re-formed a compact Golgi apparatus. However, as previously mentioned, these karyoplasts neither regenerated centrioles nor entered mitosis, even after prolonged periods (up to 1 week). The salient conclusions from this study were: (1) that the presence of a pre-existing centriole is required for the formation of a new centriole/centrosome; and (2) that the centrosome is required for cell-cycle progression and entry into mitosis.

One limitation of the Maniotis and Schliwa approach is that it works only on highly flattened cells in interphase. As a result, the stage of the cell in the cell cycle (G1, S or G2) at the time of the operation is unknown. Another problem that complicates the interpretation of any conventional microsurgery study is that, along with the centrosome, a large part of the cytoplasm is also removed. Neither of these limitations is an issue with laser microsurgery, which allows only the centrosome to be destroyed, at any time in the cell cycle. We therefore asked whether CV-1 cells continue to cycle when they are born without a centrosome; if not, where does arrest occur? For this study we destroyed one centrosome during metaphase after the spindle had formed, at the point when the centrosomes were maximally separated from the chromosomes by the intervening spindle (Figure 10.2A). This approach has two important advantages. First, there is little likelihood of inducing DNA damage during the operation, which by itself would arrest the next cell cycle in G1. Second, at the conclusion of mitosis, two independent cells are usually produced: one that lacks a centrosome, and another perfect internal control that contains a centrosome.

The data from this series of experiments [43] revealed that CV-1 cells born without a centrosome arrest in G1, before entering S phase. Although this conclusion conflicts with that reported by Maniotis and Schliwa, it is consistent with the data of Hinchcliffe and co-workers [16] which were published at the same time as our study. By combining needle microsurgery, continuous time-lapse recording, and bromo-deoxyuridine labeling, these workers were able to better define the stage of the cell cycle during which the operation was performed. They found that removal of the centrosomes from cells that are already in S or G2, did not inhibit the ensuing mitosis; however, the resultant (acentrosomal) progeny arrested during the ensuing G1. Our study and that of Hinchcliffe and co-workers clearly demonstrate that removal of the centrosome, by whatever approach, prevents green monkey kidney cells from initiating S phase. The reasons for this constitute an important area for future research. One possibility is that the centrosome acts as an essential catalytic site for activation of one or more cell cycle regulatory molecules that control the G1/S transition. Another is that cells have a checkpoint control pathway that somehow monitors the presence of the centrosome; in its absence, a checkpoint must be triggered that prohibits entry into S phase. In related recent work, the Doxsey laboratory reports that the older (maternal) centriole in the cell contains a protein, centriolin, that is required for entry into S phase [61]. If this is true, then selective destruction of only the mother, but not the daughter centriole, during early G1 should prevent the cell from entering S phase. We are currently testing this possibility using laser microsurgery.

The finding that the centrosome is not required for S or G2 cells to enter and complete mitosis has a number of ramifications. For example, it proves that this organelle is not essential for triggering entry into mitosis in vertebrate somatic cells, although it may facilitate the process (see [62, 63]).

In summary, it is evident from our laser microsurgery studies and the complementary approaches of others that the centrosome serves more functions in the cell than mere nucleation and organization of microtubules. Although such an activity is the centrosome's most visible function, this activity seems not to be essential for cell survival or reproduction [64]. This conclusion is based on the fact that, with the exception of cilia and flagella formation, most microtubule arrays can be organized in the absence of a centrosome by redundant pathways. However, it does appear that, at the level of the higher animal cell, the centrosome is essential, because its presence is somehow required for entry into S phase.

10.4.2
De Novo Centrosome Formation

Since the discovery of the centrosome in the late 1800s, finding an answer to the question of how this organelle replicates has remained a major research challenge. The answer is not simply of academic interest; it will be key to understanding one of the major routes by which genetic instability and neoplastic disease originate. This is because cells do not have a checkpoint to inhibit cell cycle progression in the presence of supernumerary centrosomes [54, 65], and extra centrosomes lead to the formation of multipolar spindles and aneuploid progeny. Instead of monitoring the number of their centrosomes, cells have evolved a mechanism(s) to ensure that this organelle is replicated once, and only once, during each cell cycle. When these mechanisms break down, extra centrosomes are produced, with deleterious results (reviewed in [47]).

We now know that the number of centrosomes is defined by the number of centriole pairs (e. g. [66, 67]). Normally, a cell inherits a single pair of centrioles at the end of mitosis; this pair is then duplicated during the subsequent S phase. One notable feature of the duplication process is that the new centriole grows directly from the wall of a pre-existing centriole. Thus, each G1 cell contains a mother centriole that has parented, and her virgin daughter offspring that has not parented. Although in many respects these two centrioles are similar structurally, they are chemically dissimilar since the more mature mother contains proteins not found in the daughter (e. g. ninein [44]; centriolin [61]; cenexin [68]; reviewed in [39]). By G2, the cell contains four centrioles: one grandmother, one mother, and two new daughters.

One long-standing idea is that centrioles duplicate precisely from one to two because new centrioles can only grow from centrioles that have been "licensed" by passing through an entire cell cycle. In this model, each centriole acquires the ability to seed the assembly of a daughter on a region of its wall, via a templated process, only after the original centriole has passed through mitosis (reviewed in [67, 69]). Another feature of this design is that the conditions for new centriole assembly are favorable only in S or G2, but not in G1. Some aspects of this model have recently been validated by cell fusion studies: centrosomes can duplicate in G2 cytoplasm, but they normally do not because they have already duplicated prior to this time, during S phase. This implies that the block to re-duplication is intrinsic to the centrosome, rather than due to unfavorable cytoplasmic conditions during G2 [70].

As a first attempt to test this model, we asked whether we could define conditions under which vertebrate cells lacking centrioles (i.e. centrosomes) can form centrioles from scratch (*de novo*). Our thinking was influenced by the report of Marshal and coworkers [71] that the green alga, *Chlamydomonas*, can form centrioles *de novo* during S but not during G2 phase. Further, the efficiency (speed) of the formation process is only 50% of the rate seen for templated assembly off the maternal centriole. Importantly, the *de novo* formation occurred only during the S period. This offered an explanation for why *de novo* assembly has not previously been seen in vertebrates: the cells lacking centrioles are arrested at G1, before the conditions allow *de novo* assembly. One point that follows from this idea is that if cells lacking a centrosome are arrested for long periods in S phase, a centrosome will ultimately form. To evaluate this, we used laser microsurgery to remove the centrosome from CHO cells that were arrested in S phase by hydroxyurea treatment [72]. We found that new γ-tubulin foci lacking centrioles reappeared in these cells 5–8 h after the operation (Figure 10.6). These foci, which were enriched for centrosomal proteins, formed even in cells in which the microtubules had been disassembled with nocodazole. Moreover, as predicted by our model, 24 h after the surgery, centrioles began to form *de novo* in association with these foci and the number of centrioles formed within this period ranged from two to 14.

Although the molecular mechanisms remain to be defined, our experiments provide some cogent information regarding the process of *de novo* centrosome formation in vertebrates. First, it requires $\sim$ 24 h, which is longer than the duration of a cell cycle in CHO cells. Second, since S phase normally takes only a few hours, a cell will not have time to form a centrosome *de novo,* unless it is arrested in S phase. Also, the first signs of *de novo* formation, which is signaled by the accumulation of γ-tubulin into a focus, require 4–5 h to appear and occur independently of the presence of microtubules. This length of time suggests the possibility that specific genes need to be expressed or proteins synthesized, to enable the *de novo* assembly pathway to operate. Finally, since as many as 14 centrioles can be formed in just 24 h, the *de novo* pathway is obviously able to support parallel production of multiple centrioles. Together, these data suggest that the templated pathway for centriole replication has evolved not because the *de novo* pathway is less efficient than the templated replication, but instead, because it sets limits on the number of centrioles produced during each cell cycle.

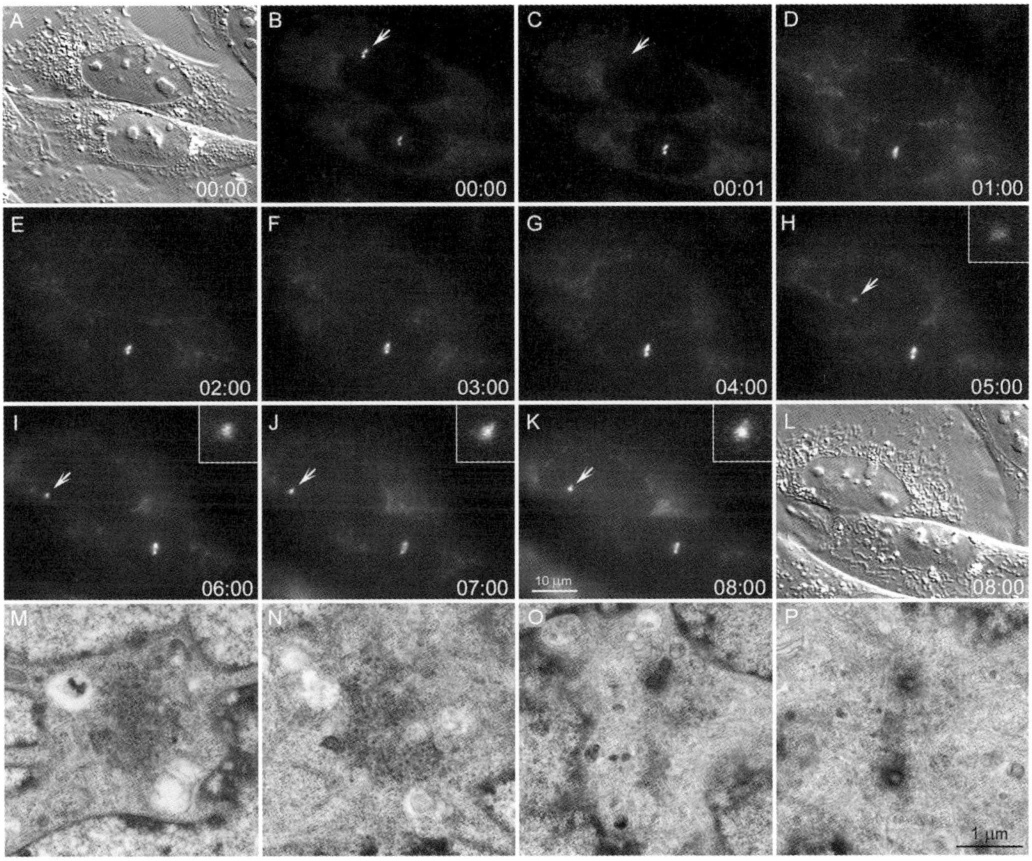

Figure 10.6 The *de novo* formation of centrosomes in CHO cells lacking a centrosome requires > 6 h in S phase. DIC (A and L) and fluorescent (B–K) micrographs of a CHO cell in which the centrosome was ablated (arrow in B and C) during S phase. An adjacent cell serves as a control. Note that a single focus of γ-tubulin subsequently reappeared 5–6 h later in the experimental cell (arrow in H–K), but on the opposite side of the nucleus from the original centrosome. A serial-section EM of the same cell revealed that this focus (M and N) lacked centrioles, which normally do not form *de novo* until 24 h post-operation. The centrosome in the non-irradiated control cell (O and P) contained normal replicated diplosome and associated pericentriolar material. Time in hours and minutes.

10.5
For the Future

The experiments and results described in this chapter illustrate the synergistic power of combining GFP-imaging with laser microsurgery. They clearly demonstrate that this combined approach can be used to differentiate between cellular functions that are supported by the centrosome as an organelle, and functions that are supported by the centrosome's individual constituents. Over the past 125 years, a number of questions have been raised regarding the function and replication of the centrosome, questions that could not be answered until a suitable method could be devised to selectively remove the centrosome from the cell. At long last such a method is available, and we see two areas of centrosome research in particular that it will positively impact in the near future.

First, the unique precision of laser microsurgery will provide the key to understanding the mysterious "two-ness" of the centrosome. As illustrated in Figure 10.1C, it is possible to specifically ablate one of the two centrioles inside a centrosome, to create a cell containing a centrosome with just half of its normal "valence" [73]. Is the presence of only the mother centriole or its daughter sufficient to support cell-cycle progression? Does the absence of the mother centriole change the structural and functional characteristics of the daughter? Our preliminary data suggest that centrosomes containing a single centriole are fairly normal and capable of supporting most centrosomal functions; however, these issues need to be explored in greater detail.

Laser microsurgery can also be conducted on cells in which individual centrosomal components have been depleted via RNAi and/or small-molecule chemical inhibitors. This very exciting development may help to identify the key molecules involved, e. g. in processes such as *de novo* centrosome formation. Will *de novo* formation occur in cells depleted of proteins that have been shown to be involved in the earliest stages of centrosome replication (e. g. centrin-2 [74])? Does the mechanism that blocks cell-cycle progression in cells lacking a centrosome depend on functional p53? These and similar questions can be answered using laser microsurgery. The answers will provide new and exciting information on the function of the centrosome, and will teach us why we need this mysterious and complex organelle.

Acknowledgments

We thank Ms Susan Nowogrodzki for helpful editorial comments.

The work in our laboratories is supported by NIH General Medical Sciences grants R01 59363 (to AK) and R37-40198 (to CLR).

References

1. Wilson, E. B. **1925**. *The Cell in Development and Heredity.* The Macmillan Co.: New York.
2. Heald, R., Tournebize R., Blank T., Sandaltzopoulos R., Becker P., Hyman A., and Karsenti E. **1996**. *Nature* **382**, 420–425.
3. Merdes, A. and Cleveland D. W. **1997**. *J Cell Biol* **138**, 953–956.
4. Heald, R., Tournebize R., Habermann A., Karsenti E., and Hyman A. **1997**. *J Cell Biol* **138**, 615–628.
5. Hyman, A. and Karsenti E. **1998**. *J Cell Sci* **111**, 2077-2083.
6. Khodjakov, A., Cole R. W., Oakley B. R., and Rieder C. L. **2000**. *Curr Biol* **10**, 59–67.
7. Oakley, B. R., Oakley C. E., Yoon Y., and Jung M. K. **1990**. *Cell* **61**, 1289–1301.
8. Joshi, H. C., Palacios M. J., McNamara L., and Cleveland D. W. **1992**. *Nature* **356**, 80-83.
9. Moudjou, M., Bordes N., Paintrand M., and Bornens M. **1996**. *J Cell Sci* **109**, 875–887.
10. Paoletti, A., Moudjou M., Paintrand M., Salisbury J. L., and Bornens M. **1996**. *J Cell Sci* **109**, 3089–3102.
11. Khodjakov, A. and Rieder C. L. **1999**. *J Cell Biol* **146**, 585–596.
12. Megraw, T. L., Kao L.-R., and Kaufman T. C. **2001**. *Curr Biol* **11**, 116–120.
13. Zorn, G. A., Lucas J. J., and Kates J. R. **1979**. *Cell* **18**, 659–672.
14. Sluder, G., Miller F. J., and Rieder C. L. **1989**. *Cell Motil Cytoskeleton* **13**, 264–273.
15. Maniotis, A. and Schliwa M. **1991**. *Cell* **67**, 495–504.
16. Hinchcliffe, E. H., Miller F. J., Cham M., Khodjakov A., and Sluder G. **2001**. *Science* **291**, 1547–1550.
17. Skibbens, R. V. and Salmon E. D. **1997**. *Exp Cell Res* **235**, 314–324.
18. Berns, M. W., Rattner J. B., Brenner S., and Meredith S. **1977**. *J. Cell Biol.* **72**, 351–367.
19. Berns, M. W. and Richardson S. M. **1977**. *J Cell Biol* **75**, 977–982.
20. Berns, M. W. **1998**. *Sci Am* 62–67.
21. Zirkle, R. E. and Bloom W. **1953**. *Science* **117**, 487–493.
22. Uretz, R. B., Bloom W., and Zirkle R. E. **1954**. *Science* **120**, 197–199.
23. Zirkle, R. E. **1970**. *J Cell Biol* **47**, 235a.
24. Forer, A. **1965**. *J Cell Biol* **25**, 95–117.
25. Bajer, A. **1959**. *Chrom Today* **3**, 63–69.
26. Walker, R. A., Inoue S., and Salmon E. D. **1989**. *J Cell Biol* **108**, 931–937.
27. Spurck, T. P., Stonington O. G., Snyder J. A., Pickett-Heaps J. D., Bajer A., and Mole-Bajer J. **1990**. *J Cell Biol* **111**, 1505–1518.
28. Izutsu, K. and Sato H. **1992**. *Biol Cell* **76**, 339–350.
29. Zirkle, R. E. **1970**. *Rad Res* **41**, 516–537.
30. Forer, A. **1991**. *J Cell Sci* **98**, 415–422.
31. Koonce, M. P., Cloney R. A., and Berns M. W. **1984**. *J Cell Biol* **98**, 1999–2010.
32. Grill, S. W., Gonczy P., Stelzer E. H.K., and Hyman A. A. **2001**. *Nature* **409**, 630–633.
33. Grill, S. W., Howard J., Schaffer E., Stelzer E. H., and Hyman A. A. **2003**. *Science* **301**, 518–521.
34. Khodjakov, A., Cole R. W., and Rieder C. L. **1997**. *Cell Motil Cytoskeleton* **38**, 311–317.
35. Greulich, K. O. and Leitz G. **1994**. *Exp Tech Physics* **40**, 1–14.
36. Berns, M. W., Aist J., Edwards J., Strahs K., Girton J., McNeill P., Rattner J. B., Kitzes M., Hammer-Wilson M., Liaw L. H., Siemens A., Koonce M., Peterson S., Brenner S., Burt J., Walter R., Bryant P. J., van Dyk D., Coulombe J., Cahill T., and Berns G. S. **1981**. *Science* **213**, 505–513.
37. Strahs, K. R. and Berns M. W. 1979. *Exp Cell Res* **119**, 31–45.
38. Cole, R. W., Khodjakov A., Wright W. H., and Rieder C. L. 1995. *J Microsc Soc Am* **1**, 203–215.
39. Bornens, M. **2002**. *Curr Opin Cell Biol* **14**, 25–34.
40. McNeill, P. A. and Berns M. W. **1981**. *J Cell Biol* **88**, 543–553.
41. Rattner, J. B. and Berns M. W. **1974**. *J Cell Biol* **62**, 526–533.

42. Peterson, S. P. and Berns M. W. **1978**. *J Cell Sci* **34**, 289–301.
43. Khodjakov, A. and Rieder C. L. **2001**. *J Cell Biol* **153**, 237–242.
44. Piel, M., Meyer P., Khodjakov A., Rieder C. L., and Bornens M. **2000**. *J Cell Biol* **149**, 317–330.
45. Piel, M., Nordberg J., Euteneuer U., and Bornens M. **2001**. *Science* **291**, 1550–1553.
46. Brinkley, B. R. **2001**. *Trends Cell Biol* **11**, 18–21.
47. Nigg, E. A. **2002**. *Nature Rev Cancer* **2**, 1–11.
48. Walczak, C. E., Vernos I., Mitchison T. J., Karsenti E., and Heald R. **1998**. *Curr Biol* **8**, 903–913.
49. Bajer, A. S. **1982**. *J Cell Biol* **93**, 33–48.
50. Ault, J. G., DeMarco A. J., Salmon E. D., and Rieder C. L. **1991**. *J Cell Sci* **99**, 701–710.
51. Wilson, P. G., Fuller M. T., and Borisy G. G. **1997**. *J Cell Sci* **110**, 451–464.
52. Sibon, O. C., Kelkar A., Lemstra W., and Theurkauf W. E. **2000**. *Nature Cell Biol* **2**, 90–95.
53. Takada, S., Kelkar A., and Theurkauf W. E. **2003**. *Cell* **113**, 87–99.
54. Sluder, G., Thompson E. A., Miller F. J., Hayes J., and Rieder C. L. **1997**. *J Cell Sci* **110**, 421–429.
55. Rusan, N. M., Tulu S., Fagerstrom C. J., and Wadsworth P. **2002**. *J Cell Biol* **158**, 997–1003.
56. Khodjakov, A., Copenagle L., Gordon M. B., Compton D. A., and Kapoor T. M. **2003**. *J Cell Biol* **160**, 671–683.
57. O'Connell, C. B. and Wang Y.-L. **2000**. *Mol Biol Cell* **11**, 1765–1774.
58. Shaw, S. L., Yeh E., Maddox P., Salmon E. D., and Bloom K. **1997**. *J Cell Biol* **139**, 985–994.
59. Mullins, J. M. and Biesele J. J. **1977**. *J Cell Biol* **73**, 672–684.
60. Rappaport, R. **1986**. *Int Rev Cytol* **105**, 245–281.
61. Gromley, A., Jurczyk A., Sillibourne J., Halilovic E., Mogensen M. M., Groisman I., Blomberg M., and Doxsey S. **2003**. *J Cell Biol* **161**, 535–545.
62. Jackman, M., Lindon C., Nigg E. A., and Pines J. **2003**. *Nature Cell Biol* **5**, 143–148.
63. Lane, H. A. and Nigg E. A. **1996**. *J Cell Biol* **135**, 1701–1713.
64. Rieder, C. L., Faruki S., and Khodjakov A. **2001**. *Trends Cell Biol* **11**, 413–419.
65. Rieder, C. L., Khodjakov A., Paliulis L. V., Fortier T. M., Cole R. W., and Sluder G. **1997**. *Proc Natl Acad Sci USA* **94**, 5107–112.
66. Bobinnec, Y., Khodjakov A., Mir L. M., Rieder C. L., Edde B., and Bornens M. **1998**. *J Cell Biol* **143**, 1575–1589.
67. Hinchcliffe, E. H. and Sluder G. **2001**. *Genes Dev* **15**, 1167–1181.
68. Lange, B. M.H. and Gull K. **1995**. *J Cell Biol* **130**, 919–927.
69. Marshall, W. F. and Rosenbaum J. L. **2000**. *Curr Topics Dev Biol* **49**, 187–205.
70. Wong, C. and Stearns T. **2003**. *Nature Cell Biol* **5**, 539–544.
71. Marshall, W. F., Vucica Y., and Rosenbaum J. L. **2001**. *Curr Biol* **11**, 308–317.
72. Khodjakov, A., Rieder C. L., Sluder G., Cassels G., Sibon O. C., and Wang C. L. **2002**. *J Cell Biol* **158**, 1171–1181.
73. Mazia, D. **1987**. *Int Rev Cytol* **100**, 49–92.
74. Salisbury, J. L., Suino K. M., Busby R., and Springett M. **2002**. *Curr Biol* **12**, 1287–1292.

11
Centrosome Regulation in Response to Environmental and Genotoxic Stress

Ody C. M. Sibon and William E. Theurkauf

11.1
Introduction

The centrosome plays a role in a remarkably broad range of cellular processes, including microtubule nucleation, organization of the spindle poles and interphase organelles, and cell cycle progression. In addition, critical cell-cycle regulators and components of the signal transduction pathways localize to the centrosome, suggesting that it may help integrate diverse signals to produce a coherent cellular response to extracellular and internal queues. Environmental and genotoxic stress trigger global changes in gene expression, cytoplasm organization, and cell cycle progression. Centrosome structure and function are disrupted in response to heat shock and DNA damage, which may play a role in coordinating the pleiotropic cellular stress response. In this chapter we review the changes in centrosomes organization and function triggered by heat shock, unfolded protein accumulation, and genotoxic lesions, and speculate about the potential biological functions of the centrosome during the cellular response to stress.

11.2
Heat Shock

Heating above physiological growth temperature induces heat shock and in a variety of cells this is accompanied by changes in centrosome organization and loss of centrosome protein localization (Table 11.1). Changes in centrosome protein localization in response to heat shock have not been systematically studied and only a subset of components has been assayed in most systems. However, loss of γ-tubulin localization is observed in a number of systems (Table 11.1). A reduced capacity of centrosomes to nucleate microtubules after heat shock has been reported, consistent with loss of γ-tubulin and the γ-tubulin ring complex (γ-TuRC), a critical microtubule-nucleating factor [1, 2], from the centrosome [3–5]. In addition, electron

Centrosomes in Development and Disease. Edited by Erich A. Nigg

ISBN 3-527-30980-2

Table 11.1 Centrosome alterations associated with heat shock.

Cell line	*Alteration*	*Reference*
HeLa cells	Hsp72 at mitotic centrosomes*; loss of γ-tubulin*, pericentrin, Hsp 73 and TCP-1*, multiple, minute centrosomes	5, 13, 14
CHO cells	Loss of pericentrin, γ-tubulin*, increased density PCM†, multipolar spindles, additional foci of PCM*	3, 6, 52, 53
Drosophila Kc cells	Disintegration of mitotic centrosomes†, more condensed PCM†, loss of Bx63/CP190 labeling at mitotic centrosomes*	4, 7, 9,
Mouse T lymphocytes	Aggregation of PCM†, loss of centriole-associated protein*	8, 54
Human blood monocytes	Loss of centrosome antigen labeling	54

*Assayed by immunocytochemistry.
†Assayed by electron microscopy.

microscopic analyses show that heat shock can lead to increased density of pericentriolar material (PCM) [6–8] or complete disintegration of centrosomes [9].

The loss of centrosome-associated protein localization after heat shock, as detected by immunolabeling, could reflect protein dissociation from a centrosome scaffold, protein degradation, protein denaturation or aggregation, or a combination of these factors. For the most part, the mechanism of heat shock-induced loss of centrosome protein labeling is not understood. Localization of the centrosomal protein CP190 is reduced after heat shock, but protein levels on Western blots do not change [4]. Therefore, at least in this instance, diminished labeling does not appear to reflect protein degradation.

Recovery of centrosome function following heat shock, by contrast, may be directly linked to changes in heat shock protein expression or function. A universal response to heat shock is the rapid and transient increase in expression of a small number of heat shock proteins or Hsps. Hsps belong to a large family of both constitutive and stress-induced "chaperones" that play an important role in mediating protein folding, transport and assembly–disassembly of polypeptide complexes (for reviews see [10–12]). Cells exposed to a relatively short heat treatment become "thermotolerant", and show increased survival in response to a second heat shock. The capacity of the centrosome to re-grow microtubules and the efficiency of centrosomal staining by anti-pericentrin antibody recover more efficiently in thermotolerant cells. Furthermore, the inducible form of Hsp72 accumulates at centrosomes following heat shock [13]. Heat shock proteins thus appear to play a role in restoring centrosome function following heat stress, perhaps by promoting assembly of multi-component centrosome protein complexes.

The ability to rapidly recover centrosome function following heat shock correlates with cell survival, and this may reflect the importance of centrosome function during mitosis. Thermotolerant cells recover centrosome function more efficiently than naïve cells. These cells also show enhanced clonogenic survival and less frequent centrosome and spindle abnormalities after heat shock [8]. By contrast, following heat shock, naïve cells often progress through an aborted division and die by a non-apoptotic pathway. Mitotic cells with multiple centrosomes and abnormal spindles are more frequently observed in heat-shocked tumor cells than in heat-shocked non-tumor controls, and the increase in abnormal spindles correlates with increased cell death by a non-apoptotic mechanism [14]. Cell death following an aborted mitotic division, termed "mitotic catastrophe", is also commonly observed following DNA damage (for a review see [15]). Cell death by mitotic catastrophe eliminates cells from the population. Centrosome disruption may therefore function to eliminate heat shock-damaged cells from a normal cell population.

Several lines of evidence suggest that heat shock proteins are also required for normal centrosome assembly. A number of heat shock proteins co-localize with centrosomes under normal growth conditions, including Hsp90, TCP-1, Hsp73, and Hsp70 [13, 16, 17]. Furthermore, isolated *Drosophila* centrosomes contain Hsp90, as determined by MALDI mass spectrometry, and Hsp90 localizes to centrosomes throughout the cell cycle, and at different stages of development. Significantly, Hsp90 mutations disrupt centrosome organization and lead to assembly of aberrant spindles and impaired chromosome segregation [18]. Hsp90 is required for the stability of Polo kinase, which regulates several aspects of cell division, including centrosome maturation and function. Hsp90 may therefore promote centrosome function by maintaining Polo kinase activity [19]. Moreover, in *Xenopus* oocyte extracts an Hsp70/Hsp90 complex appears to directly sequester centrin, which is released when the centrosome assembles on oocyte activation [20]. Heat shock proteins may therefore promote centrosome function by directly interacting with structural components of this organelle and by stabilizing key regulators of centrosome function.

Heat shock appears to have differential effects on centrosome organization, depending on cell-cycle phase. In HeLa cells, heat shock-induced Hsp72 localization to centrosomes is most pronounced during mitosis [13], and we have recently found that heat shock-induced loss of γ-tubulin-GFP from the centrosome is also more dramatic during mitosis (O. C. M. Sibon, unpublished data). Furthermore, in *Drosophila* cultured cells, heat shock does not alter the ultrastructure of interphase centrioles, but leads to severe defects in centriole organization in mitotic cells [9]. Mitotic centrosomes thus appear to be particularly sensitive to heat stress. This may function to trigger mitotic catastrophe, which then disposes of cells that sustain irreparable heat damage.

11.3
Centrosomes and the Unfolded Protein Response

A potentially related form of stress is triggered by accumulation of unfolded proteins, which leads to formation of dense protein aggregates, called "aggresomes", which localize to the vicinity of centrosomes. Under normal physiological conditions, quality control pathways catalyze proper folding of nascent proteins and degrade improperly-folded polypeptides and protein aggregates [21, 22]. However, when the quality control mechanisms are disrupted or overloaded, through a pathological condition or experimental manipulation, improperly-folded proteins accumulate in the cytoplasm and aggregates often form in the pericentrosomal area [23] (for reviews see [24, 25]). Johnston and colleagues [23] first described these structures in cells over-expressing the cystic fibrosis transmembrane conductance regulator (CFTR). In these cells, CFTR accumulated in a pericentriolar structure surrounded by a cage of vimentin filaments. Inhibition of proteasome function induces deposition of other proteins in similar aggresomes, which are also associated with centrosomes. Aggresome formation is dependent on intact microtubules, suggesting that clustering around the centrosome is an active process. Johnston and coworkers proposed that aggresome formation is a general response to accumulation of unfolded proteins, which occurs when proteasome capacity is exceeded [23].

Consistent with this hypothesis, aggresome-like clusters of insoluble/misfolded proteins that co-localize with the centrosomal marker γ-tubulin have now been observed in a number of situations, including human diseases and disease models. Intra-cytoplasmic protein aggregates (or Lewy bodies) that are found in neurodegenerative disorders such as Parkinson's disease and dementia co-localize with γ-tubulin-containing structures [26]. A mutant form of the prion protein associated with transmissible spongiform encephalopathies also forms aggresome-like clusters that co-localizes with γ-tubulin [27]. A missense mutation in αB-crystallin (αB), the cause of a desmin-related myopathy, causes aggregates of desmin and αB in aggresomes concentrating at and around centrosomes as determined using γ-tubulin as a centrosomal marker [28]. A mutation in the FERM domain of schwannomin, the product of the NF2 tumor suppressor gene, also causes misfolding and accumulation in aggresome-like structures that again co-localize with structures that are immunoreactive for γ-tubulin [29]. It is unclear if these structures contribute to disease progression or are a secondary consequence of the disease state. However, aggresome formation appears to be a common consequence of accumulation of insoluble proteins.

Active proteasomal complexes localize to centrosomes under basal conditions [30, 31]. The centrosome may therefore represent a primary site for degradation of misfolded proteins. Consistent with this speculation, centrosome-associated aggresomes enlarge when proteosome activity is inhibited by drugs [23, 26, 27, 32].

Under some condition, the distribution of γ-tubulin appears to be altered when aggresomes form. γ-Tubulin and pericentrin are present at aggresomes in cells from patients with Parkinson's disease and other dementias that produce Lewy bodies [26]. However, in some cells containing mutant schwannomin aggregates γ-tubulin labeling is absent [29]. Garcia-Mata et al. reported disruption of astral microtubule organization around the aggresomes, consistent with defects in the γ-TuRC [33]. However, at the electron microscopic level, aggresomes appear electron dense and are surrounded by a cage of intermediate filaments [23, 32–34], and the interior of the aggresome is not generally immunolabeled by antibodies

[23]. It is therefore possible that γ-TuRC is at the centrosome and biochemically active, but is sterically prevented from functioning in microtubule nucleation when dense aggresomes are present.

The link between aggresomes and human disease appears to be quite strong. By nucleating and organizing microtubules, the centrosome could promote aggresome assembly, and defects in centrosome function due to aggresome formation could contribute to pathogenesis. A molecular understanding of the link between centrosomes and the unfolded protein response could shed light on a clinically intractable group of neurodegenerative diseases.

11.4 Centrosome Disruption in Response to Genotoxic Stress

Maintenance of genomic integrity is critical to normal development and disease prevention, and conserved pathways promote damage repair or eliminate mutant cells from the population. DNA damage and replication checkpoints delay the cell cycle to allow repair of genetic lesions or completion of DNA replication. In systems ranging from mammalian tumors to early *Drosophila* embryos, checkpoint failures that allow DNA damage or incomplete replication to persist into mitosis trigger "mitotic catastrophe", a poorly understood process characterized by delays in metaphase, chromosome segregation failures, and cell death by non-apoptotic mechanisms[15, 35–37]. The molecular mechanism of mitotic catastrophe has not been analyzed in detail, but this process appears to be a significant cause of chemotherapy-induced cell death in tumors and may serve an important genome maintenance function (reviewed in[15]) Studies in *Drosophila* embryos demonstrate that checkpoint failures and "mitotic catastrophe" are linked to mitosis-specific defects in centrosome structure and function, anastral mitotic spindle assembly, and chromosome segregation failures on mitotic exit [36]. The *Drosophila* homolog of the Chk2 tumor suppressor kinase is essential to mitotic catastrophe in early embryos, demonstrating that this is a genetically programmed response to genotoxic lesions [38]. The mitotic response to DNA damage in early *Drosophila* embryos is reviewed below, followed by a discussion of recent studies in cultured mammalian cells.

11.4.1 Centrosome Inactivation in Early Embryos

The mechanism of DNA damage-induced division failure has been extensively studied in early *Drosophila* embryos, and work in this system is therefore described in some detail. *Drosophila* embryogenesis is initiated by 13 very rapid mitotic divisions that proceed without cytokinesis. These syncytial divisions, like the cleavage stage divisions in other embryos, are characterized by alternating S and M phases without intervening gap phases [39]. The first nine divisions are uniformly rapid, with S phase taking approximately 5 min and M phase 3 min. These initial divisions take place in the interior of the embryo, but the majority of nuclei migrates to the cortex and forms a monolayer by interphase of division 10. The final four syncytial blastoderm stage divisions (mitosis 10–13) take place in a cortical monolayer, and during these divisions the length of S phase progressively increases while M phase remains relatively constant. Following division 13, the cell cycle slows dramatically, gap phases are introduced, cortical nuclei are incorporated into cells, and high level zygotic gene expression is initiated [39].

ATR and Chk1 are highly conserved kinases required for G2/M phase checkpoint control. The *Drosophila* homologs of these kinases are required for DNA replication checkpoint function during early embryogenesis, and this checkpoint pathway is required to delay mitosis as S phase slows during the final syncytial blastoderm divisions [40, 41]. Late syncytial blastoderm stage embryos mutant for these replication checkpoint components thus spontaneously initiate mitosis before S phase is completed, triggering mitotic catastrophe. Time-lapse confocal microscopic analyses show that these aborted mitotic divisions are tightly linked to mitosis-specific centrosome inactivation, anastral spindle assembly, and delays in mitosis [36]. The centrosome defects correlate with loss of multiple components of the γ-TuRC from a core centrosome structure [36]. In wild-type embryos, identical mitotic defects are triggered by DNA replication inhibitors, a wide range of DNA damaging agents, and direct injection of restriction enzyme-digested DNA. Centrosome disruption and mitotic division failure thus appear to be a consistent response to the presence of genotoxic lesions at the onset of mitosis [36, 38].

Following the 13th syncytial blastoderm division, nuclei at the cortex are incorporated into the cells that will form the embryo proper [39]. The polyploid nuclei that result from damage-induced division failure invariably dissociate from the cortex and drop into the internal yolk mass, while the centrosomes associated with these nuclei remain at the cortex. The mitotic catastrophe response to DNA damage in early embryos thus prevents transmission of damaged nuclei to the post-cellularization embryo [36, 38]. As outlined above, mitotic catastrophe in mammalian cells often leads to death by a non-apoptotic mechanism. In syncytial fly embryos and mammalian cells, "mitotic catastrophe" thus leads to disposal of defective mitotic products, and appears to serve a genome-maintenance function analogous to apoptosis.

11.4.2
Chk2 is Required for DNA Damage-induced Mitotic Catastrophe

Mitotic failure following checkpoint failure could be a non-specific response to accumulated genetic lesions, or the product of a mitosis-specific DNA damage pathway. A number of genes that mediate interphase damage responses, including the homologs of p53, the Bloom's helicases, ATR and Chk1 kinase, are not required for mitotic catastrophe in early embryos. However, mutations in the *Drosophila* homolog of Checkpoint kinase 2 (Chk2), encoded by the *mnk* gene, block all aspects of mitotic catastrophe in early embryos [38]. This observation demonstrates that mitotic catastrophe, at least in fly embryos, is a genetically programmed response to genotoxic stress.

A null mutation in *mnk* does not affect viability or fertility, but disrupts DNA damage-induced apoptosis and leads to mutagen sensitivity [42, 43]. *In vivo* assays revealed profound defects in the mitotic response to DNA damage. In wild-type embryos, DNA-damaging drugs and replication inhibitors consistently trigger the mitosis-specific loss of γ-tubulin from a core centrosome complex, anastral spindle assembly, metaphase delays, and chromosome segregation failures [36, 38]. By contrast, *mnk* mutant embryos assemble cytologically normal spindles with wild-type γ-tubulin localization following treatment with a wide range of DNA-damaging agents. The *mnk* mutation also suppresses the spindle assembly and γ-TuRC localization defects that occur spontaneously in *grp* (Chk1 kinase) checkpoint mutant embryos [38].

DNA damage also triggers severe defects in anaphase chromosome movement. In *mnk* mutants, however, damaged chromosomes are stretched between the poles, leading to unequal segregation and formation of daughter nuclei that are linked by chromatin bridges [38]. A wild-type *mnk* transgene restores damage-induced loss of γ-TuRC localization and the anaphase chromosome segregation block, demonstrating that Chk2 is required for both of these aspects of the mitotic damage response.

In syncytial *Drosophila* embryos, damage-induced division failure is invariably followed by dissociation of the resulting abnormal nucleus from the cortex, and the *mnk* mutation also prevents this step in the mitotic catastrophe response [38]. Dissociation from the cortex prevents transmission of damaged nuclei to the embryo proper, and is therefore likely to be critical to the mitotic catastrophe response. Nuclear loss can be dissociated from centrosome inactivation and division failure. Wild-type embryos treated with the topoisomerase II inhibitor ICRF-193, which does not directly induce DNA damage, assemble normal astral spindles and progress through mitosis with wild-type kinetics. However, the chromosomes are topologically linked, anaphase chromosome segregation fails, and the resulting abnormal nuclei invariably drop into the interior. In *mnk* mutant embryos, by contrast, ICRF-193 blocks chromosome segregation but the abnormal mitotic products are invariably retained in the cortical monolayer [38]. Chk2 thus appears to disrupt the link between centrosomes and nuclei in response to division failure through a process that is independent of centrosome inactivation. Chk2 may therefore directly target factors that link centrosomes to nuclei.

Mitotic catastrophe, at least in the early embryo, thus appears to be triggered by a multi-step pathway that links centrosome function, spindle assembly and chromosome segregation to genome integrity. This response functions to eliminate damaged nuclei from the embryonic precursor pool, and thus fulfills a similar function to apoptosis.

11.4.3
DNA Damage and Mitosis in Mammalian Cells

Mitotic catastrophe in mammalian cells, triggered by G2/M checkpoint failures, is characterized by delays in metaphase, chromosome segregation and cytokinesis failures, and cell death. In some cells, mitotic division failure is followed by apoptosis. However, cells more commonly arrest in G1 or die by a non-apoptotic mechanism (reviewed in [15]) [44–47]. Cytologically similar mitotic catastrophe responses have been described in diverse systems, including primary mouse embryo fibroblasts, *Drosophila* embryos, and a number of cultured cells [40, 41, 44–47]. Significantly, mitotic catastrophe appears to be the primary mechanism of cell death in a number of tumor cells lines and in tumors *in vivo* following treatment with chemotherapeutic agents (reviewed by [15]).

A hallmark of damage-induced mitotic catastrophe is the accumulation of cells with large polyploidy nuclei or multiple nuclei. *In vivo* studies in human colorectal tumor cells demonstrate that these cells can be formed by mitotic division failure [45]. Following ionizing radiation, human HCT116 cells progress into mitosis and chromosomes align, but anaphase chromosome segregation and cytokinesis fail, producing polyploid cells that contain bi-lobed nuclei [45]. Following division failure, nuclei fragment into compact masses that resemble clusters of grapes. Similar nuclear morphology is observed during apoptosis, and conventional apoptosis is sometimes observed following damage-induced division failure. However, cells produced by damage-induced mitotic failure are more typically TUNEL negative and isolated DNA does not show laddering characteristic of apoptosis [15, 48, 49]. In addition, cell death following division failure is generally not blocked by apoptotic inhibitors, cells do not contract or bleb, and apoptotic bodies are not formed (reviewed in [15]). Cell death by mitotic catastrophe thus appears to be distinct from apoptosis in both cell-cycle phase and mechanism of execution.

Two recent studies have analyzed the effects of DNA damage and incomplete DNA replication on mitotic centrosome organization in mammalian cells. Hut et al. analyzed centrosomes as hamster cells progress into mitosis prior to completion of DNA replication [50], while Mikhailov et al. analyzed centrosome and spindle organization in mammalian cells when DNA damage was induced following commitment to mitosis [51]. Using GFP-γ-tubulin as a centrosome marker, Hut et al. showed that centrosomes frequently fragment when checkpoint control is disrupted and mitosis is initiated before S phase is completed. These cells often assemble multi-polar spindles and progress through an aborted mitotic division to produce a single polyploidy cell. These authors also found that cells carrying a mutation that disrupts DNA damage repair spontaneously show similar mitotic defects, indicating that centrosome fragmentation is not due to the caffeine treatments used in the replication studies. These observations strongly suggest that centrosome disruption is a conserved feature of the damage-induced mitotic-division failure. However, complete dissociation of γ-tubulin from a core centrosome scaffold, observed in *Drosophila* embryos, was not found in mammalian cells. Therefore, aspects of centrosome disruption may vary according to organism.

Mikhailov et al. [51] used laser light to induce DNA damage following commitment to mitosis. In contrast to cells that progress into mitosis with DNA damage, these cells did not show centrosome defects. However, mitosis was delayed and H2AX histone was phosphorylated, indicating that damage can be detected within the context of condensed mitotic chromosomes. Inducing DNA damage during mitosis also has no clear effect on centrosome structure in the early *Drosophila* embryos, and centrosome inactivation is not blocked by mutations in the *Drosophila* H2Av homolog (S. Kwak and W. E. Theurkauf, unpublished observations). In a wide range of systems, mitotic catastrophe and centrosome disruption thus appear to require persistence of genetic lesions through the interphase to mitosis transition. We therefore speculate that the signaling machinery which triggers centrosome inactivation cannot recognize damage within the context of condensed mitotic chromatin, and thus requires damage detection prior to mitosis.

11.5
Final Thoughts

Cellular stress is induced by a broad range of environmental factors and triggers diverse changes in cell physiology. However, alterations in centrosome organization are a common feature of the cellular stress response and may play important biological roles in this response (Figure 11.1). Cell cycle checkpoints and apoptosis represent distinct salvage and disposal responses to genotoxic stress, and we speculate that centrosome alterations may contribute to salvage and disposal responses to a wide range of stress-inducing factors. As outlined above, genotoxic lesions trigger mitosis-specific centrosome defects that are linked to chromosome segregation and cytokinesis failures and cell death by a poorly understood process that appears to be distinct from apoptosis. Mitotic centrosome disruption thus appears to contribute to a mitotic catastrophe response that removes damaged nuclei and cells from the population. Mitotic centrosome function is also sensitive to heat shock, which has been reported to trigger both thermotolerance and cell death by mitotic catastrophe. The response to thermal stress, like the response to genotoxic stress, may therefore be determined by the extent of damage. We speculate that mild thermal stress induces a salvage response that includes heat shock gene expression, which promotes repair and thermotolerance. However, more severe heat shock may induce irreparable cellular damage that triggers the disposal pathway, characterized by mitotic catastrophe and cell death. Within this framework, the centrosome is a common target during the cellular responses to heat shock and genotoxic stress. Rigorous tests of this hypothesis await further molecular, genetic, and cellular characterization of the mitotic response to diverse forms of stress.

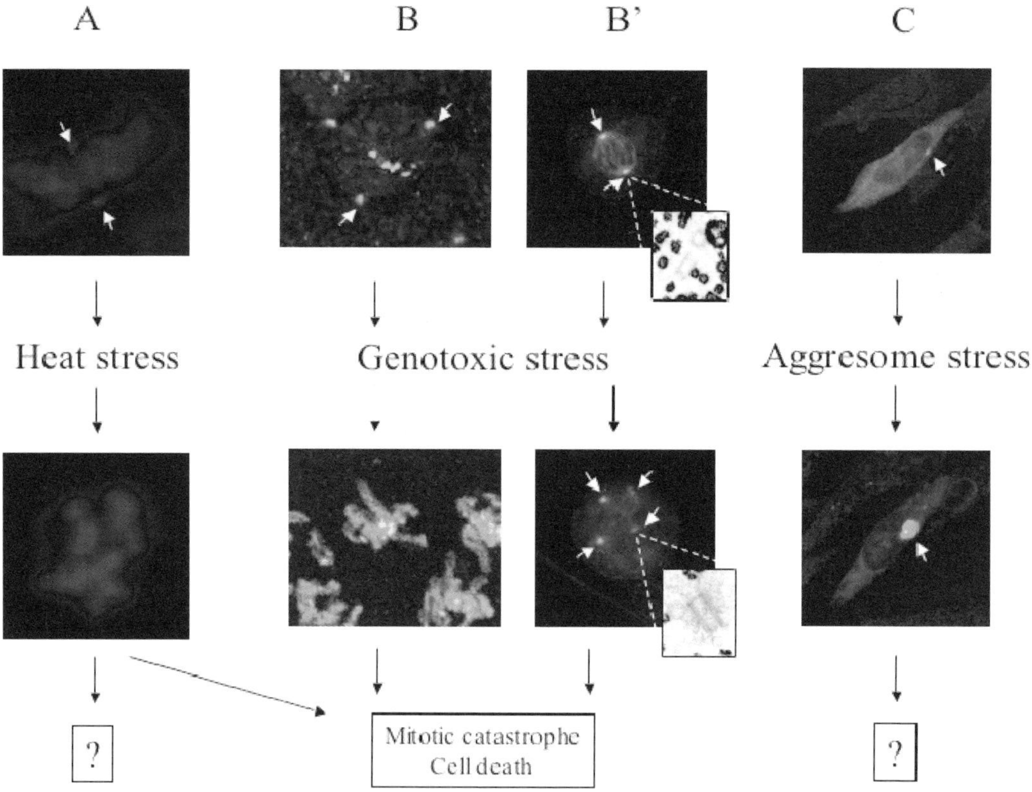

Figure 11.1 Centrosome alterations in response to heat, genotoxic and aggresome stress. In diverse systems, γ-tubulin (red) localizes to centrosomes at the mitotic spindle poles (A, B, B') and close to interphase nuclei (C). In Chinese hamster ovary (CHO) cells, heat stress (A) triggers loss of γ-tubulin localization to the poles (courtesy of H. Hut) while genotoxic stress (B') leads to mitotic centrosome fragmentation. Electron microscopic examination demonstrates that the centrosome fragments contain single centrioles (insets). In response to heat shock and genotoxic stress, centrosome disruption is associated with failures of mitotic division and mitotic catastrophe. In *Drosophila* embryos, genotoxic stress also leads to dissociation of γ-tubulin from the spindle poles (B) and mitotic catastrophe. Over-expression of a mutant form of GFP taggered the Huntingtin protein (green) in hamster cells (C), leads to aggresome formation around interphase centrosomes (courtesy of F. Salomons and M. Rujano). The significance of aggresome formation is not known, but this structure may contribute to neurodegeneration in a number of pathological conditions. In all panels, γ-tubulin is in red and DNA is in blue. In B, the kinetochore marker MeiS332 is in green. In C the Huntingtin-GFP protein is in green (see Color Plates page XXVII).

References

1. Zheng, Y., Jung, M. K., and Oakley, B. R. **1991**. Gamma-tubulin is present in *Drosophila melanogaster* and *Homo sapiens* and is associated with the centrosome. *Cell* 65, 817–823.
2. Zheng, Y., Wong, M. L., Alberts, B., and Mitchison, T. **1995**. Nucleation of microtubule assembly by a gamma-tubulin-containing ring complex. *Nature* 378, 578–583.
3. Vidair, C. A., Doxsey, S. J., and Dewey, W. C. **1995**. Thermotolerant cells possess an enhanced capacity to repair heat-induced alterations to centrosome structure and function. *J Cell Physiol* 163, 194–203.
4. Debec, A., Courgeon, A., Maingourd, M., and Maisonhaute, C. **1990**. The response of the centrosome to heat shock and related stresses in a *Drosophila* cell line. *J Cell Sci* 96, 403–412.
5. Brown, C. R., Hong-Brown, L. Q., Doxsey, S. J., and Welch, W. J. **1996**. Molecular chaperones and the centrosome. A role for HSP 73 in centrosomal repair following heat shock treatment. *J Biol Chem* 271, 883–840.
6. Barrau, M. D., Blackburn, G. R., and Dewey, W. C. **1978**. Effects of heat on the centrosomes of Chinese Hamster Ovary cells. *Cancer Res* 38, 2290–2294.
7. Marcaillou, C., Debec, A., Lauverjat, S., and Saihi, A. **1993**. The effect of the heat shock response on ultrastructure of the centrosome of *Drosophila* cultured cells in interphase: possible relation with changes in the chemical state of calcium. *Biochem Cell Biol* 71, 507–517.
8. Knox, J. D., Mitchell, R. E., and Brown, D. L. **1991**. Effects of hyperthermia on microtubule organization and cytolytic activity of murine cytotoxic T lymphocytes. *Exp Cell Res* 194, 275–283.
9. Debec, A., and Marcaillou, C. **1997**. Structural alterations of the mitotic apparatus induced by the heat shock response in *Drosophila* cells. *Biol Cell* 89, 76–78.
10. Morimoto, R. I., and Milarski, K. L. **1990**. Expression and function of vertebrate hsp70 genes. In *Stress Proteins in Biology and Medicine*, R. R. Morimoto, A. Tissieres and C. Georgopoulos (Eds). CSH Laboratory Press: Cold Spring Harbor, NY, 323–359.
11. Nollen, E. A., and Morimoto, R. I. **2002**. Chaperoning signaling pathways: molecular chaperones as stress-sensing "heat shock" proteins. *J Cell Sci* 115, 2809–2816.
12. Liang P., and MacRae T. H. **1997**. Molecular chaperones and the cytoskeleton. *J Cell Sci* 110, 1431–1440.
13. Rattner, J. B. **1991**. Hsp70 is localized to the centrosome of dividing HeLa cells. *Exp Cell Res* 195, 110–113.
14. Nakahata, K., Miyakoda, M., Suzuki, K., Kodama, S., and Watanabe, M. **2002**. Heat shock induces centrosomal dysfunction, and causes non-apoptotic mitotic catastrophe in human tumor cells. *Int J Hyperthermia* 18, 332–343.
15. Roninson, I. B., Broude, E. V., and Chang, B. D. **2001**. If not apoptosis, then what? Treatment-induced senescence and mitotic catastrophe in tumor cells. *Drug Resist Update* 4, 303–313.
16. Brown, C. R., Doxsey, S. J., Hong-Brown, L. Q., Martin, R. L., and Welch, W. J. **1996**. Molecular chaperones and the centrosome. A role for TCP-1 in microtubule nucleation. *J Biol Chem* 271, 824–832.
17. Perret, E., Moudjou, M., Geraud, M. L., Derancourt, J., Soyer-Gobillard, M., and Bornens, M. **1995**. Identification of an HSP-related protein associated with centrosomes from dinoflagellates to human cells. *J Cell Sci* 108, 711–725.
18. Lange, B. M., Bachi, A., and Wilm, G. **2000**. Hsp90 is a core centrosomal component and is required at different stages of the centrosomal cycle in *Drosophila* and vertebrates. *EMBO J* 18, 1252–1262.
19. De Carcer, G., do Carmo Avides, M., Lallena, M. J., Glover, D. M., and Gonzales, C. **2001**. Requirement of

Hsp90 for centrosomal function reflects its regulation of Polo kinase stability. *EMBO J* 20, 2878–2884.

20. Uzawa, M., Grams, J., Madden, B., Toft, D., and Salisbury, J. F. **1995**. Identification of a complex between centrin and heat shock proteins in CSF-arrested *Xenopus* oocytes and dissociation of the complex following oocyte activation. *Dev Biol* 171, 51–59.

21. Ellgaard, L., Molinari, M., and Helenius A. **1999**. Setting the standards: quality control in the secretory pathway. *Science* 286, 1882–1888.

22. Wickner, S., et al., Maurizi, M. R., and Gottesmann, S. **1999**. Posttranslational quality control: folding, refolding, and degrading proteins. *Science* 286, 1888–1893.

23. Johnston, J. A., Ward, C. L., and Kopito, R. R. **1998** Aggresomes: a cellular response to misfolded proteins. *J Cell Biol* 143, 1883–1898. 1998.

24. Wójcik, C., and DeMartino, G. N. **2003**. Intracellular localization of proteasomes (review). *Int J Biochem Cell Biol* 35, 579–589.

25. Kopito, R. R. **2000**. Aggresomes, inclusion bodies and protein aggregation. *Trends Cell Biol* 10, 524–530.

26. McNaught, K., Shashidharan, P., Perl, D. P., Jenner, P., and Olanow, C. W. **2002**. Aggresome-related biogenesis of Lewy bodies. *Eur J Neurosci* 16, 2136–2148.

27. Ma, J., and Lindquist, S. **2001**. Wild-type PrP and a mutant associated with prion disease are subject to retrograde transport and proteasome degradation. *Proc Natl Acad Sci USA* 98, 14955–14960.

28. Zobel, A. T. C., Loranger, A., Marceau, N., Thériault, J. R., Lambert, H., and Landry, J. **2003**. Distinct chaperone mechanisms can delay the formation of aggresomes by myopathy-causing R120G αB-crystallin mutant. *Hum Mol Gen* 12, 1609–1620.

29. Gautreau, A., Fievet, B. T., Brault, E., Antony, C., Houdusse, A., Louvard, D., and Arpin, M. 2003. Isolation and characterization of an aggresome determinat in the NF2 tumor suppressor. *J Biol Chem* 278, 6235–6242.

30. Fabunmi, R., Wigley, W. C., Thomas, P. J., and DeMartino, G. N. **2000**. Activity and regulation of the centrosome-associated proteasome. *J Biol Chem* 275, 409–413.

31. Wigley, W. C., Fabunmi, R. P., Lee, M. G., Marino, C. R. Muallem, S., DeMartino, G. N., and Thomas P. J. **1999**. Dynamic association of proteasomal machinery with the centrosome. *J Cell Biol* 145, 481–490.

32. Waelter, S., Boeddrich, A., Lurz, R., Scherzinger, E., Lueder, G., Lehrach, H., and Wanker E. E. **2001**. Accumulation of mutant huntingtin fragments in aggresome-like inclusion bodies as a result of insufficient protein degradation. *Mol Biol Cell* 12, 1393–1407.

33. García-Mata, R., Bebök, Z., Sorscher, E. J., and Sztul, E. S. **1999**. Charecterization and dynamics of aggresome formation by a cytosolic GFP-chimera. *J Cell Biol* 146, 1239–1254.

34. Saliba, R. S., Munro, P. M.G., Luthert, P. J., and Cheetham, M. E. **2002**. The cellular fate of mutant rhodopsin: quality control, degradation and aggresome formation. *J Cell Sci* 115, 2907–2918.

35. Canman, C. E. **2001**. Replication checkpoint: preventing mitotic catastrophe. *Curr Biol* 11, R121–R124.

36. Sibon, O. C., Kelkar, A., Lemstra, W., and Theurkauf, W. E. **2000**. DNA-replication/DNA-damage-dependent centrosome inactivation in *Drosophila* embryos. *Nature Cell Biol* 2, 90–95.

37. Andreassen, P. R., Lacroix, F. B., Lohez, O. D., Margolis, R. L. **2001**. Neither p21WAF1 nor 14-3-3sigma prevents G2 progression to mitotic catastrophe in human colon carcinoma cells after DNA damage, but p21WAF1 induces stable G1 arrest in resulting tetraploid cells. *Cancer Res* 61, 7660–7668.

38. Takada, S., Kelkar, A., and Theurkauf, W. E. **2003**. Drosophila checkpoint kinase 2 couples centrosome function and spindle assembly to genomic integrity. *Cell* 113, 87–99.

39. Foe, V., Odell, G., and Edgar, B. **1993**. In *The Development of* Drosophila melanogaster, M. A. Bate (Ed.). Cold Spring Harbor Laboratory Press: New York, 149–300.

40. Sibon, O. C., Laurencon, A., Hawley, R., and Theurkauf, W. E. **1999**. The *Drosophila* ATM homologue Mei-41 has an essential checkpoint function at the midblastula transition. *Curr Biol* 9, 302–312.

41. Sibon, O. C., Stevenson, V. A., and Theurkauf, W. E. **1997**. DNA-replication checkpoint control at the *Drosophila* midblastula transition. *Nature* 388, 93–97.

42. Xu, J., Xin, S., and Du, W. **2001**. *Drosophila* Chk2 is required for DNA damage-mediated cell cycle arrest and apoptosis. *FEBS Lett* 508, 394–398.

43. Peters, M., DeLuca, C., Hirao, A., Stambolic, V., Potter, J., Zhou, L., Liepa, J., Snow, B., Arya, S., Wong, J., Bouchard, D., Binari, R., Manoukian, A. S., and Mak, T. W. **2002**. Chk2 regulates irradiation-induced, p53-mediated apoptosis in *Drosophila. Proc Natl Acad Sci USA* 99, 11305–11310. E-pub. 2002 August.

44. Brown, E. J., and Baltimore, D. **2000**. ATR disruption leads to chromosomal fragmentation and early embryonic lethality. *Genes Dev* 14, 397–402.

45. Bunz, F., Dutriaux, A., Lengauer, C., Waldman, T., Zhou, S., Brown, J. P., Sedivy, J. M., Kinzler, K. W., and Vogelstein, B. **1998**. Requirement for p53 and p21 to sustain G2 arrest after DNA damage. *Science* 282, 1497–1501.

46. Chan, T. A., Hermeking, H., Lengauer, C., Kinzler, K. W., and Vogelstein, B. **1999**. 14-3-3sigma is required to prevent mitotic catastrophe after DNA damage. *Nature* 401, 616–620.

47. Liu, Q., Guntuku, S., Cui, X. S., Matsuoka, S., Cortez, D., Tamai, K., Luo, G., Carattini-Rivera, S., DeMayo, F., Bradley, A., Donehower, L. A., and Elledge, S. **2000**. Chk1 is an essential kinase that is regulated by Atr and required for the G(2)/M DNA damage checkpoint. *Genes Dev* 14, 1448–1459.

48. Lock, R. B., and Stribinskiene, L. **1996**. Dual modes of death induced by etoposide in human epithelial tumor cells allow Bcl-2 to inhibit apoptosis without affecting clonogenic survival. *Cancer Res* 56, 4006–4012.

49. Nabha, S. M., Mohammad, R. M., Dandashi, M. H., Coupaye-Gerard, B., Aboukameel, A., Pettit, G. R., and Al-Katib, A. M. **2002**. Combretastatin-A4 prodrug induces mitotic catastrophe in chronic lymphocytic leukemia cell line independent of caspase activation and poly(ADP-ribose) polymerase cleavage. *Clin Cancer Res* 8, 2735–2741.

50. Hut, H. M., Lemstra, W., Blaauw, E. H., Van Cappellen, G. W., Kampinga, H. H., and Sibon, O. C. **2003**. Centrosomes split in the presence of impaired DNA integrity during mitosis. *Mol Biol Cell* 14, 1993–2004.

51. Mikhailov, A., Cole, R. W., and Rieder, C. L. **2002**. DNA damage during mitosis in human cells delays the metaphase/anaphase transition via the spindle-assembly checkpoint. *Curr Biol* 12, 1797–1806.

52. Vidair, C. A., Huang, R. N., and Doxsey, S. J. **1996**. Heat shock causes protein aggregation and reduced protein solubility at the centrosome and other cytoplasmic locations. *Int J Hyperthermia* 12, 681–695.

53. Vidair, C. A., Doxsey, S. J., and Dewey, W. C. **1993**. Heat shock alters centrosome organization leading to mitotic dysfunction and cell death. *J Cell Physiol* 154, 443–455.

54. Malawista, S. E., Boisfleury-Chevance, A., Maunoury, R., and Bessis, M. **1983**. Heat as a probe of centrosomal function: a phase contrast and immunofluorescent study of human blood monocyes *Blood Cells* 9, 443–448.

Part III
The Centrosome in Development and Tissue Architecture

Centrosomes in Development and Disease. Edited by Erich A. Nigg

ISBN 3-527-30980-2

12
The *C. elegans* Centrosome during Early Embryonic Development

Laurence Pelletier, Thomas Müller-Reichert, Martin Srayko, Nurhan Özlü, Anne-Lore Schlaitz and Anthony A. Hyman

Abbreviations

γ-TuRC, γ-tubulin ring complexes; γ-TuSC, γ-tubulin small complexes; AP, anterior–posterior axis; dsRNA, double-stranded RNA; EM, electron microscope; FRAP, fluorescence recovery after photobleaching; GFP, green fluorescent protein; LR, left-right axis; MT, microtubules; MTOC, microtubule organizing center; PCM, pericentriolar material; RNAi, RNA-mediated interference; WT, wild-type

12.1
Introduction

Although more than 100 years have passed since Theodor Boveri gave the centrosome its name [1], the understanding of its structure, composition and function still remains in a state of infancy. This chapter will focus on recent advances made in understanding centrosome function and biogenesis using the nematode *C. elegans* as an experimental system. The role of the centrosome in determining the asymmetry of the early *C. elegans* embryo will also be discussed. We will first focus on the reasons that motivated the choice of *C. elegans* as a system to study centrosome function. This will be followed by a detailed description of the first cell division of the *C. elegans* zygote, the stage at which most of the research on the *C. elegans* centrosomes to date has been performed.

12.1.1
C. elegans as a Tool to Study Centrosome Biogenesis

The use of *C. elegans* as a powerful genetic system has long been recognized [2] culminating in the 2002 Nobel Prize in medicine to Brenner, Horvitz and Sulston. The *C. elegans* adult hermaphrodite is approximately 1 mm in length and is composed of 959 somatic cells, each having a defined cell lineage, and a variable num-

Centrosomes in Development and Disease. Edited by Erich A. Nigg

ISBN 3-527-30980-2

ber of germ cells [3]. Its genome has been completely sequenced since 1998, and is now skillfully annotated (http://www.wormbase.org, [4]). In order to effectively study centrosome function and biogenesis one needs an experimental system such as *C. elegans* that is easy to manipulate, in which centrosomes can be easily visualized and where gene inactivation works efficiently. *C. elegans* is an hermaphroditic species that can be routinely cultured in the laboratory, feeding primarily on *E. coli*. *C. elegans* is capable of internal self-fertilization, each hermaphrodite being able to generate hundreds of embryos. Once the embryos are laid, their survival in the environment depends on their chitinous eggshell. This, in turn, renders them amenable to experimental manipulation in the laboratory. When choosing *C. elegans* as a system for biological research, Brenner recognized the advantage of a translucent embryo, where Nomarski optics allow the visualization of cellular components such as nuclei, centrosomes and chromosomes. The embryos are relatively large ellipsoid structures approximately 25 by 50 μm (Figure 12.1). The position of the centrosomes and the mitotic spindle can be followed over time, making the identification and characterization of genes required for centrosome functions possible (Figure 12.1, arrowheads). In addition, GFP (Green fluorescent protein) technology is now well established in *C. elegans* and transgenic worms expressing various proteins tagged with GFP are readily available. This makes it possible to directly visualize, in real-time, various cellular processes [5]. Over the last decade or so, the use of RNAi (for RNA-mediated interference) has proven to be a powerful experimental and gene discovery tool [6, 7]. This technique relies on the introduction of gene-specific double-stranded RNA (dsRNA) into the germ line (or soma) of adult hermaphrodites. Unfortunately, it seems that RNAi does not work efficiently in the sperm, making it difficult to identify genes with strict paternal requirements. However, for most maternally provided proteins, the introduction of dsRNA results in the formation of oocytes containing cytoplasm essentially depleted, after 20–30 h, of the target gene's protein [7]. Upon fertilization, the effect of the RNAi can be observed in the developing embryo. Large scale RNAi screens, where the dsRNA was delivered either by soaking the worms in a solution containing dsRNA [8], feeding them with bacteria synthesizing dsRNA [9–12] or by direct microinjection of the dsRNA into the gonad of hermaphrodites [13, 14], have been performed. This reverse genetics approach resulted in the identification of genes required for specific processes, such as spindle assembly and centrosome function.

12.1.2 The First Cell Division of the *C. elegans* Embryo

One of the features of *C. elegans*' development that lends itself well to genome-wide analysis is that the cellular events are largely invariant between different embryos [3, 15–17]. Prior to fertilization, *C. elegans* oocytes are arrested in meiotic prophase I. Soon after fertilization, an acentriolar meiotic spindle forms to ensure chromosome segregation and polar body extrusion [18]. Subsequently, a second meiotic spindle forms and the remaining haploid female pronucleus and male pronucleus, typically located at opposite ends of the embryo, become visible. Shortly afterwards,

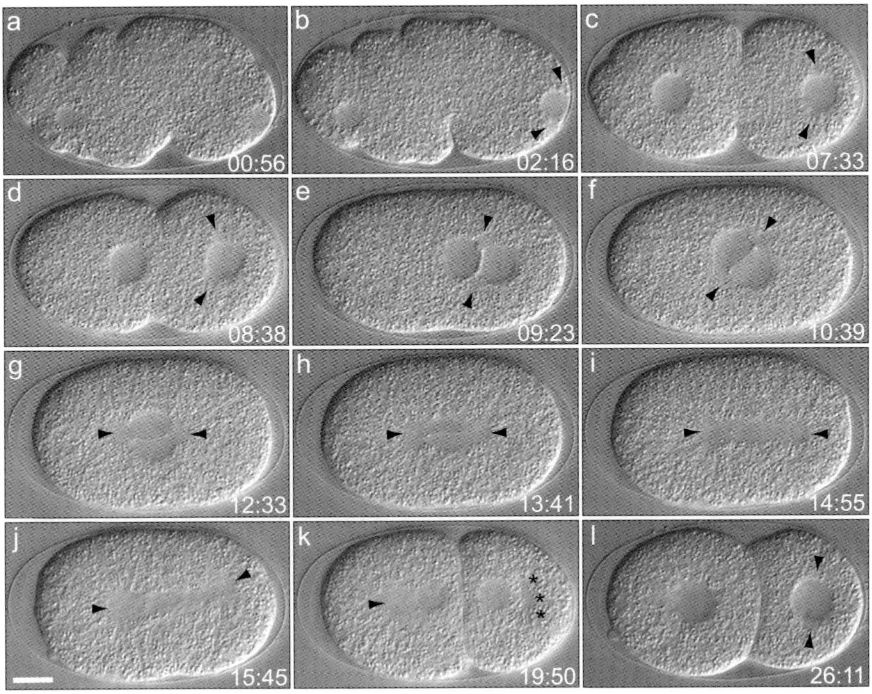

Figure 12.1 The first cell division of the *C. elegans* embryo. (a) Shortly after completing female meiosis, both male and female pronuclei become visible, a process referred to as pronuclear appearance. At this point membrane ruffling can be observed in both the anterior and posterior part of the embryo. (b) Shortly afterwards, the pronuclei become more defined, two centrosomal poles can be observed on each side of the male pronucleus (arrowheads) and ruffling becomes less abundant in the posterior part of the embryo of the embryo. (c) A pseudocleavage furrow then forms roughly at mid-egg length and ruffling becomes restricted to the anterior part. (d) The female pronucleus begins migrating towards the male pronucleus. (e) The male and female pronuclei meet. (f) After the female and male pronuclei have met, the centrosome–pronucleus complex begins to rotate. (g) After having rotated 90°, the nuclear envelopes break down and spindle formation begins. (h–j) Elongation of the mitotic spindle. (j) The spindle positions itself more towards the posterior part of the embryo so that the first division is asymmetric, yielding two daughter cells of different size. (k) During cytokinesis, the centrosomal poles disassemble, the posterior pole adopting a flattened morphology (asterisks). (l) Soon after the completion of cytokinesis, two defined centrosomal poles re-form in the daughter cells. In all images shown, the anterior of the embryo is to the left and the posterior to the right. Arrowheads indicate the position of the centrosomes and asterisks indicate the flattened posterior centrosome. Images were acquired at the indicated time points (min : s) using Nomarski optics. Scale bar = 10 μm.

cortical contractions, a process also referred to as "ruffling", can be observed over the whole embryo cortex (Figure 12.1a) and both the female and male pronucleus become more visible (Figure 12.1b). The time at which the two pronuclei become apparent is referred to as "pronuclear appearance". The female pronucleus is commonly located near the cortex opposite the male pronucleus, which is located in the posterior part of the embryo (Figure 12.1b). The position of the male pronucleus defines the posterior, the female pronucleus the anterior (Figure 12.1b). A pseudocleavage furrow then forms at about mid-egg distance (Figure 12.1c). At this time, the first signs of asymmetry become evident as cortical ruffling becomes restricted to the anterior part of the embryo (Figure 12.1c). During pseudocleavage, the male pronucleus lifts off the posterior cortex while the female pronucleus migrates towards it (Figure 12.1c–d). The pronuclei meet and subsequently move to the center of the embryo where they rotate 90° before the first mitotic spindle forms (Figure 12.1e–i). Although it is possible to visualize the rotation of the centrosome/pronucleus complex and spindle assembly using Nomarski optics, it is best observed by time-lapse microscopy of transgenic embryos expressing GFP-tagged versions α-, β- or γ-tubulin [19–21]. The metaphase spindle forms at the center of the embryo, parallel to the long axis (Figure 12.1h). During anaphase, the posterior pole of the spindle is displaced towards the posterior end of the embryo (Figure 12.1i) while it oscillates (Figure 12.1j). During cytokinesis, the centrosome at the posterior end of the embryo displays a flattened morphology (asterisks, Fig. 1k). The posterior displacement of the mitotic spindle causes the first cell division to be asymmetric, generating a larger anterior daughter, (AP), and a smaller posterior daughter, (P_1), after cytokinesis (Figure 12.1l) [3].

12.2
The *C. elegans* Centrosome

The centrosome, or the microtubule-organizing center (MTOC), is a non-membrane bound organelle that can be divided into at least three distinct domains in *C. elegans* based on the sub-cellular distribution of its resident proteins. It is composed of a centriole pair surrounded by an electron-dense proteinaceous lattice termed the pericentriolar material (PCM). Some components of the PCM, referred to in this chapter as peripheral PCM components, are found surrounding the PCM, giving them a doughnut-shaped appearance by immunofluorescence (Figure 12.4). The *C. elegans* centrosome is usually found in close proximity to the nucleus (arrowheads, Figure 12.1), its anchoring mediated by the Hook family member ZYG-12 that localizes to both the centrosome and the nuclear envelope (Table 12.1; Christian J. Malone, personal communication). In this section, we will discuss the structure and composition of the *C. elegans* centrosome. The *C. elegans* centrosomal proteins identified so far are listed in Table 12.1 along with their proposed function and homologs in other species if applicable. The function of individual centrosomal proteins will be discussed in further detail in Sections 12.3 and 12.4.

Table 12.1 Components of the *C. elegans* centrosome.

Gene name	*Function*	*Homolog in other species*	*References*
air-1	Centrosome maturation	Aurora-A	20, 49
grp-1	MT nucleation	Spc97p/Spc98p family member	21
sas-4	Centriole duplication	Weak homology to mammalian CPAP	25, 34
spd-2	Spindle assembly and polarity	None identified so far	Kevin O'Connell (personal communication), 47
spd-5	Spindle assembly and polarity	None identified so far	42
tac-1	Required for long astral and spindle microtubules	*Drosophila* TACC	67–69
tbg-1	MT nucleation	γ-Tubulin	19, 21, 37
zyg-1	Centriole duplication	Similarity to Nek2	33
zyg-9	Required for long astral and spindle microtubules	XMAP215/ch-TOG/ Msps family	66–69
zyg-12	Nuclear attachment	Hook family member	Christian Malone (personal communication)

12.2.1
The Centrioles

The centrioles in the *C. elegans* early embryo possess a simpler morphology, compared to those of mammals. They are composed of nine singlet microtubules surrounding a central tube, rather than the usual nine triplet microtubules (Figure 12.2A) [22–25]. Singlet microtubules are also observed in the early *Drosophila melanogaster* embryo [23, 26]. Although they differ morphologically, it appears likely that centrioles in *C. elegans* duplicate and function in a conventional manner [27]. It should be noted however, that the basal bodies of the inner and outer labial and cephalic neurons of *C. elegans* appear to be composed of doublet microtubules [28]. In the early embryo, each of the cylindrical centrioles is approximately 200–250 nm in length and 175 nm in diameter (Figure 12.2B–C). From EM analysis it appears that the appendages found on the mature, or mother centriole, are considerably less prominent in *C. elegans* than in mammalian cells [29]. Furthermore, whereas in mammals proteins such as centriolin, ε-tubulin, ninein and cenexin have been shown to specifically associate with the mother centriole, no such proteins have been identified so far in *C. elegans* [30–32]. However, most of the detailed descriptions of the centriolar structure have been derived from early embryos, where the speed of division is such that the centriole pair may not have enough time to reach a mature stage [26]. Another possibility is that higher

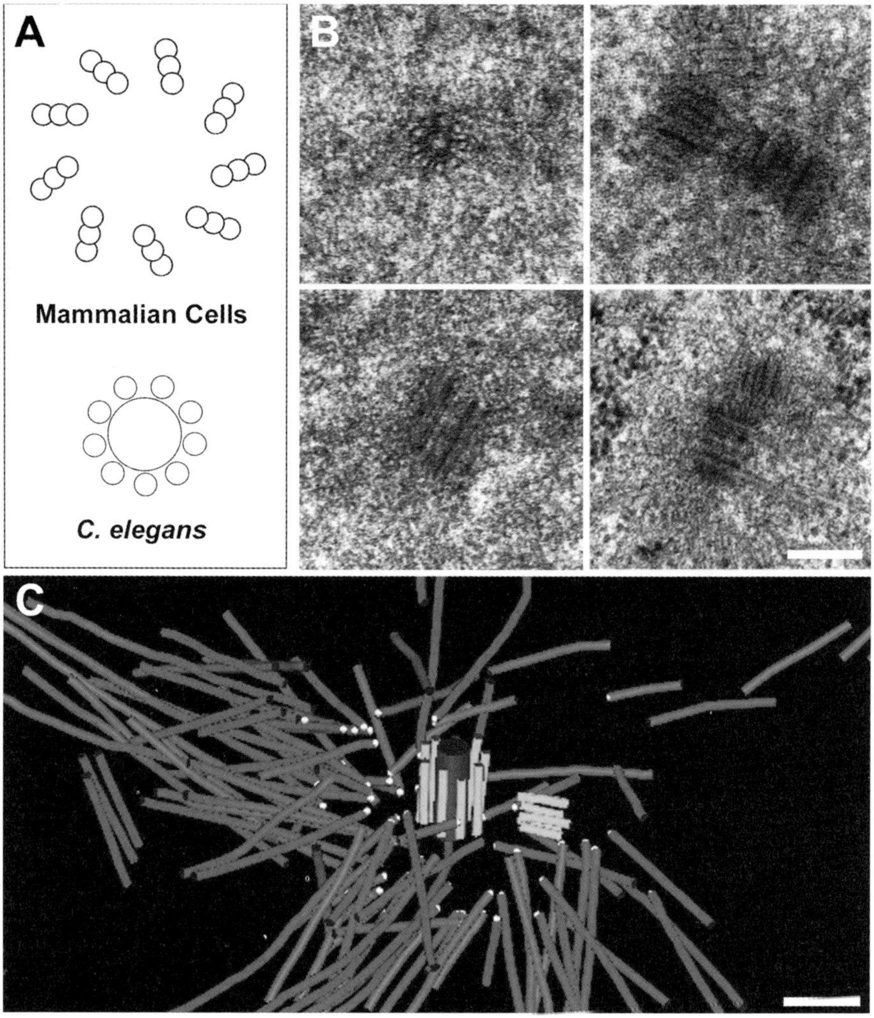

Figure 12.2 The ultrastructure of the *C. elegans* centrosome. (A) Schematic representation of the triplet structure of centrioles found in mammalian cells (top) and the singlet structure observed in *C. elegans* (bottom). (B) Electron micrographs of wild-type centrioles in cross-section and longitudinal orientation (left) and wild-type centriole pairs in orthogonal orientation (right). (C) 3-D model of a centriole pair during prometaphase derived from a tomographic reconstruction. Microtubules (red) are organized mainly around one centriole (blue), referred to as the mother centriole. Note that the minus ends of the microtubules do not come in contact with this centriole. Scale bars = 250 nm (see Color Plates page XXVIII).

eukaryotic centrioles have acquired appendages for specialized functions not utilized by the *C. elegans* early embryo.

To date, two *C. elegans* proteins, SAS-4 and ZYG-1, display centriole-specific localization patterns, and both play a role in centriole duplication (Table 12.1) [24, 25, 33–35]. In addition, SAS-4 appears to play a role in regulating centrosome size [25, 36]. These two proteins will be discussed further in Section 12.3.1.

12.2.2
The Pericentriolar Material (PCM)

Ultrastructurally, the PCM is often described as an "amorphous cloud" of electron-dense material that surrounds the centrioles. In *C. elegans* the size of the PCM varies during the cell cycle, reaching a maximum of about 1–2 μm in diameter during the metaphase to anaphase transition as judged by the localization of γ-tubulin (Figure 12.3e–f) [21, 37]. An array of microtubules emanating from the PCM can be observed in *C. elegans* , but these microtubules do not come into contact with the centriole pair, a situation also observed in other organisms (Figures 12.2C and 12.3) [29, 38, 39]. The analysis of tomographic volumes has shown that spindle microtubules appear to interact preferentially with the PCM in the vicinity of one of the two centrioles (Figure 12.2C) [38]. Whether or not this represents the same "mother" centrioles described in other systems (containing proteins such as centriolin and ninein) remains to be determined.

A functional definition of a PCM component could be that of a protein whose localization to the centrosome is not lost upon microtubule depolymerization. A well-established marker for PCM in *C. elegans* is γ-tubulin (Figures 12.3 and 12.4). The *C. elegans* gene *tbg-1* was shown to encode a protein homologous to γ-tubulin of other species (Table 12.1) [37]. During oogenesis, γ-tubulin can be detected on the nuclei of the germ cells present in the distal part of the gonad [37]. In meiotic cells at pachytene stage, located in the middle of the gonad, γ-tubulin is more difficult to observe, while only a faint cytoplasmic staining of γ-tubulin is found in oocytes arrested in diakinesis of prophase I [37]. The fact that distinct foci of γ-tubulin are not observed in mature oocytes reflects the elimination of the centrioles during oogenesis, but the mechanism behind this disappearance is not yet understood. Although immuno-EM using anti γ-tubulin antibodies has not been performed, some images suggest that, as in mammalian cells, γ-tubulin is also present at the core of the centriole [37, 40]. In *Paramecium*, γ-tubulin is required to initiate daughter centriole formation [41]. In *C. elegans*, it still remains unclear whether or not γ-tubulin is required to initiate daughter centriole formation [19, 21].

SPD-5 is another well-characterized PCM component that displays a similar localization pattern to that of γ-tubulin by immunofluorescence, surrounding the centriole pair (Table 12.1) [42]. It appears that SPD-5 acts upstream of other known PCM components since it is required for the recruitment of all the PCM components analyzed so far, including γ-tubulin, AIR-1 and ZYG-9 [42]. The failure to recruit PCM components leads to the inability of the centrosomes to

Figure 12.3 PCM recruitment and spindle assembly in *C. elegans*. Early embryos at different stages of the cell cycle were fixed and labeled for DNA (blue), microtubules (green) and γ-tubulin (red). Z-stacks through entire embryos were acquired, the images deconvolved and shown as two-dimensional projections. Scale bar = 10 μm. The anterior is to the left in all the images. (a) An acentrosomal meiotic spindle can be observed soon after fertilization (arrow). At this stage the centrosome contributed by the sperm has yet to separate. (b) At the beginning of pronuclear migration, the sperm-derived centrosome has separated and recruited some γ-tubulin therefore increasing the amount of microtubules it is able to nucleate. (c) At the time when the pronucleus–centrosome complex begins rotating, the DNA is condensed, the nuclear envelope broken down, and the levels of γ-tubulin and the nucleating capacity begin increasing rapidly, a process termed centrosome maturation. (d) After rotation, spindle assembly begins. (e) Late metaphase spindle with aligned chromosomes. (f) Late anaphase. (g) During telophase, the nuclear envelope re-forms and the posterior centrosome adopts a flattened morphology. At this time the γ-tubulin staining becomes more diffuse. (h) After cytokinesis, the centrosomes separate again in preparation for the next cellular division (see Color Plates page XXIX). ▶

nucleate microtubules further highlighting the role of SPD-5 as a key regulator of PCM function [42].

The distribution of AIR-1 on the centrosome is distinct from other PCM components like γ-tubulin and SPD-5, as it is more peripheral and also decorates microtubules [20] (Figures 12.3 and 12.4). Upon the addition of nocodazole, AIR-1 is no longer found on microtubules and in the periphery of γ-tubulin but instead appears to have collapsed onto PCM components like γ-tubulin and SPD-5 [20]. This observation suggests that an intact array of microtubules emanating from the centrosomes is required to maintain the localization of peripheral PCM components. In summary, the *C. elegans* centrosome appears to be composed of at least three sets of components: the core centriolar components, the PCM components and the more peripheral PCM components, such as AIR-1.

12.3
The Centrosome Cycle in *C. elegans* Embryos

In this section we will describe the centrosome cycle of the first cell division of the *C. elegans* zygote. During oogenesis, the oocyte loses its centriole pair and as a consequence of this, the centriole pair of the zygote is solely of paternal origin, contributed by the sperm upon fertilization (Figure 12.3a). Because of the absence of most PCM components on the sperm centrosome it is thought to be largely inactive in the mature spermatocyte but quickly becomes re-activated once liberated in the oocyte cytoplasm [27, 37, 42]. After entering the egg, the sperm-derived centrosome duplicates, yielding two small centrosomes that separate and position themselves between the sperm pronucleus and the cell cortex (Figure 12.3b), each of them labeling for centriolar components (Figure 12.4a–b). This suggests that each of the small centrosomes contains either a single centriole, or a centriole pair. Unfortunately, it is difficult do discriminate between these two possibilities under the light microscope. Although centriole separation is relatively easy to see, centriole duplication itself is hard to visualize without ultrastructural studies, and for this

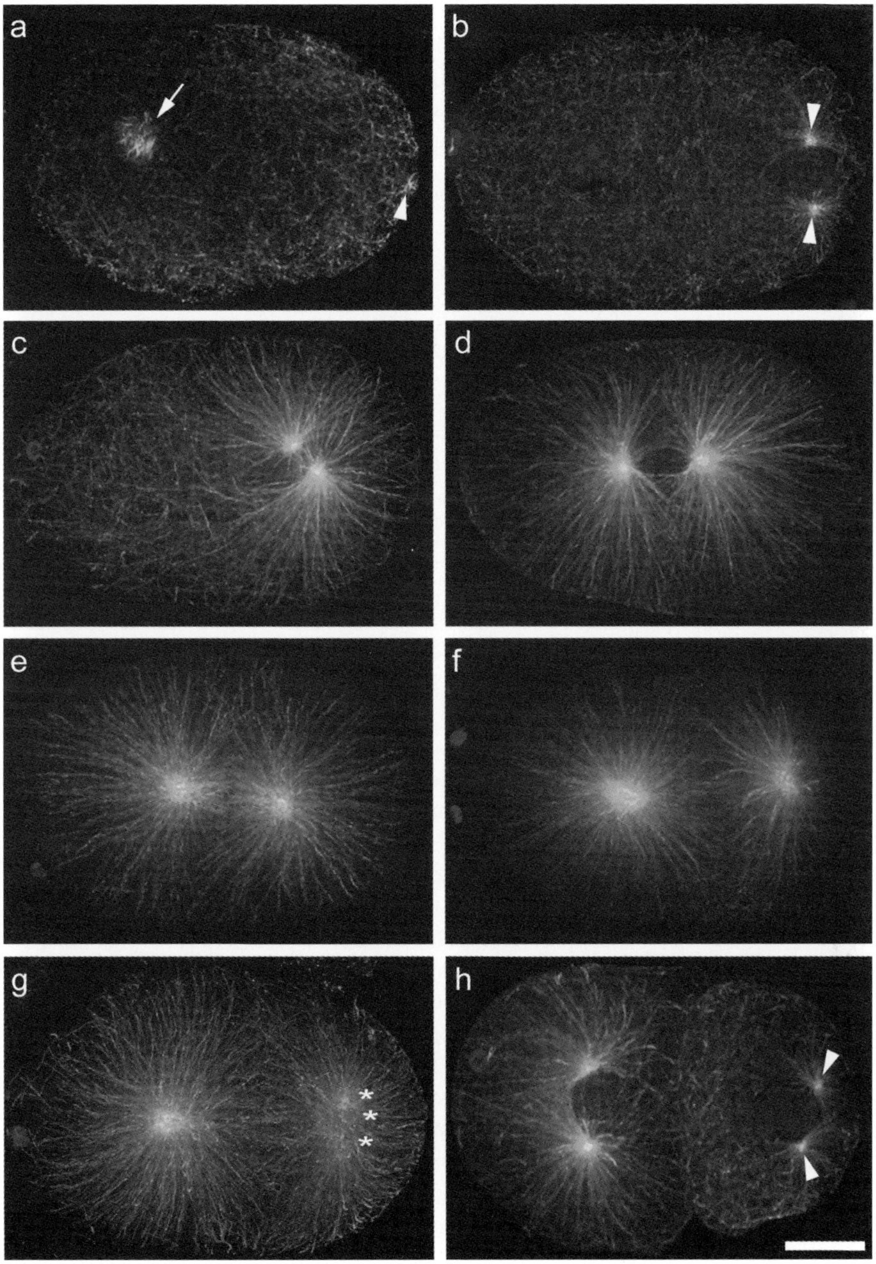
a
b
c
d
e
f
g
h

reason, the exact time at which the centrioles duplicate in *C. elegans* has not been determined. Concurrently, the centrosomes begin recruiting PCM components such as γ-tubulin (Figure 12.3b) and AIR-1 (Figure 12.4a–b) and the number of microtubules emanating from the centrosome increases slightly (Figure 12.3a–b). This phase will be referred to in this chapter as *PCM recruitment*. At the onset of mitosis, once the female and male pronuclei have met, the rate of PCM recruitment on the centrosomes increases dramatically, PCM levels reaching a maximum at the metaphase to anaphase transition [20] (Figure 12.3c–g). This increase correlates with the number of microtubules nucleated by the centrosomes, and is termed *centrosome maturation* (Figure 12.3c–f). In order to ensure that each daughter cell inherits a centriole pair, the centrioles contributed by the sperm need to duplicate prior to cytokinesis, a process known as *centriole duplication* (Figures 12.3h and 12.4e). Below we will describe each of these phases in more detail.

12.3.1
Centriole Duplication

Much akin to DNA, the centrosome needs to duplicate once and only once during the cell cycle. Failure to do so leads to the formation of aberrant spindles which, in turn, leads to defects in chromosome segregation and hence is a major cause of genomic instability and cancer progression (for a review see [43]). In this section, we will focus on the centriolar components ZYG-1 and SAS4 and their roles in centriole duplication in *C. elegans* (Table 12.1).

zyg-1 was first identified over 20 years ago as a zygote-defective mutation, but the molecular basis of this defect was only recently elucidated [33, 44] (for a review see [24]). The *zyg-1* gene encodes a protein kinase that localizes to the centrosome during anaphase, but is undetectable at other stages of the cell cycle [33]. One interesting possibility would be that the temporal recruitment of ZYG-1 to centrosomes represents the signal that initiates the centriole duplication/separation process. In *zyg-1* mutants, the centrosome contains a single centriole that appears to possess normal nucleation capacity but fails to assemble bipolar spindles. O'Connell and colleagues showed that paternal ZYG-1 is required for the formation of a bipolar spindle in the first division while maternal ZYG-1 is required for that process in the subsequent divisions. The fact that the nucleation capacity of the centrosomes in *zyg-1* mutant embryos appears to be normal suggests that it is directly involved in centriole duplication rather than being required for other centrosome-associated functions. Although *in vitro* experiments have shown that the ZYG-1 protein is able to phosphorylate itself, the existence of other substrates of the ZYG-1 kinase still remains to be demonstrated [24, 33], the identification of which will help us better understand the role of this key regulator of centriole duplication in *C. elegans*.

A second protein, named SAS-4 (for Spindle ASsembly defective-4), also involved in centriole duplication has been characterized in two separate studies. The *sas-4* gene was initially identified during the course of an RNAi-based screen aimed at identifying the genes on the *C. elegans* chromosome III that are required for cell division [13]. *SAS-4* encodes a polypeptide, with a predicted molecular weight of 92 kDa [25, 34]. It contains a central coiled-coil region and is localized to the centrioles [25, 34]. The centriole duplication defect in *sas-4(RNAi)* is similar to that caused by the lack of maternal *zyg-1*, the daughter cells failing to establish a bipolar spindle during the second cell division [25, 33, 34]. In *sas-4(RNAi)* embryos, the monopolar spindles formed during the second division accumulate normal amounts of PCM components, such as γ-tubulin, suggesting that, like ZYG-1, SAS-4 is directly required for centrosome duplication [25, 34]. Kirkham and colleagues used ultrastructural analysis to show that the defect was specifically due to the inability of centrioles to duplicate and that each pole of the bipolar spindles formed during the first division contained a single centriole. One issue that needs to be reconciled is to what degree SAS-4 remains associated with the centrioles. On the one hand, the work by Kirkham and colleagues suggests that cytoplasmic SAS-4 becomes incorporated into the daughter centriole as it forms during centriole duplication and remains stably associated with it thereafter. On the other hand, Leidel and colleagues argue, using results from FRAP experiments, that SAS-4 is a stable component of the centriole before the end of mitosis, and that shortly afterwards it is capable of rapidly exchanging with a cytoplasmic pool of SAS-4 at the time when centriole duplication is thought to occur [34]. The further characterization of SAS-4 and ZYG-1, along with the identification of other proteins required for centriole duplication, is needed in order to better understand the molecular mechanisms that govern centriole duplication in *C. elegans*.

12.3.2
PCM Recruitment

Soon after fertilization, the sperm centrosome liberated into the oocyte begins recruiting PCM components. As already mentioned above, the centrosomal protein SPD-5 was shown to be required for this process [42]. The centrosomes are unable to recruit any PCM in *spd-5* mutants, and consequently are unable to nucleate microtubules and assemble a mitotic spindle. This is the first clear evidence that centrosomes in *C. elegans* are required for spindle assembly [42]. SPD-5 is an 1198-amino acid protein with a predicted molecular weight of 135 kDa which contains 11 predicted coiled-coil domains [42]. The localization of SPD-5 to centrosomes persists in the presence of the microtubule depolymerizing agent nocodazole suggesting that SPD-5 is a bona fide PCM component. Interestingly it is largely absent in mature spermatocytes but is recruited soon after fertilization [42]. The *spd-5 (or213ts)* mutant was first identified as a conditional maternal-effect mutation that displayed severe defects in cell division at the restrictive temperature [45]. *spd-5* mutant embryos exhibit pronuclear migration defects, are unable to assemble a mitotic spindle and fail at cytokinesis [42]. However, meiosis is not impaired in *spd-5* mutants and acentrosomal meiotic spindles form and correctly extrude DNA of maternal origin into two polar bodies [42]. The inability of *spd-5* embryos to assemble a mitotic spindle can be explained by the incapacity of centrosomes to recruit PCM components such as γ-tubulin, ZYG-9 and AIR-1 in the absence of SPD-5 [42].

Another protein involved in PCM recruitment that localizes to the centrosome is SPD-2 (Kevin O'Connell, Eva Hannak and Laurence Pelletier, unpublished observations). The *spd-2(oj29)* mutation was first identified in a screen for mutants defective in cell division [46]. O'Connell and colleagues showed that *spd-2* embryos are defective in mitotic spindle assembly and that the ability of the centrosome to recruit normal amounts of γ-tubulin is severely impaired [47]. This interesting observation suggests that SPD-2 also plays a role in PCM recruitment but further experiments will be required to better understand the role of SPD-2 in PCM recruitment.

Perhaps one of the most interesting characteristics of SAS-4, a protein already discussed above for its role in centriole duplication, is that it appears to control the size of the centrosome by directing the amount of PCM it recruits [25]. Partial depletion of SAS-4 leads to mitotic spindle poles that differ in the amount of PCM recruited to the centrosomes during the second cell division and in the number of microtubules they nucleate [25]. Analysis of these asymmetric spindle poles revealed that one pole contains normal amounts of γ-tubulin and a single centriole while the other contains reduced amounts of γ-tubulin and a single centriole that is considerably smaller in size [25]. The importance of this observation is that it provides evidence that the centrioles themselves may dictate how much PCM is recruited to the centrosome [25].

12.3.3 Centrosome Maturation

Centrosome maturation corresponds to the drastic increase in the rate of PCM recruitment at the onset of mitosis that is thought to be a prerequisite for spindle assembly. In vertebrate cells, centrosome maturation occurs at the G2/M transition, at which time, the amount of centrosome-associated γ-tubulin rapidly increases 3–5-fold [48]. In *C. elegans*, centrosome maturation also occurs as the cells enter mitosis, the levels of γ-tubulin also increasing rapidly (Figure 12.3) [20]. The protein kinase AIR-1, shown to be a regulator of this process in *C. elegans*, was first identified as a component of the centrosome required for normal spindle assembly and chromosome segregation [49]. It was later shown that the spindle assembly defect observed in *air-1(RNAi)* embryos is due to a failure to undergo centrosome maturation [20]. AIR-1 belongs to the Aurora-A family of protein kinases (Table 12.1) [50, 51]. So far, Aurora-A kinases have been identified in fly, yeast, worm, frog and human [49, 52–56]. As mentioned previously, the localization pattern of AIR-1 on the centrosome is different from other PCM components (Figure 12.4). AIR-1 is peripheral to γ-tubulin and can also be found extending away from the PCM, decorating microtubules (Figure 12.4) [20]. Like other PCM components, AIR-1 localizes to the centrosome throughout the cell cycle, but the levels increase drastically at the onset of mitosis, reaching a maximum at the metaphase to anaphase transition (Figure 12.4a–e) [20]. In *air-1(RNAi)* embryos, spindle assembly does not occur, instead, two closely opposed centrosomal asters are

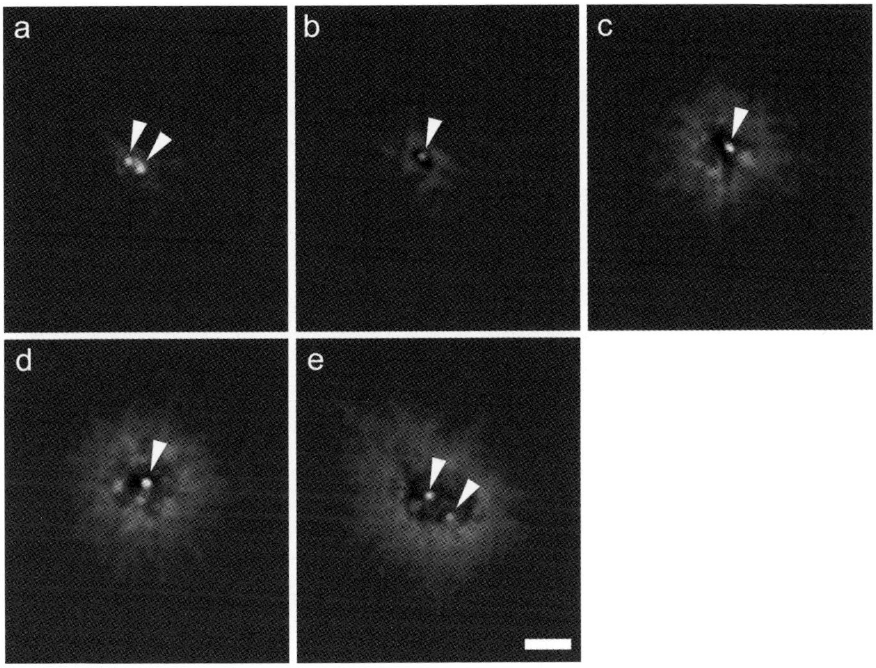

Figure 12.4 Layers of the *C. elegans* centrosome. Early embryos at different stages of the cell cycle were fixed and triple labeled for γ-tubulin (light blue), AIR-1 (green) and the centriolar protein SAS-4 (red). Z-stacks through entire embryos were acquired and the images deconvolved. Individual images or two-dimensional projections of two sections are shown. (a) Centrosome during female meiosis. Note the presence of two SAS-4 positive structures, indicating that the two centrioles contributed by the sperm have separated. (b) One pole after centrosome separation. (c) One pole during the rotation of the pronucleus–centrosome complex. (d) One pole during metaphase. (e) One pole during telophase. Note that at this stage two SAS-4 positive structures can be seen at each pole, suggesting that centriole duplication has occurred (see Color Plates page XXX).

observed. In addition, these asters contain significantly less α-tubulin than wild-type centrosomes [20]. These observations are consistent with the role in spindle assembly previously ascribed to Aurora-A in *Drosophila* and *Xenopus* [52, 53].

Using time-lapse imaging, followed by the measurement of the distance between centrosomes, Hannak and colleagues have established that although centrosomes undergo separation prior to mitosis, AIR-1 was required to maintain centrosome separation at the onset of mitosis [20]. Centrosomes in *air-1(RNAi)* embryos are unable to increase their levels of PCM components such as CeGrip-1, γ-tubulin and ZYG-9 at the onset of mitosis [20]. This suggests that AIR-1 is required for the centrosomal accumulation of these proteins during maturation. Spindle assembly also fails in *air-1(RNAi)*, one possibility being that centrosome maturation is a prerequisite for spindle assembly in *C. elegans* [20]. The reduced levels of PCM components on the centrosome of *air-1(RNAi)* embryos is not caused by the lack of microtubules since WT embryos are capable of recruiting normal amounts of PCM after treatment with nocodazole [20]. This observation points to a direct role for AIR-1 in the maturation process. The depletion of AIR-1 also causes a delay in nuclear envelope breakdown (NEBD). One interesting possibility is that centrosome maturation and subsequent spindle assembly, regulated by AIR-1, somehow controls the timing of NEBD [20]. In *spd-5* embryos, γ-tubulin, ZYG-9 and AIR-1 cannot be detected on centrosomes at any stage during the cell cycle [42]. In contrast, *air-1(RNAi)* embryos are able to undergo the initial phase of PCM recruitment and the above proteins can be detected [20]. Together these results suggest that AIR-1 is required to carry out a PCM recruitment pathway during centrosome maturation. This event, which appears to be distinct from early PCM recruitment, is regulated, at least in part, by the AIR-1 kinase.

12.4 Centrosome Functions

In this section we will discuss the various functions of the centrosome in the early *C. elegans* embryo. First, we will concentrate on the role of the centrosome in microtubule nucleation and spindle assembly. This will be followed by a section dedicated to the role of the centrosome in determining the anterior–posterior polarity of the embryo and a section on the role of the centrosomes in positioning the mitotic spindle during mitosis.

12.4.1 Spindle Assembly and Microtubule Nucleation

Perhaps the most obvious characteristic of the centrosome is that it constitutes the main source of microtubule nucleation in the cell. In embryonic systems such as the *C. elegans* zygote and the *Xenopus laevis* oocyte, the transition between interphase and mitosis is marked by a drastic change in the number of microtubules growing out of the centrosome (Figure 12.3). Microtubule nucleation at the centrosome in *C. elegans* increases significantly at the onset of mitosis, reaching a maximum at the metaphase to anaphase transition (Figure 12.3e–f). This increase appears to be related to the amount of PCM present on the centrosome, for example the amount of γ-tubulin being the highest at that time (Figure 12.3c–g) [21]. Consistent with this, depletion of γ-tubulin leads to a decrease in the nucleation capacity of centrosomes [21].

It appears that the kinetically dominant assembly pathway for centrosomal asters in *C. elegans* is dependent upon γ-tubulin, although MT asters presumably can also form in a γ-tubulin-independent fashion [19, 21]. Even though this awaits biochemical verification, it is thought that γ-tubulin in *C. elegans* also functions within larger protein complexes [21]. In all organisms examined so far, γ-tubulin appears to exist in a heteromeric complex that contains two members of the Spc97/Spc98 protein family, conserved from *Drosophila* to humans [57, 58]. This complex, termed the "small" γ-tubulin complex (γ-TuSC), is a subunit of a larger complex referred to as the γ-tubulin ring complex (γ-TuRC) that possesses microtubule capping and nucleating activity *in vitro* [59, 60]. *C. elegans* contains two open reading frames (H04J21.3 and C45G3.3) encoding respectively CeGrip1 and CeGrip2 that likely represent Spc97/Spc98 family members [21]. Both proteins contain two γ ring protein (grip) domains, a characteristic shared by many γ-tubulin interacting proteins [57]. Depletion of CeGrip1 leads to the inability to recruit γ-tubulin to the centrosome, leading to subsequent defects in microtubule nucleation [21]. This observation is consistent with these proteins being the *C. elegans* equivalent of γ-TuSC but this still needs to be verified experimentally through biochemical assays. To date no γ-TuRC components have been identified in *C. elegans*.

In *C. elegans*, it appears that functional centrosomes are required to assemble a mitotic spindle. This is best supported by the fact that *spd-5* mutants fail to assemble a mitotic spindle, presumably because their centrosomes are unable to recruit any PCM components. This is in sharp contrast with work in other organisms, such as fly and human, or *in vitro* experiments that suggest, on the contrary, that centrosomes are not necessary for spindle formation [61–64] (for a review see the Chapter 10). One possibility is that the chromatin-Ran-GTP-mediated microtubule nucleation pathway does not operate with sufficient efficiency, if present at all, in the early *C. elegans* embryo to compensate for the absence of functional centrosomes [65]. Further experiments will be required to clarify this point.

Two other proteins, ZYG-9 and TAC-1, that localize to the *C. elegans* centrosome and do not appear to affect the number of microtubules they nucleate, but rather the length of astral microtubules, have been identified (Table 12.1) [66–69]. The XMAP215/ch-TOG/Msps family of proteins consists of microtubule-associated proteins (MAPs) that modulate the dynamic instability of microtubules by directly binding to them [70–73]. *C. elegans zyg-9* encodes the ortholog of XMAP215 [66]. One-cell *zyg-9* mutants exhibit both meiotic and mitotic defects [66]. Consistent with this meiotic phenotype, ZYG-9 localizes to the meiotic spindle where it presumably participates in the formation of the acentrosomal spindle, and upon depletion of ZYG-9, meiotic spindles become disorganized. During mitosis, the spindles are smaller than normal and composed of unusually short microtubules [66, 67, 74]. A known physical interactor of the *Drosophila* protein Msps is D-TACC (transforming acidic coiled-coil) [75, 76]. The efficient localization of Msps to the centrosome requires D-TACC and vice versa [75, 76]. The *C. elegans* ortholog of D-TACC is TAC-1 [67–69]. This protein also localizes to the centrosome and physically interacts with ZYG-9, the Msps homolog. ZYG-9 and TAC-1 depend on each other for their localization to the centrosome. The depletion of TAC-1 by RNAi resulted in microtubule-based defects very similar to that of *zyg-9(RNAi)*, both being required for the generation of long astral and spindle microtubules [67–69]. Interestingly, it does not appear that either ZYG-9 or TAC-1 is required for microtubule nucleation since the amount of α-tubulin near the centrosomes is indistinguishable from that of WT embryos [67].

12.4.2
Determination of Anterior–Posterior Polarity

C. elegans oocytes are apolar with respect to the future body axes. Following fertilization, however, the polarity of the zygote soon becomes apparent, even before the completion of the first cellular division, approximately 30 min after fertilization [77, 78]. The first axis of polarity seems to be directed by the sperm since the location of the sperm pronucleus correlates with the posterior pole of the embryo [79, 80]. The establishment of the anterior–posterior (AP) axis is evident when contractions throughout the embryo cortex suddenly stop in the region adjacent to the sperm pronucleus (Figure 12.1a). Coincident with this cortical "smoothing", internal cytoplasm begins to flow towards and peripheral cytoplasm away from this area, a process know as fountain flows [81]. These flows are required to segregate germline fate determinants, such as P-granules [82, 83], and the polarity of the cytoplasmic flows appears to be dictated by the centrosomes associated with the sperm pronucleus [81]. Other markers of cell polarity in *C. elegans*, the PAR proteins, become asymmetric at this time. Prior to fertilization, PAR proteins are uniformly distributed but quickly redistribute in response to the sperm signal. PAR-1 and PAR-2 become localized to a posterior cortical domain while PAR-3 and PAR-6 redistribute to a non-overlapping anterior domain [84–87].

What sperm-supplied component is required to establish anterior–posterior polarity? The mature *C. elegans* sperm contributes the paternal genetic material, a centriole pair, an RNA pool, cytoplasm and mitochondria. Anucleate *C. elegans* sperm can still determine the anterior–posterior axis of the zygote suggesting that the sperm DNA itself is not required to establish polarity [88]. Certain lines of evidence support the hypothesis that the centrosome is the important contribution of the sperm with regard to polarity. Although not proven directly, it has been proposed that the polarization of the embryo is a microtubule-directed process [89]. The fact that the centrosome is the major site of microtubule nucleation is consistent with a role for the sperm asters in determining polarity. As described previously, both *spd-2* and *spd-5* mutants are defective in PCM recruitment, and hence contain centrosomes that do not nucleate microtubules adequately. Interestingly, both these mutants have defects in polarity [42, 47, 90]. In *spd-2* mutants cytoplasmic flows are not observed, P-granules fail to segregate to the posterior part of the embryo and the accurate partitioning of PAR proteins does not occur [47]. In *spd-5* mutants, the partitioning of PAR-2 is affected, and the anteriorly directed cytoplasmic flows and the segregation of P-granules to the posterior are slowed considerably [90]. Taken together, these results demonstrate a clear correlation between centrosomes and the establishment of polarity in the one-cell *C. elegans* embryo.

12.4.3
Spindle Positioning

Prior to mitotic spindle assembly, the centrosome–pronucleus complex is typically in a position orthogonal to the AP axis (Figure 12.1e). Since the mitotic spindle in *C. elegans* must form along the AP axis and be posteriorly displaced during anaphase of the first cell division of the zygote (Figures 12.1i and 12.3g), mechanisms must exist to ensure that both these criteria are met prior to cytokinesis. The centrosome–pronucleus complex first needs to rotate 90 ° in order for the mitotic spindle to form in the AP axis. Initial experiments addressing the rotation process were performed in early blastomeres of the P cell lineage. During the rotation in the P_1 cell, one of the two centrosomes (termed the leading centrosome) moves towards the anterior part of the embryo while the other centrosome (called the lagging centrosome) serves as a pivoting point for the centrosome–pronucleus complex [91]. The transient disruption of either the leading centrosome or the microtubules that emanate from it and reach the anterior cortex, temporarily halts rotation [92]. These data are consistent with the idea that the centrosome rotates towards its correct position by shortening the connections, most likely microtubules, between a centrosome and a defined site on the cortex of the embryo [92]. Components of the dynein/dynactin complex are localized to cortical sites and are required for the rotation raising the interesting possibility that dynein may be involved in tethering astral microtubules and, through microtubule depolymerization, may provide the tension that mediates the rotation of the centrosome–pronucleus complex [93, 94]. Whether or not a similar mechanism operates during the first mitotic division, in P_0, to ensure that the centrosome–pronucleus complex rotates, is still unknown.

The generation of daughter cells of different size through unequal cell division is crucial for ensuring cell diversity during development [95]. This can be exemplified by the ability of stem cells, through asymmetric cell divisions, to generate an exact copy of themselves (other stem cells) and new types of cells that will eventually differentiate into the mature cells of the tissue [96, 97]. To ensure that cells are capable of dividing asymmetrically, molecular mechanisms are in place to ensure the mitotic spindle is properly positioned before the end of anaphase [98]. The microtubule cytoskeleton and the associated centrosomes (or spindle-pole body in yeast) are required to achieve this task in flies and yeast. Furthermore, evidence suggests that one of the mechanisms behind spindle positioning involves the stabilization of microtubule plus-ends and cortical anchors [99–101]. But what is the direct evidence so far in favor of interactions between the mitotic spindle and the cell cortex that would mediate spindle positioning in *C. elegans*? The answer to this question came from experiments performed by Grill and colleagues where a high-power UV laser was used to physically ablate the central spindle of embryos during the first cell division [102]. Their results show, by measuring the maximal peak velocity of both spindle poles after the ablation of the central spindle that external forces of different strengths act on each pole of the mitotic spindle, the net force being greater on the posterior pole than the on anterior pole [102]. This imbalance of forces could explain why the spindle is displaced posteriorly during the first cell division. Interestingly, the unequal distribution of the forces acting on the spindle poles appears to be under the control of the *par* genes since it is no longer observed in *par-2(RNAi)* or *par-3(RNAi)* embryos [102]. The imbalance of forces acting on the mitotic spindle poles is actually caused by an asymmetric distribution of the number of active force generators that differentially affect the pulling forces acting on the anterior and posterior centrosomal asters [103]. The activation of heterotrimeric G protein α-subunits is required for the generation of these astral pulling forces [103]. The GoLoco-containing proteins GPR-1 and GPR-2, and the two α-subunits GOA-1 and GPA-16 are all required in order to generate proper pulling forces in the early embryo [102, 104]. GPR-1 and GPR-2 interact specifically with the guanosine diphosphate-bound GOA-1 and are localized to the posterior cortex, the localization being *par-2* and *par-3* dependent [104]. Together, these results suggest that the extent of cortical pulling forces exerted on the spindle depends on the cortically localized Gα activity that is regulated through an anterior–posterior signal regulated, at least in part, through GPR-1 and GPR-2.

The left–right asymmetry (LR) in *C. elegans*, or "handedness choice", first becomes apparent between the four-cell to six-cell stage of the early embryo, and from then on persists throughout development [3, 105, 106]. The *spn-1* mutant phenotype is due to a loss-of-function mutation in GPA-16 that leads to the impaired recruitment of Gβ subunits to the centrosome and leads to defects in LR asymmetry determination [106]. Genetic evidence also suggests that *gpa-16* interacts genetically with *par* genes [106]. It thus appears that the mechanisms that define spindle positioning in the one-cell embryo and later in the four-cell embryo are both based on G-protein-mediated force generation through cortical anchors.

12.5
Concluding Remarks

Much has been learned over the last decades about the *C. elegans* centrosome. Many novel genes required for proper centrosome function have been identified, some of them through standard genetic methods, others through large-scale RNAi screens. Further analysis of the available data will provide us with a new challenge: identifying all the components that constitute a functional centrosome able to duplicate, mature and assemble a mitotic spindle. To do so, large-scale protein interaction screens will have to be used to identify non-essential components of the centrosome missed in the RNAi screens due to the lack of phenotype, and perhaps other essential genes that were missed [107]. The fact that RNAi does not appear to function in sperm brings about another problem that may be solved by two-hybrid screens, that of the RNAi screens missing most, if not all, the genes that have a strict paternal requirement. Biochemical assays such as those used routinely in *Xenopus* for centriole duplication, spindle assembly and centrosome maturation will need to be developed in order to further dissect the molecular mechanisms behind centrosome function in *C. elegans*. A better understanding of the *C. elegans* centrosome will provide us with useful insights that, in turn, will lead to a better understanding of the centrosomes of other species, such as humans, where the core mechanisms behind their function and biogenesis are expected to be similar.

Acknowledgments

The authors would like to thank all the members of the Hyman laboratory for many stimulating conversations, Kevin O'Connell and Christian Malone for sharing unpublished results, and Mike Boxem, Alexandru Constantinescu, Carrie Cowan, Sophie Quintin and Tim Nötzel for their judicious comments on the manuscript.

Laurence Pelletier is supported by postdoctoral fellowships from EMBO and Martin Srayko by a postdoctoral fellowship from the HFSP.

References

1. Boveri, T. **1901**. Zellenstudien IV. Ueber die Nature der Centrosomen (Jena).
2. Wood, W. B. **1988**. The Nematode *Caenorhabditis elegans*. Cold Spring Harbour Laboratory.
3. Sulston, J. E., Schierenberg, E., White, J. G., and Thomson, J. N. **1983**. The embryonic cell lineage of the nematode *Caenorhabditis elegans*. *Dev Biol* 100, 64–119.
4. The *C. elegans* sequencing consortium. **1998**. Genome sequence of the nematode *C. elegans*: a platform for investigating biology. *Science* 282, 2012–2018.
5. Chalfie, M., Tu, Y., Euskirchen, G., Ward, W. W., and Prasher, D. C. **1994**. Green fluorescent protein as a marker for gene expression. *Science* 263, 802–805.
6. Fire, A., Albertson, D., Harrison, S. W., and Moerman, D. G. **1991**. Production of antisense RNA leads to effective and specific inhibition of gene expression in *C. elegans* muscle. *Development* 113, 503–514.
7. Fire, A., Xu, S., Montgomery, M. K., Kostas, S. A., Driver, S. E., and Mello, C. C. **1998**. Potent and specific genetic interference by double-stranded RNA in *Caenorhabditis elegans*. *Nature* 391, 806–811.
8. Maeda, I., Kohara, Y., Yamamoto, M., and Sugimoto, A. **2001**. Large-scale analysis of gene function in *Caenorhabditis elegans* by high-throughput RNAi. *Curr Biol* 11, 171–176.
9. Simmer, F., Moorman, C., Van Der Linden, A. M., Kuijk, E., Van Den Berghe, P. V., Kamath, R., Fraser, A. G., Ahringer, J., and Plasterk, R. H. **2003**. Genome-wide RNAi of *C. elegans* using the hypersensitive rrf-3 strain reveals novel gene functions. *PLoS Biol* 1, E12.
10. Kamath, R. S., and Ahringer, J. **2003**. Genome-wide RNAi screening in *Caenorhabditis elegans*. *Methods* 30, 313–321.
11. Zipperlen, P., Fraser, A. G., Kamath, R. S., Martinez-Campos, M., and Ahringer, J. **2001**. Roles for 147 embryonic lethal genes on *C.elegans* chromosome I identified by RNA interference and video microscopy. *EMBO J* 20, 3984–3992.
12. Fraser, A. G., Kamath, R. S., Zipperlen, P., Martinez-Campos, M., Sohrmann, M., and Ahringer, J. **2000**. Functional genomic analysis of *C. elegans* chro-

mosome I by systematic RNA interference. *Nature* 408, 325–330.

13. Gonczy, P., Echeverri, C., Oegema, K., Coulson, A., Jones, S. J., Copley, R. R., Duperon, J., Oegema, J., Brehm, M., Cassin, E., Hannak, E., Kirkham, M., Pichler, S., Flohrs, K., Goessen, A., Leidel, S., Alleaume, A. M., Martin, C., Ozlu, N., Bork, P., and Hyman, A. A. **2000**. Functional genomic analysis of cell division in *C. elegans* using RNAi of genes on chromosome III. *Nature* 408, 331–336.
14. Piano, F., Schetter, A. J., Mangone, M., Stein, L., and Kemphues, K. J. **2000**. RNAi analysis of genes expressed in the ovary of *Caenorhabditis elegans*. *Curr Biol* 10, 1619–1622.
15. Krieg, C., Cole, T., Deppe, U., Schierenberg, E., Schmitt, D., Yoder, B., and con Ehrenstein, G. **1978**. The cellular anatomy of embryos of the nematode *Caenorhabditis elegans*. Analysis and reconstruction of serial section electron micrographs. *Dev Biol* 65, 193–215.
16. Deppe, U., Schierenberg, E., Cole, T., Krieg, C., Schmitt, D., Yoder, B., and von Ehrenstein, G. **1978**. Cell lineages of the embryo of the nematode *Caenorhabditis elegans*. *Proc Natl Acad Sci USA* 75, 376–380.
17. Gonczy, P., Schnabel, H., Kaletta, T., Amores, A. D., Hyman, T., and Schnabel, R. **1999**. Dissection of cell division processes in the one cell stage *Caenorhabditis elegans* embryo by mutational analysis. *J Cell Biol* 144, 927–946.
18. Albertson, D. G., and Thomson, J. N. **1993**. Segregation of holocentric chromosomes at meiosis in the nematode, *Caenorhabditis elegans*. *Chromosome Res* 1, 15–26.
19. Strome, S., Powers, J., Dunn, M., Reese, K., Malone, C. J., White, J., Seydoux, G., and Saxton, W. **2001**. Spindle dynamics and the role of γ-tubulin in early *Caenorhabditis elegans* embryos. *Mol Biol Cell* 12, 1751–1764.
20. Hannak, E., Kirkham, M., Hyman, A. A., and Oegema, K. **2001**. Aurora-A kinase is required for centrosome maturation in *Caenorhabditis elegans*. *J Cell Biol* 155, 1109–1116.
21. Hannak, E., Oegema, K., Kirkham, M., Gonczy, P., Habermann, B., and Hyman, A. A. **2002**. The kinetically dominant assembly pathway for centrosomal asters in *Caenorhabditis elegans* is γ-tubulin dependent. *J Cell Biol* 157, 591–602.
22. Albertson, D. G. **1984**. Formation of the first cleavage spindle in nematode embryos. *Dev Biol* 101, 61–72.
23. Wolf, N., Hirsh, D., and McIntosh, J. R. **1978**. Spermatogenesis in males of the free-living nematode, *Caenorhabditis elegans*. *J Ultrastruct Res* 63, 155–169.
24. O'Connell, K. F. **2002**. The ZYG-1 kinase, a mitotic and meiotic regulator of centriole replication. *Oncogene* 21, 6201–6208.
25. Kirkham, M., Muller-Reichert, T., Oegema, K., Grill, S., and Hyman, A. A. **2003**. SAS-4 is a *C. elegans* centriolar protein that controls centrosome size. *Cell* 112, 575–587.
26. Moritz, M., Braunfeld, M. B., Fung, J. C., Sedat, J. W., Alberts, B. M., and Agard, D. A. **1995**. Three-dimensional structural characterization of centrosomes from early *Drosophila* embryos. *J Cell Biol* 130, 1149–1159.
27. O'Connell, K. F. **2000**. The centrosome of the early *C. elegans* embryo: inheritance, assembly, replication, and developmental roles. *Curr Top Dev Biol* 49, 365–384.
28. Ward, S., Thomson, N., White, J. G., and Brenner, S. **1975**. Electron microscopical reconstruction of the anterior sensory anatomy of the nematode *Caenorhabditis elegans*. *J Comp Neurol* 160, 313–337.
29. O'Toole, E. T., K. L. McDonald, J. Mäntler, J. R. McIntosh, A. A. Hyman, and T. Müller-Reichert **2003**. Morphologically distinct microtubule ends in the mitotic centrosome of *Caenorhabditis elegans*. *J Cell Biol* 163, 451–456.
30. Lange, B. M., and Gull, K. **1995**. A molecular marker for centriole maturation in the mammalian cell cycle. *J Cell Biol* 130, 919–927.
31. Chang, P., Giddings, T. H., Winey, M., and Stearns, T. **2003**. ε-Tubulin is required for centriole duplication and

microtubule organization. *Nature Cell Biol* 5, 71–76.

32. Gromley, A., Jurczyk, A., Sillibourne, J., Halilovic, E., Mogensen, M., Groisman, I., Blomberg, M., and Doxsey, S. **2003**. A novel human protein of the maternal centriole is required for the final stages of cytokinesis and entry into S phase. *J Cell Biol* 161, 535–545.

33. O'Connell, K. F., Caron, C., Kopish, K. R., Hurd, D. D., Kemphues, K. J., Li, Y., and White, J. G. **2001**. The *C. elegans* zyg-1 gene encodes a regulator of centrosome duplication with distinct maternal and paternal roles in the embryo. *Cell* 105, 547–558.

34. Leidel, S., and Gonczy, P. **2003**. SAS-4 is essential for centrosome duplication in *C. elegans* and is recruited to daughter centrioles once per cell cycle. *Dev Cell* 4, 431–439.

35. Wong, C., and Stearns, T. **2003**. Centrosome biology: a SAS-sy centriole in the cell cycle. *Curr Biol* 13, R351–R352.

36. Salisbury, J. L. **2003**. Centrosome size is controlled by centriolar SAS-4. Trends *Cell Biol* 13, 340–343.

37. Bobinnec, Y., Fukuda, M., and Nishida, E. **2000**. Identification and characterization of *Caenorhabditis elegans* γ-tubulin in dividing cells and differentiated tissues. *J Cell Sci* 113(Pt 21), 3747–3759.

38. Muller-Reichert, T., Hohenberg, H., O'Toole, E. T., and McDonald, K. **2003**. Cryoimmobilization and three-dimensional visualization of *C. elegans* ultrastructure. *J Microsc* 212, 71–80.

39. Gould, R. R., and Borisy, G. G. **1977**. The pericentriolar material in Chinese hamster ovary cells nucleates microtubule formation. *J Cell Biol* 73, 601–615.

40. Fuller, S. D., Gowen, B. E., Reinsch, S., Sawyer, A., Buendia, B., Wepf, R., and Karsenti, E. **1995**. The core of the mammalian centriole contains γ-tubulin. *Curr Biol* 5, 1384–1393.

41. Ruiz, F., Beisson, J., Rossier, J., and Dupuis-Williams, P. **1999**. Basal body duplication in *Paramecium* requires γ-tubulin. *Curr Biol* 9, 43–46.

42. Hamill, D. R., Severson, A. F., Carter, J. C., and Bowerman, B. **2002**. Centrosome maturation and mitotic spindle assembly in *C. elegans* require SPD-5, a protein with multiple coiled-coil domains. *Dev Cell* 3, 673–684.

43. Nigg, E. A. **2002**. Centrosome aberrations: cause or consequence of cancer progression? *Nature Rev Cancer* 2, 815–825.

44. Wood, W. B., Hecht, R., Carr, S., Vanderslice, R., Wolf, N., and Hirsh, D. **1980**. Parental effects and phenotypic characterization of mutations that affect early development in *Caenorhabditis elegans*. *Dev Biol* 74, 446–469.

45. Encalada, S. E., Martin, P. R., Phillips, J. B., Lyczak, R., Hamill, D. R., Swan, K. A., and Bowerman, B. **2000**. DNA replication defects delay cell division and disrupt cell polarity in early *Caenorhabditis elegans* embryos. *Dev Biol* 228, 225–238.

46. O'Connell, K. F., Leys, C. M., and White, J. G. **1998**. A genetic screen for temperature-sensitive cell-division mutants of *Caenorhabditis elegans*. *Genetics* 149, 1303–1321.

47. O'Connell, K. F., Maxwell, K. N., and White, J. G. **2000**. The spd-2 gene is required for polarization of the anteroposterior axis and formation of the sperm asters in the *Caenorhabditis elegans* zygote. *Dev Biol* 222, 55–70.

48. Khodjakov, A., and Rieder, C. L. **1999**. The sudden recruitment of γ-tubulin to the centrosome at the onset of mitosis and its dynamic exchange throughout the cell cycle, do not require microtubules. *J Cell Biol* 146, 585–596.

49. Schumacher, J. M., Ashcroft, N., Donovan, P. J., and Golden, A. **1998**. A highly conserved centrosomal kinase, AIR-1, is required for accurate cell cycle progression and segregation of developmental factors in *Caenorhabditis elegans* embryos. *Development* 125, 4391–4402.

50. Nigg, E. A. **2001**. Mitotic kinases as regulators of cell division and its checkpoints. *Nature Rev Mol Cell Biol* 2, 21–32.

51. Dutertre, S., Descamps, S., and Prigent, C. **2002**. On the role of aurora-A in centrosome function. *Oncogene* 21, 6175–6183.

52. Roghi, C., Giet, R., Uzbekov, R., Morin, N., Chartrain, I., Le Guellec, R., Couturier, A., Doree, M., Philippe, M., and Prigent, C. **1998**. The *Xenopus* protein kinase pEg2 associates with the centrosome in a cell cycle-dependent manner, binds to the spindle microtubules and is involved in bipolar mitotic spindle assembly. *J Cell Sci* 111(Pt 5), 557–572.

53. Glover, D. M., Leibowitz, M. H., McLean, D. A., and Parry, H. **1995**. Mutations in aurora prevent centrosome separation leading to the formation of monopolar spindles. *Cell* 81, 95–105.

54. Chan, C. S., and Botstein, D. **1993**. Isolation and characterization of chromosome-gain and increase-in-ploidy mutants in yeast. *Genetics* 135, 677–691.

55. Petersen, J., Paris, J., Willer, M., Philippe, M., and Hagan, I. M. **2001**. The *S. pombe* aurora-related kinase Ark1 associates with mitotic structures in a stage dependent manner and is required for chromosome segregation. *J Cell Sci* 114, 4371–4384.

56. Bischoff, J. R., and Plowman, G. D. **1999**. The Aurora/Ipl1p kinase family: regulators of chromosome segregation and cytokinesis. *Trends Cell Biol* 9, 454–459.

57. Gunawardane, R. N., Martin, O. C., Cao, K., Zhang, L., Dej, K., Iwamatsu, A., and Zheng, Y. **2000**. Characterization and reconstitution of *Drosophila* γ-tubulin ring complex subunits. *J Cell Biol* 151, 1513–1524.

58. Murphy, S. M., Urbani, L., and Stearns, T. **1998**. The mammalian γ-tubulin complex contains homologues of the yeast spindle pole body components spc97p and spc98p. *J Cell Biol* 141, 663–674.

59. Zheng, Y., Wong, M. L., Alberts, B., and Mitchison, T. **1995**. Nucleation of microtubule assembly by a γ-tubulin-containing ring complex. *Nature* 378, 578–583.

60. Oegema, K., Wiese, C., Martin, O. C., Milligan, R. A., Iwamatsu, A., Mitchison, T. J., and Zheng, Y. **1999**. Characterization of two related *Drosophila* γ-tubulin complexes that differ in their ability to nucleate microtubules. *J Cell Biol* 144, 721–733.

61. Bonaccorsi, S., Giansanti, M. G., and Gatti, M. **2000**. Spindle assembly in *Drosophila* neuroblasts and ganglion mother cells. *Nature Cell Biol* 2, 54–56.

62. Khodjakov, A., Cole, R. W., Oakley, B. R., and Rieder, C. L. **2000**. Centrosome-independent mitotic spindle formation in vertebrates. *Curr Biol* 10, 59–67.

63. Vaizel-Ohayon, D., and Schejter, E. D. **1999**. Mutations in centrosomin reveal requirements for centrosomal function during early *Drosophila* embryogenesis. *Curr Biol* 9, 889–898.

64. Heald, R., Tournebize, R., Blank, T., Sandaltzopoulos, R., Becker, P., Hyman, A., and Karsenti, E. **1996**. Self-organization of microtubules into bipolar spindles around artificial chromosomes in *Xenopus* egg extracts. *Nature* 382, 420–425.

65. Karsenti, E., and Vernos, I. **2001**. The mitotic spindle: a self-made machine. *Science* 294, 543–547.

66. Matthews, L. R., Carter, P., Thierry-Mieg, D., and Kemphues, K. **1998**. ZYG-9, a *Caenorhabditis elegans* protein required for microtubule organization and function, is a component of meiotic and mitotic spindle poles. *J Cell Biol* 141, 1159–1168.

67. Srayko, M., Quintin, S., Schwager, A., and Hyman, A. A. **2003**. *Caenorhabditis elegans* TAC-1 and ZYG-9 form a complex that is essential for long astral and spindle microtubules. *Curr Biol* 13, 1506–1511.

68. Bellanger, J. M., and Gonczy, P. **2003**. TAC-1 and ZYG-9 form a complex that promotes microtubule assembly in *C. elegans* embryos. *Curr Biol* 13, 1488–1498.

69. Le Bot, N., Tsai, M. C., Andrews, R. K., and Ahringer, J. **2003**. TAC-1, a regulator of microtubule length in the *C. elegans* embryo. *Curr Biol* 13, 1499–1505.

70. Gard, D. L., and Kirschner, M. W. **1987**. A microtubule-associated protein from *Xenopus* eggs that specifically promotes assembly at the plus-end. *J Cell Biol* 105, 2203–2215.

71. Tournebize, R., Popov, A., Kinoshita, K., Ashford, A. J., Rybina, S., Pozniakovsky, A., Mayer, T. U., Walczak, C. E., Karsenti, E., and Hyman, A. A. **2000**. Control of microtubule dynamics by the antagonistic activities of XMAP215 and XKCM1 in *Xenopus* egg extracts. *Nature Cell Biol* 2, 13–19.

72. Kinoshita, K., Habermann, B., and Hyman, A. A. **2002**. XMAP215: a key component of the dynamic microtubule cytoskeleton. *Trends Cell Biol* 12, 267–273.

73. Ohkura, H., Garcia, M. A., and Toda, T. **2001**. Dis1/TOG universal microtubule adaptors – one MAP for all? *J Cell Sci* 114, 3805–3812.

74. Kemphues, K. J., Wolf, N., Wood, W. B., and Hirsh, D. **1986**. Two loci required for cytoplasmic organization in early embryos of *Caenorhabditis elegans*. *Dev Biol* 113, 449–460.

75. Cullen, C. F., and Ohkura, H. 2001. Msps protein is localized to acentrosomal poles to ensure bipolarity of *Drosophila* meiotic spindles. *Nature Cell Biol* 3, 637–642.

76. Lee, M. J., Gergely, F., Jeffers, K., Peak-Chew, S. Y., and Raff, J. W. **2001**. Msps/XMAP215 interacts with the centrosomal protein D-TACC to regulate microtubule behaviour. *Nature Cell Biol* 3, 643–649.

77. Hill, D. P., and Strome, S. **1988**. An analysis of the role of microfilaments in the establishment and maintenance of asymmetry in *Caenorhabditis elegans* zygotes. *Dev Biol* 125, 75–84.

78. Hill, D. P., and Strome, S. **1990**. Brief cytochalasin-induced disruption of microfilaments during a critical interval in 1-cell *C. elegans* embryos alters the partitioning of developmental instructions to the 2-cell embryo. *Development* 108, 159–172.

79. Goldstein, B., and Hird, S. N. **1996**. Specification of the anteroposterior axis in *Caenorhabditis elegans*. *Development* 122, 1467–1474.

80. Goldstein, B., Hird, S. N., and White, J. G. **1993**. Cell polarity in early *C. elegans* development. *Development* (Suppl.), 279–287.

81. Hird, S. N., and White, J. G. **1993**. Cortical and cytoplasmic flow polarity in early embryonic cells of *Caenorhabditis elegans*. *J Cell Biol* 121, 1343–1355.

82. Strome, S., and Wood, W. B. **1983**. Generation of asymmetry and segregation of germ-line granules in early *C. elegans* embryos. *Cell* 35, 15–25.

83. Hird, S. N., Paulsen, J. E., and Strome, S. **1996**. Segregation of germ granules in living *Caenorhabditis elegans* embryos: cell-type-specific mechanisms for cytoplasmic localisation. *Development* 122, 1303–1312.

84. Etemad-Moghadam, B., Guo, S., and Kemphues, K. J. **1995**. Asymmetrically distributed PAR-3 protein contributes to cell polarity and spindle alignment in early *C. elegans* embryos. *Cell* 83, 743–752.

85. Hung, T. J., and Kemphues, K. J. **1999**. PAR-6 is a conserved PDZ domain-containing protein that colocalizes with PAR-3 in *Caenorhabditis elegans* embryos. *Development* 126, 127–135.

86. Guo, S., and Kemphues, K. J. **1995**. par-1, a gene required for establishing polarity in *C. elegans* embryos, encodes a putative Ser/Thr kinase that is asymmetrically distributed. *Cell* 81, 611–620.

87. Boyd, L., Guo, S., Levitan, D., Stinchcomb, D. T., and Kemphues, K. J. **1996**. PAR-2 is asymmetrically distributed and promotes association of P granules and PAR-1 with the cortex in *C. elegans* embryos. *Development* 122, 3075–3084.

88. Sadler, P. L., and Shakes, D. C. **2000**. Anucleate *Caenorhabditis elegans* sperm can crawl, fertilize oocytes and direct anterior-posterior polarization of the 1-cell embryo. *Development* 127, 355–366.

89. Wallenfang, M. R., and Seydoux, G. **2000**. Polarization of the anterior–posterior axis of *C. elegans* is a microtubule-directed process. *Nature* 408, 89–92.

90. Severson, A. F., and Bowerman, B. **2003**. Myosin and the PAR proteins polarize microfilament-dependent forces that shape and position mitotic spindles in *Caenorhabditis elegans*. *J Cell Biol* 161, 21–26.

91. Hyman, A. A., and White, J. G. **1987**. Determination of cell division axes in the early embryogenesis of *Caenorhabditis elegans*. *J Cell Biol* 105, 2123–2135.

92. Hyman, A. A. **1989**. Centrosome movement in the early divisions of *Caenorhabditis elegans*: a cortical site determining centrosome position. *J Cell Biol* 109, 1185–1193.

93. Skop, A. R., and White, J. G. **1998**. The dynactin complex is required for cleavage plane specification in early *Caenorhabditis elegans* embryos. *Curr Biol* 8, 1110–1116.

94. Gonczy, P., Pichler, S., Kirkham, M., and Hyman, A. A. **1999**. Cytoplasmic dynein is required for distinct aspects of MTOC positioning, including centrosome separation, in the one cell stage *Caenorhabditis elegans* embryo. *J Cell Biol* 147, 135–150.

95. Horvitz, H. R., and Herskowitz, I. **1992**. Mechanisms of asymmetric cell division: two Bs or not two Bs, that is the question. *Cell* 68, 237–255.

96. Lin, H. **2003**. Stem cells: to be and not to be. *Nature* 425, 353–355.

97. Yamashita, Y. M., Jones, D. L., and Fuller, M. T. **2003**. Orientation of asymmetric stem cell division by the APC tumor suppressor and centrosome. *Science* 301, 1547–1550.

98. Gonczy, P. **2002**. Mechanisms of spindle positioning: focus on flies and worms. *Trends Cell Biol* 12, 332–339.

99. Korinek, W. S., Copeland, M. J., Chaudhuri, A., and Chant, J. **2000**. Molecular linkage underlying microtubule orientation toward cortical sites in yeast. *Science* 287, 2257–2259.

100. Lee, L., Tirnauer, J. S., Li, J., Schuyler, S. C., Liu, J. Y., and Pellman, D. **2000**. Positioning of the mitotic spindle by a cortical-microtubule capture mechanism. *Science* 287, 2260–2262.

101. Miller, R. K., Cheng, S. C., and Rose, M. D. **2000**. Bim1p/Yeb1p mediates the Kar9p-dependent cortical attachment of cytoplasmic microtubules. *Mol Biol Cell* 11, 2949–2959.

102. Grill, S. W., Gonczy, P., Stelzer, E. H., and Hyman, A. A. **2001**. Polarity controls forces governing asymmetric spindle positioning in the *Caenorhabditis elegans* embryo. *Nature* 409, 630–633.

103. Grill, S. W., Howard, J., Schaffer, E., Stelzer, E. H., and Hyman, A. A. **2003**. The distribution of active force generators controls mitotic spindle position. *Science* 301, 518–521.

104. Colombo, K., Grill, S. W., Kimple, R. J., Willard, F. S., Siderovski, D. P., and Gonczy, P. **2003**. Translation of polarity cues into asymmetric spindle positioning in *Caenorhabditis elegans* embryos. *Science* 300, 1957–1961.

105. Wood, W. B. **1991**. Evidence from reversal of handedness in *C. elegans* embryos for early cell interactions determining cell fates. *Nature* 349, 536–538.

106. Bergmann, D. C., Lee, M., Robertson, B., Tsou, M. F., Rose, L. S., and Wood, W. B. **2003**. Embryonic handedness choice in *C. elegans* involves the Gα protein GPA-16. *Development* 130, 5731–5740.

107. Reboul, J., Vaglio, P., Rual, J. F., Lamesch, P., Martinez, M., Armstrong, C. M., Li, S., Jacotot, L., Bertin, N., Janky, R., Moore, T., Hudson, J. R., Jr., Hartley, J. L., Brasch, M. A., Vandenhaute, J., Boulton, S., Endress, G. A., Jenna, S., Chevet, E., Papasotiropoulos, V., Tolias, P. P., Ptacek, J., Snyder, M., Huang, R., Chance, M. R., Lee, H., Doucette-Stamm, L., Hill, D. E., and Vidal, M. **2003**. *C. elegans* ORFeome version 1.1: experimental verification of the genome annotation and resource for proteome-scale protein expression. *Nature Genet* 34, 35–41.

13
Centrosomes in a Developing Organism: Lessons from *Drosophila*

Jordan W. Raff

13.1
Introduction

The fruit fly *Drosophila melanogaster* has been a popular organism for developmental biologists for more than 100 years. The enormous power of fly genetics has made it possible to identify many of the most important proteins that orchestrate the development of an animal from a fertilized egg. More recently, the combination of genetics, biochemistry, and time-lapse microscopy in living embryos has made *Drosophila* an attractive system to study fundamental problems in cell biology. In this chapter, I first review the *Drosophila* life cycle, highlighting the developmental stages that have been most useful for studying centrosome function. I then summarize what has been learned about centrosome structure and function in the fly. Finally, I discuss the lessons that *Drosophila* has taught us about how centrosomes and centrosomal proteins function within the context of a multicellular organism.

13.2
Centrosome and Microtubule Organisation during the *Drosophila* Life Cycle

13.2.1
Oogenesis

In flies, oogenesis begins with the asymmetric division of a germline stem cell to produce a cystoblast and another stem cell [1]. The cystoblast undergoes four rounds of incomplete cell division to generate a cyst of 16 cells, which remain interconnected by large intracellular bridges called ring canals (Figure 13.1A). Due to the pattern of cell divisions, two of the 16 cells contain four ring canals, and one of these invariably becomes the oocyte; the remaining 15 cells become supporting nurse cells. During the initial four divisions, one centrosome of each mitotic spindle remains connected to the "fusome", an amorphous structure that passes through the ring canals, connecting all of the cells in the cyst [1]. When the four

Centrosomes in Development and Disease. Edited by Erich A. Nigg

ISBN 3-527-30980-2

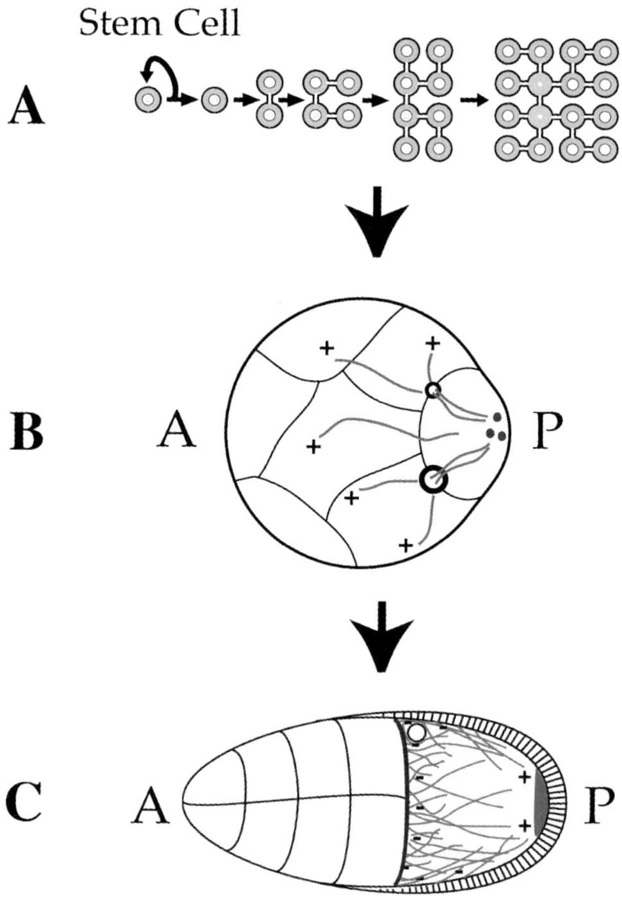

Figure 13.1 A schematic summary of oogenesis in *Drosophila*. (A) The division of a stem cell gives rise to another stem cell and a cystoblast that goes through four rounds of mitosis to generate a cyst of 16 cells which remain interconnected by ring canals. Two cells (orange) contain four ring canals, and one of these invariably becomes positioned at the posterior of the cyst and becomes the oocyte, while the other cells become nurse cells. (B) After mitosis is finished, the centrioles (blue) of the nurse cells lose all PCM markers and migrate into the oocyte. As the oocyte enlarges, the centrioles move to the posterior and an MTOC is assembled in this region, although it is not clear that the centrioles are required to form this MTOC. The microtubules (green) extend away from the MTOC and spread through the ring canals to the nurse cells (only two ring canals are shown here). (C) At later stages of oogenesis, the centrioles disappear, and the posterior MTOC is disassembled. The minus ends of the microtubules now associate with a diffuse region spread along the anterior cortex. The exact organization of microtubules at this stage of oogenesis is controversial (see, for example [151]) and long microtubules, like those depicted here, are rarely visible at this stage of oogenesis. A microtubule plus end-directed kinesin, however, can accumulate at the posterior pole (red), while a minus end-directed kinesin can accumulate at the anterior cortex (blue) suggesting that the microtubules have an overall polarity [152]. In this figure, anterior is on the left and posterior on the right, as indicated. This figure was adapted from a figure provided by Daniel St Johnston (see Color Plates page XXXI).

divisions are complete, all of the centrioles in the 16-cell cyst migrate into the presumptive oocyte [2–4]. This migration depends on the fusome, but not on microtubules [5]. Although the centrioles appear to lack all pericentriolar material (PCM) markers[1], they come to lie at the centre of a poorly defined microtubule organizing centre (MTOC) that forms at the posterior of the oocyte, which organizes an array of microtubules that spreads throughout the 16-cell cyst [6] (Figure 13.1B). This array plays an important part in guiding the transport of materials from the nurse cells into the oocyte.

As oogenesis proceeds, the oocyte enlarges, and the centrioles mysteriously disappear. As the centrioles disappear, the posterior MTOC is disassembled, and the microtubules re-organize, with their minus ends clustered diffusely along the anterior margins of the oocyte cortex, and their plus ends clustered at the posterior tip of the oocyte (Figure 13.1C) [6]. This polarized array of microtubules plays a crucial role in organizing positional cues within the oocyte, and thereby in establishing the two major body axes of the organism (anterior/posterior and dorsal/ventral) [7].

As the oocyte grows in size and eventually matures into an egg, the female pronucleus enters meiosis I. Because the centrioles have disappeared by this stage, the meiosis I spindle forms in the absence of centrosomes. It is now clear that many cell types have the ability to organize a bi-polar spindle in the absence of centrosomes (reviewed in [8]), but the *Drosophila* female meiosis I spindle is perhaps the best studied example of a naturally-occurring acentrosomal spindle. These spindles also lack astral microtubules, and most PCM markers, including γ-tubulin, are not concentrated at the spindle poles [9–11]. Elegant live imaging experiments have shown how this spindle is constructed: microtubules initially polymerize around the chromatin and then the co-ordinated action of various microtubule motors, organize the microtubules into a bipolar spindle [9].

Mature eggs remain arrested in metaphase of meiosis I until fertilization, when they exit meiosis I and immediately enter meiosis II. The two meiosis II spindles are arranged in a line and share one pole [12]. The non-shared poles appear to be similar to the poles of the meiosis I spindle, in that they are anastral and lack most PCM markers. The shared pole, however, organizes a robust array of astral microtubules and contains the centrosomal markers γ-tubulin and CP190, although it does not contain centrioles [12]. The linear arrangement of the meiosis II spindles ensures that one female pronucleus is moved into the interior of the embryo, and it is usually this pronucleus that interacts with microtubules emanating from the centrosome associated with the fertilizing male pronucleus. The female pronucleus migrates along these microtubules, and joins the male pronucleus on the first mitotic spindle formed after the sperm-derived centrosome has replicated [13].

[1] Although there is one report of an anti-γ-tubulin monoclonal antibody that stains oocyte centrioles [3], we have found that these centrioles are not stained with any other anti-γ-tubulin antibodies, or with different batches of the same monoclonal antibody ascites fluid used in this study. It seems likely, therefore, that a batch of this monoclonal antibody ascites fluid contained another antibody that cross-reacted with *Drosophila* centrioles.

13.2.2 Spermatogenesis

Like oogenesis, spermatogenesis begins with the asymmetric division of a germ-line stem cell to produce a gonialblast and another stem cell. The gonialblast proceeds through four rounds of mitosis to generate a cyst of 16 primary spermatocytes, which remain interconnected by intracellular bridges [14]. Unlike oogenesis, centrosomes are retained in spermatocytes, and the meiosis I and II spindles contain canonical centrosomes. Because these cells are very large, the study of male meiosis I and II has been a favored system for studying cell division in flies. In particular, these cells have provided important insights into the respective roles of the centrosomes and the central spindle in organizing cytokinesis [15].

At the end of meiosis II, a cyst containing 64 mature sperm is produced. The sperm tail is organized at its base by a centriole/basal body, which will ultimately organize the centrosome in the zygote [16]. Thus, as in most species, centrioles are paternally inherited in *Drosophila melanogaster.* This is not always the case in insects, however. In the hymenopteran *Nasonia vitripennis,* for example, all males are normally produced by the parthenogenetic development of the egg. In unfertilized eggs, a large number of asters appear to be formed *de novo* at the egg cortex, and these then migrate into the egg interior where two of them will become stably associated with the female pronucleus [17]. These two maternally-derived asters form the centrosomes used for subsequent parthenogenetic development, while the other asters disappear. A similar process can also occur in a laboratory strain of *Drosophila mercatorum* [18], suggesting that many insect eggs may have the ability to form centrosomes *de novo* but are normally inhibited from doing so. There is increasing evidence that this is also the case in other species [19, 20].

13.2.3 Early Embryogenesis

The early *Drosophila* embryo offers several important advantages for studying centrosomal proteins. It starts life with a single nucleus that proceeds through a rapid series of synchronous nuclear divisions, which all occur within a common cytoplasm, producing a syncytium. These early nuclear divisions are remarkably fast: the entire genome is replicated in under 6 min, and the spindle is assembled, chromosomes segregated, and the spindle disassembled in less than 3 min [21]. Thus, the embryo has to form several thousand centrosomes in less than 2.5 h. Moreover, this stage of development does not require zygotic transcription, so all the components needed to build several thousand centrosomes and spindles are laid down in the egg by the mother. This makes this early embryo an excellent source of centrosome and spindle components for biochemical analysis.

During these early rounds of mitosis, two processes are responsible for positioning the nuclei within the embryo. The zygotic nucleus produced at fertilization is invariably positioned towards the anterior end of the embryo, and a process termed "axial expansion" spreads the dividing nuclei evenly throughout the syncytial embryo. When viewed by light microscopy, axial expansion is seen to occur by a highly co-ordinated series of contractile movements, and both actin and myosin are essential for this process [22, 23]. Once spread evenly along the A/P axis of the embryo, the majority of the nuclei co-ordinately migrate to the embryo cortex. This process of "nuclear migration" depends on the very large astral arrays of microtubules that are nucleated from the centrosomes at late stages of mitosis and in early interphase [24].

The migrating nuclei first reach the posterior cortex at nuclear cycle 9, where they quickly pinch off from the rest of the embryo to form "pole cells", which are the future germ cells of the organism. The rest of the migrating nuclei reach the cortex at nuclear cycle 10, where they proceed through four more rounds of mitosis. Because these divisions occur at the embryo surface and are synchronous, they provide a unique opportunity to follow centrosome and microtubule behavior in multiple spindles at the same time (see, for example, Figure 13.2). Moreover, it is easy to generate stable transgenic lines that express GFP-tagged versions of proteins, or to inject fluorescently-labelled proteins or antibodies into the early *Drosophila* embryo [25]. Because the embryo is remarkably resistant to photodamage, one can follow the behavior of fluorescently-tagged proteins through several rounds of mitosis at high resolution. As we shall see, this approach has provided important insights into the dynamic behavior of centrosomes and centrosomal proteins.

The early fly embryo also provides a powerful system to investigate the interaction between the microtubule and actin/myosin cytoskeletal networks. Although the nuclei divide within a common cytoplasm, once at the cortex each nucleus organizes its own domain of actin [26] In interphase, the cortical actin forms a "cap" that lies above each nucleus, while, in mitosis, the actin re-organizes into a "pseudo-cleavage furrow" that surrounds each spindle. This furrow prevents the microtubules from one spindle interacting with the chromosomes of another spindle in the cramped environment of the surface monolayer. Centrosomes are essential for organizing the cortical actin. In embryos in which DNA synthesis has been blocked with drugs, the centrosomes still migrate to the embryo cortex,

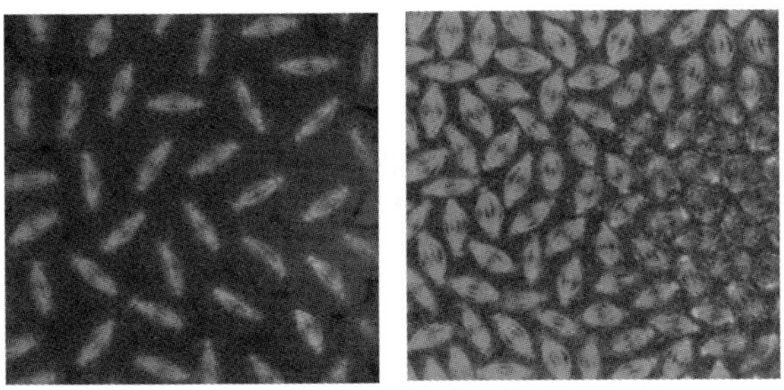

Figure 13.2 Perturbing centrosome function by antibody injection in syncytial embryos. Embryos that express a tubulin-GFP fusion protein were injected with Texas Red-labeled anti-CP190 (A) or anti-D-TACC (B) antibodies and imaged on a confocal microscope. The anti-CP190 antibodies bind to the centrosomes closest to the injection site. The behavior of the microtubules, however, is not perturbed, suggesting that CP190 is not involved in regulating centrosome or microtubule behavior (see Section 13.3.4). The anti-D-TACC antibodies precipitate the endogenous D-TACC protein and form large lumps in the cytoplasm, effectively depleting D-TACC from the spindles around the injection site. As a result, the spindles closest to the injection site are noticeably shorter than those away from the injection site, suggesting that D-TACC is involved in stabilizing spindle microtubules. Scale bar = 10 μm (see Color Plates page XXXII).

although without the nuclei [27]. Once at the cortex, the centrosomes at the posterior pole can form pole cells without nuclei; the centrosomes elsewhere at the cortex organize only actin caps. It remains a mystery how centrosomes stimulate cell formation at the posterior pole but only actin cap formation elsewhere.

After 13 rounds of division, the syncytial nuclei cellularize, as the plasma membrane invaginates around each nucleus to create a monolayer of columnar epithelial cells. Cellularization involves complex interactions between the centrosomes and the microtubule, actin, and myosin networks [26]. After cellularization is complete, gastrulation begins, and the cells no longer divide in synchrony [28].

13.2.4
Asymmetric Divisions of Embryonic Neuroblasts

Cell division in the gastrulating epithelial cells is invariably symmetric, producing two daughters of equal size. Embryonic neuroblast cells, however, delaminate from the epithelial sheet and undergo an asymmetric division to generate two cells of different size and of different developmental fates: a large neuroblast cell, and a smaller ganglion mother cell (GMC) [29, 30]. In most asymmetric divisions, the mitotic spindle is positioned in the cell through interactions between astral microtubules and specific cues at the cell cortex [31, 32]. These interactions align the spindle with information in the cortex, and they usually lead to the displacement of the metaphase spindle within the cell. As a result, division produces two cells of unequal size, and cortically placed developmental cues are differentially segregated between the two daughter cells. In *Drosophila* neuroblasts, the spindle aligns with cortical cues but it remains symmetrically localized within the cell during metaphase [33]. Only during anaphase does an asymmetry become apparent, as the microtubule aster in the large nascent neuroblast cell becomes much larger than that in the smaller nascent GMC. The large aster appears to "push" the cleavage furrow away, toward the small aster, making the GMC smaller than the neuroblast.

During these asymmetric neuroblast divisions, the centrosomal proteins γ-tubulin, CP190, and CP60 are symmetrically distributed on the centrosomes in metaphase [33]. By anaphase, however, all these proteins are present at higher levels on the centrosome in the large nascent neuroblast, presumably explaining why it nucleates a larger array of astral microtubules. Thus, the basis for this asymmetric division appears to be the development of an asymmetry between the two centrosomes during anaphase. Many of the proteins involved in setting up this asymmetric division in flies have now been identified [32]. Of particular importance is the localization to the apical cortex of two protein complexes, one containing Par6, Bazooka/Par3, and an atypical protein kinase C (aPKC) and the other containing Partner of Inscutable (Pins) and the Gα subunit of a heterotrimeric G protein [34]. Both of these complexes are highly conserved and play an important part in organizing various aspects of cell polarity in many cell types [35]. It remains a major challenge to understand how these cortically localized complexes influence both the positioning of the spindle within the cell and the recruitment of proteins to the centrosome in anaphase.

13.2.5
Larval Development

Apart from the neuroblasts, most embryonic cells go through only a few more rounds of cell division after cellularization. During the latter part of embryonic development and most of larval development, the majority of cells become polyploid through repeated rounds of S-phase without intervening rounds of M-phase. Although not well studied, it seems that most of these non-dividing cells lose their centrosomes during this phase of development [36]. In the larval stages, exceptions to this generalization are the cells of the larval brain and the imaginal discs. These discs are clusters of cells that will ultimately form most tissues of the adult fly, and they continue to divide throughout the larval stages of development, although they are not essential for the survival of the larva. A mutation in a gene that is essential for cell division will, therefore, often only produce a noticeable cell division phenotype at later stages of larval development, as the maternally supplied protein and mRNA stored in the egg are sufficient to drive all cell divisions in the embryo. Thus, animals carrying mutations in genes encoding many centrosomal proteins eventually die as late larvae, and their mutant phenotypes are often studied in larval brains, which are particularly mitotically active [37, 38].

At the end of larval development, the larva pupates, and a large pulse of the hormone ecdyosone stimulates the imaginal discs to undergo the complex tissue re-arrangements that ultimately produce the fully formed adult fly. In most adult fly tissues there is little, if any, cell division and, with a few exceptions [39, 40], there has been little study of centrosome and microtubule organization.

13.3
Drosophila Centrosomal Proteins

13.3.1
Microtubule Nucleation from Centrosomes: γ-Tubulin and the γ-TuRC

Until 15 years ago, how centrosomes interacted with microtubules was a complete mystery. The discovery of γ-tubulin provided an essential clue [41]. Although the exact mechanism remains controversial [42], there is general agreement that γ-tubulin ring complexes (γ-TuRCs) concentrated in the PCM nucleate centrosomal microtubules [43–45], and ring-like structures containing γ-tubulin were first identified in the PCM of purified *Drosophila* centrosomes [46].

When purified from *Drosophila* embryos, γ-tubulin exists in two forms: a large (~ 2.2 mDa) complex (the γ-TuRC) and a smaller (~ 220 kDa) complex (the γ-tubulin small complex, γ-TuSC) [47, 48]. The *Drosophila* γ-TuRC consists of six to eight proteins (called *Drosophila* γ-tubulin ring proteins, Dgrips), most of which have now been identified [49, 50]. Interestingly, apart from γ-tubulin and Dgrip71WD, these proteins all share two "grip" motifs, the function of which is unclear. *In vitro* reconstitution experiments have identified a complex web of interactions between the components of the γ-TuRC. In contrast, the γ-TuSC contains only two molecules of γ-tubulin and one each of D-grip84 and D-grip91 [48]. The budding yeast *S. cerevisiae* contains only this smaller complex [51, 52], suggesting that the small complex is sufficient to nucleate microtubule growth and may act as a building block to assemble the larger complex. In support of this idea, the γ-TuSC can nucleate microtubules *in vitro*, although it is much less efficient than the γ-TuRC [48].

While studies on unicellular eukaryotes have demonstrated that γ-tubulin is essential for proper microtubule organization, γ-tubulin mutations in flies have revealed some surprising aspects of γ-tubulin function. There are two γ-tubulin genes in *Drosophila*, *γ-tubulin 37C* (*γt37C*), which is expressed only during oogenesis and early embryogenesis, and *γ-tubulin 23C* (*γt23C*), which is expressed at all other stages of development [11]. The two isoforms appear to be functionally equivalent, as the 23C isoform can rescue the defects associated with mutations in *γt37C* when it is expressed from the 37C promoter (C. Gonzalez, personal communication). Flies homozygous for a mutation in the *γt23C* gene die as late larvae, and, although microtubules are often disorganized during mitosis in mutant cells, in some cells, centrosomes can still interact with substantial numbers of microtubules [53]. Surprisingly, the centrosomal localization of the centrosomal protein CP190 is strongly disrupted in mutant cells, while the centrosomal localization of another protein, Centrosomin (CNN) is not. This suggests that γ-tubulin plays a structural role in organizing specific components of the centrosome. Flies with a mutation in the *Dgrip81* gene (called *discs degenerate 4* (*dd4*)) have a similar phenotype, and γ-tubulin is no longer concentrated at centrosomes in *dd4* mutants [54].

Elegant studies of spermatocytes in living *Drosophila* testes have confirmed that centrosomes associate with robust microtubule asters in γ-tubulin mutants [55]. Surprisingly, however, mutant cells completely fail to form a meiotic spindle, and the separated centrosomes rapidly collapse together to form a large monopolar aster as soon as meiosis is initiated. A similar phenotype has been observed in *C. elegans* embryos depleted of γ-tubulin by RNA interference (RNAi) [56, 57]. Thus, centrosomes may interact with spindle microtubules and astral microtubules in different ways, and it is only the interaction with spindle microtubules that absolutely requires γ-tubulin.

13.3.2
The Recruitment of the γ-TuRC to Centrosomes: The Potential Roles of Asp, Polo, CNN, Aurora A, and CP309/D-PLP

Semi-purified centrosomes can be readily prepared from *Drosophila* embryo extracts, which can nucleate microtubule asters when mixed with pure tubulin [58]. These centrosomes can be depleted of many components of the PCM, including γ-tubulin, by treatment with high salt [47], leaving behind a core "centromatrix" [59]. Such salt-stripped centrosomes can no longer nucleate microtubules when mixed with purified tubulin, but they can do so if mixed with embryo extracts, which allows γ-tubulin and other factors to re-load onto the centrosomes [47]. If the extracts are depleted of γ-tubulin, they can no longer restore the microtubule nucleating potential of the centrosomes, but mixing salt-stripped centrosomes with purified γ-TuRCs is insufficient to restore their activity. Thus, there is at least one other component required for microtubule nucleation, and it was suggested that this factor might "load" the γ-TuRC onto the stripped centrosomes [47].

Extracts depleted of the Abnormal spindle (Asp) protein, also cannot restore the microtubule nucleating potential of salt-stripped centrosomes, suggesting that it may be the γ-TuRC "loading" factor [60]. Moreover, Asp is phosphorylated by Polo [61], a mitotic kinase that is concentrated in centrosomes [62] and has been implicated in recruiting PCM components to the centrosome upon the entry into mitosis [63, 64]. Extracts prepared from *polo* mutants are unable to restore microtubule nucleation to stripped centrosomes, but this can be rescued by the addition of purified Polo, or of partially-purified Asp that has been phosphorylated by Polo [61]. *In vivo*, however, Asp is most strongly concentrated around the minus ends of the spindle microtubules that are clustered near the centrosome, and a mutational analysis suggests that it is not required to load the γ-TuRC onto centrosomes, but is required to maintain the link between centrosomes and spindle microtubules and for the proper formation of the central spindle [65, 66].

In *Drosophila* and *C. elegans*, the centrosomal Aurora A kinase is essential for the mitotic recruitment of several proteins to the PCM, including γ-tubulin [67, 68]. Like Polo, Aurora A regulates many aspects of mitosis [69], and it has recently been shown to interact with another centrosomal protein, CNN [70]. CNN is required for the efficient recruitment of γ-tubulin and CP190 to centrosomes in embryos, and the formation of astral microtubules is severely impaired in *cnn* mutant embryos [71, 72]. Aurora A can bind directly to the C-terminal region of CNN, and each protein is required for the centrosomal localization of the other [70]. Moreover, the N-terminal region of CNN can interact with γ-TuRCs, although it is not clear if this interaction is direct. Thus, CNN may help link Aurora A to the γ-TuRC and thereby help regulate the recruitment of the γ-TuRC to centrosomes during mitosis. Although there are no known homologs of CNN in vertebrates, the overexpression of fly CNN in human cells causes the formation of large cytoplasmic aggregates of CNN that bind to endogenous γ-tubulin and are associated with large microtubule asters [70].

Pericentrin is involved in recruiting γ-tubulin to centrosomes in vertebrates [73], and it shares a conserved domain – the pericentrin/AKAP450 centrosomal targeting (PACT) domain – with several other proteins that have been implicated in recruiting specific proteins to the centrosome or to the spindle pole body (SPB) in yeast [74]. In *Drosophila*, there is only one PACT domain protein [74], and embryo extracts depleted of this protein (called CP309 or *Drosophila* pericentrin-like protein, D-PLP) can no longer restore the microtubule nucleating capacity of salt stripped centrosomes [75]. Surprisingly, a mutation in this gene disrupts the centrosomal recruitment of not just γ-tubulin, but of all other centrosomal proteins tested, including γ-tubulin, CNN, Aurora A, D-TACC, Msps, CP190, and CP60 [153]. Thus, CP309/D-PLP is required for the efficient recruitment of most, if not all, proteins to the PCM. Interestingly, this protein is a component of both the centrioles and the PCM in flies, suggesting that it may function to link the centrioles and the PCM during centrosome assembly [153].

13.3.3
The Interaction between Centrosomes and Microtubules: The Role of D-TACC and Msps

The genetic dissection of γ-TuRC function in flies has shown that the interaction between centrosomes and microtubules is more complicated than just a simple nucleation event from γ-TuRCs embedded in the PCM. One strategy to identify other proteins potentially involved in the interaction between centrosomes and microtubules, has been to purify proteins from *Drosophila* embryo extracts on the basis of their ability to interact with microtubules and then to raise antibodies against them to identify those that are concentrated at centrosomes [76].

One such centrosomal microtubule-associated protein (MAP) is the *Drosophila* transforming acidic coiled-coil-containing (D-TACC) protein, which is required to stabilize centrosomal microtubules in embryos [77] (see Figure 13.2). D-TACC interacts with microtubules indirectly via another centrosomal MAP, Minispindles (Msps) [78]. Msps is the *Drosophila* homolog of the highly conserved XMAP215/ch-Tog family of microtubule stabilizing proteins [79]. In frog egg extracts, XMAP215 appears to be the major microtubule stabilizing factor [80], and Msps and Zyg9 (the *C. elegans* homolog) have a similar function in fly and worm embryos, respectively [81] (M. Lee, T. Barros and J. Raff, unpublished observations). Paradoxically, these proteins stabilize microtubules *in vitro* mainly by stabilizing microtubule plus ends, yet, *in vivo*, they are highly concentrated at centrosomes, where the minus ends of the microtubules are concentrated [82]. In flies, D-TACC is required to efficiently recruit Msps to centrosomes [78], and this is also true in worms [83–85]. It has been proposed that the two proteins co-operate to stabilize centrosomal microtubules, either by binding to the minus ends of microtubules released from their nucleating sites or by loading onto the plus ends of microtubules that grow from the centrosomes [78]. In support of the latter possibility, small particles of D-TACC-GFP and Msps-GFP can be seen oscillating to and fro from the centrosome, as if they were binding to the growing and shrinking plus ends of the microtubules. Such centrosomal "flares" have also been observed with a CNN-GFP fusion protein [86].

13.3.4
Centrosomes and the Organization of the Actin/Myosin Cytoskeleton in Early Fly Embryos: The Role of Scrambled, Nuf, and CP190

As described in Section 13.2.3, there is strong evidence to suggest that centrosomes influence the behavior of the actin/myosin cytoskeleton in the early fly embryo. 3-D reconstructions of the actin, myosin and microtubule networks in fixed embryos before and after various drug treatments have suggested that actin and myosin

II filaments can be transported along microtubules from the minus ends at centrosomes to the plus ends [87]. This transport, together with assumed differential affinities of actin and myosin for the cell cortex and a regulation of these properties during the cell cycle, could explain the distribution and interdependencies of the cytoskeletal networks. Studies on the *Drosophila* Scrambled (Sced) protein, however, suggest a different model of how centrosomes could influence actin [88]. Sced localizes to both centrosomes and to the pseudo-cleavage furrows that normally surround the spindles during mitosis in the syncytial embryos. Mutations in *sced* do not interfere with actin cap formation in interphase, but block the recruitment of actin to the pseudo-cleavage furrows in mitosis. Surprisingly, injecting microtubule depolymerizing drugs into embryos does not disrupt the localization of Sced or the organization of actin in caps or furrows. These observations suggest that Sced is essential to help centrosomes recruit actin into the furrow and that centrosomes can influence actin recruitment even in the absence of microtubules.

Studies on the protein Nuclear fall out (Nuf) have suggested that centrosomes may also direct the recruitment of membranes to both pseudo-cleavage furrows and the furrows that invaginate around the nuclei during cellularization [89]. Nuf is recruited to centrosomes specifically during prophase. In *nuf* mutant embryos, actin is not properly recruited to the furrows, even though other furrow components are recruited. Nuf is a member of the arfophilin family, which has been implicated in regulating membrane trafficking. Moreover, Nuf is required to recruit the membrane-associated protein Discontinuous actin hexagon (Dah) to furrows, providing a potential mechanism whereby centrosomes could guide the transport of membrane components to the furrow [90, 91].

One of the most studied *Drosophila* centrosomal proteins, CP190, has an unexpected role in regulating myosin in the early fly embryo. CP190 was the first centrosomal protein to have its cDNA cloned in *Drosophila* [92] and it is widely used as a centrosomal marker. CP190, together with its binding partner CP60, cycles between the nucleus in interphase and the centrosomes in mitosis, although CP60 levels do not peak at the centrosome until anaphase/telophase [93–95]. Both proteins interact directly with microtubules *in vitro*, and the ability of CP60 to bind microtubules is abolished when it is phosphorylated by cdc2 (also known as Cdk1)/cyclin B kinase. These observations suggested that CP190 and CP60 were involved in regulating microtubule behavior specifically during late stages of mitosis, when cdc2/cyclin B levels are in decline. A mutation in the *cp190* gene, however, does not detectably affect centrosomal microtubules or any aspect of mitosis, even though CP190 and CP60 are no longer detectable at centrosomes in the mutant cells [96].

The *cp190* mutation, however, is lethal [96]. The lethality can be rescued by the expression of a deleted form of CP190 (CP190ΔM) that no longer binds to centrosomes or microtubules, demonstrating that CP190 must have some critical function (possibly in the nucleus) that is independent of its ability to bind to centrosomes or microtubules. In early embryos that lack CP190 function, axial expansion fails (S. Chodagam, W. Whitfield, and J. Raff, unpublished observations). As described in Section 13.2.2, axial expansion is an actin/myosin-dependent process that spreads the nuclei throughout the early embryo [22, 23]. Recent studies have shown that cycles of myosin II accumulation occur in the region of the embryo cortex that is directly above the migrating nuclei, even before the nuclei reach the cortex [23]. These cycles stimulate a cortical contraction that appears to drive axial expansion. In *cp190* mutant embryos, the cycles are severely diminished, and axial expansion fails. Moreover, this failure can be rescued by the expression of a constitutively activated myosin light chain, strongly suggesting that CP190 is somehow involved in regulating myosin function in the early embryo. Although the axial expansion defect of *cp190* mutant embryos is rescued by the expression of CP190, it is not rescued by the expression of CP190ΔM, suggesting that the association of CP190 with centrosomes and microtubules is required to regulate myosin II function (S. Chodagam, W. Whitfield, and J. Raff, unpublished observations).

These observations serve as a salutary lesson for those of us who rely on biochemical techniques to identify proteins of interest. The CP190/CP60 complex had many properties suggestive of a role in regulating centrosomal microtubules during mitosis, but the genetic analysis reveals that, *in vivo*, this is not the case.

13.3.5
Centrosomes and Cytokinesis: Studies on asl, cnn, and γ-Tubulin Mutant Spermatocytes

Interactions between the centrosome/microtubule cytoskeleton and the actin/myosin cytoskeleton also play an important part in positioning the contractile ring that mediates cytokinesis [97]. There has been a long-standing controversy over the relative contributions of the chromosomes, the astral microtubules, and the central spindle (midbody) microtubules in positioning the ring. Studies on the large fly spermatocyte cells during meiosis I and II have provided important insights. In particular, in *asterless* (*asl*) mutant fly spermatocytes, the centrosomes nucleate very few, if any, astral microtubules during meiosis [98]. Surprisingly, however, these spindles can undergo a relatively normal looking cytokinesis, suggesting that astral microtubules are not required for the formation and proper positioning of the contractile ring. This finding supports earlier studies that emphasized the importance of the central spindle in organizing cytokinesis in fly spermatocytes. Indeed, there seems to be an interdependency between the central spindle and the contractile ring: mutations that disrupt one, invariably disrupt the other [99, 100].

As mentioned above, in spermatocytes that are deficient in γ-tubulin or Dgrip81, spindle formation is prevented, but large numbers of astral microtubules emanate from the single spindle pole that forms in these cells [55, 101]. This pole is usually asymmetrically positioned in the cell, and long microtubules project from the pole in the direction of the chromosomes, but then continue on past them to the cell cortex. Remarkably, a functional contractile ring assembles close to the distal (plus) ends of these long microtubules. This finding is at odds with most models of contractile ring positioning, but a similar phenomenon has recently been observed in mammalian tissue culture cells that have been forced to enter anaphase with a monopolar spindle [102]. In the latter study, it was postulated that the astral microtubules which pass close to the chromosomes gain an as yet unidentified factor that stabilizes the microtubules and also allows them to stimulate contractile activity when they reach the cortex. The astral microtubules that do not pass close to the chromosomes seem to be less stable, and may actively suppress contractile activity at other regions of the cortex. It has recently been shown, however, that *Drosophila* spermatocytes can undergo cytokinesis even when there are no chromosomes present in the cell [103]. Clearly, the mechanism of contractile ring positioning is likely to remain controversial for some time.

Although this section has focused on the role of centrosomes in organizing the actin/myosin cytoskeleton, the actin/myosin cytoskeleton can also influence centrosome/microtubule behavior. Centrosome separation, for example, often fails in *Drosophila* spermatocytes that are mutant for *chickadee* or *twinstar* (*Drosophila* profilin and cofilin, respectively), and this is also the case in cells treated with actin-depolymerizing drugs [100, 104].

13.3.6 Centrosomes and the Cell Cycle

It has been known for many years that many of the most important regulators of the cell cycle are concentrated at centrosomes, at least transiently, during some stages of the cycle. These include the Polo and Aurora A kinases discussed above, and cdc2/cyclin B kinase. While some of these proteins are probably concentrated at centrosomes specifically to influence centrosomal events during the cell cycle (see Section 13.3.2), others may simply use centrosomes as a convenient meeting place to bring enzymes and their substrates together to maximize efficiency and co-ordination (see Chapter 8).

Examples of the latter case may be the proteins Cdc20 (Fizzy in flies) and Cdh1 (Fizzy-related in flies). These proteins bind to the anaphase promoting complex/cyclosome (APC/C) and are essential for triggering the destruction of proteins such as cyclin B and securin, which enables cells to exit from mitosis [105, 106]. It is thought that APC/C^{Cdc20} is responsible for initiating mitotic exit, but its activity is held in check by an inhibitory signal from unattached kinetochores that prevents anaphase onset until all chromosomes are properly aligned at the metaphase plate. This inhibitory mechanism is commonly referred to as the spindle assembly checkpoint (reviewed in [107]). Once activated, APC/C^{Cdc20} degrades cyclin B and securin, thereby initiating the exit from mitosis. As cyclin B levels fall, APC/C^{Cdh1} is activated to initiate a second phase of destruction that degrades many additional proteins, including Cdc20, which effectively shuts down the first phase of cyclin B and securin destruction [106]. Studies of the dynamic properties of Cdc20 and Cdh1 in flies have shown that both proteins are concentrated at centrosomes but rapidly exchange with a cytoplasmic pool [108]. Thus, the centrosomal concentration of these proteins may simply ensure the efficient switching between the Cdc20- and Cdh1-mediated phases of protein destruction.

There is also evidence that the negative signaling system from unattached kinetochores discussed above, is in some way linked to centrosomes. Many of the Mad and Bub proteins that are essential for the operation of this spindle checkpoint are concentrated both at kinetochores and centrosomes, and several have been observed to travel along kinetochore microtubules toward the centrosomes [109]. For example, two such proteins initially identified in flies, Zeste white 10 (ZW10) and Rough deal (Rod), form a complex that recruits dynein to the kinetochore: both proteins are essential for the spindle checkpoint to operate and both travel along microtubules from the kinetochore to the centrosome [110, 111]. Intriguingly, the destruction of cyclin B is spatially regulated in many cell types [112, 113], and, in flies, it is initiated at centrosomes [113, 114]. Moreover, in human cells, the initial activation of cdc2/cyclin B kinase also occurs at centrosomes [115].

Another conserved centrosomal component, Hsp83 (Hsp90 in vertebrates), regulates cell-cycle events in a novel way. It was initially identified as a centrosomal protein using mass spectroscopy techniques on centrosomes partially purified from *Drosophila* embryos [116]. Hsp83 is a molecular chaperone that is a core component of the centrosome and is essential for proper centrosome function in both flies and vertebrates. It functions at the centrosome, at least in part, by stabilizing the kinase activity of Polo: if Hsp83/90 function is perturbed, Polo kinase activity is dramatically reduced [117]. Thus, Hsp83 is essential for the proper folding of Polo.

13.3.7
Centrosome Dynamics: Inactivation and Flares

As discussed in Section 13.2.3, the rapid and synchronous nuclear cycles of the early fly embryo offer an unusual opportunity to follow the dynamic behavior of specific centrosomal proteins at high resolution in real time. Here, I briefly focus on two aspects of centrosome dynamics that may have important implications for centrosome function – centrosome inactivation and centrosomal "flares".

Centrosome inactivation occurs when DNA is damaged during the rapid nuclear divisions that occur at the cortex of the syncytial fly embryo [118] (see also Chapter 11). When DNA is damaged in a typical somatic cell, the cell usually responds by arresting the cell cycle prior to mitosis; if the DNA cannot be repaired quickly enough, the cells undergo apoptosis, thereby eliminating cells with potentially harmful mutations [119]. In syncytial embryos, where the nuclei divide synchronously in a common cytoplasm, a damaged nucleus cannot arrest. Instead, the centrosomes associated with the damaged nucleus become inactivated as the embryos enter mitosis [118]. The γ-TuRCs dissociate from the centrosome, the spindle fails to assemble normally, and, as a result, the chromosomes fail to segregate properly. Although the γ-TuRCs re-associate with the inactivated centrosomes after mitosis is complete, a nucleus that has not proceeded through mitosis correctly is somehow recognized as abnormal, and falls into the interior of the embryo [120]. As internal nuclei are all destined to form yolk nuclei, which do not contribute to the adult fly, the defective nucleus is effectively eliminated. Thus, centrosome inactivation ensures that nuclei that enter mitosis with damaged DNA are not propagated. Although the γ-TuRCs dissociate from the centrosome during centrosome inactivation, other centrosomal proteins such as CP190 do not, suggesting that a core structure remains intact in the inactivated mitotic centrosome [118] (see also Chapter 11).

In most somatic cells DNA damage is monitored by a number of protein kinases, including ATM and ATR. These kinases then usually activate a cascade of phosphorylations via the downstream protein kinases Chk1 and Chk2 [121]. In *Drosophila*, Chk2 is recruited to the centrosomes associated with damaged nuclei and is required for centrosome inactivation, suggesting that it directly phosphorylates one or more centrosomal proteins [122]. It remains unclear whether centrosome inactivation only occurs in the *Drosophila* syncytial embryo, or whether it can occur in other cells with damaged DNA that fail to arrest their cycle prior to the entry into mitosis. Whatever its general significance, centrosome inactivation demonstrates that different centrosomal proteins interact with centrosomes in different ways and that the association can be highly dynamic.

Centrosomal "flares" were first described when the behavior of a CNN-GFP fusion protein was observed in living *Drosophila* embryos [86]. Fluorescent protein particles appeared to be ejected from the centrosome and then move to and fro close to the centrosome, in a microtubule-dependent manner. D-TACC-GFP and Msps-GFP form similar flares, which are thought to represent complexes of the two proteins bound to the growing and shrinking plus ends of microtubules emanating from the centrosome [78]. It seems that only a subset of centrosomal proteins form flares (N. Peel and J. Raff, unpublished observations) and that there is at least a partial overlap between the flares that contain CNN and those that contain D-TACC/Msps [86]. Many questions remain about the nature of centrosomal flares. Is there more than one type of flare complex? How are the complexes bound to the growing and shrinking end of a microtubule? Are the flares of any functional significance?

13.3.8
Microtubule Motors and Plus-end Tracking Proteins at the Centrosome

Several microtubule motors appear to function at centrosomes in *Drosophila* (see, for example, [123–126]). In most cases, however, these motors are also located elsewhere in the cell and the extent to which the overall function of the motor depends on the centrosomally-localized fraction of the protein is unclear. For this reason, I will not discuss the function of specific microtubule motors in any detail here. In a recent study, however, all of the known kinesin motor proteins were depleted from *Drosophila* S2 cells in culture using RNAi, and nine of the 25 kinesins were found to be essential for some aspect of mitosis [127]. This study illustrates another advantage of *Drosophila*: RNAi is extremely efficient in *Drosophila* tissue culture cells (see also [128]).

Drosophila also has a number of proteins that bind to the growing plus ends of microtubules, the so-called +TIPS [129]. Fractions of some of these proteins, such as EB1 and Mast/Orbit, are found at centrosomes. As with the motor proteins discussed above, these proteins can have important functions in regulating microtubule behavior, but the significance of their localisation at centrosomes is unclear [130–133].

13.3.9 The Interphase Centrosome in Flies: Missing in Action?

Virtually all of the centrosomal proteins discussed so far are recruited to centrosomes only during mitosis in most *Drosophila* somatic cells. This fact is easily overlooked as most of these proteins bind to centrosomes throughout the nuclear cycle in syncytial embryos, probably because there are no "Gap" phases in these rapid nuclear cycles and only a few minutes separate each round of mitosis. In somatic cells, such as those found in larval brains, however, most of these proteins are not detectable on interphase centrosomes. Moreover, interphase centrosomes do not seem to function as major MTOCs in most non-embryonic interphase cells in the fly (our unpublished observations).

This may explain why *Drosophila* seems to lack a number of conserved vertebrate centrosomal proteins such as ninein, ninein-like protein (NLP), PCM-1, and ε-tubulin. These proteins appear to be involved in organizing centrosomal microtubules during interphase [134–138]. Perhaps, flies do not need these proteins because the centrosome usually does not function as a major MTOC in interphase cells.

13.4 The Role of Centrosomes and Centrosomal Proteins *In Vivo*

In this section, I will discuss two specific areas where *Drosophila* has provided important insights into centrosome function within the context of a multicellular organism: (1) the role of centrosomes in the development of a multicellular organism and (2) the role of centrosomal proteins in organizing microtubules in cells that do not contain canonical centrosomes.

13.4.1 The Essential Role of Centrosomes In *Drosophila*

As discussed in Section 13.3.5, *asl* mutant spermatocytes appear to lack astral microtubules, yet they are able to undergo relatively normal cytokinesis, suggesting that astral microtubules are not essential for positioning the contractile ring [98]. Surprisingly, the asymmetric divisions of the larval neuroblast cells also occur relatively normally in *asl* mutants [139]. This result is unexpected, as interactions between the astral microtubules and the cortex are thought to be essential for properly positioning the spindle during asymmetric divisions [140]. Although this result calls into question the role of astral microtubules in asymmetric division, its significance is unclear as *asl* mutations are lethal. Thus, Asl must provide some essential function to the fly.

Mutations in another *Drosophila* centrosomal gene, *cnn*, have a similar phenotype to *asl* mutants. As discussed in Section 13.3.2, CNN is thought to link Aurora A to the γ-TuRC [70], and the centrosomal recruitment of γ-tubulin and CP190 is severely disrupted during mitosis in *cnn* mutant flies [71]. Like *asl* mutants, *cnn* mutants appear to lack astral microtubules in mitosis but their neuroblasts can undergo apparently normal asymmetric divisions [141]. Unlike *asl* mutants, however, *cnn* null mutants are viable: they develop at normal rates and form almost perfectly normal adult flies, which are born at normal Mendelian ratios [141]. The finding that a complex multicellular organism like a fly can apparently complete (in the authors words) "zygotic development without functional mitotic centrosomes" [141] was an unexpected revelation.

Cnn mutant flies, however, are invariably sterile. In females, the sterility arises because astral microtubules play an essential part in preventing mitotic spindles from colliding with one another during mitosis in the syncytial embryo [142]. Thus, *cnn* mutants are viable because homozygous mutant embryos (laid by heterozygous mothers) proceed through early development using the maternally supplied stockpile of CNN. When this runs out, further development does not require CNN, so homozygous mutant animals are viable. Homozygous mutant females are sterile because the embryos they lay do not contain CNN and so die during early embryonic development due to an accumulation of mitotic defects [71, 72]. In *cnn* mutant males, sterility reflects cytokinesis defects during meiosis I and II, and sperm tail defects, suggesting that CNN normally plays a part in centriole function, as well as in centrosome function [143]. It is unclear why, in *cnn* mutants, there are cytokinesis defects during male meiosis but not during mitosis.

At one extreme, the *asl* and *cnn* mutant phenotypes could be interpreted to suggest that the only essential function of centrosomes during mitosis in flies is to generate astral microtubules during the syncytial divisions. At all other stages of the fly life cycle, mitotic centrosomes and astral microtubules would seem dispensable. However, as discussed at greater length elsewhere [8], the problem with this interpretation is that it is not clear that these mutants completely lack functional mitotic centrosomes and astral microtubules. While the formation of astral microtubules is clearly compromised, it is not clear that they are completely absent. Recent live imaging of tubulin-GFP in *asl* mutant spermatocytes (an analysis that was not possible previously) has shown that a small number of astral microtubules are present (H. Varmark and C. Gonzalez, unpublished observations). Moreover, while asymmetric divisions clearly can occur in both *asl* and *cnn* mutants, they are not entirely normal. In both cases, the normal alignment of the mitotic spindle with asymmetrically localized cell-fate determinants in neuroblasts is at least partially disrupted [139, 141], suggesting that robust astral microtubules may be essential for properly positioning the spindle, even if they are not essential for asymmetric division *per se*. Thus, the normal development of *cnn* mutant flies may be more a testament to the ability of the developing nervous system to cope with mistakes in asymmetric divisions, rather than to the ability of cells to properly position spindles without astral microtubules.

Nonetheless, the finding that *cnn* null mutant flies can develop so normally with clearly dysfunctional mitotic centrosomes must be taken into account when considering centrosomal function *in vivo*. It is interesting that CNN, which recruits proteins such as γ-tubulin and Aurora A to centrosomes during mitosis, is not essential for the development of the fly. The same is true for two other proteins that appear to function by recruiting proteins to mitotic centrosomes. CP309/D-PLP (the *Drosophila* homolog of pericentrin; see Section 13.3.2) is required for the efficient recruitment of most, if not all, PCM components to the centrosome during mitosis, yet mitosis occurs largely normally in *d-plp* null mutant larval brain cells [153]. Similarly, D-TACC is required to recruit the microtubule stabilizing protein Msps to centrosomes, yet *d-tacc* null mutants are viable: as with CNN, D-TACC is only essential for mitosis in the syncytial embryo [77]. These studies suggest that centrosomal proteins that simply recruit other proteins to the centrosome may only be essential for the rapid nuclear divisions of the syncytial embryo. Centrosomal proteins such as γ-tubulin, Msps, or Aurora A, however, are essential for fly viability, perhaps because they are required to stabilize microtubules even when they are not concentrated at centrosomes, or because they fulfill some other function.

Perhaps, these results should not be so surprising. After all, animal centrosomes have centrioles at their core, and centrioles are themselves complex microtubule structures. Centrioles, therefore, presumably evolved some time after microtubules were organizing cell division, and, if so, there must have been a more primitive mechanism of animal cell division that did not require canonical centrosomes (see also Chapter 6). Indeed, plant cells may still use such a mechanism. Perhaps, this primitive mechanism can be re-activated in flies, explaining why efficient centrosome function appears to be so easily dispensed with during mitosis. Perhaps, vertebrates will prove to be as adaptable as *Drosophila*, as microtubules can self organize into bi-polar spindles in the absence of centrosomes in several vertebrate cell types in culture [144, 145]. Although vertebrate cells that have had their centrosomes removed by microsurgery or laser ablation can proceed through mitosis, cytokinesis is inefficient, and they invariably arrest during G1 of the next cell cycle. Understanding the molecular basis of this cell-cycle arrest constitutes a major challenge for centrosome research.

13.4.2
The Role of Centrosomal Proteins in Oogenesis

As discussed in Section 13.2.1, microtubules play an essential role in organizing many aspects of oogenesis in flies, yet centrosomes disappear early in oocyte development. Oogenesis in flies therefore offers a powerful system with which to study how microtubules are organized in the absence of typical centrosomes and which "centrosomal" proteins are involved instead.

At least two different arrays of polarized microtubules are established at different stages of oocyte development (Figure 13.1; see Section 13.2.1). In somatic cells, γ-tubulin is essential for the proper organization of microtubules, probably even when it is not strongly concentrated at centrosomes (see Section 13.4.1), so it seems likely that it plays a role in organizing microtubules within the oocyte. Only γ-tubulin 37C is expressed in the developing oocyte [11], but, surprisingly, it is not detectably enriched at the MTOC that forms around the centrioles at the posterior pole of the oocyte, nor at the MTOC that forms at the anterior pole of the oocyte later in oogenesis [10, 11]. Recently, however, *γt37C* and *dgrip75* (a gene encoding a component of the γ-TuRC) were both identified in a screen for mutations that failed to localize *bicoid* (the morphogen that specifies anterior fates in *Drosophila*) to the anterior pole of the oocyte at later stages of development [146]. This localization of *bicoid* is microtubule dependent, and, in this study, both γ-tubulin 37C and Dgrip75 proteins were detectably (although weakly) enriched at the anterior cortex during late stages of oogenesis. Mutations in the genes that encode either protein, however, did not alter the gross morphology or organization of microtubules in the oocyte. Thus, γ-tubulin 37C is apparently essential for the organization of only a specific subset of oocyte microtubules.

More recently, oogenesis has been studied in mutant flies in which both γ-tubulin genes were mutated at the same time [147]. These "double" mutant oocytes had strong, but relatively pleiotropic, abnormalities, suggesting that γ-tubulin is essential for oocyte development but the two isoforms are partially redundant. In support of this possibility, preliminary data suggest that strong mutations in the γ-tubulin 23C gene lead to the upregulation of the γ-tubulin 37C gene (P. Sampaio and C. E. Sunkel, personal communication).

Although it may have only a limited role in organizing microtubules in the oocyte, γ-tubulin 37C seems to be essential for organizing the meiosis I spindle [10] (but see also [11]). γ-Tubulin is not detectable at the acentrosomal poles of these spindles, but the spindles are highly disorganized in *γt37C* mutants. That γ-tubulin functions in these spindles without being detectable there, serves as a reminder that the inability to detect a protein in a particular location does not prove its absence or its lack of involvement in a process.

D-TACC and Msps are the only known centrosomal proteins to be identified at the poles of the meiosis I spindle [148]. If either protein is inactivated by mutation, the spindles often become tripolar or quadrapolar, suggesting that both proteins are required to maintain spindle bipolarity. This phenotype is reminiscent of that observed in both *Drosophila* and human somatic cells that have been depleted of Msps [79] or ch-TOG (the human homolog of Msps) [149]. How these proteins function to maintain spindle bipolarity is unclear, but their ability to form higher order polymers when overexpressed in tissue culture cells suggests that they may form a structural lattice at the spindle poles that maintains the integrity of the poles [150].

During earlier stages of oogenesis, D-TACC is diffusely concentrated in both the region of the MTOC that initially forms at the posterior pole of the oocyte and the region of the MTOC that forms slightly later at the anterior pole (J. Raff and

W. Theurkauf, unpublished observations; [5]). Although this distribution suggests that D-TACC may play a part in stabilizing microtubules in the oocyte, *d-tacc* mutations do not detectably disrupt either the organization or function of the non-spindle microtubule arrays in the oocyte (B. Cha and W. Theurkauf, personal communication). The centrosomal protein CNN is also concentrated in both the posterior and anterior MTOCs during oogenesis, but, like D-TACC, the significance of this distribution is unclear [4]. Perhaps, like γ-tubulin, D-TACC and CNN are involved in organizing only a subset of the microtubules in the oocyte, or they play redundant roles in organizing microtubules within the oocyte.

These studies reveal that many "centrosomal" proteins play a part in organizing microtubules in cells that do not contain canonical centrosomes. The exact details of how they do so remains to be established, but further studies in *Drosophila* will doubtless provide important clues.

13.5
Summary

Using *Drosophila* allows the combination of biochemical, cell biological and genetic approaches to study centrosome biology within the context of an intact animal. New approaches, such as the use of photoactivatable GFPs, fluorescence recovery after photobleaching (FRAP), and fluorescence resonance energy transfer (FRET), will make it possible to study the dynamic behavior and interactions of many centrosomal proteins in the early *Drosophila* embryo in exquisite detail. The well annotated fly genome sequence, together with increasingly powerful mass spectroscopy techniques for protein identification, should make it possible to identify all of the proteins of the fly centrosome (see Chapter 7). By inactivating all of the genes or proteins one by one and in combination, both *in vivo* and *in vitro*, the mysteries of centrosome function will slowly, but surely, be resolved.

Acknowledgments

I thank the many colleagues quoted in this chapter who allowed me to discuss their unpublished observations; Daniel St Johnston for providing the building blocks for Figure 13.1, and Martin Raff, Renata Basto, Teresa Barros, and Eliana Lucas for many helpful comments on the manuscript.

Work in my laboratory is supported by the Wellcome Trust and by Cancer Research, UK.

References

1. Deng, W., and Lin, H. **2001**. Asymmetric germ cell division and oocyte determination during *Drosophila* oogenesis. Int Rev Cytol *203*, 93–138.
2. Mahowald, A. P., and Strassheim, J. M. **1970**. Intercellular migration of centrioles in the germarium of Drosophila melanogaster. An electron microscopic study. J Cell Biol *45*, 306–320.
3. Grieder, N. C., de Cuevas, M., and Spradling, A. C. **2000**. The fusome organizes the microtubule network during oocyte differentiation in *Drosophila*. Development *127*, 4253–4264.
4. Megraw, T. L., and Kaufman, T. C. **2000**. The centrosome in *Drosophila* oocyte development. Curr Top Dev Biol *49*, 385–407.
5. Bolivar, J., Huynh, J. R., Lopez-Schier, H., Gonzalez, C., St Johnston, D., and Gonzalez-Reyes, A. **2001**. Centrosome migration into the *Drosophila* oocyte is independent of BicD and egl, and of the organization of the microtubule cytoskeleton. Development *128*, 1889–1897.
6. Theurkauf, W. E., Smiley, S., Wong, M. L., and Alberts, B. M. **1992**. Reorganization of the cytoskeleton during *Drosophila* oogenesis: implications for axis specification and intercellular transport. Development *115*, 923–936.
7. Cooley, L., and Theurkauf, W. E. **1994**. Cytoskeletal functions during *Drosophila* oogenesis. Science *266*, 590–596.
8. Raff, J. W. **2001**. Centrosomes: central no more? Curr Biol *11*, R159–R161.
9. Matthies, H. J., McDonald, H. B., Goldstein, L. S., and Theurkauf, W. E. **1996**. Anastral meiotic spindle morphogenesis: role of the non-claret disjunctional kinesin-like protein. J Cell Biol *134*, 455–464.
10. Tavosanis, G., Llamazares, S., Goulielmos, G., and Gonzalez, C. **1997**. Essential role for gamma-tubulin in the acentriolar female meiotic spindle of Drosophila. EMBO J *16*, 1809–1819.
11. Wilson, P. G., Zheng, Y., Oakley, C. E., Oakley, B. R., Borisy, G. G., and Fuller, M. T. **1997**. Differential expression of two gamma-tubulin isoforms during gametogenesis and development in *Drosophila*. Dev Biol *184*, 207–221.
12. Riparbelli, M. G., and Callaini, G. **1996**. Meiotic spindle organization in fertilized Drosophila oocyte: presence of centrosomal components in the meiotic apparatus. J Cell Sci *109*(Pt 5), 911–918.
13. Callaini, G., and Riparbelli, M. G. **1996**. Fertilization in *Drosophila melanogaster*: centrosome inheritance and organization of the first mitotic spindle. Dev Biol *176*, 199–208.
14. Fuller, M. **1993**. Spermatogenesis. In *The Development of* Drosophila melanogaster, M. Bate, and A. Martinez-Arias (Eds). Cold Spring Harbour Laboratory Press: New York, 71–148.
15. Giansanti, M. G., Bonaccorsi, S., Bucciarelli, E., and Gatti, M. **2001**. *Drosophila* male meiosis as a model system for the study of cytokinesis in animal cells. Cell Struct Funct *26*, 609–617.
16. Gonzalez, C., Tavosanis, G., and Mollinari, C. **1998**. Centrosomes and microtubule organisation during Drosophila development. J Cell Sci *111*(Pt 18), 2697–2706.
17. Tram, U., and Sullivan, W. **2000**. Reciprocal inheritance of centrosomes in the parthenogenetic hymenopteran *Nasonia vitripennis*. Curr Biol *10*, 1413–1419.
18. Riparbelli, M. G., and Callaini, G. **2003**. *Drosophila* parthenogenesis: a model for *de novo* centrosome assembly. Dev Biol *260*, 298–313.
19. Marshall, W. F., Vucica, Y., and Rosenbaum, J. L. **2001**. Kinetics and regulation of *de novo* centriole assembly. Implications for the mechanism of centriole duplication. Curr Biol *11*, 308–317.
20. Khodjakov, A., Rieder, C. L., Sluder, G., Cassels, G., Sibon, O., and Wang, C. L. **2002**. *De novo* formation of centrosomes in vertebrate cells arrested

during S phase. J Cell Biol *158*, 1171–1181.

21. Foe, V. E., and Alberts, B. M. 1983. Studies of nuclear and cytoplasmic behaviour during the five mitotic cycles that precede gastrulation in *Drosophila* embryogenesis. J Cell Sci *61*, 31–70.

22. von Dassow, G., and Schubiger, G. **1994**. How an actin network might cause fountain streaming and nuclear migration in the syncytial *Drosophila* embryo. J Cell Biol *127*, 1637–1653.

23. Royou, A., Sullivan, W., and Karess, R. **2002**. Cortical recruitment of nonmuscle myosin II in early syncytial *Drosophila* embryos: its role in nuclear axial expansion and its regulation by Cdc2 activity. J Cell Biol *158*, 127–137.

24. Baker, J., Theurkauf, W. E., and Schubiger, G. **1993**. Dynamic changes in microtubule configuration correlate with nuclear migration in the preblastoderm *Drosophila* embryo. J Cell Biol *122*, 113–121.

25. Francis-Lang, H., Minden, J., Sullivan, W., and Oegema, K. **1999**. Live confocal analysis with fluorescently labeled proteins. Methods Mol Biol *122*, 223–239.

26. Foe, V. E., Odell, G. M., and Edgar, B. A. **1993**. Mitosis and morphogenesis in the Drosophila embryo: point and counterpoint. In *The Development of* Drosophila melanogaster, M. Bate, and A. Martinez-Arias (Eds). Cold Spring Harbour Laboratory Press: New York, 149–300.

27. Raff, J. W., and Glover, D. M. **1989**. Centrosomes, and not nuclei, initiate pole cell formation in *Drosophila* embryos. Cell *57*, 611–619.

28. Foe, V. E. 1989. Mitotic domains reveal early commitment of cells in *Drosophila* embryos. Development *107*, 1–22.

29. Doe, C. Q., and Skeath, J. B. **1996**. Neurogenesis in the insect central nervous system. Curr Opin Neurobiol *6*, 18–24.

30. Chia, W., and Yang, X. **2002**. Asymmetric division of *Drosophila* neural progenitors. Curr Opin Genet Dev *12*, 459–464.

31. Kaltschmidt, J. A., and Brand, A. H. **2002**. Asymmetric cell division: microtubule dynamics and spindle asymmetry. J Cell Sci *115*, 2257–2264.

32. Ahringer, J. **2003**. Control of cell polarity and mitotic spindle positioning in animal cells. Curr Opin Cell Biol *15*, 73–81.

33. Kaltschmidt, J. A., Davidson, C. M., Brown, N. H., and Brand, A. H. **2000**. Rotation and asymmetry of the mitotic spindle direct asymmetric cell division in the developing central nervous system. Nature Cell Biol *2*, 7–12.

34. Cai, Y., Yu, F., Lin, S., Chia, W., and Yang, X. **2003**. Apical complex genes control mitotic spindle geometry and relative size of daughter cells in Drosophila neuroblast and pI asymmetric divisions. Cell *112*, 51–62.

35. Doe, C. Q., and Bowerman, B. **2001**. Asymmetric cell division: fly neuroblast meets worm zygote. Curr Opin Cell Biol *13*, 68–75.

36. Mahowald, A. P., Caulton, J. H., Edwards, M. K., and Floyd, A. D. **1979**. Loss of centrioles and polyploidization in follicle cells of *Drosophila melanogaster*. Exp Cell Res *118*, 404–410.

37. Gatti, M., and Baker, B. S. 1989. Genes controlling essential cell-cycle functions in *Drosophila melanogaster*. Genes Dev *3*, 438–453.

38. Glover, D. M. 1989. Mitosis in *Drosophila*. J Cell Sci *92*(Pt 2), 137–146.

39. Mogensen, M. M., and Tucker, J. B. **1987**. Evidence for microtubule nucleation at plasma membrane-associated sites in Drosophila. J Cell Sci *88*(Pt 1), 95–107.

40. Mogensen, M. M., Tucker, J. B., and Baggaley, T. B. **1993**. Multiple plasma membrane-associated MTOC systems in the acentrosomal cone cells of *Drosophila* ommatidia. Eur J Cell Biol *60*, 67–75.

41. Oakley, C. E., and Oakley, B. R. **1989**. Identification of gamma-tubulin, a new member of the tubulin superfamily encoded by mipA gene of *Aspergillus nidulans*. Nature *338*, 662–664.

42. Erickson, H. P. **2000**. Gamma-tubulin nucleation: template or protofilament? Nature Cell Biol *2*, E93–E96.

43. Zheng, Y., Wong, M. L., Alberts, B., and Mitchison, T. **1995**. Nucleation of microtubule assembly by a gamma-tubulin-containing ring complex. Nature *378*, 578–583.

44. Wiese, C., and Zheng, Y. **1999**. Gamma-tubulin complexes and their interaction with microtubule-organizing centers. Curr Opin Struct Biol *9*, 250–259.

45. Moritz, M., and Agard, D. A. **2001**. Gamma-tubulin complexes and microtubule nucleation. Curr Opin Struct Biol *11*, 174–181.

46. Moritz, M., Braunfeld, M. B., Sedat, J. W., Alberts, B., and Agard, D. A. **1995**. Microtubule nucleation by gamma-tubulin-containing rings in the centrosome. Nature *378*, 638–640.

47. Moritz, M., Zheng, Y., Alberts, B. M., and Oegema, K. **1998**. Recruitment of the gamma-tubulin ring complex to *Drosophila* salt-stripped centrosome scaffolds. J Cell Biol *142*, 775–786.

48. Oegema, K., Wiese, C., Martin, O. C., Milligan, R. A., Iwamatsu, A., Mitchison, T. J., and Zheng, Y. **1999**. Characterization of two related *Drosophila* gamma-tubulin complexes that differ in their ability to nucleate microtubules. J Cell Biol *144*, 721–733.

49. Gunawardane, R. N., Martin, O. C., Cao, K., Zhang, L., Dej, K., Iwamatsu, A., and Zheng, Y. **2000**. Characterization and reconstitution of *Drosophila* gamma-tubulin ring complex subunits. J Cell Biol *151*, 1513–1524.

50. Gunawardane, R. N., Martin, O. C., and Zheng, Y. **2003**. Characterization of a new gammaTuRC subunit with WD repeats. Mol Biol Cell *14*, 1017–1026.

51. Geissler, S., Pereira, G., Spang, A., Knop, M., Soues, S., Kilmartin, J., and Schiebel, E. **1996**. The spindle pole body component Spc98p interacts with the gamma-tubulin-like Tub4p of *Saccharomyces cerevisiae* at the sites of microtubule attachment. EMBO J *15*, 3899–3911.

52. Knop, M., Pereira, G., Geissler, S., Grein, K., and Schiebel, E. **1997**. The spindle pole body component Spc97p interacts with the gamma-tubulin of *Saccharomyces cerevisiae* and functions in microtubule organization and spindle pole body duplication. EMBO J *16*, 1550–1564.

53. Sunkel, C. E., Gomes, R., Sampaio, P., Perdigao, J., and Gonzalez, C. **1995**. Gamma-tubulin is required for the structure and function of the microtubule organizing centre in *Drosophila* neuroblasts. EMBO J *14*, 28–36.

54. Barbosa, V., Yamamoto, R. R., Henderson, D. S., and Glover, D. M. **2000**. Mutation of a *Drosophila* gamma tubulin ring complex subunit encoded by discs degenerate-4 differentially disrupts centrosomal protein localization. Genes Dev *14*, 3126–3139.

55. Sampaio, P., Rebollo, E., Varmark, H., Sunkel, C. E., and Gonzalez, C. **2001**. Organized microtubule arrays in gamma-tubulin-depleted *Drosophila* spermatocytes. Curr Biol *11*, 1788–1793.

56. Strome, S., Powers, J., Dunn, M., Reese, K., Malone, C. J., White, J., Seydoux, G., and Saxton, W. **2001**. Spindle dynamics and the role of gamma-tubulin in early *Caenorhabditis elegans* embryos. Mol Biol Cell *12*, 1751–1764.

57. Hannak, E., Oegema, K., Kirkham, M., Gonczy, P., Habermann, B., and Hyman, A. A. **2002**. The kinetically dominant assembly pathway for centrosomal asters in *Caenorhabditis elegans* is gamma-tubulin dependent. J Cell Biol *157*, 591–602.

58. Moritz, M., and Alberts, B. M. **1999**. Isolation of centrosomes from *Drosophila* embryos. Methods Cell Biol *61*, 1–12.

59. Schnackenberg, B. J., Hull, D. R., Balczon, R. D., and Palazzo, R. E. **2000**. Reconstitution of microtubule nucleation potential in centrosomes isolated from *Spisula solidissima* oocytes. J Cell Sci *113*(Pt 6), 943–953.

60. do Carmo Avides, M., and Glover, D. M. **1999**. Abnormal spindle protein, Asp, and the integrity of mitotic centrosomal microtubule organizing centers. Science *283*, 1733–1735.

61. do Carmo Avides, M., Tavares, A., and Glover, D. M. **2001**. Polo kinase and Asp are needed to promote the mitotic organizing activity of centrosomes. Nature Cell Biol *3*, 421–424.

62. Moutinho-Santos, T., Sampaio, P., Amorim, I., Costa, M., and Sunkel, C. E. **1999**. *In vivo* localisation of the mitotic POLO kinase shows a highly dynamic association with the mitotic apparatus during early embryogenesis in *Drosophila*. Biol Cell *91*, 585–596.

63. Lane, H. A., and Nigg, E. A. **1996**. Antibody microinjection reveals an essential role for human polo-like kinase 1 (Plk1) in the functional maturation of mitotic centrosomes. J Cell Biol *135*, 1701–1713.

64. Donaldson, M. M., Tavares, A. A., Ohkura, H., Deak, P., and Glover, D. M. **2001**. Metaphase arrest with centromere separation in polo mutants of Drosophila. J Cell Biol *153*, 663–676.

65. Wakefield, J. G., Bonaccorsi, S., and Gatti, M. **2001**. The drosophila protein asp is involved in microtubule organization during spindle formation and cytokinesis. J Cell Biol *153*, 637–648.

66. Riparbelli, M. G., Callaini, G., Glover, D. M., and Avides Mdo, C. **2002**. A requirement for the Abnormal Spindle protein to organize microtubules of the central spindle for cytokinesis in *Drosophila*. J Cell Sci *115*, 913–922.

67. Hannak, E., Kirkham, M., Hyman, A. A., and Oegema, K. **2001**. Aurora-A kinase is required for centrosome maturation in *Caenorhabditis elegans*. J Cell Biol *155*, 1109–1116.

68. Berdnik, D., and Knoblich, J. A. **2002**. Drosophila Aurora-A is required for centrosome maturation and actin-dependent asymmetric protein localization during mitosis. Curr Biol *12*, 640–647.

69. Dutertre, S., Descamps, S., and Prigent, C. **2002**. On the role of aurora-A in centrosome function. Oncogene *21*, 6175–6183.

70. Terada, Y., Uetake, Y., and Kuriyama, R. **2003**. Interaction of Aurora-A and centrosomin at the microtubule-nucleating site in *Drosophila* and mammalian cells. J Cell Biol *162*, 757–763.

71. Megraw, T. L., Li, K., Kao, L. R., and Kaufman, T. C. **1999**. The centrosomin protein is required for centrosome assembly and function during cleavage in *Drosophila*. Development *126*, 2829–2839.

72. Vaizel-Ohayon, D., and Schejter, E. D. **1999**. Mutations in centrosomin reveal requirements for centrosomal function during early *Drosophila* embryogenesis. Curr Biol *9*, 889–898.

73. Dictenberg, J. B., Zimmerman, W., Sparks, C. A., Young, A., Vidair, C., Zheng, Y., Carrington, W., Fay, F. S., and Doxsey, S. J. **1998**. Pericentrin and gamma-tubulin form a protein complex and are organized into a novel lattice at the centrosome. J Cell Biol *141*, 163–174.

74. Gillingham, A. K., and Munro, S. **2000**. The PACT domain, a conserved centrosomal targeting motif in the coiled-coil proteins AKAP450 and pericentrin. EMBO Rep *1*, 524–529.

75. Kawaguchi, S. I., and Zheng, Y. **2004**. Characterization of a *Drosophila* centrosome protein CP309 that shares homology with Kendrin and CG-NAP. Mol Biol Cell *15*, 37–45.

76. Kellogg, D. R., Field, C. M., and Alberts, B. M. **1989**. Identification of microtubule-associated proteins in the centrosome, spindle, and kinetochore of the early *Drosophila* embryo. J Cell Biol *109*, 2977–2991.

77. Gergely, F., Kidd, D., Jeffers, K., Wakefield, J. G., and Raff, J. W. **2000**. D-TACC: a novel centrosomal protein required for normal spindle function in the early *Drosophila* embryo. EMBO J *19*, 241–252.

78. Lee, M. J., Gergely, F., Jeffers, K., Peak-Chew, S. Y., and Raff, J. W. **2001**. Msps/XMAP215 interacts with the centrosomal protein D-TACC to regulate microtubule behaviour. Nature Cell Biol *3*, 643–649.

79. Cullen, C. F., Deak, P., Glover, D. M., and Ohkura, H. **1999**. Mini spindles: a gene encoding a conserved microtubule-associated protein required for the integrity of the mitotic spindle in *Drosophila*. J Cell Biol *146*, 1005–1018.

80. Kinoshita, K., Habermann, B., and Hyman, A. A. **2002**. XMAP215: a key component of the dynamic microtubule cytoskeleton. Trends Cell Biol *12*, 267–273.

81. Matthews, L. R., Carter, P., Thierry-Mieg, D., and Kemphues, K. **1998**. ZYG-9, a *Caenorhabditis elegans* protein required for microtubule organization and function, is a component of meiotic and mitotic spindle poles. J Cell Biol *141*, 1159–1168.

82. Ohkura, H., Garcia, M. A., and Toda, T. **2001**. Dis1/TOG universal microtubule adaptors – one MAP for all? J Cell Sci *114*, 3805–3812.

83. Bellanger, J. M., and Gonczy, P. **2003**. TAC-1 and ZYG-9 form a complex that promotes microtubule assembly in C. elegans embryos. Curr Biol *13*, 1488–1498.

84. Le Bot, N., Tsai, M. C., Andrews, R. K., and Ahringer, J. **2003**. TAC-1, a regulator of microtubule length in the *C. elegans* embryo. Curr Biol *13*, 1499–1505.

85. Srayko, M., Quintin, S., Schwager, A., and Hyman, A. A. **2003**. *Caenorhabditis elegans* TAC-1 and ZYG-9 form a complex that is essential for long astral and spindle microtubules. Curr Biol *13*, 1506–1511.

86. Megraw, T. L., Kilaru, S., Turner, F. R., and Kaufman, T. C. **2002**. The centrosome is a dynamic structure that ejects PCM flares. J Cell Sci *115*, 4707–4718.

87. Foe, V. E., Field, C. M., and Odell, G. M. **2000**. Microtubules and mitotic cycle phase modulate spatiotemporal distributions of F-actin and myosin II in *Drosophila* syncytial blastoderm embryos. Development *127*, 1767–1787.

88. Stevenson, V. A., Kramer, J., Kuhn, J., and Theurkauf, W. E. **2001**. Centrosomes and the Scrambled protein coordinate microtubule-independent actin reorganization. Nature Cell Biol *3*, 68–75.

89. Rothwell, W. F., Fogarty, P., Field, C. M., and Sullivan, W. **1998**. Nuclear-fallout, a *Drosophila* protein that cycles from the cytoplasm to the centrosomes, regulates cortical microfilament organization. Development *125*, 1295–1303.

90. Rothwell, W. F., Zhang, C. X., Zelano, C., Hsieh, T. S., and Sullivan, W. **1999**. The *Drosophila* centrosomal protein Nuf is required for recruiting Dah, a membrane-associated protein, to furrows in the early embryo. J Cell Sci *112*(Pt 17), 2885–2893.

91. Riggs, B., Rothwell, W., Mische, S., Hickson, G. R., Matheson, J., Hays, T. S., Gould, G. W., and Sullivan, W. **2003**. Actin cytoskeleton remodeling during early *Drosophila* furrow formation requires recycling endosomal components Nuclear-fallout and Rab11. J Cell Biol *163*, 143–154.

92. Whitfield, W. G., Millar, S. E., Saumweber, H., Frasch, M., and Glover, D. M. **1988**. Cloning of a gene encoding an antigen associated with the centrosome in *Drosophila.* J Cell Sci *89*(Pt 4), 467–480.

93. Kellogg, D. R., and Alberts, B. M. **1992**. Purification of a multiprotein complex containing centrosomal proteins from the *Drosophila* embryo by chromatography with low-affinity polyclonal antibodies. Mol Biol Cell *3*, 1–11.

94. Kellogg, D. R., Oegema, K., Raff, J., Schneider, K., and Alberts, B. M. **1995**. CP60: a microtubule-associated protein that is localized to the centrosome in a cell cycle-specific manner. Mol Biol Cell *6*, 1673–1684.

95. Oegema, K., Marshall, W. F., Sedat, J. W., and Alberts, B. M. **1997**. Two proteins that cycle asynchronously between centrosomes and nuclear structures: *Drosophila* CP60 and CP190. J Cell Sci *110*(Pt 14), 1573–1583.

96. Butcher, R. D. J., Chodagam, S., Basto, R., Wakefield, J. G., Henderson, D. S., Raff, J. W., and Whitfield, W. G. F. **2004**. The Drosophila centrosome-associated protein CP190 is essential for viability, but not for cell division. J Cell Sci (in press).

97. Robinson, D. N., and Spudich, J. A. **2000**. Towards a molecular understanding of cytokinesis. Trends Cell Biol *10*, 228–237.

98. Bonaccorsi, S., Giansanti, M. G., and Gatti, M. **1998**. Spindle self-organization and cytokinesis during male meiosis in asterless mutants of Drosophila melanogaster. J Cell Biol *142*, 751–761.

99. Williams, B. C., Riedy, M. F., Williams, E. V., Gatti, M., and Goldberg, M. L. **1995**. The *Drosophila* kinesin-like protein KLP3A is a midbody component required for central spindle assembly and initiation of cytokinesis. J Cell Biol *129*, 709–723.

100. Giansanti, M. G., Bonaccorsi, S., Williams, B., Williams, E. V., Santolamazza, C., Goldberg, M. L., and Gatti, M. **1998**. Cooperative interactions between the central spindle and the contractile ring during *Drosophila* cytokinesis. Genes Dev *12*, 396–410.

101. Barbosa, V., Gatt, M., Rebollo, E., Gonzalez, C., and Glover, D. M. **2003**. Drosophila dd4 mutants reveal that gammaTuRC is required to maintain juxtaposed half spindles in spermatocytes. J Cell Sci *116*, 929–941.

102. Canman, J. C., Cameron, L. A., Maddox, P. S., Straight, A., Tirnauer, J. S., Mitchison, T. J., Fang, G., Kapoor, T. M., and Salmon, E. D. **2003**. Determining the position of the cell division plane. Nature *424*, 1074–1078.

103. Bucciarelli, E., Giansanti, M. G., Bonaccorsi, S., and Gatti, M. **2003**. Spindle assembly and cytokinesis in the absence of chromosomes during Drosophila male meiosis. J Cell Biol *160*, 993–999.

104. Gunsalus, K. C., Bonaccorsi, S., Williams, E., Verni, F., Gatti, M., and Goldberg, M. L. **1995**. Mutations in twinstar, a *Drosophila* gene encoding a cofilin/ADF homologue, result in defects in centrosome migration and cytokinesis. J Cell Biol *131*, 1243–1259.

105. Zachariae, W., and Nasmyth, K. **1999**. Whose end is destruction: cell division and the anaphase-promoting complex. Genes Dev *13*, 2039–2058.

106. Kramer, E. R., Scheuringer, N., Podtelejnikov, A. V., Mann, M., and Peters, J. M. **2000**. Mitotic regulation of the APC activator proteins CDC20 and CDH1. Mol Biol Cell *11*, 1555–1569.

107. Shah, J. V., and Cleveland, D. W. **2000**. Waiting for anaphase: Mad2 and the spindle assembly checkpoint. Cell *103*, 997–1000.

108. Raff, J. W., Jeffers, K., and Huang, J. Y. **2002**. The roles of Fzy/Cdc20 and Fzr/Cdh1 in regulating the destruction of cyclin B in space and time. J Cell Biol *157*, 1139–1149.

109. Howell, B. J., McEwen, B. F., Canman, J. C., Hoffman, D. B., Farrar, E. M., Rieder, C. L., and Salmon, E. D. **2001**. Cytoplasmic dynein/dynactin drives kinetochore protein transport to the spindle poles and has a role in mitotic spindle checkpoint inactivation. J Cell Biol *155*, 1159–1172.

110. Basto, R., Gomes, R., and Karess, R. E. **2000**. Rough deal and Zw10 are required for the metaphase checkpoint in Drosophila. Nature Cell Biol *2*, 939–943.

111. Chan, G. K., Jablonski, S. A., Starr, D. A., Goldberg, M. L., and Yen, T. J. **2000**. Human Zw10 and ROD are mitotic checkpoint proteins that bind to kinetochores. Nature Cell Biol *2*, 944–947.

112. Clute, P., and Pines, J. **1999**. Temporal and spatial control of cyclin B1 destruction in metaphase. Nature Cell Biol *1*, 82–87.

113. Huang, J., and Raff, J. W. **1999**. The disappearance of cyclin B at the end of mitosis is regulated spatially in *Drosophila* cells. EMBO J *18*, 2184–2195.

114. Wakefield, J. G., Huang, J. Y., and Raff, J. W. **2000**. Centrosomes have a role in regulating the destruction of cyclin B in early *Drosophila* embryos. Curr Biol *10*, 1367–1370.

115. Jackman, M., Lindon, C., Nigg, E. A., and Pines, J. **2003**. Active cyclin B1-Cdk1 first appears on centrosomes in prophase. Nature Cell Biol *5*, 143–148.

116. Lange, B. M., Bachi, A., Wilm, M., and Gonzalez, C. **2000**. Hsp90 is a core centrosomal component and is required at different stages of the centrosome cycle in *Drosophila* and vertebrates. EMBO J *19*, 1252–1262.

117. de Carcer, G., do Carmo Avides, M., Lallena, M. J., Glover, D. M., and Gonzalez, C. **2001**. Requirement of Hsp90 for centrosomal function reflects its regulation of Polo kinase stability. EMBO J *20*, 2878–2884.

118. Sibon, O. C., Kelkar, A., Lemstra, W., and Theurkauf, W. E. **2000**. DNA-replication/DNA-damage-dependent centrosome inactivation in *Drosophila* embryos. Nature Cell Biol *2*, 90–95.

119. Dasika, G. K., Lin, S. C., Zhao, S., Sung, P., Tomkinson, A., and Lee, E. Y. **1999**. DNA damage-induced cell cycle checkpoints and DNA strand break repair in development and tumorigenesis. Oncogene *18*, 7883–7899.

120. Sullivan, W., Daily, D. R., Fogarty, P., Yook, K. J., and Pimpinelli, S. **1993**. Delays in anaphase initiation occur in individual nuclei of the syncytial *Drosophila* embryo. Mol Biol Cell *4*, 885–896.

121. Taylor, W. R., and Stark, G. R. **2001**. Regulation of the G2/M transition by p53. Oncogene *20*, 1803–1815.

122. Takada, S., Kelkar, A., and Theurkauf, W. E. **2003**. *Drosophila* checkpoint kinase 2 couples centrosome function and spindle assembly to genomic integrity. Cell *113*, 87–99.

123. Endow, S. A., and Komma, D. J. **1996**. Centrosome and spindle function of the *Drosophila* Ncd microtubule motor visualized in live embryos using Ncd-GFP fusion proteins. J Cell Sci *109*(Pt 10), 2429–2442.

124. Sharp, D. J., Yu, K. R., Sisson, J. C., Sullivan, W., and Scholey, J. M. **1999**. Antagonistic microtubule-sliding motors position mitotic centrosomes in *Drosophila* early embryos. Nature Cell Biol *1*, 51–54.

125. Sharp, D. J., Brown, H. M., Kwon, M., Rogers, G. C., Holland, G., and Scholey, J. M. **2000**. Functional coordination of three mitotic motors in *Drosophila* embryos. Mol Biol Cell *11*, 241–253.

126. Sharp, D. J., Rogers, G. C., and Scholey, J. M. **2000**. Cytoplasmic dynein is required for poleward chromosome movement during mitosis in *Drosophila* embryos. Nature Cell Biol *2*, 922–930.

127. Goshima, G., and Vale, R. D. **2003**. The roles of microtubule-based motor proteins in mitosis: comprehensive RNAi analysis in the *Drosophila* S2 cell line. J Cell Biol *162*, 1003–1016.

128. Somma, M. P., Fasulo, B., Cenci, G., Cundari, E., and Gatti, M. **2002**. Molecular dissection of cytokinesis by RNA interference in *Drosophila* cultured cells. Mol Biol Cell *13*, 2448–2460.

129. Carvalho, P., Tirnauer, J. S., and Pellman, D. **2003**. Surfing on microtubule ends. Trends Cell Biol *13*, 229–237.

130. Inoue, Y. H., do Carmo Avides, M., Shiraki, M., Deak, P., Yamaguchi, M., Nishimoto, Y., Matsukage, A., and Glover, D. M. **2000**. Orbit, a novel microtubule-associated protein essential for mitosis in *Drosophila melanogaster.* J Cell Biol *149*, 153–166.

131. Lemos, C. L., Sampaio, P., Maiato, H., Costa, M., Omel'yanchuk, L. V., Liberal, V., and Sunkel, C. E. **2000**. Mast, a conserved microtubule-associated protein required for bipolar mitotic spindle organization. EMBO J *19*, 3668–3682.

132. Maiato, H., Sampaio, P., Lemos, C. L., Findlay, J., Carmena, M., Earnshaw, W. C., and Sunkel, C. E. **2002**. MAST/Orbit has a role in microtubule-kinetochore attachment and is essential for chromosome alignment and maintenance of spindle bipolarity. J Cell Biol *157*, 749–760.

133. Rogers, S. L., Rogers, G. C., Sharp, D. J., and Vale, R. D. **2002**. *Drosophila* EB1 is important for proper assembly, dynamics, and positioning of the mitotic spindle. J Cell Biol *158*, 873–884.

134. Mogensen, M. M., Malik, A., Piel, M., Bouckson-Castaing, V., and Bornens, M. **2000**. Microtubule minus-end anchorage at centrosomal and non-centrosomal sites: the role of ninein. J Cell Sci *113*(Pt 17), 3013–3023.

135. Dammermann, A., and Merdes, A. **2002**. Assembly of centrosomal proteins and microtubule organization depends on PCM-1. J Cell Biol *159*, 255–266.

136. Casenghi, M., Meraldi, P., Weinhart, U., Duncan, P. I., Korner, R., and Nigg, E. A. **2003**. Polo-like kinase 1 regulates Nlp, a centrosome protein involved in microtubule nucleation. Dev Cell *5,* 113–125.
137. Chang, P., Giddings, T. H., Jr., Winey, M., and Stearns, T. **2003**. Epsilon-tubulin is required for centriole duplication and microtubule organization. Nature Cell Biol *5,* 71–76.
138. Kubo, A., and Tsukita, S. **2003**. Non-membranous granular organelle consisting of PCM-1: subcellular distribution and cell-cycle-dependent assembly/disassembly. J Cell Sci *116,* 919–928.
139. Giansanti, M. G., Gatti, M., and Bonaccorsi, S. **2001**. The role of centrosomes and astral microtubules during asymmetric division of *Drosophila* neuroblasts. Development *128,* 1137–1145.
140. Gonczy, P. **2002**. Mechanisms of spindle positioning: focus on flies and worms. Trends Cell Biol *12,* 332–339.
141. Megraw, T. L., Kao, L. R., and Kaufman, T. C. **2001**. Zygotic development without functional mitotic centrosomes. Curr Biol *11,* 116–120.
142. de Saint Phalle, B., and Sullivan, W. **1998**. Spindle assembly and mitosis without centrosomes in parthenogenetic *Sciara* embryos. J Cell Biol *141,* 1383–1391.
143. Li, K., Xu, E. Y., Cecil, J. K., Turner, F. R., Megraw, T. L., and Kaufman, T. C. **1998**. *Drosophila* centrosomin protein is required for male meiosis and assembly of the flagellar axoneme. J Cell Biol *141,* 455–467.
144. Khodjakov, A., Cole, R. W., Oakley, B. R., and Rieder, C. L. **2000**. Centrosome-independent mitotic spindle formation in vertebrates. Curr Biol *10,* 59–67.
145. Hinchcliffe, E. H., Miller, F. J., Cham, M., Khodjakov, A., and Sluder, G. **2001**. Requirement of a centrosomal activity for cell cycle progression through G1 into S phase. Science *291,* 1547–1550.
146. Schnorrer, F., Luschnig, S., Koch, I., and Nusslein-Volhard, C. **2002**. Gamma-tubulin37C and gamma-tubulin ring complex protein 75 are essential for bicoid RNA localization during drosophila oogenesis. Dev Cell *3,* 685–696.
147. Tavosanis, G., and Gonzalez, C. **2003**. gamma-Tubulin function during female germ-cell development and oogenesis in *Drosophila.* Proc Natl Acad Sci USA *100,* 10263–10268.
148. Cullen, C. F., and Ohkura, H. **2001**. Msps protein is localized to acentrosomal poles to ensure bipolarity of *Drosophila* meiotic spindles. Nature Cell Biol *3,* 637–642.
149. Gergely, F., Draviam, V. M., and Raff, J. W. **2003**. The ch-TOG/XMAP215 protein is essential for spindle pole organization in human somatic cells. Genes Dev *17,* 336–341.
150. Gergely, F., Karlsson, C., Still, I., Cowell, J., Kilmartin, J., and Raff, J. W. **2000**. The TACC domain identifies a family of centrosomal proteins that can interact with microtubules. Proc Natl Acad Sci USA *97,* 14352–14357.
151. Cha, B. J., Serbus, L. R., Koppetsch, B. S., and Theurkauf, W. E. **2002**. Kinesin I-dependent cortical exclusion restricts pole plasm to the oocyte posterior. Nature Cell Biol *4,* 592–598.
152. Clark, I. E., Jan, L. Y., and Jan, Y. N. **1997**. Reciprocal localization of Nod and kinesin fusion proteins indicates microtubule polarity in the *Drosophila* oocyte, epithelium, neuron and muscle. Development *124,* 461–470.
153. Martinez-Campos, M., Basto, R., Baker, J., Kernan, M., Raff, J. W. **2004**. The *Drosophila* pericentrin-like protein is essential for cilia/flagella function, but appears to be dispensable for mitosis. J Cell Biol *165,* 673–683.

14
Centrosome Inheritance during Human Fertilization and "Therapeutic" Cloning: Reproductive and Developmental Diseases and Disorders Caused by Centrosome Dysfunction

C. S. Navara, C. Simerly and G. Schatten

14.1
Introduction

Over a century ago, van Beneden [1] and Boveri [2] discovered that the centrosome is vital for successful fertilization and the beginning of embryonic development. In the 1925 third edition of his pioneering monograph *The Cell in Development and Heredity*, sadly dedicated to the memory of his dear friend Theodor Boveri, E. B. Wilson writes:

> *"The essential postulates of (Boveri's) theory were (1) that the central body ('centrosome') is the fertilizing element proper; (2) it is actually imported into the egg by the sperm; (3) that the cleavage centers arise directly by division of the sperm-center. As outlined by Boveri the theory took the following form: during the somatic divisions the center (centrosome) is continuously handed on by division from generation to generation of cells. This process comes to an end in the mature egg after extrusion of the second polocyte (polar body), when the egg-center degenerates or becomes physiologically ineffective; further cell-division is thus inhibited and the occurrence of parthenogenesis is avoided. The ripe egg possesses all of the elements necessary for development save an active division-center (centrosome). The sperm, on the other hand, possesses such a center but lacks the protoplasmic substratum in which to operate. In this respect the egg and sperm are complementary structures; their union in syngamy thus restores to each the missing element necessary for further development (p. 155 in [3]). Accepting this it follows that the nuclei of the embryo are derived equally from the two parents; the central bodies (centrosomes) are purely of paternal origin; and to this it might be added that the general cytoplasm of the embryo seems to be almost wholly of maternal origin." (From [4] p. 440; parentheses added).*

More than a century ago, Boveri and Van Beneden recognized that the sperm contributes the centrosome, the structure that organizes the mitotic spindle poles (cleavage centers). Even in our time, these observations are profound, extraordinary for their clarity and elegant simplicity; modern centrosome biologists should be humbled by the brilliance of our great-great-grand-professors. While much of the past quarter-century has witnessed significant progress in the molecular dissec-

Centrosomes in Development and Disease. Edited by Erich A. Nigg

ISBN 3-527-30980-2

tion of the centrosome, as well as discoveries from laboratory experiments and natural reproduction, Boveri's theory remains pioneering, even from the vantage of an early 21st century centrosome biologist.

Even the minor criticism of Boveri's postulate on the uniparental origin of the centrosome ("the central bodies (centrosomes) are purely of paternal origin; and to this it might be added that the general cytoplasm of the embryo seems to be almost wholly of maternal origin") can be countered by the possibility of Wilson's slight mistranslation. The term "protoplasm" was coined by Purkinje when he first discovered the germinal vesicle [5], to refer to the unique cytoplasm that is found within an unfertilized egg. The egg's cytoplasm or protoplasm, unlike other cytoplasm, is uniquely capable of supporting the development of the next generation. When Boveri wrote "The sperm, on the other hand, possesses such a center but lacks the protoplasmic substratum in which to operate", perhaps he was indeed aware that while the egg cytoplasm is capable of initiating parthenogenetic development, the sperm centrosome recruits the maternal proteins that are essential for fertilization and early development.

A century after Boveri's theory, human *in vitro* fertilization (IVF) was achieved [6]. Now more than 1 million IVF babies have been born. The discarded specimens from IVF clinics have provided a precious and unique research resource for centrosome biologists, and a new breed of centrosome physician–scientist is emerging. Reproductive mistakes including polyspermy (fertilization by more than one sperm) and parthenogenesis (development beginning in an activated egg without any sperm) subtly challenge aspects of the unipaternal centrosome inheritance theory.

Here, we focus on the centrosome during fertilization, with special attention to human reproduction and development. We also consider the centrosome during nuclear transfer which represents research frontiers for the next generation of centrosome biologists. For obvious ethical reasons, experimental results are obtained and/or corroborated by studying non-human primate development. This is essential, ironically, because fertilization in mouse and other rodents represents rare exceptions to Boveri's theory [7]. In this chapter, we consider:

- Centrosomes during human fertilization
- Centrosome dysfunction as causes of human infertility
- Centrosome functional assays for diagnosing male infertility
- Polyspermy in humans
- "Dispermy hypothesis" for the origins of genomic imprinted disorders
- Maternal centrosome anomalies and birth defects
- Resolving the special problem of parthenogenetic development: roles of cytoplasmic motors and NuMA
- Centrosomes during cloning, and centrosomes in embryonic stem cells derived after nuclear transfer
- Research challenges for centrosome developmental biologists: developmental centrosomopathies

14.2 Centrosomes during Human Fertilization

Centrosome inheritance during human fertilization [7–10] mirrors the inheritance pathway found in most animals (reviewed in [7]). Simply stated, the human sperm contributes the proximal centriole, which recruits egg proteins including γ-tubulin, centrin, pericentrin, NuMA, and microtubule motors, to the sperm centriolar complex. Within the fertilized egg, the now enlarged "sperm centrosome", which more properly should be called the "zygote centrosome", nucleates microtubules that assemble the first microtubule-based structure in the fertilized egg – the sperm aster. The sperm aster in fertilized human oocytes (Figure 14.1A) is the typical radially-arrayed monaster juxtaposed to the sperm nucleus (which is called the "male pronucleus"; Figure 14.1B, M) after the sperm chromatin has decondensed within the egg cytoplasm. As in most animal eggs, the sperm tail enters the egg and one or two punctate structures, detectable with γ-tubulin imaging, are found at the center of the sperm aster exactly at the junction between the sperm axoneme and the male pronuclear surface (Figure 14.1C and D). These have been shown by Sathanathan et al. [11] to be the sperm centriole(s).

Additional evidence supporting the sperm contribution of the centrosome and centriolar complex in humans comes from studies on polyspermic fertilization and parthenogenesis [8, 9, 12]. As shown in Figure 14.1E, because two sperm enter the oocyte (i. e. dispermic fertilization), each paternal centrosome organizes a sperm aster at the base of the sperm head (Figure 14.1F, M). Conversely, artificial activation of oocytes (i. e. parthenogenesis), in which no contribution of the paternal centrosome is provided, leads to a random, disarrayed microtubule pattern at interphase, as is shown in this activated oocyte in which the sperm failed to penetrate (Figure 14.1G and H). Collectively, these data reinforce the observation that the centrosome is paternally inherited in humans [7, 9].

14.3 Centrosome Dysfunction as Causes of Human Infertility

The inheritance of the centrosome during human reproduction has crucial implications for the diagnosis and treatment of human infertility, especially male-factor infertility. Male infertility may be the only example in medicine in which one patient carries a disorder (e. g. defective sperm centrosome), but another (his wife or partner) undergoes the surgical procedure – and perhaps without enjoying a successful outcome.

During the past decade, several teams of investigators around the world have made seminal discoveries concerning the pattern of centrosome inheritance in human and non-human primate fertilization [8, 9, 10, 13]. For obvious reasons, these studies were all carried out in gametes discarded from human infertility clinics employing assisted reproductive technologies (ART). Because couples seek ART treatment for reasons including male-factor infertility, certain forms of

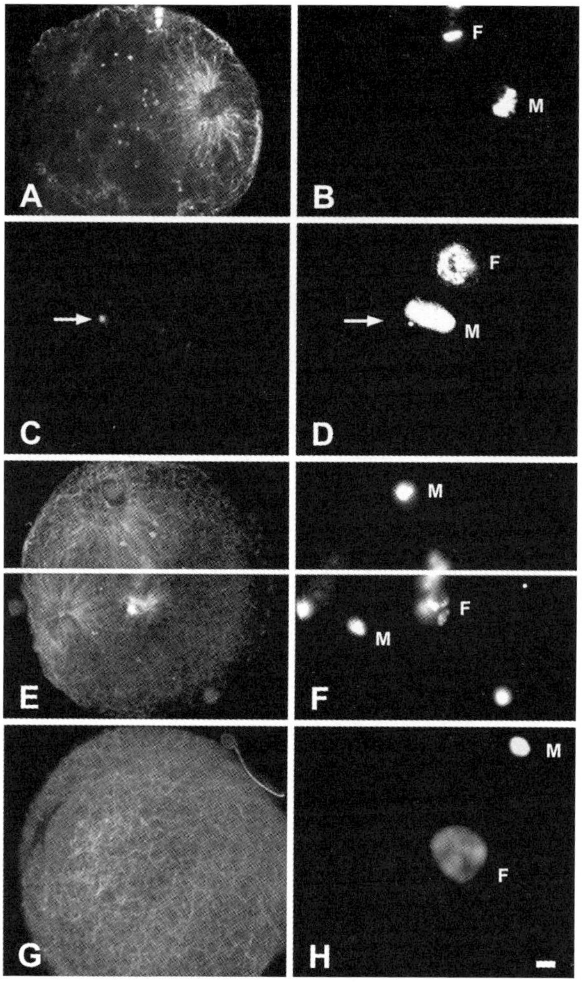

Figure 14.1 Microtubule organization during human fertilization, dispermy and parthenogenesis. The incorporated sperm nucleates a radially-arrayed monaster (A) adjacent to the male pronucleus (B) in fertilized human oocytes. At the base of the aster the centrosomal protein γ-tubulin is observed as one or two tight foci (C). Dispermy results in a sperm aster being organized by each paternal centrosome (E) whereas activation without sperm penetration results in a random, disarrayed microtubule pattern at interphase (Figure 14.1G). A, E, G, = microtubules; C = γ-tubulin and B, D, F, H = DNA. Arrow, γ-tubulin foci; M, paternal DNA; and F, maternal DNA. All images reprinted with permission from Simerly et al. [9], except C and D which are from [12]. Bar = 10 μm.

fertilization failures found by ART clinics appear to result from defects in the sperm centrosome. Centrosome dysfunction during human reproduction, in which the sperm enters the oocyte but the zygote is unable to divide, is being recognized as a new cause of male-factor infertility (Figure 14.2) [9, 13–17]. During human fertilization, if the assembled sperm centrosome is dysfunctional in microtubule assembly or organization, the sperm aster fails to form or does so in a manner inconsistent with the ability to promote pronuclear apposition. Very small asters (Figure 14.2A) may be observed adjacent to incorporated sperm heads (Figure 14.2B) which fail to activate the oocyte and stimulate the exit from metaphase arrest. Instead of a single sperm aster the incorporated sperm may nucleate multiple asters (Figure 14.2C, D), or may fail to organize the microtubules after sperm incorporation resulting in a microtubule pattern reminiscent of partheno-

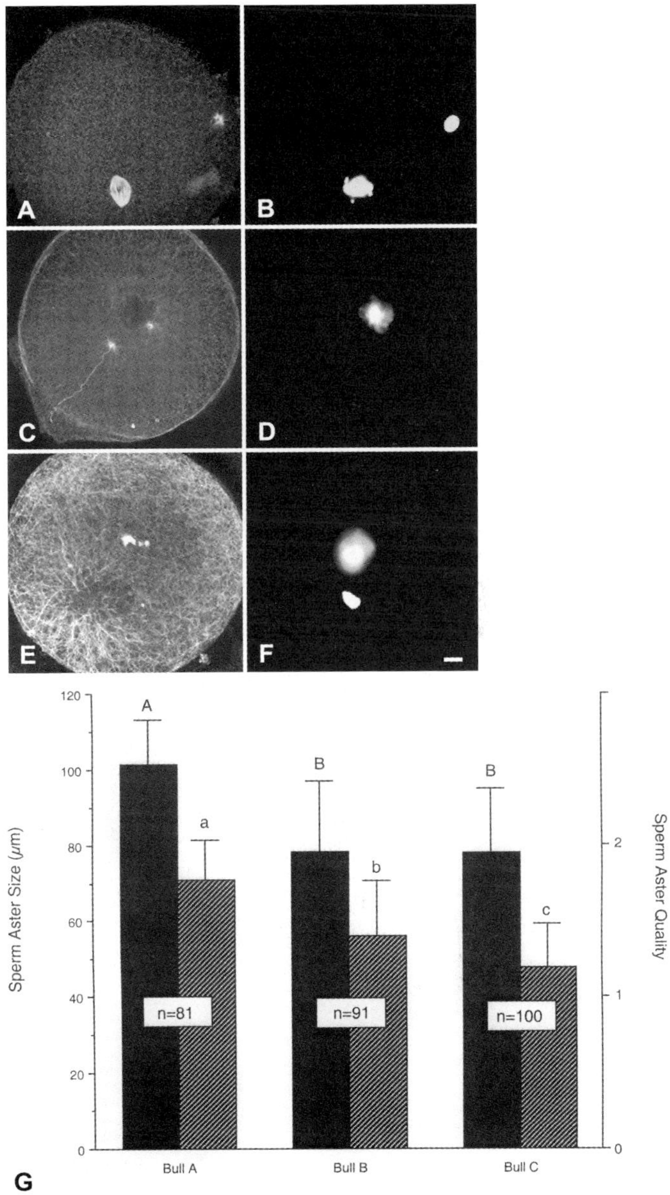

Figure 14.2 Phenotypic variations among paternal centrosomes. A small aster or asters (A) adjacent to the incorporated sperm head (B) is observed in oocytes remaining at metaphase. A single sperm (D, F) may nucleate multiple asters (C), or may fail to organize the microtubules resulting in a disorganized microtubule pattern (E). (G) Comparison of sperm aster size and quality in three bulls of known fertility. Diameter of the sperm aster at its largest plane was measured using the confocal microscope (left axis). A quality score was also given to each aster (right axis). The bull with the highest field fertility and *in vitro* fertility (bull A) also had the largest and best organized sperm asters with averages of 101.4 and 1.8 µm respectively. Bull B had an average sperm aster diameter of 78.2 µm and an aster quality score of 1.4. The bull with the worst *in vitro* fertility (bull C) had the smallest (77.9 µm) and most poorly organized sperm asters (1.2). These data represent five repetitions: comparisons were made using the protected means of the least squares method. Different letters indicate significant differences ($p = 0.025$); bars indicate standard error. A, C, E = microtubules; B, D, F = DNA. Images A–E reprinted with permission from Asch et al. [18]. Graph reprinted from Navara et al. [37]. Bar = 10 µm.

genesis (Figure 14.2E, F). Such oocytes fail to complete the fertilization process and arrest in early development [9, 18, 19].

The clinical and fundamental importance of these observations in humans has been reinforced by animal studies on the phenotypic expression of centrosomes in bulls [20]. Using randomized bovine oocytes and sperm from bulls proven in the field or after *in vitro* fertilization to be superb, average, or sub-fertile, the organization and size of the sperm aster was shown to vary according to the father (Figure 14.2G). This suggests that the quality or quantity of the sperm centrosome has a direct affect on the success and speed of fertilization and is correlated with the frequency of live births. It is possible that there are variations in centrosomal vigor as is found in other inherited components.

14.4 Centrosome Functional Assays for Diagnosing Male Infertility

Centrosome reconstitution is a multi-step process occurring between the end of second meiosis and the transition into interphase of the first cell cycle. Microtubule nucleation and organization capabilities must function properly and quickly to form the sperm aster that directs pronuclear migration. This is a critical step in accurately completing the fertilization process, defined as when the male and female genomes can intermix at first mitotic metaphase [7]. A functional centrosome also defines the site of first bipolar mitotic spindle assembly within the activated cytoplasm and participates in spindle organization by serving as a dominant microtubule organizing center (MTOC) at the spindle poles.

Centrosome reconstitution during fertilization is a unique model for exploring the molecular components necessary for determining centrosome parental origin and function [7, 12]. Pioneering work in cell-free systems in *Xenopus laevis* have successfully explored the molecular events leading to centrosome reconstitution and microtubule assembly [21–23]. These studies demonstrated how exposure of demembranated *X. laevis* sperm exposed to *X. laevis* cytostatic factor (CSF)-arrested egg extracts leads to the binding of vital microtubule nucleating components such as γ-tubulin and MPM-2 phosphorylated epitopes. These sperm centrosomes thus become competent for nucleating microtubule growth into sperm asters *in vitro* [22]. Furthermore, this process is microtubule and microfilament independent, but egg extract and ATP dependent [22].

Analysis of human and bovine sperm in *X. laevis* CSF-arrested extracts provides the basis for studying the assembly of a zygotic centrosome capable of nucleating and organizing microtubules *in vitro* [12, 24]. Mammalian sperm exposed to increased calcium levels, plasma membrane destabilization, and disulfide bond reduction unveils paternal γ-tubulin and other centrosomal protein binding sites, concomitant with the onset of pronuclear decondensation. This "procentrosome" structure is thus "primed" to attract and bind maternal γ-tubulin from the egg's cytoplasmic pool. Conversely, other paternal centrosomal proteins predicted to be critical for the reorganization of the sperm centrosomal complex following insemination (i. e. centrin; [25–27]) are modified following exposure to the egg's cytoplasm. Exposure to an elevated kinase activity within the meiotic cytoplasm then shifts the microtubule dynamics to a state conducive with nucleation and polymerization.

Clinical assays can now be posed based on the initial molecular characterization of the human sperm centrosome (reviewed in [7]). In addition to studies dissecting the relative contributions of the male and female constituents to the human zygotic centrosome, microtubule assembly *in vitro* from the sperm centrosome can be assayed using *Xenopus* egg extracts in combination with polymerization-competent rhodamine-tagged tubulin protein [12]. The advantage of such an assay is that it can be a prospective test to investigate sperm microtubule nucleation ability in the sperm of men of varying fertility *in vitro*, prior to assigning couples to the arduous procedures of ART methods such as intracytoplasmic sperm injection (ICSI).

Examining microtubule assembly and centrosome functioning after microinjecting human sperm into mature bovine oocytes is also a potentially useful prospective centrosomal assay system [15]. The major advantage of this test is the ability to observe pronuclear apposition mediated by a functional human sperm centrosome within a living egg, as opposed to just microtubule assembly as currently observed in egg extract models. Past research has shown that the recipient oocyte must be from a species that follows the paternal inheritance of the centrosome, which thus excludes rodent oocytes. For instance, the zona-free hamster oocyte sperm penetration test, a commonly used mammalian oocyte for assaying human male infertility, is a uniquely inappropriate model for the investigation and diagnosis of impaired sperm centrosome function of human sperm [28]. Instead, oocytes from animals like the rabbit or cow, which support paternal centrosomal functioning, are more relevant models to investigate human centrosome reconstitution and sperm aster formation [15, 29].

Since defective centrosomes can be responsible for fertilization arrest and initiate certain types of male infertility, centrosome microinjection therapy has been proposed as a method to correct this defect [7, 8]. However, only centrosomes introduced from intact sperm prove capable of completing the fertilization process and correctly segregate their chromosomes at cell division [30], indicating that the position of the centrosome is a critical parameter for completion of fertilization.

14.5
Polyspermy in Humans

Polyspermy represents an experimental test of the relative parental contributions to the centrosome, since the paternal contribution is multiplied. In many animal systems, polyspermy introduces two centrosomes that duplicate and separate at mitosis to form a tetrapolar spindle. Typically, dispermic insemination leads to aneuploid embryos because the triploid chromosome complement is abnormally divided into the resultant four blastomeres at the end of cell division. Such evidence has provided the rationale that the inheritance of only one centrosome is critical to forming bipolar spindles that can accurately segregate the chromosomes [31, 32].

Evidence from polyspermy experiments in mammals, however, argues against this central dogma on the universal contribution of the sperm centrosome. First, mice violate the notion of paternal centrosome contribution at fertilization since both the distal and proximal sperm centrioles are highly degenerated in the mature sperm (reviewed in [33]) and no sperm aster is detected at the base of the sperm head following sperm incorporation [34]. Furthermore, dispermic or trispermic mouse zygotes do not display sperm asters, and they go on to divide from one to two [35], suggesting that in this system the sperm do not contribute the dominant centrosomal foci. Secondly, most other mammals, including marsupials [36], cows [37], sheep [38], pigs [39], rabbits [15], monkeys [40], and humans [9, 41, 42] form supernumerary sperm asters after polyspermy. But, as shown in the study of human fertilization, such dispermic zygotes in these mammals may divide from one cell into two, three, or even four [41–43]. Analysis of dispermic human and rhesus zygotes at mitosis (Figure 14.3) demonstrates the assembly of bipolar metaphase spindles in the presence of supernumerary centrosomes, consistent with other reports on human polyspermy [8, 10, 41]. Tripolar or tetrapolar mitotic spindles were rarely observed in dispermic zygotes; instead, two γ-tubulin foci for each incorporated sperm are observed at first mitosis. In the case of dispermy this results at prophase of first mitosis in a disorganized multipolar spindle (Figure 14.3A) with four foci of γ-tubulin (Figure 14.3B). This disorganized struc-

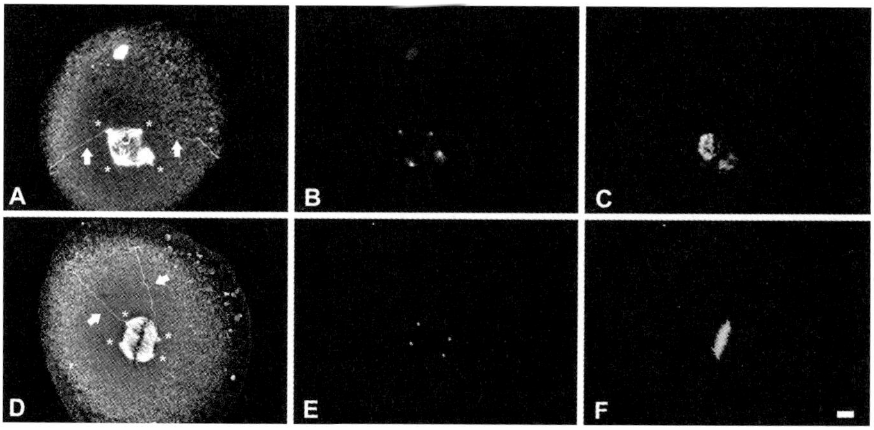

Figure 14.3 Mitotic spindle formation in dispermic human zygotes. Dispermy in human zygotes results in a disorganized prophase mitotic spindle (A) with four foci of γ-tubulin (B; location denoted by asterisks in A). This disorganized prophase spindle resolves at metaphase resulting in a bipolar spindle (D) with well-organized chromosomes (F). Four foci of γ-tubulin can still be observed on the spindle (E; location denoted by asterisks in D). A, D = microtubules; B, E = γ-tubulin; and C, F = DNA. Arrows indicate incorporated sperm axonemes and asterisks highlight γ-tubulin foci. Reprinted with permission from Simerly et al. [12]. Bar = 10 μm.

ture surprisingly resolves itself at metaphase of mitosis resulting in a bipolar spindle (Figure 14.3D) with well-organized chromosomes (Figure 14.3F) dividing from one to two. The four foci of γ-tubulin are still observed on the spindle (Figure 14.3E). These observations in humans may well be due to requirements of non-centrosomal components (i. e. molecular motors and spindle matrix proteins) necessary for bipolar spindle assembly in somatic cells. As discussed in a subsequent section, analysis of parthenogenetic development in primate oocytes supports this view.

14.6 "Dispermy Hypothesis" for the Origins of Genomic Imprinted Disorders

Since the now-infamous report of the Fol brothers [44] on the "Quadrille of Centers", dispermy holds a special place in the hearts and minds of centrosome biologists. The first descriptions of the mitotic apparatus at first division referred to tetrapolar structures, and the Fols believed that both the sperm and the egg each contributed two sets of centrosomes, thereby generating the tetrapolar structure (see Figure 14.3D: dispermic human oocyte). Wilson [4], in a stinging publication, recognized that the "quadrille" was the result of dispermy – not monospermy – and that two sperm each contribute two centrosomes so that four mitotic poles are found at first mitosis after dispermy.

Dispermic fertilizations in humans are observed frequently (see Figure 14.3D) [43], but unlike sea urchins, first cleavage may result in two, three, or rarely four blastomeres. Diandric triploidy is one of the consequences, although a variety of other mis-segregations leading to chromosome mosaicism can also result. 2N/3N mixoploidies have resulted in live births with a variety of developmental disorders [45–47] (reviewed in [48–50]). Diploid sperm, the result of male meiotic error, can also produce triploidy [51, 52].

Golubovsky [50] presents a hypothesis stating that postzygotic diploidization of triploid embryos may be the origin of disorders resulting from genomic imprinting errors. In mammals, the paternal and maternal chromosomes are specifically modified by DNA-methylation of "imprinted" genes, so that certain maternal genes are silenced, as are complementary paternal genes. Angelman (AS) and Prader–Willi syndromes (PWS; reviewed [53–56]) are uniparental disomy disorders (UPD) with a spectrum of developmental, neurological, and behavioral consequences. AS results from the loss of two maternally-expressed genes, whereas PWS is the result of the reciprocal loss of two paternally-expressed genes.

Since the centrosomes in human zygotes robustly form bipolar spindles even in the case of dispermy, it is possible that daughter blastomeres after cell division inherit two or more paternal or maternal chromosome sets. While enormous strides have been made in understanding the molecular consequences of these imprinting errors in both humans and now mouse models, the origins for uniparental disomies remain mysterious.

14.7
Maternal Centrosome Anomalies and Birth Defects

The aging of the maternal centrosome is perhaps the primary cause of reproductive failure in older women [57], since the rates of aneuploidy in pre-implantation human embryos has been reported to be as high as 52–61 % [58]. The majority of these aneuploidies result from errors in female meiosis, particularly meiosis I. This has led to the clinically useful FISH analysis of the first polar body [59], the product of first meiosis, as a preconception genetic diagnostic (in contrast to the more typical embryonic blastomere test: pre-implantation genetic diagnosis or PGD; [60–62]).

Triploidy has been estimated to occur at ~ 1 % in humans; greater than 10 % of spontaneous miscarriages are triploid. The majority of triploidies are due to errors in female meiosis, especially in polar body extrusion. However, a significant number of triploidies result from accurate female meiosis and dispermic fertilization.

Chromosome non-disjunction is well known to result in tragic birth defects [63, 64]. The majority of these errors occur in female meiosis, and first meiosis appears particularly prone since the stringency of metaphase–anaphase checkpoint controls is very low. While the precise phase at which mammalian oocytes reduce and lose their traditional, replicating centrosomes is not well known, extrapolations from invertebrates [65–67] suggest that the maternal centrosome is reduced before the completion of the meiotic divisions [68].

In humans, all the mitotic divisions of the oogonia are completed by mid-gestation within female fetuses, i. e. 4–5 months prior to birth. As reproduction cannot start for more than a decade and the potential of eggs from women older than 35 years declines swiftly, the nature of the maternal centrosome and its stability over decades in these arrested oocytes is of both fundamental and clinical importance. With the advent of oocyte donation, women well beyond menopause [69] can deliver healthy children – but only if the oocyte is obtained from a younger woman.

For the centrosome biologist, questions persist regarding the capacity of the maternal centrosome inside an oocyte (which is reproductively viable) one-third of a century old versus one that is half-century old (which produces aberrant meiotic spindles and generates aneuploid embryos incapable of either implantation or embryogenesis). In addition, it is not yet clear why the nature of female meiosis is so different from that of male meiosis. Female meiosis in mammals begins *in utero*, is arrested until puberty, and then once per month, is reinitiated in a small number of arrested oocytes until they are depleted and menopause ensues. In contrast, mitosis in the male germ line [70] restarts at puberty with continuous waves of male meiosis as well as on-going proliferation of male germ cells. This permits the remarkable ability to generate transgenic sperm as well as heterologous sperm (e. g. rat sperm in mouse testes) by male germ cell transplantation [70, 71].

14.8 Resolving the Special Problem of Parthenogenesis: Roles of Cytoplasmic Motors and NuMA

Centrosomes have traditionally been described as critical for spindle microtubule nucleation and spindle bipolarity (reviewed in [32]). As exemplified in rodents, the dominant microtubule organizing center (MTOC) is derived strictly from maternal sources and remains active during meiosis as well as after insemination for directing the motility events crucial to the completion of the fertilization process in these species [35, 72] (reviewed in [7]). However, for all other mammals, the theory of uniparental contribution of the centrosome is observed and reinforced by the presence of multiple sperm asters during polyspermic insemination, as well as by the lack of cytoplasmic asters after parthenogenetic activation. Most species that follow a paternal method of centrosome inheritance also support parthenogenic activation (reviewed in [7]), often leading to the development of cleavage stage embryos [73–76].

Immunostaining and serial transmission electron microscopy (TEM) investigations on spindle pole organization in unfertilized primate and bovine oocytes show a lack of any maternal centrosomes at their poles, although they still form functional bipolar meiotic spindles [10, 77, 78]. These findings in mature oocytes are consistent with prior reports suggesting that the loss or masking of the maternal centrosome occurs before the onset of meiosis in mammals [68]. However, parthenogenetic rabbit blastocysts display *de novo* centrioles, structures not observed in early development in this species [73, 79, 80]. This poses a dilemma of how the maternal centrosome might be restored in the unfertilized egg in a manner that permits it to duplicate and split accurately for assembling bipolar mitotic spindles after parthenogenetic activation [7].

During human [9] rhesus (Figure 14.4A), and bovine parthenogenesis [20] no sperm astral microtubules are nucleated. Disarrayed cytoplasmic microtubules are formed, mainly cortically derived, but these microtubules do not participate in any nuclear positioning event within the cytoplasm. However, a bipolar spindle is formed at the time of first mitotic metaphase, though its spindle poles lack specific centrosomal foci containing γ-tubulin or pericentrin (Figure 14.4C and D). This suggests that the maternal centrosome is not necessarily resurrected after activation of the mature oocyte to initially assemble a mitotic bipolar spindle [7].

Recent evidence suggests that meiotic and mitotic spindles maintain the ability to self-assemble and direct cell division without functional MTOCs [78, 81–85]. The opposing activities of Eg5 and HSET kinesins have been implicated in bipolarity establishment in somatic cells by maintaining centrosome separation during mitosis, a property closely tied to bipolar spindle formation in many somatic cells, and cross-linking both parallel and antiparallel microtubules to produce opposing forces within the spindle lattice [86–90]. Analysis of the spindle kinesin proteins Eg5 and HSET during mouse meiotic maturation demonstrates their presence in meiotic spindles while microinjection of function-blocking antibodies to Eg5 and HSET showed their requirement for bipolar spindle assembly in this system

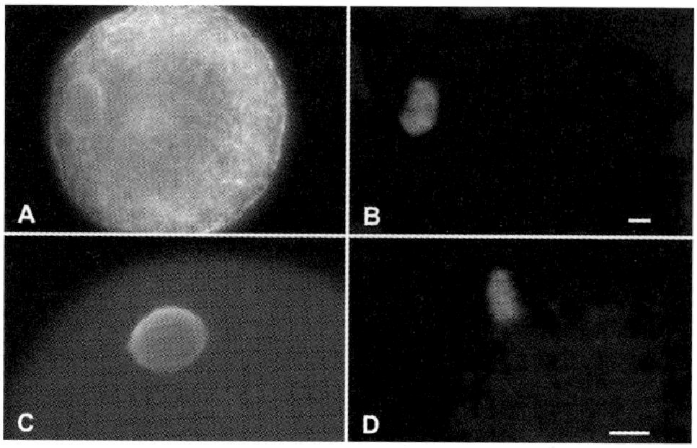

Figure 14.4 Parthenogenesis in rhesus monkey oocytes. Parthenogenetically activated rhesus oocytes display disorganized microtubules during interphase (A) with no preferential association with the female pronucleus (B). At mitosis, however, a bipolar well-organized spindle is formed (C) centering the chromosomes on the metaphase plate (D). A, C = microtubules; B, D = DNA. Reprinted with permission from Wu et al. [40]. Bar = 10 μm.

[90]. In addition, these kinesin proteins along with the spindle matrix component, NuMA (nuclear mitotic apparatus protein), have been identified in primate oocytes, zygotes, and parthenogenotes [91]. Microinjection studies using function blocking antibodies to these kinesins and NuMA have shown that these proteins play critical roles in spindle assembly in unfertilized primate oocytes and mitotic parthenogenotes (C. Simerly and G. Schatten, unpublished observations). Similar crucial roles for minus end-directed kinesin motors in organizing female meiotic spindles have been described for *Drosophila* mutants deficient in the minus-end kinesin, non-claret disjunctional (ncd) [92]. Taken together, these data suggest that the centrosome may be dispensible for meiotic and first mitotic spindle formation after parthenogenetic activation in primate eggs. Rather, it is the combined action of microtubule cross-linking and oppositely oriented motor activity that is required for spindle bipolarity in early primate eggs [90, 93].

14.9
Centrosomes during Cloning, and Centrosomes in Embryonic Stem Cells Derived after Nuclear Transfer

Nuclear transfer (NT) cloning, sporadically successful in mammals [94–97], challenges fundamental tenets of developmental biology, including the precise requirement for exactly two parents of opposite sexes during natural reproduction, and its mechanisms still remain inexplicable. Unlike fertilization, when two haploid genomes unite within the activated egg's cytoplasm as directed by the introduced sperm centrosome, the maternal chromosomes on the meiotic spindle are removed during NT cloning, a diploid nucleus is inserted, and the egg is activated as during parthenogenesis. By adapting NT-cloning procedures, successful in domestic species and mice, to non-human primates, however, unexpected problems have been identified in the extranuclear inheritance of the centrosome.

Imaged live, monkey NT-cloned embryos appear normal, yet no pregnancies resulted from 33 embryo transfers into 16 surrogates [91]. Microtubule and DNA immunostaining of NT-cloned embryos at interphase demonstrated multiple nuclei and inappropriate numbers of centrosomes during interphase compared to fertilized controls and bovine clones produced by similar NT techniques (Figure 14.5A) [37, 98]. Malformed first mitotic spindles after nuclear transfer were also observed using a variety of somatic or embryonic cell types transferred into enucleated rhesus oocytes [91]. The majority of mitotic spindles in rhesus cloned embryos was multipolar and had misaligned chromosomes (Figure 14.5A). Despite the assembly of dysfunctional mitotic spindles, cleavage continues but with unequal chromosome segregation, giving rise to aneuploid embryos. These results are distinct from the normal bipolar mitotic spindles observed in cows after nuclear transfer using similar methods (Figure 14.5A, inset) [37]. The abnormal failure rate of correctly assembled bipolar mitotic spindles in rhesus clones is surprising, given the ability of rhesus polyspermic and parthenogenetic eggs to assemble bipolar spindles with aligned chromosomes (Figure 14.4C). How-

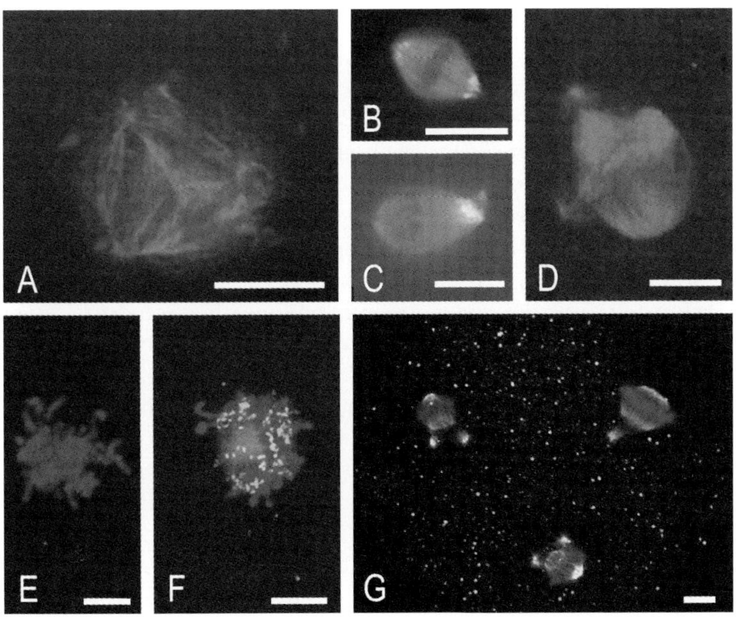

Figure 14.5 Faulty mitotic spindles produce aneuploid embryos after primate nuclear transfer. (A) Defective NT mitotic spindle with misaligned chromosomes. Centrosomal NuMA at meiosis (B) and mitosis (C), but not in mitotic spindles after NT (D). The centrosomal kinesin HSET is also missing after NT (E), but not centromeric Eg5 (F). Bipolar mitotic spindles with aligned chromosomes and centrosomal NuMA after NT into fertilized eggs (G). Blue, DNA; red, B-tubulin; green, NuMA in B, C, D, and G; HSET in E; and Eg5 in F. Reprinted with permission from Simerly et al. [91]. Bars = 10 µm (see Color Plates page XXXII).

ever, evidence suggests that primate NT cloning fails due to imbalances between the chromosome sets, spindle pole numbers, microtubule-based molecular motors, and the acquired somatic cell centrosome [91].

NuMA concentrates exclusively at the spindle poles in unfertilized meiotic and fertilized mitotic oocytes (Figure 14.5B and C), but is not detected in cloned mitotic spindles after nuclear transfer (Figure 14.5D). Likewise, the M-phase kinesin motor HSET, which is routinely detected on meiotic and mitotic spindle poles [90], is not detected on NT-cloned spindles (Figure 14.5E). Conversely, the oppositely-oriented kinesin motor Eg5 binds centromere pairs in meiotic and mitotic spindles, including those on NT-cloned mitotic spindles (Figure 14.5G). Collectively, these data suggest that both NuMA and HSET are disrupted from participating in mitotic bipolar spindle assembly, perhaps because the majority of these proteins are removed with the meiotic spindle during the initial enucleation step [91]. Supporting evidence for meiotic spindle removal as the source for NT-clone mitotic errors is found in experiments combining fertilization with nuclear transfer. In such instances, the resulting tetraploid spindles are assembled with properly aligned chromosomes on bipolar spindles and NuMA detected at the spindle poles (Figure 14.5G).

Primate nuclear transfer appears challenged by molecular requirements in assembling the first mitotic spindle which appear more stringent than those in mammals where nuclear transfer succeeds. In cattle, the somatic centrosome transferred during nuclear transfer organizes a large, well-formed microtubule aster within the recipient's cytoplasm [37], and mice rely exclusively on the oocyte's maternal MTOCs [7]. Successful reproductive cloning is achieved routinely only in those well-studied species that provide vast numbers of both oocytes and surrogates, systems often capable of multiple deliveries and amenable to transfers of supernumerary embryos.

Recent success in deriving a pluripotent human embryo stem cell line from a human blastocyst generated by somatic cell nuclear transfer [99] underscores the challenges in centrosome biology during cloning. While this accomplishment may well pave the path to embryonic stem cell derivations after cloning, the single successful line derived from the 242 oocytes was suggested by the authors to be due in part to the aneuploidy generated by centrosome protein loss during enucleation.

14.10
Research Challenges for Centrosome Developmental Biologists: Developmental Centrosomopathies

Reproduction is error-prone: clinically, fewer than one-quarter of natural conceptions succeed [100]. Even in favorable systems, nuclear transfer is at least 10-fold worse. Any number of defects can account for nuclear transfer failures [101]: e. g. nuclear reprogramming; cell cycle asynchrony; gene misexpression during development; genomic imprinting errors; placental dysfunction; and technical damage. Included in this growing list are epigenetic obstacles like cytoplasmic incompatibilities in mitotic spindle assembly. Notwithstanding their seeming normalcy as imaged live and non-invasively, oocytes reconstituted after primate nuclear transfer have limited reproductive potential due to aneuploidy. A greater understanding of these abnormalities may facilitate the production of genetically identical nonhuman primate models for the study of human diseases and for assessing the promise of innovative stem cell therapies [102], goals which cannot be met with current NT-cloning approaches due to chromosome imbalances resulting from spindle defects.

As the process that spans the fields of reproductive and developmental biology, a full understanding of the cellular and molecular events during fertilization is critical. Centrosomal biologists are progressing rapidly in determining the molecular constituents that comprise the centrosome and the minimal structure needed to promote microtubule nucleation from this structure. As described throughout this book, terrific advances in understanding centrosomal inheritance, composition, assembly, duplication, and separation in a variety of cell types are providing important insights into how the centrosome mediates intracellular motility events and the many other cellular processes linked to centrosomal activities. The importance of such advances in basic centrosomal biology has now translated to molecular medicine, where clinical problems ranging from sophisticated infertility treatments [7] and contraception, hinge on a more complete knowledge underlying this crucial cellular structure. Questions remain about gametogenesis and whether the process has placed special restrictions on centrosome restorative phases in each gamete, especially after injecting immature spermatogenic cells (spermatids, spermatid nuclei, spermatocytes) for the clinical treatment of male infertility (reviewed in [7]).

References

1. Van Beneden PJ. *Commensaux et les Parasites dans le Règne Animal* **1876** D. Appleton and Co.: New York.
2. Boveri T. **1901**. *Zellen-Studien: Ueber die Natur der Centrosomen. IV.* Fischer: Jena, Germany.
3. Boveri T. **1887**. *Ueber die "Befruchtung der Eier von* Ascaris megalocephala". Fischer: Jena, Germany.
4. Wilson EB. **1925**. *The Cell in Development and Heredity* vol. 3. Macmillan: New York.
5. Purkinje JE. **1840**. "Uber die Analogien in den Strukturelementen des thierischen and pflanzlichen Organismus", *Übersicht der Arbeiten und Veränderungen der Schlesischen Gesellschaft für Vaterländische Cultur* im Jahre 1839, 81–82, Breslau.
6. Steptoe PC, Edwards RG. **1978**. Birth after the reimplantation of a human embryo. *Lancet* **2**:366.
7. Schatten G. **1994**. The centrosome and its mode of inheritance: the reduction of the centrosome during gametogenesis and its restoration during fertilization *Dev Biol* **165**:299–335.
8. Palermo GD, Colombero LT, Rosenwaks Z. 1997. The human sperm centrosome is responsible for normal syngamy and early embryonic development. *Rev Reprod* **2**:19–27.
9. Simerly C, Wu GJ, Zoran S, Ord T, Rawlins R, Jones J, Navara C,

Gerrity M, Rinehart J, Binor Z. **1995**. The paternal inheritance of the centrosome, the cell's microtubule-organizing center, in humans, and the implications for infertility. *Nature Med* **1**:47–52.

10. Sathananthan AH. **1997**. Mitosis in the human embryo: the vital role of the sperm centrosome (centriole). *Histol Histopathol* **12**:827–856.
11. Sathananthan AH, Kola I, Osborne J, Trounson A, Ng SC, Bongso A, Ratnam SS. 1991. Centrioles in the beginning of human development. *Proc Natl Acad Sci USA* **88**:4806–4810.
12. Simerly C, Zoran SS, Payne C, Dominko T, Sutovsky P, Navara CS, Salisbury JL, Schatten G. **1999**. Biparental inheritance of gamma-tubulin during human fertilization: molecular reconstitution of functional zygotic centrosomes in inseminated human oocytes and in cell-free extracts nucleated by human sperm. *Mol Biol Cell* **10**:2955–2969.
13. Van Blerkom J. **1996**. Sperm centrosome dysfunction: a possible new class of male factor infertility in the human. *Mol Hum Reprod* **2**:349–354.
14. Hewitson L, Simerly C, Schatten G. **1997**. Inheritance defects of the sperm centrosome in humans and its possible role in male infertility. *Int J Androl* **20**(Suppl. 3):35–43.
15. Terada Y, Simerly CR, Hewitson L, Schatten G. **2000**. Sperm aster formation and pronuclear decondensation during rabbit fertilization and development of a functional assay for human sperm *Biol Reprod* **62**:557–563.
16. Sutovsky P, Terada Y, Schatten G. **2001**. Ubiquitin-based sperm assay for the diagnosis of male factor infertility. *Hum Reprod* **16**:250–258.
17. Rawe VY, Terada Y, Nakamura S, Chillik CF, Olmedo SB, Chemes HE. **2002**. A pathology of the sperm centriole responsible for defective sperm aster formation, syngamy and cleavage. *Hum Reprod* **17**:2344–2349.
18. Asch R, Simerly C, Ord T, Ord VA, Schatten G. **1995**. The stages at which human fertilization arrests: microtubule and chromosome configurations in inseminated oocytes which failed to complete fertilization and development in humans. *Hum Reprod* **10**:1897–1906.
19. Rawe VY, Olmedo SB, Nodar FN, Doncel GD, Acosta AA, Vitullo AD. **2000**. Cytoskeletal organization defects and abortive activation in human oocytes after IVF and ICSI failure. *Mol Hum Reprod* **6**:510–516.
20. Navara CS, First NL, Schatten G. **1996** Phenotypic variations among paternal centrosomes expressed within the zygote as disparate microtubule lengths and sperm aster organization: Correlations between centrosome activity and developmental success. *Proc Natl Acad Sci USA* **93**:5384–5388.
21. Doxsey SJ, Stein P, Evans L, Calarco PD, Kirschner M. **1994**. Pericentrin, a highly conserved centrosome protein involved in microtubule organization. *Cell* **76**:639–650.
22. Stearns T, Kirschner M. **1994**. *In vitro* reconstitution of centrosome assembly and function: the central role of γ-tubulin. *Cell* **76**:623–637.
23. Félix M-A, Antony C, Wright M, Maro B. **1994**. Centrosome assembly *in vitro*. *J Cell Biol* **124**:19–31.
24. Fouquet JP, Kann ML, Combeau C, Melki R. **1998**. Gamma-tubulin during the differentiation of spermatozoa in various mammals and man. *Mol Hum Reprod* **4**:1122–1129.
25. Salisbury JL, Suino KM, Busby R, Springett M. **2002**. Centrin-2 is required for centriole duplication in mammalian cells. *Curr Biol* **12**:1287–1292.
26. Bornens M. **2002**. Centrosome composition and microtubule anchoring mechanisms. *Curr Opin Cell Biol.* **14**:25–34.
27. Beisson J, Wright, M. **2003**. Basal body/centriole assembly and continuity. *Curr Opin Cell Biol* **15**: 96–104.
28. Hewitson L, Haavisto A, Simerly C, Jones J, Schatten G. **1997**. Microtubule organization and chromatin configurations in hamster oocytes during fertilization and parthenogenetic acti-

vation, and after insemination with human sperm. *Biol Reprod* **57**:967–975.

29. Nakamura S, Terada Y, Horiuchi T, Emuta C, Murakami T, Yaegashi N, Okamura K. **2002**. Analysis of the human sperm centrosomal function and the oocyte activation ability in a case of globozoospermia, by ICSI into bovine oocytes. *Hum Reprod* **17**:2930–2934.

30. Moomjy M, Colombero LT, Veeck LL, Rosenwaks Z, Palermo GD. **1999**. Sperm integrity is critical for normal mitotic division and early embryonic development. *Mol Hum Reprod* **5**:836–844.

31. Hyman AA. **2000**. Centrosomes: *Sic transit gloria centri*. *Curr Biol* **10**:R276–R278.

32. Whittman T, Hyman A, Desai A. **2001**. The spindle: a dynamic assembly of microtubules and motors. *Nature Cell Biol* **3**:E28–E34.

33. Manandhar G, Simerly C, Schatten G. **2000**. Centrosome reduction during mammalian spermiogenesis. *Curr Top Dev Biol* **49**:343–363.

34. Schatten H, Schatten G, Mazia D, Balczon R, Simerly C. **1986**. Behavior of centrosomes during fertilization and cell division in mouse oocytes and in sea urchin eggs. *Proc Natl Acad Sci USA* **83**:105–109.

35. Schatten G, Simerly C, Schatten H. **1991**. Maternal inheritance of centrosomes in mammals? Studies on parthenogenesis and polyspermy in mice. *Proc Natl Acad Sci USA* **88**:6785–6789.

36. Breed W, Simerly C, Navara C, Vanderberg J, Schatten G. **1994**. Distribution of microtubules in eggs and early embryos of marsupial, *Monodelphis domestica*. *Dev Biol* **164**:230–240.

37. Navara CS, First NL, Schatten G. **1994**. Microtubule organization in the cow during fertilization, polyspermy, parthenogenesis, and nuclear transfer: the role of the sperm aster. *Develop Biol* **162**:29–40.

38. Crozet N, Dahirel M, Chesne P. **2000**. Centrosome inheritance in sheep zygotes: centrioles are contributed by the sperm. *Microsc Res Tech* **49**:445–450.

39. Kim N-H, Simerly C, Funahashi H, Schatten G, Day BN. **1996**. Microtubule organization in porcine oocytes during fertilization and parthenogenesis. *Biol Reprod* **54**:1397–1404.

40. Wu GJ, Simerly C, Zoran SS, Funte LR, Schatten G. **1996**. Microtubule and chromatin dynamics during fertilization and early development in rhesus monkeys, and regulation by intracellular calcium ions. *Biol Reprod* **55**:260–270.

41. Kola, I. and Trounson, A. **1989**. Dispermic human fertilization: violation of expected cell behavior. In *The Cell Biology of Fertilization*, Schatten H, Schatten G. (Eds). Academic Press, Inc.: San Diego.

42. Van Blerkom J, Henry G. **1991**. Dispermic fertilization of human oocytes. *J Electr Micros Tech* **17**:437–449.

43. Plachot M, Crozet N. **1992**. Fertilization abnormalities in human *in-vitro* fertilization. *Hum Reprod* **7**(Suppl. 1):89–94.

44. Fol H. 1891. Die "Centrenquadrille", eine neue Episode aus der Befruchtungsgeschichte. *Arch Des Sci Phys Nat* **VI**:266–274.

45. Dewald G, Alvarez MN, Clouter MD, Kelatis PP, Gordon H. **1975**. A diploid–triploid human mosaic with cytogenetic evidenced of double fertilization. *Clin Genet* **8**:142–160.

46. Tharapel AT, Wilroy RS, Martens PR, Holbert JM, Summitt RL. **1983**. Diploid–triploid mosaicism: delineation of the syndrome. *Ann Genet* **26**:229–233.

47. Carakushansky G, Teich E, Ribeiro MG, Horowitz D. **1993**. Diploid/triploid mosaicism: further delineation of the phenotype. *Am J Med Genet* **52**:399–401.

48. Wulfsberg EA, Wassel WC, Polo CA. **1991**. Monozygotic twin girls with diploid/triploid chromosome mosaicism and cutaneous pigmentary dysplasia. *Clin Genet* **39**:370–375.

49. Phelan MC, Rogers CR, Michaelis RC, Moore LC, Blackburn W. **2001**. Prenatal diagnosis of mosaicism for triploidy and trisomy 13. *Prenat Diagn* **21**:457–460.

50. Golubovsky MD. **2003**. Postzygotic diploidization of triploids as a source of unusual cases of mosaicism, chimerism and twinning. *Hum Reprod* **18**:236–242.
51. Macas E, Imthurn B, Keller PJ. **2001**. Increased incidence of numerical chromosome abnormalities in spermatozoa injected into human oocytes by ICSI. *Hum Reprod* **16**:115–120.
52. Egozcue S, Blanco J, Vidal F, Egozcue J. **2002**. Diploid sperm and origin of triploidy. *Hum Reprod* **17**:5–7.
53. Vogels A, Fryns JP. **2002**. The Prader–Willi syndrome and the Angelman syndrome. *Genet Counsel* **13**:385–396.
54. Bestor TH. **2003**. Imprinting errors and developmental asymmetry. *Phil Trans R Soc Lond B Biol Sci* **358**:1411–1415.
55. Walter J, Paulsen M. **2003**. Imprinting and disease. *Semin Cell Dev Biol* **14**:101–110.
56. Verona RI, Mann MR, Bartolomei MS. **2003**. Genomic imprinting: intricacies of epigenetic regulation in clusters. *Annu Rev Cell Dev Biol* **19**:237–259.
57. Battaglia DE, Goodwin P, Klein NA, Soules MR. **1996**. Influence of maternal age on meiotic spindle assembly in oocytes from naturally cycling women. *Hum Reprod* **11**:2217–2222.
58. Munne S, Cohen J. **1998**. Chromosome abnormalities in human embryos. *Hum Reprod Update* **4**:842–855.
59. Verlinsky Y, Cieslak J, Kuliev A. **2002**. Preimplantation FISH diagnosis of aneuploidies. *Methods Mol Biol* **204**:259–273.
60. Handyside AH, Scriven PN, Ogilvie CM. **1998**. The future of preimplantation genetic diagnosis. *Hum Reprod* **13**(Suppl. 4):249–255.
61. Werlin L, Rodi I, DeCherney A, Marello E, Hill D, Munne S. **2003**. Preimplantation genetic diagnosis as both a therapeutic and diagnostic tool in assisted reproductive technology. *Fertil Steril* **80**:467–468.
62. Munne S. **2003**. Preimplantation genetic diagnosis and human implantation – a review. *Placenta* **24**(Suppl. B):S70–S76.
63. Hassold T, Hunt P. **2001**. To err (meiotically) is human: the genesis of human aneuploidy. *Nature Rev Genet* **2**:280–291.
64. Hunt PA, Hassold TJ. **2002**. Sex matters in meiosis. *Science* **296**:2181–2183.
65. Sluder G, Miller FJ, Lewis K. **1993**. Centrosome inheritance in starfish zygotes. II: Selective suppression of the maternal centrosome during meiosis. *Dev Biol* **155**:58–67.
66. Palazzo RE, Vogel JM, Schnackenberg BJ, Hull DR, Wu X. **2000**. Centrosome maturation. *Curr Top Dev Biol* **49**:449–470.
67. Uetake Y, Kato KH, Washitani-Nemoto S, Nemoto Si S. **2002**. Nonequivalence of maternal centrosomes/centrioles in starfish oocytes: selective casting-off of reproductive centrioles into polar bodies. *Dev Biol* **247**:149–164.
68. Briggs D, Miller D, Gosden,R. **1999**. Molecular biology of female gametogenesis. In *Molecular Biology in Reproductive Medicine,* Fauser BCJM, Rutherford AJ, Strauss III JF, Van Steirteghem A. (Eds). Parthenon Publishing: New York.
69. Paulson RJ, Boostanfar R, Saadat P, Mor E, Tourgeman DE, Slater CC, Francis MM, Jain JK. **2002**. Pregnancy in the sixth decade of life: obstetric outcomes in women of advanced reproductive age. *JAMA* **288**:2320–2323.
70. Brinster RL, Nagano M. **1998**. Sprematogonial stem cell transplantation, cryopreservation and culture. *Semin Cell Dev Biol* **9**:401–409.
71. Orwig KE, Ryu BY, Avarbock MR, Brinster RL. **2002**. Male germ-line stem cell potential is predicted by morphology of cells in neonatal rat testes. *Proc Natl Acad Sci* **99**:11706–11711. E-pub Aug 15.
72. Schatten G, Simerly C, Asai DJ, Szöke E, Cooke P, Schatten H. **1988**. Acetylated α-tubulin in microtubules during mouse fertilization and early development. *Devel Biol* **130**:74–86.
73. Szöllösi D, Ozil JP. **1991**. *De novo* formation of centrioles in partheno-

genetically activated diploidized rabbit embryos. *Biol Cell* **72**:61–66.

74. Cibelli JB, Grant KA, Chapman KB, Cunniff K, Worst T, Green HL, Walker SJ, Gutin PH, Vilner L, Tabar V, Dominko T, Kane J, Wettstein PJ, Lanza RP, Studer L, Vrana KE, West MD. **2002**. Parthenogenetic stem cells in nonhuman primates. *Science* **295**:819.

75. Vrana KE, Hipp JD, Goss AM, McCool BA, Riddle DR, Walker SJ, Wettstein PJ, Studer LP, Tabar V, Cunniff K, Chapman K, Vilner L, West MD, Grant KA, Cibelli JB. **2003**. Nonhuman primate parthenogenetic stem cells. *Proc Natl Acad Sci USA* Suppl. 1:11911–11916.

76. Prather RS. **2001**. Basic mechanisms of fertilization and parthenogenesis in pigs. *Reprod Suppl* **58**:105–112.

77. George MA, Pickering SJ, Braude PR, Johnson MH. **1996**. The distribution of alpha- and gamma-tubulin in fresh and aged human and mouse oocytes exposed to cryoprotectant. *Mol Hum Reprod* **2**:445–456.

78. Lee J, MiyanoT, Moore RM. **2000**. Spindle formation and dynamics of γ-tubulin and nuclear mitotic apparatus protein distribution during meiosis in pig and mouse oocytes. *Biol Reprod* **62**:1184–1192.

79. Szöllösi D, Calarco P, Donahue RP. **1972**. Absence of centrioles in the first and second meiotic spindles of mouse oocytes. *J Cell Sci* **11**:521–541.

80. Magnuson T, Epstein CJ. **1984**. Oligosyndactly: a lethal mutation in the mouse that results in mitotic arrest very early in development. *Cell* **38**:823–833.

81. Megraw TL, Kao L-R, Kaufman T. **2001**. Zygotic development without functional mitotic centrosomes. *Curr Biol* **11**:116–120.

82. Bonaccorsi S, Giansanti MG, Gatti M. **2000**. Spindle assembly in *Drosophila* neuroblasts and ganglion mother cells. *Nature Cell Biol* **2**:54–56.

83. Khodjakov A, Cole RW, Oakley BR, Rieder CL. **2000**. Centrosome-independent mitotic spindle formation in vertebrates. *Curr Biol* **10**:59–67.

84. Khodjakov A, Rieder CL. **2001**. Centrosomes enhance the fidelity of cytokinesis in vertebrates and are required for cell cycle progression. *J Cell Biol* **153**:237–242.

85. Raff JW. **2001**. Centrosomes: central no more? *Curr Biol* **11**:R159–R161.

86. Blangy A, Lane HA, d'Hérin P, Harper M, Kress M, Nigg EA. **1995**. Phosphorylation by $p34^{cdc2}$ regulates spindle association of human Eg5, a kinesin-related motor essential for bipolar spindle formation *in vivo*. *Cell* **83**:1159–1169.

87. Heck MMS, Pereira A, Pesavento P, Yannoni Y, Spradling AC, Goldstein LSB. **1993**. The kinesin-like protein KLP61F is essential for mitosis in *Drosophila*. *J Cell Biol* **123**:665–679.

88. Walczak CE, Vernos I, Mitchison TJ, Karsenti E, Heald R. **1998**. A model for the proposed roles of different microtubule-based motor proteins in establishing spindle bipolarity. *Curr Biol* **8**:903–913.

89. Whitehead CM, Rattner JB. **1998**. Expanding the role of HsEg5 within the mitotic and post-mitotic phases of the cell cycle. *J Cell Sci* **111**:2551–2561.

90. Mountain V, Simerly C, Howard L, Ando A, Schatten G, Compton DA. **1999**. The kinesin-related protein, HSET, opposes the activity of Eg5 and cross-links microtubules in the mammalian mitotic spindle. *J Cell Biol* **147**:351–365.

91. Simerly C, Dominko T, Navara C, Payne C, Capuano S, Gosman G, Chong KY, Takahashi D, Chace C, Compton D, Hewitson L, Schatten G. **2003**. Molecular correlates of primate nuclear transfer failures. *Science* **300**:297.

92. Endow SA. **1999**. Microtubule motors in spindle and chromosome motility. *Eur J Biochem* **262**:12–18.

93. Compton DA. **1998**. Focusing on spindle poles. *J Cell Sci* **111**:1477–1481.

94. Campbell KHS. **1999**. Nuclear equivalence, nuclear transfer, and the cell cycle. *Cloning* **1**:3–15.

95. Pennisi E, Vogel G. **2000**. Clones: a hard act to follow. *Science* **288**:1722–1727.

96. Wilmut I, Paterson L. **2003**. Somatic cell nuclear transfer. *Oncol Res* **13**:303–307.

97. Alberio R, Campbell KH. **2003**. Epigenetics and nuclear transfer. *Lancet* **361**:1239–1240.

98. Dominko T, Simerly C, Martinovich C, Schatten G. **2003**. Cloning in nonhuman primates. In *Principles of Cloning*, Cibelli J, Lanza RP, Campbell KHS, West MD. (Eds). Academic Press: San Diego.

99. Hwang W-S, Ry YJ, Park JH, Park ES, Lee EG, Koo JM, Chun' HY, Lee BC, Kang SK, Kim SJ, Ahn C, Hwang JH, Park JY, Cibelli JB, Shin Yong Moon SY. **2004**. Pluripotent human embryonic stem cell line derived from a cloned blastocyst. *Science* (in press).

100. Edwards RG. **2001**. IVF and the history of stem cells. *Nature* **413**:349–351.

101. Humpherys D, Eggan K, Akutsu H, Freidman A, Hochedlinger K, Yanagimachi R, Lander ES, Golub TR, Jaenisch R. **2002**. Abnormal gene expression in cloned mice derived from embryonic stem cell and cumulus cell nuclei. *Proc Natl Acad Sci USA* **99**:12889–12894.

102. McLaren A. **2001**. Stem cells: therapy for the 21st century. *Trans Med Soc Lond* **117**:78–79.

15
Microtubule Organizing Centers in Polarized Epithelial Cells

Mette M. Mogensen

15.1
Introduction

Microtubules are essential for many cellular functions including cell motility, polarity, division and the targeting of vesicles and signaling molecules. The precise and intricate microtubule pattern deployed in cells is crucial for many of these microtubule-dependent processes and is mainly defined by the microtubule organizing center (MTOC). Many different morphologies of the MTOC exist, from the spindle pole body (SPB) of yeast [1], the nucleus-associated body (NAB) of the cellular slime molds [2], floating nucleating sites in plants [3] and multiple plasma membrane associated nucleating sites in *Drosophila* epidermal cells [4] to the classic centrosome present in most animal cells. The centrosome consists of a pair of centrioles and pericentriolar material (PCM) and it is responsible not only for the nucleation of microtubules, but also for their anchorage [5, 6]. The minus ends of the microtubules are anchored at the centrosome and elongation occurs by distal plus-end addition of tubulin units [7]. However, the centrosome is not only a microtubule organizer but also an anchor for many regulatory molecules and a key player in the control of cell cycle progression and of cell function generally [5, 8]. Based on evidence from centriole-induced disassembly studies, which suggest that the centrioles are important for centrosome integrity and for the recruitment and assembly of the PCM, it seems that when present the centrioles act as a dominant focus for and organizer of the PCM [9]. In extracts of *Xenopus* eggs, the centriole-associated protein kinase Nek2 has proved important for efficient recruitment of PCM to the centrosome, with depletion delaying accumulation of γ-tubulin and microtubule assembly [10, 11]. The recent discovery of the centriolar compound SAS-4, which appears to dictate centrosome size in *Caenorhabditis elegans*, lends further support for centrioles as organizers of the PCM [12, 13].

Centrosomes in Development and Disease. Edited by Erich A. Nigg

ISBN 3-527-30980-2

15.2 Centrosomal Microtubule Nucleation

The microtubule nucleating power of the centrosome resides within the PCM [14, 15], a fibrous scaffolding lattice composed of coiled-coil proteins able to anchor signaling molecules and components of the γ-tubulin complex. Overwhelming evidence points to the γ-tubulin complex as the most efficient nucleator of microtubules and as essential for nucleation at the centrosome [16–21]. Immunoelectron microscopic tomography studies have identified γ-tubulin ring complexes (γ-TuRC) within the PCM of *Drosophila* and *Spisula*, which seem to act as templates for the initiation of microtubule assembly [17, 22]. So far six components of the γ-TuRC, in addition to γ-tubulin, have been identified [23–25]. Similar complexes have also been observed in the cytoplasm, where they have been suggested to act as a store for centrosomal recruitment [20, 21, 26]. So far most evidence points to the γ-TuRC acting as a template for the assembly of microtubules but it still remains to be determined whether an uncoiled γ-TuRC forming a stable protofilament is adapted in certain circumstances [27–32]. The requirement of γ-TuRCs for nucleation at the centrosome is elegantly demonstrated in a series of experiments on salt-extracted centrosomes. Salt-extracted centrosomes lack γ-TuRCs and are unable to nucleate microtubules, but subsequent recruitment of γ-tubulin complexes from cytoplasmic extracts make them nucleation competent [33, 34].

However, the γ-TuRC alone is not sufficient to restore the nucleating capacity of salt-extracted centrosomes, other components for anchoring the γ-TuRCs to the centrosome are needed. In budding yeast, Spc110p anchors the γ-TuRC by interacting with Spc97p and Spc98p of the γ-TuRC [35, 36]. In *Drosophila* the addition of Asp together with γ-TuRCs can restore the nucleation ability of salt-stripped centrosomes and Asp may be responsible for anchoring γ-TuRC to the PCM lattice, although it has also been implicated in tethering microtubules [37]. In mammalian cells large coiled-coil A-kinase anchoring proteins such as AKAP450/CG-NAP and pericentrin/kendrin seem to provide an important supporting network for the binding of γ-TuRCs. AKAP450 and pericentrin, which exist in two isoforms as pericentrin A and pericentrin B (also known as kendrin), interact with the γ-TuRC via the GCP2 and/or GCP3 (orthogs of Spc97p and Spc98p respectively) [25, 38] and pericentrin has been suggested to be implicated in the recruitment of γ-tubulin to the centrosome [39–47]. Cep135 is another coiled-coil protein of the PCM and when it is depleted microtubule organization but not nucleation, is affected. It has been suggested to form part of the PCM scaffold and to be important for the maintenance of the PCM structure [48]. Recently, Nlp (ninein-like protein) has been suggested to be a docking protein for the γ-TuRC during interphase [49]. Nlp interacts with two components of the γ-TuRC, recruits γ-tubulin and hGCP4 (human γ-tubulin complex protein 4) and stimulates microtubule nucleation during interphase, while it is released from the centrosome prior to mitosis. Nlp is a strong candidate for a γ-tubulin binding protein and may thus be the missing link (adaptor), which directly anchors the γ-TuRC to the PCM lattice during interphase. The small GTPase Ran, which is best known as a key regulator of nucleocytoplasmic transport, is also involved in microtubule assembly. A fraction of Ran is associated with the centrosome throughout the cell cycle. Ran forms a complex with AKAP450, which excludes γ-tubulin, and its delocalization from the centrosome affects microtubule assembly although γ-tubulin remains at the centrosome [50].

Although there is little doubt that the γ-TuRC is responsible for microtubule nucleation and in principle could cap and anchor microtubules, it is unlikely that it plays a major role in microtubule minus-end anchorage *in vivo*. It is becoming increasingly clear that other complexes are responsible for microtubule minus-end anchorage. The novel concept of two functionally distinct centrosomal complexes, a γ-tubulin nucleating complex and an anchoring complex responsible for the anchorage of microtubule minus ends released from the γ-tubulin complex, was proposed following studies of polarized cochlear epithelial cells, where nucleation and anchoring sites are spatially separated [4, 51]. The centrosomal protein ninein has emerged as a strong candidate for being part of the anchoring complex [52, 53].

15.3 Non-centrosomal Microtubule Arrays

A radial array of microtubules anchored at a centrally located centrosome predominates in many animal cells (Figure 15.1). However, non-centrosomal microtubules also occur and their proportion depends on the cell type. In many differentiated cells including polarized cells such as intestinal, kidney, hepatocyte and retinal pigmented epithelial cells and certain mammalian cochlear cells as well as neuronal and skeletal muscle cells, the majority of the microtubules are non-centrosomal and a non-radial array is evident. Non-centrosomal microtubule arrays are critical for many of the specialized functions of differentiated cells. For example, in intest-

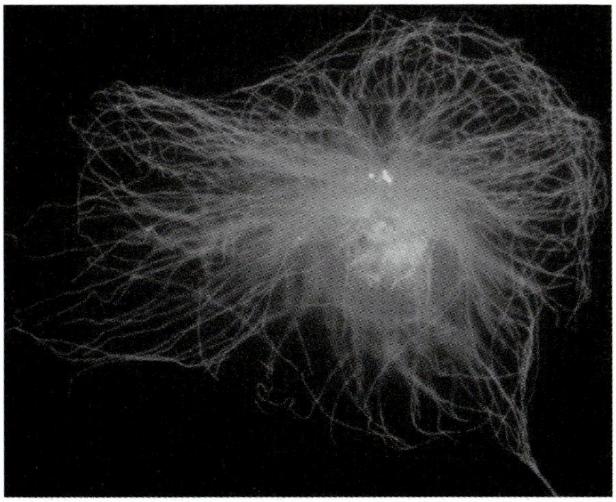

Figure 15.1 Cell with a radial array of microtubules focused on a juxta-nuclear centrosome. A wide field fluorescent image of a cell triple labeled for microtubules with an antibody to α-tubulin (green), for the centrosome with an antibody to γ-tubulin (red or yellow where co-localized with α-tubulin) and for the nucleus with DAPI (blue). Images by courtesy of Gemma Bellett (see Color Plates page XXXIII).

inal and kidney epithelial cells the apico-basal microtubules are essential for proper sorting of membrane components and for directing vesicle traffic [54]. In the organ of Corti in the inner ear large non-centrosomal apico-basal microtubule arrays in the supporting cells provide physical resilience to its cellular architecture and contribute to the efficient transmission of mechanical vibrations to the sensory hair cells, which is critical for auditory perception [55].

How non-centrosomal microtubules are generated and which molecular mechanisms govern the dramatic reorganization of a radial microtubule array into a non-centrosomal, apico-basal array has not been fully established. In some cells the centrosome disassembles and new MTOCs are generated. The classic example of redistribution of nucleating material occurs during vertebrate myogenesis when microtubules are nucleated from sites associated with the nuclear envelope while parallel arrays form in myotubes [56]. In polarized epithelial cells, redistribution of the nucleating material occurs in certain polarized epithelia of *Drosophila* during late pupal morphogenesis, where microtubule minus ends are anchored and elongate from multiple plasma membrane-associated MTOCs following the loss of the centrioles [57–59]. In lens epithelial cells the centrosome apparently disassembles only to relocate and reassemble *de novo* at a later stage in lens fiber differentiation [4, 60, 61]. However, in most polarized epithelial cells the centrosome remains, but relatively few microtubules radiate from it and, instead, the vast majority of microtubule minus ends are non-centrosomal. The question then arises, as to whether the centrosome is still active and whether it is the predominant site of microtubule nucleation. In rat Sertoli cells the centrosome, which remains juxta-nuclear and basal during polarization, apparently no longer acts as the main site of nucleation. Here the minus ends of the apico-basal microtubules are anchored at apical non-centrosomal sites and re-grow from these sites following nocodazole removal, which suggests non-centrosomal nucleation [62].

Establishment of polarity in most epithelial cells involves migration of the centrioles to the apical cell surface, loss of radial microtubule organization and the formation of a predominantly apico-basal array [63–66]. In many epithelial cells one of the centrioles becomes a basal body, which gives rise to a primary cilium, but also supports a small cytoplasmic fascicle of microtubules. The majority of the microtubules are non-centrosomal and in a few epithelial cell types, such as retinal photoreceptor cells and sensory hair cells of the mammalian inner ear, they remain free in the cytoplasm [67–70], while in most others they are anchored at apical non-centrosomal sites [63–65, 71–73]. So is the apical centrosome still the dominant nucleating site in these polarized epithelial cells? Data from Madin–Darby canine kidney (MDCK) cells [71] and from cochlear supporting cells [51] suggest that the non-centrosomal microtubules originate from the centrosome and that they are not nucleated at the non-centrosomal apical sites where the minus ends of the apico-basal microtubules are anchored.

Inner ear epithelial cells located in the cochlea, which retain their centriole-containing centrosome, have proved extremely useful as a model system for studying microtubule nucleation and anchorage in a tissue system where non-centrosomal microtubules predominate. This is particularly the case for the supporting inner and outer pillar cells of the organ of Corti, which assemble large apico-basal arrays of several thousand microtubules during tissue morphogenesis [51, 73, 74] (Figures 15.2 and 15.3). Microtubule polarity investigations show that the minus ends of the microtubules are associated with the apical anchoring sites in the pillar cells [75]. This provides a large target for microtubule-associated proteins which facilitates their unambiguous detection. In addition, the distinct and morphologically well-defined temporal and spatial differentiation process of the pillar cells enables assessment of the distribution of such proteins during assembly of the apico-basal arrays. Further, it provides an opportunity to study centrosomal and microtubule organization during normal tissue morphogenesis *in situ*.

γ-Tubulin and pericentrin are concentrated at the centrosome in the pillar cells throughout development and they have not been detected at the apical sites, where most of the microtubule minus-ends are concentrated (Figure 15.2b). This

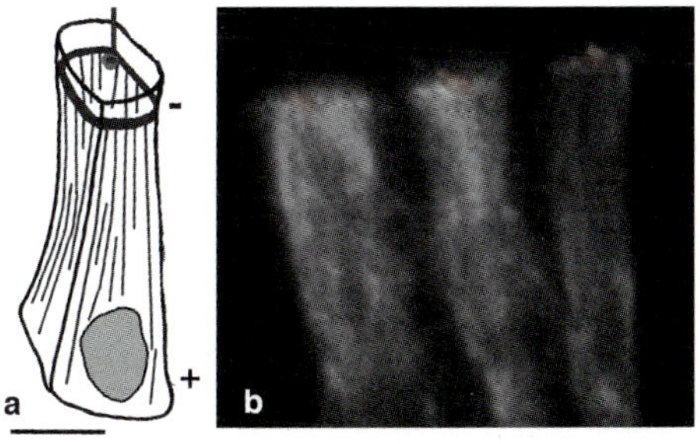

Figure 15.2 Apico-basal microtubule arrays in cochlear inner pillar epithelial cells. (a) Schematic 3-D representation of the microtubule organization (blue) in an inner pillar cell during assembly of the apico-basal array. The centrosome (red with centriole and basal body/primary cilium in blue) is located at the apex and most of the microtubule minus-ends are anchored at a peripheral apical ring (grey). The nucleus is shown in yellow. Adapted from [52]. (b) Wide field fluorescent image of the apical halves of three inner pillar cells showing an apico-basal array of microtubules (labeled with an antibody to α-tubulin in green) and γ-tubulin (red) concentrated at the apical centrosome. Images by courtesy of Gemma Bellett. Bar = 10 μm (see Color Plates page XXXIII).

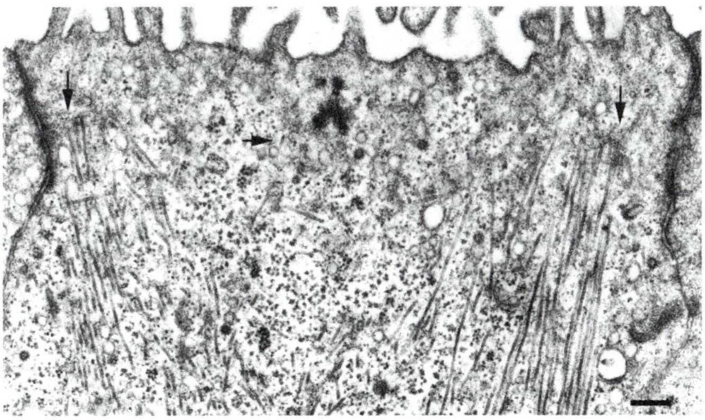

Figure 15.3 Microtubules at the apical non-centrosomal site in an inner pillar cell. Transmission electron microscopic image of a longitudinal section through the apical region of an inner pillar cell during assembly of the apico-basal microtubule array. The majority of the microtubules are associated with dense material at the apical periphery (large arrows) while a few microtubules can be seen near the centrosome (small arrow). Bar = 0.5 μm.

suggests that microtubule nucleation is confined to the centrosome. Similarly, γ-tubulin is concentrated at a single apical focal point in other polarized epithelial cells such as MDCK and retinal pigment epithelia cells [71, 76]. The microtubule minus end-anchoring protein ninein, on the other hand, is evident at the centrosome and at the apical non-centrosomal sites where several thousand microtubule minus-ends are concentrated (Figure 15.4). Furthermore, a transit population of microtubules and ninein speckles are evident between the centrosome and the apical sites, while microtubule assembly is progressing. This correlates with the observed gradual shift in microtubules from the centrosome to the apical sites and the simultaneous accumulation of ninein at the apical sites during development. Evidence from studies on polarized MDCK cells also supports microtubule nucleation at the centrosome and anchorage at the apical sites. Microtubule re-growth analyses, following nocodazole-induced microtubule depolymerization, reveal an initial radial centrosomal array, which is subsequently replaced by an apico-basal non-centrosomal array. In addition dynactin, which has been suggested to be important for microtubule minus-end anchorage at the centrosome, is located at the cell apex in polarized MDCK cells [71, 77, 78]. Analyses of microtubule assembly in these cells led to the proposal of a novel microtubule release and capture mechanism for the generation of non-centrosomal apico-basal arrays in polarized epithelial cells. The studies also suggested that the centrosome contains two functionally distinct complexes, one concerned with nucleation and the other with anchorage, which in these highly specialized cells are located at different sites [4, 51, 52, 75].

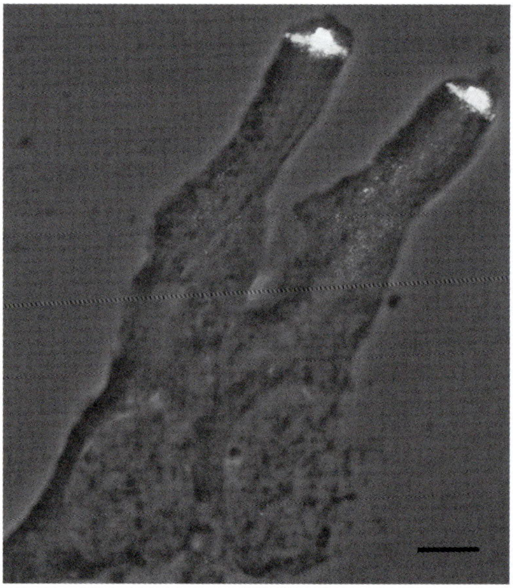

Figure 15.4 Ninein localization in inner pillar cells. Projection of confocal optical sections through two isolated inner pillar cells at a stage when some 3000 microtubule minus-ends are concentrated at the apical sites. The pillar cells have been labeled with an antibody for ninein (yellow). Ninein is concentrated at the centrosome and at the apical sites but ninein speckles are also evident within the apical half of the cytoplasm. From [52]. Bar = 5 μm (see Color Plates page XXXIV).

15.4
Microtubule Minus-end Anchorage at Centrosomal and Non-centrosomal Sites

The concept of two functionally distinct centrosomal complexes was conjectured by De Brabander and co-workers [79] who speculated that microtubule nucleation and anchorage may be separate events involving different parts of the centrosome. This has now become highly relevant and is clearly central to the control of microtubule assembly and organization in cells generally. Two functionally distinct microtubule minus-end-associated complexes, one nucleating and one anchoring, thus seem to be fundamental components of the centrosome (Figure 15.5). The assembly of non-radial arrays, typical of many differentiated cells, is likely to depend on the redeployment of either nucleating or anchoring complexes or both to other sites. The assembly of the apico-basal arrays in, for example, the pillar cells involves the relocation of anchoring complexes to apical sites, while in *Drosophila* wing cells it seems that both nucleating and anchoring complexes become associated with multiple apical sites.

The centrosomal protein ninein is a strong contender as a component of the anchoring complex. Ninein is a coiled-coil protein which exists as several splice var-

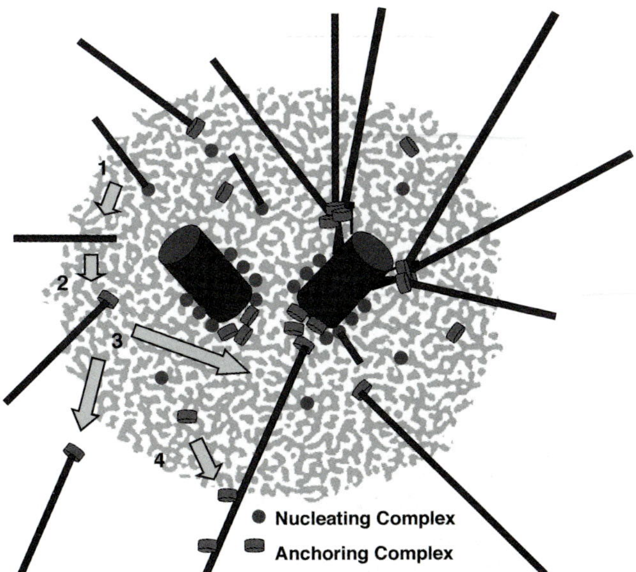

Figure 15.5 Two centrosomal complexes: a nucleating and an anchoring complex. Schematic diagram showing the organization of nucleating (γ-TURC) and anchoring (ninein) complexes within the centrosome based on recent findings. The possible fates of a microtubule nucleated by a γ-TURC and centrosomal anchoring complexes are outlined in 1–4. (1) Microtubule release from the γ-TURC following nucleation; (2) microtubule minus-end capping by a capping/anchoring complex; (3) microtubule release from the centrosome or firm anchorage within the PCM closely associated with the mother centriole; (4) release of anchoring complexes from the centrosome and their transport along a microtubule (see Color Plates page XXXIV).

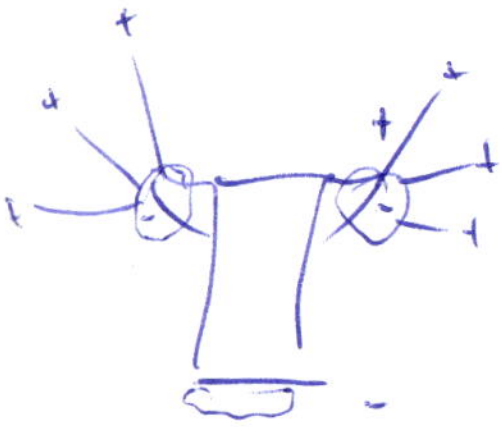

iants and localizes to the centrosome in most cells displaying a radial microtubule array [80–82]. Ninein localizes together with γ-tubulin at the centrosome in cochlear pillar cells, but is evident also at the apical sites where the majority of the microtubule minus-ends accumulate [52]. Immunofluorescent imaging reveals that ninein is concentrated around one of the centrioles, often organized into several dots, while only a single, usually faintly stained, dot is associated with the other centriole [52, 53, 83]. Immunoelectron microscopy has pinpointed ninein to the tip of the subdistal appendages of the mother (older) centriole and to the minus ends of both centrioles, which provides evidence for a role in minus-end anchorage [52]. The tips of the subdistal appendages are the focus for the minus ends of a fascicle of microtubules and the proximal ends of the centrioles, which contain the minus ends of the centriolar microtubules, are regions with strong sequestering affinity for minus-end binding proteins [84, 85]. Ninein is also present at the minus ends of microtubules within the PCM [52]. Affinity for the minus end is further supported by co-localization of ninein with C-Nap1, which is known to associate with the minus ends of the centrioles [83]. Microtubule re-growth, overexpression and depletion studies, in particular, support the role of ninein in microtubule minus-end anchorage. γ-Tubulin is usually closely associated with the centrioles [85] and while both mother and daughter centrioles nucleate similar numbers of microtubules following nocodazole removal, only the mother with its high concentration of ninein is able to maintain a radial array [53]. Ninein overexpression, which leads to the formation of highly ordered aggregates consisting of nodes and fibers, causes an increase in microtubule anchorage, and a marked decrease in microtubule release. Nocodazole re-growth experiments in cells overexpressing ninein show no effect on microtubule nucleation or elongation but result in a dramatic fall in non-centrosomal microtubules [86]. Striking evidence for a role in anchorage comes from ninein inhibition and depletion studies, which affect microtubule organization but not nucleation. Microinjection of anti-ninein antibodies or RNAi depletion of ninein results in loss of the radial array [87].

Other centrosomal proteins such as cenexin/ODF2, centriolin, Cep110 and ε-tubulin also localize to the subdistal appendices [83, 88–90]. ε-Tubulin, like ninein does not appear to play a role in microtubule nucleation as depletion does not affect the ability of sperm centrosomes to nucleate. However, it does seem to influence microtubule organization. ε-Tubulin together with ninein and γ-tubulin is present at the focus of non-centrosomal DMSO-induced asters in *Xenopus* extracts and ε-tubulin depletion disrupts the radial arrangement [88]. Centriolin, which has proved to be important for the normal progression of cytokinesis, does not seem to play a central role in microtubule minus-end anchorage as RNAi depletion does not affect microtubule organization [89]. However, centriolin localizes to the apical anchoring sites in the pillar cells, which may reflect a role in microtubule stabilization rather than anchorage.

Dynactin has also been reported to have a key role in microtubule anchorage, independently of dynein, at the centrosome and at apical non-centrosomal sites in polarized MDCK cells. Dynactin shows preferential association with the mother centriole and overexpression of dynactin subunits affects centrosome integrity and induces microtubule disorganization [77, 78, 91]. Dynein/dynactin has recently been shown to be needed for the recruitment of ninein to the centrosome, and PCM-1 seems to play a role in this process [87].

15.5
Centrosomal Release of Microtubules and Anchoring Complexes

Centrosomal microtubule nucleation and release is not a new concept but was originally proposed as part of the conveyer belt hypothesis in which microtubules are continuously nucleated and released from the centrosome [92]. Release has been visualized in epithelial cell cultures [93, 94] and has been shown convincingly to be the main mechanism for producing non-centrosomal, free microtubules in neurons [95]. In polarized epithelial cells, unlike in neurons, microtubules are subsequently captured and anchored at apical non-centrosomal sites following their release. Evidence is now emerging which reveals that anchoring complexes are also released from the centrosome [5, 52, 86].

The mechanism responsible for centrosomal microtubule release in epithelial cells is not known, but the microtubule-severing protein katanin is a strong candidate. Katanin is a well-established microtubule-severing protein which has been shown to be important for microtubule release from their centrosomal attachment and for neuronal differentiation [95–98]. Evidence also suggests that katanin is responsible for M-phase severing activity at the spindle poles in mammalian cells, which is thought to play a critical role in minus-end depolymerization of spindle microtubules for pole-ward flux during mitosis [99, 100]. Interestingly, high resolution electron tomography analyses of microtubule minus-ends at mitotic spindle poles show a high proportion of open-ended kinetochore microtubules which is consistent with the action of katanin [101]. A katanin-like protein, MEI-1, in *Caenorhabditis elegans* also severs microtubules and is essential for meiotic spindle function [102, 103]. Katanin also seems to mediate the severing of axonemal microtubules during deflagellation in *Chlamydomonas* [104] and a katanin-like protein in plants is important for microtubule organization and dynamics [105].

In epithelial cells not only microtubules but also anchoring complexes are released from the centrosome [52, 86] (Figure 15.5). Release and positioning of microtubule minus-end anchoring complexes is likely to be central to the assembly and stabilization of the apico-basal arrays in polarized epithelial cells. Analyses of ninein distribution during assembly of the apico-basal arrays in the pillar cells suggest that ninein is released from the centrosome and relocates to non-centrosomal apical microtubule minus-end anchoring sites in a microtubule-dependent manner [52]. Ninein release and bi-directional movement along microtubules has been visualized in living epithelial cells expressing green fluorescent protein-ninein (D. Moss and M. M. Mogensen, unpublished observations). The mechanism responsible for release of anchoring complexes from the centrosome during assembly of the apico-basal arrays is not known but phosphorylation is likely to be involved. Ninein is possibly phosphorylated by Aurora A and PKA (cAMP-dependent protein kinase), which could be responsible for its displacement from the centrosome during mitosis [106]. The ninein related protein, Nlp has been also shown to disassociate from the centrosome prior to mitosis, in this case in response to phosphoration by Plk1 (Polo-like kinase 1) [49]. Ninein and other anchoring components may also be released together with the microtubules.

Relocation of ninein to the apical sites is likely to be microtubule dependent and involve microtubule-based motors since nocodazole disperses cytoplasmic speckles from the transit region in the pillar cells and inhibits ninein dynamics [52]. Intriguingly, findings from studies on PCM-1 suggests that PCM-1 granules (also known as pericentriolar satellites) may act as a cargo vehicle for the transport of ninein and other proteins such as centrin and pericentrin to the centrosome [87]. PCM-1 granules move along microtubules towards the centrosome in a dynein-dependent manner [107, 108]. Inhibition or RNAi depletion of PCM-1 does not affect microtubule nucleation but dramatically decreases the amount of ninein at the centrosome and indirectly affects microtubule anchorage at the centrosome [87]. This further supports the idea that microtubule nucleation and organization are controlled by separate complexes. Whether PCM-1 is also responsible for ninein translocation from the centrosome to the apical sites will be interesting to determine. PCM-1 has been observed at the apical anchoring sites in the pillar cells (G. Bellett and M. M. Mogensen, unpublished observations) and this may be indicative of a role in ninein accumulation at non-centromal sites. Alternatively, PCM-1 may have a more direct role in microtubule anchorage and form a docking platform for ninein, possibly via its interaction with dynein/dynactin. Future studies will need to resolve the composition of the anchorage sites with variants of the AKAPs being strong candidate components. In fact AKAP350 has been localized to apical foci in polarized MDCK cells [109] although a connection with microtubule minus-end anchorage remains to be determined.

15.6
Stabilization of Non-centrosomal Microtubules

Anchorage of microtubules at the centrosome ensures stabilization of microtubule minus-ends, which might otherwise be subject to depolymerization. Stabilization of the minus ends of free microtubules is an essential step towards the creation of non-centrosomal microtubule arrays. The γ-TuRC complex has been suggested to act as a microtubule minus-end cap [31, 32, 110]. However, γ-tubulin is not associated with the transit microtubule population or present at the microtubule minus-ends at the apical sites in the pillar cells. Similarly, γ-tubulin does not appear to cap the minus end of free microtubules released from the centrosome in neurons [111]. In addition, γ-tubulin has not been detected at the end of free microtubules in epithelial cells [112]. The fate of the released microtubules is likely to depend on the microenvironment that the new microtubule ends are exposed to. Treadmilling, which is the continuous polymerization at the plus ends and depolymerization at the minus end, is frequently observed in cytoplasts prepared from melanocytes or fibroblasts while rarely in cytoplasts from epithelial cells [112, 113]. Instead, in epithelial centrosome-free cytoplasts the microtubules show dynamic instability [114]. This suggests that non-centrosomal microtubules in epithelial cells may be stabilized especially at their minus ends by an end-capping complex and/or by microtubule-associated proteins (MAPs), which prevent their depolymerization. Ninein is also a potential microtubule minus-end capping protein as it associates with the transit microtubule population in the pillar cells and with free microtubules in cultured epithelial cells. Furthermore, ninein associates with the minus end of paclitaxel-induced microtubule bundles [86]. Ninein may thus perform an additional role in capping the minus end and providing microtubule stability whilst in transit.

Cell differentiation seems to involve increased microtubule stability. Polarization and formation of cell contacts in MDCK cells, for example, doubles the half-life time for microtubule turnover [115]. Cadherin-mediated signaling seems to influence the control of microtubule dynamics and stabilization of the minus ends of non-centrosomal microtubules [116, 117]. Expression of cadherins and cell-to-cell contacts in centrosome-free cytoplasts induces a switch in microtubule behavior from treadmilling to dynamic instability suggesting stabilization of the minus end. Signaling from cadherins to microtubules has been suggested to involve the small GTPases and Ran has been proposed as a potential mediator of minus-end stability [50, 116]. Interestingly, a fraction of Ran localizes to the centrosome and in the pillar cells it is also present at the apical anchoring sites [50]. Plus-end dynamic instability is also suppressed in fully contacted newt lung epithelial cells, which implies plus-end capping [117]. Selective stabilization of microtubules has been shown to involve the Rho GTPases and the downstream effector mDia (mouse diaphanous-related formins) [118, 119]. mDia binds microtubules along their length, induces stability and caps both ends without apparently binding to them. So mDia seems to protect free non-centrosomal microtubules. Interestingly, the microtubules within the apical region of the pillar cells become detyrosinated and acetylated relatively early in the assembly process, which is indicative of stabilization [120]. Several potential candidates for microtubule plus-end capping exist, including CLIP-170, EB1, dynein/dynactin and APC (adenomatis polyposis coli) protein. APC is a potential plus-end-associated protein and is likely to play an important role in microtubule stabilization in polarized epithelial cells. The role of APC in microtubule plus-end anchorage and stabilization is highlighted in studies of the pillar cells from the min mouse (APC heterozygote mutant), which reveal marked reductions in microtubule numbers [75]. The pillar cells are subjected to considerable mechanical stress during the process of hearing and microtubule stability is likely to be a prerequisite of their survival.

15.7
Release and Capture

Release and capture is likely to be the main mechanism for generating non-centrosomal apico-basal arrays in most polarized epithelial cells which retain their centrosome. The role of microtubule minus-end anchoring complexes is likely to be central to this process with not only microtubules but also anchoring complexes being released, translocated and captured at non-centrosomal sites. Indirect evidence for translocation of microtubules from the centrosome to the apical sites comes from studies of nocodazole re-growth sequences in MDCK, WIF-B and Caco-2 cells, where a centrosomal aster is initially formed but subsequently replaced by an apico-basal array [71]. Evidence also comes from analyses of microtubule assembly in cochlea pillar cells during tissue morphogenesis. Here both microtubules and ninein gradually accumulate at the apical sites and only appear in the transit region while assembly of the apico-basal arrays is proceeding. It could be argued that the transit microtubule population and apico-basal arrays are not a result of centrosomal release but due to cytoplasmic microtubule assembly. Cytoplasmic microtubule assembly has been observed in several cell types [29, 121–123] but whether it is a spontaneous process induced by locally high tubulin concentration or is mediated by free γ-TuRCs is not clear. It may be that non-centrosomal cytoplasmic microtubule nucleation contributes to the transit microtubule population especially in systems such as the pillar cells where large numbers of free microtubules may increase turnover and create microenvironments favorable for self assembly.

The release and capture mechanism may be operating in one of two ways, depending on the cell type, and is probably governed by the distance between the centrioles (Figure 15.6). Control of the distance between the centrioles could potentially be affected by the Nek2/C-Nap1/PP1 complex as has been suggested previously [5, 124–127]. The daughter centriole is not able to retain microtubules and tends to release them, while the mother centriole has a strong capacity for anchorage [52, 53]. In cells where the daughter centriole tends to be relatively far from the basal body (mother centriole), which is apparent in the pillar cells, the classic release and capture mechanism is likely to operate. Here evidence suggests that microtubules are nucleated at the centrosome and short microtubules together with anchoring complexes (ninein) are released, translocated and captured at the apical sites (Figure 15.6). How microtubule translocation is effected in these cells is not known but an extensive actin-based filament system is present in cochlear epithelial cells and a sliding interaction between actin filaments and microtubules could enable dual translocation of these cytoskeletal filaments. The final cytoskeletal apico-basal array in the pillar cells consists of microtubules and interdigitating actin filaments [72].

Substantial evidence has emerged to show that free microtubules are translocated through the cytoplasm towards the cell periphery. Microtubules have been directly observed to move by translocation in PtK1 cells [93]. Studies of microtubule release in cultured cells during re-growth following nocodazole removal reveal that

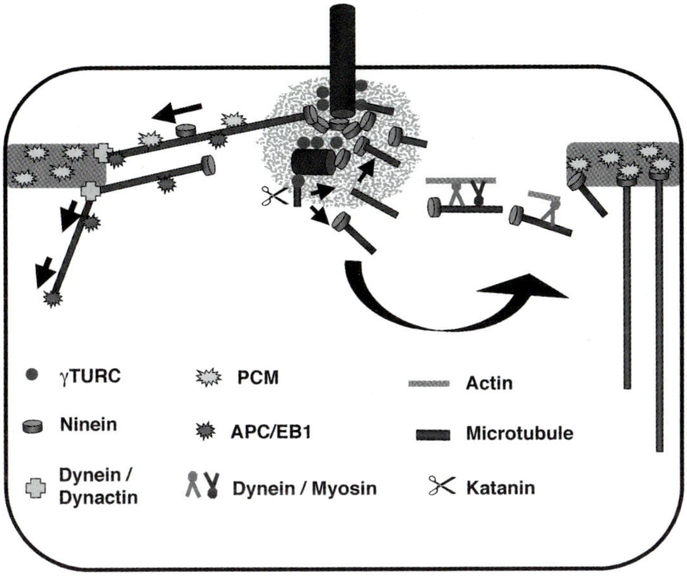

Figure 15.6 Release and capture. Two variants of the release and capture mechanism are illustrated to show the assembly of non-centrosomal apico-basal microtubule arrays in polarized epithelial cells. The right-hand side shows the classic release and capture version involving the release of relatively short microtubules and their subsequent translocation and capture at apical sites, whereas the left-hand side shows the modified version involving microtubule plus-end capture, release, translocation and minus-end capture. The two variant models are not mutually exclusive (see Color Plates page XXXV).

dynein is involved in the mediation of translocation towards the cell periphery [86]. Furthermore, it seems that translocation of short microtubules does not depend on a microtubule network, instead, movement is suggested to be mediated by dynein motors linked to actin or intermediate filaments. This has previously been shown for neurons. In neurons dynein and dynactin are responsible for microtubule translocation from the centrosome in the cell body into the axon and microtubules have been proposed to move relative to actin filaments [128, 133]. MyosinV is also a strong candidate for the translocation process. Very recent findings show that myosinVa binds to microtubules and mechanochemically couples microtubules to actin filaments [129]. It is tempting to speculate that factors such as mDia may facilitate the sliding interaction as overexpression of mDia results in co-alignment of microtubules and actin filaments [119] and that dual translocation of these filaments is mediated by dynein and myosinV (Figure 15.6).

In polarized epithelial cells, where the centrioles remain fairly close to each other, release of microtubules from the daughter centriole is likely to lead to their capture by anchoring complexes associated with the mother centriole. The consequence is likely to be the assembly of an initial extended radial array anchored at the centrosome, and this may be the case for MDCK cells. Plus-end capture of microtubules at the lateral cell surface, possibly mediated by plus-end-associated dynactin- and actin-anchored dyneins at the cell surface, may trigger microtubule release. The subsequent downward translocation and minus-end capture and anchorage at the apical site are likely to be mediated by anchored dynein. EB1 associates with the growing plus ends of microtubules and has been shown specifically to interact with the p150^{Glued} complex at the tip of microtubules and may deliver the dynein/dynactin complex to the cell cortex [130–132]. APC is likely to anchor microtubule plus-ends at cell surface sites and mediate interactions with actin at adherens junctions via β-catenin or via Asef at other sites. So a modified release and capture mechanism, involving plus-end capture, release, translocation and minus-end capture, may be operating in these polarized cells (Figure 15.6). Translocation of anchoring and docking complexes is likely to be mediated by motor proteins, microtubules and possibly carriers such as PCM-1.

The assembly of apico-basal microtubule arrays in most polarized epithelial cells, which have retained their centrosome is evidently controlled by the release and positioning of microtubules and minus-end anchoring complexes. The future holds exciting prospects for resolving the molecular basis for the release and translocation of microtubules and their minus-end anchoring complexes.

Acknowledgments

I would like to thank Martin Jackson and Kevin Tyler for helpful comments on the manuscript and Gemma Bellett for providing some of the images. I would also like to thank Jane Carter, David Moss and Gemma Bellett for useful discussions.

Notes

New exciting findings from Bornen's lab, reveal that ninein not only is critical for microtubule minus-end anchorage but also for the docking of γ-TuRCs at the centrosome. In addition ninein is shown to interact with the p150^{glued} subunit of dynactin and seems to be a major structural protein of the subdistal appendages of the mother centriole (communication Delgehyr and Bornens).

References

1. Knop,M., G. Pereira, and E. Schiebel. **1999**. Microtubule organization by the budding yeast spindle pole body. *Biol Cell* 91:291–304.
2. Daunderer, C., M. Schliwa, and R. Graf. **1999**. *Dictyostelium discoideum*: a promising centrosome model system. *Biol Cell* 91:313–320.
3. Chan, J., G. M. Calder, J. H. Doonan, and C. W. Lloyd. **2003**. EB1 reveals mobile microtuble nucleation sites in *Arabidopsis. Nature Cell Biol* 5:967–971.
4. Mogensen, M. M. **1999**. Microtubule release and capture in epithelial cells. *Biol Cell* 91:331–341.
5. Bornens, M. **2002**. Centrosome composition and microtubule anchoring mechanisms. *Curr Opin Cell Biol* 14:25–34.
6. Bornens, M., and E. Karsenti. **1984**. The centrosome. In *Membrane Structure and Function*, E. E. Bittar, (Ed.). John Wiley and Sons: New York, 99–172.
7. McIntosh, J. R., and U. Euteneuer. **1984**. Tubulin hooks as probes for microtubule polarity: an analysis of the method and an evaluation of data on microtubule polarity in the mitotic spindle. *J Cell Biol* 98:525–533.
8. Doxsey, J. **2001**. Re-evaluating centrosome function. *Nature Rev Mol Cell Biol* 2:688–698.
9. Bobinnec, Y., A. Khodjakov, L. M. Mir, C. L. Rieder, B. Edde', and M. Bornens. **1998**. Centriole disassembly *in vitro* and its effects on centrosome structure and function in vertebrate cells. *J. Cell Biol* 143:1575–1589.
10. Fry, A. M., P. Descombes, C. Twomey, R. Bacchieri, and E. A. Nigg. **2000**. The NIMA-related kinase X-Nek2B is required for efficient assembly of the zygotic centrosome in *Xenopus laevis. J Cell Sci* 113(Pt 11):1973–1984.
11. Uto, K., and N. Sagata. **2000**. Nek2B, a novel maternal form of Nek2 kinase, is essential for the assembly or maintenance of centrosomes in early *Xenopus* embryos. *EMBO J* 19:1816–1826.
12. Kirkham, M., T. Muller-Reichert, K. Oegema, S. Grill, and A. A. Hyman. **2003**. SAS-4 is a C. elegans centriolar protein that controls centrosome size. *Cell* 112:575–587.
13. Leidel, S., and P. Gonczy. **2003**. SAS-4 is essential for centrosome duplication in *C. elegans* and is recruited to daughter centrioles once per cell cycle. *Dev Cell* 4:431–439.
14. Berns, M. W., and S. M. Richardson. **1977**. Continuation of mitosis after selective laser microbeam destruction of the centriolar region. *J Cell Biol* 75:977–982.
15. Gould, R. R., and G. G. Borisy. **1977**. The pericentriolar material in Chinese hamster cells nucleates microtubule formation. *J Cell Biol* 73:601–615.
16. Hannak, E., K. Oegema, M. Kirkham, P. Gonczy, B. Habermann, and A. A. Hyman. **2002**. The kinetically dominant assembly pathway for centrosomal asters in *Caenorhabditis elegans* is γ-tubulin dependent. *J Cell Biol* 157:591–602.
17. Moritz, M., M. B. Braunfeld, J. W. Sedat, B. Albert, and D. A. Agard. **1995**. Microtubule nucleation by γ-tubulin-containing rings in the centrosome. *Nature* 378:638–640.
18. Oakley, C. E., and B. R. Oakley. **1989**. Identification of γ-tubulin, a new member of the tubulin superfamily encoded by mipA gene of *Aspergillus nidulans. Nature* 338:662–664.
19. Shu, H. B., and H. C. Joshi. **1995**. γ-Tubulin can both nucleate microtubule assembly and self-assemble into novel tubular structures in mammalian cells. *J Cell Biol* 130:1137–1147.
20. Stearns, T., and M. Kirschner. **1994**. *In vitro* reconstitution of centrosome assembly and function: the central role of γ-tubulin. *Cell* 76:623–637.
21. Zheng, Y., M. L. Wong, B. Alberts, and T. Mitchison. **1995**. Nucleation of microtubule assembly by a γ-tubulin-

containing ring complex. *Nature* 378:578–583.

22. Vogel, J. M., T. Stearns, C. L. Rieder, and R. E. Palazzo. **1997**. Centrosomes isolated from *Spisula solidissima* oocytes contain rings and an unusual stoichiometric ratio of α/β tubulin. *J Cell Biol* 137:193–202.
23. Gunawardane, R. N., O. C. Martin, and Y. Zheng. **2003**. Characterization of a new γ-TuRC subunit with WD repeats. *Mol Biol Cell* 14:1017–1026.
24. Murphy, S. M., A. M. Preble, U. K. Patel, K. L. O'Connell, D. P. Dias, M. Moritz, D. Agard, J. T. Stults, and T. Stearns. **2001**. GCP5 and GCP6: two new members of the human gamma-tubulin complex. *Mol Biol Cell* 12:3340–3352.
25. Tassin, A. M., C. Celati, M. Moudjou, and M. Bornens. **1998**. Characterization of the human homologue of the yeast spc98p and its association with gamma-tubulin. *J Cell Biol* 141:689–701.
26. Moudjou, M., N. Bordes, M. Paintrand, and M. Bornens. **1996**. γ-Tubulin in mammalian cells: the centrosomal and the cytosolic forms. *J Cell Sci* 109:875–887.
27. Erickson, H. P. **2000**. γ-Tubulin nucleation: template or protofilament? *Nature Cell Biol* 2:E93–E96.
28. Erickson, H. P., and D. Stoffler. **1996**. Protofilaments and rings, two confirmations of the γ-tubulin family conserved from bacterial FtsZ to α/β- and γ-tubulin. *J Cell Biol* 135:5–8.
29. Keating, T. J., and G. G. Borisy. **1999**. Centrosomal and non-centrosomal microtubules. *Biol Cell* 91:321–329.
30. Moritz, M., and D. A. Agard. **2001**. γ-tubulin complexes and microtubule nucleation. *Curr Opin Struct Biol* 11:174–181.
31. Moritz, M., M. B. Braunfeld, V. Guenebaut, J. Heuser, and D. A. Agard. **2000**. Structure of the γ-tubulin ring complex: a template for microtubule nucleation. *Nature Cell Biol* 2:365–370.
32. Wiese, C., and Y. Zheng. **2000**. A new function for the γ-tubulin ring complex as a microtubule minus-end cap. *Nature Cell Biol* 2:358–364.
33. Moritz, M., Y. Zheng, B. M. Alberts, and K. Oegema. **1998**. Recruitment of the γ-tubulin ring complex to *Drosophila* salt-stripped centrosome scaffolds. *J Cell Biol* 142:775–786.
34. Schnackenberg, B. J., A. Khodjakov, C. L. Rieder, and R. E. Palazzo. **1998**. The disassembly and reassembly of functional centrosomes *in vitro*. *Proc Natl Acad Sci USA* 95:9295–9300.
35. Knop, M., and E. Schiebel. **1997**. Spc98p and Spc97p of the yeast γ-tubulin complex mediate binding to the spindle pole body via their interaction with Spc110p. *EMBO J* 16:6985–6995.
36. Schiebel, E. **2000**. Gamma-tubulin complexes: binding to the centrosome, regulation and microtubule nucleation. *Curr Opin Cell Biol* 12:113–118.
37. do Carmo Avides, M., and D. M. Glover. **1999**. Abnormal spindle protein, Asp, and the integrity of mitotic centrosomal microtubule organizing centers. *Science* 283:1733–1735.
38. Murphy, S., L. Urbani, and T. Stearns. **1998**. The mammalian gamma-tubulin complex contains homologues of yeast spindle pole body components spc97p and spc98p. *J Cell Biol* 141:663–674.
39. Dictenberg, J. B., W. Zimmerman, C. A. Sparks, A. Young, C. Vidair, Y. Zheng, W. Carrington, F. S. Fay, and J. Doxsey. **1998**. Pericentrin and γ-tubulin form a protein complex and are organised into a novel lattice at the centrosome. *J Cell Biol* 141:163–174.
40. Doxsey, S. J., P. Stein, L. Evans, P. D. Calarco, and M. Kirschner. **1994**. Pericentrin, a highly conserved centrosome protein involved in microtubule organization. *Cell* 76:639–650.
41. Flory, M. R., and T. N. Davis. **2003**. The centrosomal proteins pericentrin and kendrin are encoded by alternatively spliced products of one gene. *Genomics* 82:401–405.
42. Gillingham, A. K., and S. Munro. **2000**. The PACT domain, a conserved centrosomal targeting motif in the coiled-coil proteins AKAP450 and pericentrin. *EMBO Rep* 1:524–529.
43. Kawaguchi, S. I., and Y. Zheng. **2004**. Characterization of a *Drosophila* cen-

trosome protein CP309 that shares homology with Kendrin and CG-NAP. *Mol Biol Cell* 15:37–45.

44. Keryer, G., O. Witczak, A. Delouvee, W. A. Kemmner, D. Rouillard, K. Tasken, and M. Bornens. **2003**. Dissociating the centrosomal matrix protein AKAP450 from centrioles impairs centriole duplication and cell cycle progression. *Mol Biol Cell* 14:2436–2446.

45. Li, Q., D. Hansen, A. Killilea, H. C. Joshi, R. E. Palazzo, and R. Balczon. **2001**. Kendrin/pericentrin-B, a centrosome protein with homology to pericentrin that complexes with PCM-1. *J Cell Sci* 114:797–809.

46. Steadman, B. T., P. H. Schmidt, R. A. Shanks, L. A. Lapierre, and J. R. Goldenring. **2002**. Transforming acidic coiled-coil-containing protein 4 interacts with centrosomal AKAP350 and the mitotic spindle apparatus. *J Biol Chem* 277:30165–30176.

47. Takahashi, M., A. Yamagiwa, T. Nishimura, H. Mukai, and Y. Ono. **2002**. Centrosomal proteins CG-NAP and kendrin provide microtubule nucleation sites by anchoring gamma-tubulin ring complex. *Mol Biol Cell* 13:3235–3245.

48. Ohta, T., R. Essner, J. H. Ryu, R. E. Palazzo, Y. Uetake, and R. Kuriyama. **2002**. Characterization of Cep135, a novel coiled-coil centrosomal protein involved in microtubule organization in mammalian cells. *J Cell Biol* 156:87–99.

49. Casenghi, M., P. Meraldi, U. Weinhart, P. I. Duncan, R. Korner, and E. A. Nigg. **2003**. Polo-like kinase 1 regulates Nlp, a centrosome protein involved in microtubule nucleation. *Dev Cell* 5:113–125.

50. Keryer, G., B. Di Fiore, C. Celati, K. F. Lechtreck, M. Mogensen, A. Delouvee, P. Lavia, M. Bornens, and A. M. Tassin. **2003**. Part of Ran is associated with AKAP450 at the centrosome: involvement in microtubule-organizing activity. *Mol Biol Cell* 14:4260–4271.

51. Mogensen, M. M., J. B. Mackie, S. J. Doxsey, T. Stearns, and J. B. Tucker. **1997**. Centrosomal deployment of gamma-tubulin and pericentrin: evidence for a microtubule-nucleating domain and a minus-end docking domain in certain mouse epithelial cells. *Cell Motil Cytoskeleton* 36: 276–290.

52. Mogensen, M. M., A. Malik, M. Piel, V. Bouckson-Castaing, and M. Bornens. **2000**. Microtubule minus-end anchorage at centrosomal and non-centrosomal sites: the role of ninein. *J Cell Sci.* 113(Pt 17):3013–3023.

53. Piel, M., P. Meyer, A. Khodjakov, C. L. Rieder, and M. Bornens. **2000**. The respective contributions of the mother and daughter centrioles to centrosome activity and behavior in vertebrate cells. *J Cell Biol* 149:317–330.

54. Gilbert, T., A. Le Bivic, A. Quaroni, and E. Rodriguez-Boulan. **1991**. Microtubular organization and its involvement in the biogenetic pathways of plasma membrane proteins in Caco-2 intestinal epithelial cells. *J Cell Biol* 113: 275–288.

55. Patuzzi, R. **1996**. *Cochlear Micromechanics and Macromechanics.* Springer-Verlag: Berlin, 186–257.

56. Tassin, A.-M., B. Maro, and M. Bornens. 1985. Fate of microtubules-organizing centers during myogenesis *in vitro*. *J Cell Biol* 100:35–46.

57. Mogensen, M. M., and J. B. Tucker. **1987**. Evidence for microtubule nucleation at plasma membrane-associated sites in *Drosophila*. *J Cell Sci* 88:95–107.

58. Mogensen, M. M., J. B. Tucker, and T. Baggaley. **1993**. Multiple plasma membrane-associated MTOC systems in the acentrosomal cone cells of *Drosophila* ommatidia. *Eur J Cell Biol* 60:67–75.

59. Mogensen, M. M., J. B. Tucker, and H. Stebbings. **1989**. Microtubule polarities indicate that nucleation and capture of microtubules occurs at cell surfaces in *Drosophila*. *J Cell Biol* 108:1445–1452.

60. Millar, A., A. Hooper, L. Copeland, F. Cummings, and A. Prescott. **1997**. Reorganisation of the microtubule cytoskeleton and centrosomal loss during lens fibre cell differentiation. *Nova Acta Leopoldina* 75:169–183.

61. Prescott, A. R., S. Webb, D. Rawlins, P. Shaw, and R. M. Warn. **1991**. Microtubules rich in post-translationally modified alpha-tubulin form distinct arrays in frog lens epithelial cells. *Exp Eye Res* 52:743–746.

62. Vogl, A. W., M. Weis, and D. C. Pfeiffer. **1995**. The perinuclear centriole-containing centrosome is not the major microtubule organizing center in Sertoli cells. *Eur J Cell Biol* 66:165–179.

63. Bacallao, R., C. Antony, C. Dotti, E. Karsenti, E. H.K. Stelzer, and K. Simons. **1989**. The subcellular organization of Madin–Darby Canine Kidney cells during the formation of a polarized epithelium. *J Cell Biol* 109:2817–2832.

64. Bré, M.-H., R. Pepperkok, A. M. Hill, N. Levilliers, W. Ansorge, E. H.K. Stelzer, and E. Karsenti. **1990**. Regulation of microtubule dynamics and nucleation during polarization in MDCK II cells. *J Cell Biol* 111:3013–3021.

65. Bré, M.-H., T. E. Kreis, and E. Karsenti. **1987**. Control of microtubule nucleation and stability in Madin–Darby kidney cells: the occurrence of noncentrosomal, stable detyrosinated microtubules. *J Cell Biol* 105:1283–1296.

66. Buendia, B., M.-H. Bré, G. Griffiths, and E. Karsenti. **1990**. Cytoskeletal control of centrioles movement during establishment of polarity in MDCK cells. *J Cell Biol* 110:1123–1135.

67. Furness, D. N., C. M. Hackney, and P. S. Streyger. **1990**. Organisation of microtubules in cochlear hair cells. *J Electron Microscopy Techn* 15:261–279.

68. Troutt, L., E. Wang, K. Pagh-Roehl, and B. Burnside. **1990**. Microtubule nucleation and organization in teleost photoreceptors: microtubule recovery after elimination by cold. *J Neurocytol* 19:213–223.

69. Troutt, L. L., and B. Burnside. **1988**. Microtubule polarity and distribution in teleost photoreceptors. *J Neurosci* 8:2371–2380.

70. Troutt, L. L., and B. Burnside. **1988**. The unusual microtubule polarity in teleost retinal pigment epithelial cells. *J Cell Biol* 107:1461–1464.

71. Meads, T., and T. A. Schroer. **1995**. Polarity and nucleation of microtubules in polarized epithelial cells. *Cell Motil Cytoskeleton* 32:273–288.

72. Tucker, J. B., C. C. Paton, G. P. Richardson, M. M. Mogensen, and I. J. Russell. **1992**. A cell surface-associated centrosomal layer of microtubule-organizing material in the inner pillar cell of the mouse cochlea. *J Cell Sci* 102:215–226.

73. Tucker, J. B., M. M. Mogensen, C. C. Paton, J. B. Mackie, C. G. Henderson, and L. M. Leckie. **1995**. Formation of two microtubule-nucleating sites which perform differently during centrosomal reorganization in a mouse cochlear epithelial cell. *J Cell Sci* 108:1333–1345.

74. Henderson, C. G., J. B. Tucker, M. A. Chaplin, J. B. Mackie, S. N. Maidment, M. M. Mogensen, and C. C. Paton. **1994**. Reorganisation of the centrosome and associated microtubules during the morphogenesis of a mouse cochlear epithelial cell. *J. Cell Sci* 107:589–600.

75. Mogensen, M. M., J. B. Tucker, J. B. Mackie, A. R. Prescott, and I. S. Nathke. **2002**. The adenomatous polyposis coli protein unambiguously localizes to microtubule plus ends and is involved in establishing parallel arrays of microtubule bundles in highly polarized epithelial cells. *J Cell Biol* 157:1041–1048.

76. Rizzolo, L. J., and H. C. Joshi. **1993**. Apical orientation of the microtubule organizing center and associated γ-tubulin during the polarization of the retinal pigment epithelium *in vivo*. *Dev Biol* 157:147–156.

77. Quintyne, N. J., S. R. Gill, D. M. Eckley, C. L. Crego, D. A. Compton, and T. A. Schroer. **1999**. Dynactin is required for microtubule anchoring at centrosomes. *J Cell Biol* 147: 321–334.

78. Schroer, T. A. **2001**. Microtubule don and doff their caps: dynamic attachments at plus and minus ends. *Curr Opin Cell Biol* 13:92–96.

79. De Brabander, M., G. Geuens, R. Nuydens, R. Willebrords, and J. De

May. **1982**. Microtubule stability and assembly in living cells: The influence of metabolic inhibitors, taxol and pH. *Cold Spring Harbor Symp Quant Biol* 46:227–240.

80. Bouckson-Castaing, V., M. Moudjou, D. J.P. Ferguson, S. Mucklow, Y. Belkaid, G. Milon, and P. R. Crocker. **1996**. Molecular characterisation of ninein, a new coiled-coil protein of the centrosome. *J Cell Sci* 109:179–190.

81. Hong, Y.-R., C.-H. Chen, J.-H. Chang, S.-K. Wang, W.-D. Sy, C.-K. Chou, and S.-L. Howng. **2000**. Cloning and characterization of a novel human ninein protein that interacts with the glycogen synthase kinase 3b1. *Biochem Biophys Acta* 1492:513–516.

82. Hong, Y.-R., C.-H. Chen, C.-K. Chou, S.-Y. Liou, and S.-L. Howng. **2000**. Genomic organization and molecular characterization of the human ninein gene. *Biochem Biophys Res Comm* 279:989-995.

83. Ou, Y. Y., G. J. Mack, M. Zhang, and J. B. Rattner. **2002**. CEP110 and ninein are located in a specific domain of the centrosome associated with centrosome maturation. *J Cell Sci* 115:1825–1835.

84. Paintrand, M., M. Moudjou, H. Delacroix, and M. Bornens. **1992**. Centrosome organisation and centriole architecture: their sensitivity to divalent cations. *J Struct Biol* 108:107–128.

85. Tassin, A.-M., and M. Bornens. **1999**. Centrosome structure and microtubule nucleation in animal cells. *Biol Cell* 91:343–354.

86. Abal, M., M. Piel, V. Bouckson-Castaing, M. Mogensen, J. B. Sibarita, and M. Bornens. **2002**. Microtubule release from the centrosome in migrating cells. *J Cell Biol* 159:731–737.

87. Dammermann, A., and A. Merdes. **2002**. Assembly of centrosomal proteins and microtubule organization depends on PCM-1. *J Cell Biol* 159:255–266.

88. Chang, P., T. H. Giddings Jr, M. Winey, and T. Stearns. **2003**. ε-Tubulin is required for centriole duplication and microtubule organization. *Nature Cell Biol.* 5:71–76.

89. Gromley, A., A. Jurczyk, J. Sillibourne, E. Halilovic, M. Mogensen, I. Groisman, M. Blomberg, and S. Doxsey. **2003**. A novel human protein of the maternal centriole is required for the final stages of cytokinesis and entry into S phase. *J Cell Biol* 161:535–545.

90. Nakagawa, Y., Y. Yamane, T. Okanoue, and S. Tsukita. **2001**. Outer dense fiber 2 is a widespread centrosome scaffold component preferentially associated with mother centrioles: its identification from isolated centrosomes. *Mol Biol Cell* 12:1687–1697.

91. Quintyne, N. J., and T. A. Schroer. **2002**. Distinct cell cycle-dependent roles for dynactin and dynein at centrosomes. *J Cell Biol* 159:245–254.

92. Vorobjev, I. A., and E. S. Nadezhdina. **1987**. The centrosome and its role in the organization of microtubules. *Int Rev Cytol* 106:227–293.

93. Keating, T. J., J. G. Peloquin, V. I. Rodionov, D. Momcilovic, and G. G. Borisy. **1997**. Microtubule release from the centrosome. *Proc Natl Acad Sci USA* 94:5078–5083.

94. Rusan, N. M., C. J. Fagerstrom, A. M. Yvon, and P. Wadsworth. **2001**. Cell cycle-dependent changes in microtubule dynamics in living cells expressing green fluorescent protein-alpha tubulin. *Mol Biol Cell* 12:971–980.

95. Ahmad, F. J., W. Yu, F. J. McNally, and P. W. Baas. **1999**. An essential role for katanin in severing microtubules in the neuron. *J Cell Biol* 145:305–315.

96. Baas, P. W. **1999**. Microtubules and neuronal polarity: lessons from mitosis. *Neuron* 22:23–31.

97. McNally, F. J., and R. D. Vale. **1993**. Identification of katanin, an ATPase that severs and disassembles stable microtubules. *Cell* 75:419–429.

98. McNally, K. P., O. A. Bazirgan, and F. J. McNally. **2000**. Two domains of p80 katanin regulate microtubule severing and spindle pole targeting by p60 katanin. *J Cell Sci* 113:1623–1633.

99. McNally, F. J., K. Okawa, A. Iwamatsu, and R. D. Vale. **1996**. Katanin, the microtubule-severing ATPase, is concentrated at centrosomes. *J Cell Sci* 109:561–567.

100. McNally, F. J., and S. Thomas. **1998**. Katanin is responsible for the M-phase microtubule-severing activity in *Xenopus* eggs. *Mol Biol Cell* 9:1847–1861.

101. O'Toole, E. T., K. L. McDonald, J. Mantler, J. R. McIntosh, A. A. Hyman, and T. Muller-Reichert. **2003**. Morphologically distinct microtubule ends in the mitotic centrosome of *Caenorhabditis elegans. J Cell Biol* 163:451–456.

102. Srayko, M., D. W. Buster, O. A. Bazirgan, F. J. McNally, and P. E. Mains. **2000**. MEI-1/MEI-2 katanin-like microtubule severing activity is required for *Caenorhabditis elegans* meiosis. *Genes Dev* 14:1072–1084.

103. Yang, H. Y., K. McNally, and F. J. McNally. **2003**. MEI-1/katanin is required for translocation of the meiosis I spindle to the oocyte cortex in *C. elegans. Dev Biol* 260:245–259.

104. Lohret, T. A., F. J. McNally, and L. M. Quarmby. **1998**. A role for katanin-mediated axonemal severing during *Chlamydomonas* deflagellation. *Mol Biol Cell* 9:1195–1207.

105. Burk, D. H., B. Liu, R. Zhong, W. H. Morrison, and Z. H. Ye. **2001**. A katanin-like protein regulates normal cell wall biosynthesis and cell elongation. *Plant Cell* 13:807–827.

106. Chen, C. H., S. L. Howng, T. S. Cheng, M. H. Chou, C. Y. Huang, and Y. R. Hong. **2003**. Molecular characterization of human ninein protein: two distinct subdomains required for centrosomal targeting and regulating signals in cell cycle. *Biochem Biophys Res Commun* 308:975–983.

107. Kubo, A., H. Sasaki, A. Yuba-Kubo, S. Tsukita, and N. Shiina. **1999**. Centriolar satellites: molecular characterization, ATP-dependent movement toward centrioles and possible involvement in ciliogenesis. *J Cell Biol* 147:969–980.

108. Kubo, A., and S. Tsukita. **2003**. Non-membranous granular organelle consisting of PCM-1: subcellular distribution and cell-cycle-dependent assembly/disassembly. *J Cell Sci* 116:919–928.

109. Schmidt, P. H., D. T. Dransfield, J. O. Claudio, R. G. Hawley, K. W. Trotter, S. L. Milgram, and J. R. Goldenring. **1999**. AKAP350, a multiply spliced protein kinase A-anchoring protein associated with centrosomes. *J Biol Chem* 274:3055–3066.

110. Keating, T. J., and G. G. Borisy. **2000**. Immunostructural evidence for the template mechanism of microtubule nucleation. *Nature Cell Biol* 2:352–357.

111. Baas, P. W., and H. C. Joshi. **1992**. γ-Tubulin distribution in the neuron: Implications for the origin of neuritic microtubules. *J Cell Biol* 119:171–178.

112. Rodionov, V., E. Nadezhdina, and G. Borisy. **1999**. Centrosomal control of microtubule dynamics. *Proc Natl Acad Sci USA* 96:115–120.

113. Rodionov, V. I., and G. G. Borisy. **1997**. Microtubule treadmilling *in vivo. Science* 275:215–218.

114. Mitchison, T., and M. Kirschner. **1984**. Dynamic instability of microtubule growth. *Nature* 312:237–242.

115. Pepperkok, R., M. H. Bre, J. Davoust, and T. E. Kreis. **1990**. Microtubules are stabilized in confluent epithelial cells but not in fibroblasts. *J Cell Biol* 111:3003–3012.

116. Chausovsky, A., A. D. Bershadsky, and G. G. Borisy. **2000**. Cadherin-mediated regulation of microtubule dynamics. *Nature Cell Biol* 2:797–804.

117. Waterman-Storer, C. M., W. C. Salmon, and E. D. Salmon. **2000**. Feedback interactions between cell–cell adherens junctions and cytoskeletal dynamics in newt lung epithelial cells. *Mol Biol Cell* 11:2471–2483.

118. Gundersen, G. G. **2002**. Evolutionary conservation of microtubule-capture mechanisms. *Nature Rev Mol Cell Biol* 3:296–304.

119. Palazzo, A. F., T. A. Cook, A. S. Alberts, and G. G. Gundersen. **2001**. mDia mediates Rho-regulated formation and orientation of stable microtubules. *Nature Cell Biol* 3:723–729.

120. Tannenbaum, J., and N. B. Slepecky. **1997**. Localization of microtubules containing posttranslationally modified tubulin in cochlear epithelial cells during development. *Cell Motil Cytoskeleton* 38:146–162.

121. Prescott, A. R., P. G. Dowrick, and R. M. Warn. **1992**. Stable and slow-turning-over microtubules characterize the processes of motile epithelial cells treated with scatter factor. *J Cell Sci* 102(Pt 1):103–112.

122. Vorobjev, I. A., T. M. Svitkina, and G. G. Borisy. **1997**. Cytoplasmic assembly of microtubules in cultured cells. *J Cell Sci* 110:2635–2645.

123. Yvon, A. C., and P. Wadsworth. **1997**. Non-centrosomal microtubule formation and measurement of minus end microtubule dynamics in A498 cells. *J Cell Sci* 110:2391–2401.

124. Faragher, A. J., and A. M. Fry. **2003**. Nek2A kinase stimulates centrosome disjunction and is required for formation of bipolar mitotic spindles. *Mol Biol Cell* 14:2876–2889.

125. Fry, A. M., P. Meraldi, and E. A. Nigg. **1998**. A centrosomal function for the human Nek2 protein kinase, a member of the NIMA family of cell cycle regulators. *EMBO J* 17:470–481.

126. Helps, N. R., X. Luo, H. M. Barker, and P. T. Cohen. **2000**. NIMA-related kinase 2 (Nek2), a cell-cycle-regulated protein kinase localized to centrosomes, is complexed to protein phosphatase 1. *Biochem J* 349:509–518.

127. Mayor, T., Y. D. Stierhof, K. Tanaka, A. M. Fry, and E. A. Nigg. **2000**. The centrosomal protein C-Nap1 is required for cell cycle-regulated centrosome cohesion. *J Cell Biol* 151:837–846.

128. Baas, P. W. **1998**. The role of motor proteins in establishing the microtubule arrays of axons and dendrites. *J Chem Neuroanat* 14:175–180.

129. Cao, T. T., W. M. Chang, S. E., and M. S. Mooseker. **2003**. Myosin-Va binds to and mechanochemically couples microtubules to actin filaments. *Mol Biol Cell* Epress E03-07-0504.

130. Allan, V., and I. S. Nathke. **2001**. Catch and pull a microtubule: getting a grasp on the cortex. *Nature Cell Biol* 3:E226–E228.

131. Askham, J. M., K. T. Vaughan, H. V. Goodson, and E. E. Morrison. **2002**. Evidence that an interaction between EB1 and p150(Glued) is required for the formation and maintenance of a radial microtubule array anchored at the centrosome. *Mol Biol Cell* 13:3627–3645.

132. Dikovskaya, D., J. Zumbrunn, G. A. Penman, and I. S. Nathke. **2001**. The adenomatous polyposis coli protein: in the limelight out at the edge. *Trends Cell Biol* 11:378–384.

133. Ahmad, F.J., Echeverri, C. J., Vallee, R. B., Baas, P. W. **1998**. Cytoplasmic dynein and dynactin are required for the transport of microtubules into the axon. *J Cell Biol* 140:391–401.

Part IV
Centrosomes in Disease

Centrosomes in Development and Disease. Edited by Erich A. Nigg

ISBN 3-527-30980-2

16
Centrosome Anomalies in Cancer: From Early Observations to Animal Models

Thea M. Goepfert and William R. Brinkley

16.1
Early Observations

Early in the last century, Theodor Boveri proposed that the characteristics seen in malignant tumors, such as loss of cell polarity and chromosomal segregation abnormalities (aneuploidy), result from defects in centrosome function [1]. Boveri first coined the terms "centrosome" and "centriole", described their location in the cytoplasm, and identified their role in mitosis and aneuploidy:

> *The centrosome generally lies outside but near the nucleus. In its passive condition it is a minute speck in the cell; but in its active phase it becomes the centre of an "attraction sphere", with radiating fibers, constituting the aster. Along these fibers the split chromosomes travel in cell-division to the opposite poles of the achromatic spindle. [2, 3]*

As early as 1887, Boveri determined that inheritance and heritable elements were intimately associated with chromosomes. In joint studies with Edouard van Beneden on the horse nematode, *Ascaris megalocephala*, Boveri observed that chromosomes (chromatic elements) migrated toward one of the two poles in normally dividing eggs. Boveri reasoned that "the nuclei of the resulting daughter cells must receive different qualities, in case we are to ascribe different qualities to the individual chromatic elements" [2]. Ultimately, he realized that Mendel's heredity laws from 35 years earlier were in perfect accordance with his own cytological and developmental discoveries.

Boveri and van Beneden also observed abnormal cell division in *Ascaris* eggs. Occasionally, cells divided with multiple spindle poles (tetrasters) when two sperms fertilized one egg (dispermic egg). An example is shown in Figure 16.1, which demonstrates the resulting unequal distribution of chromosomes to four poles. From this observation, they theorized that "the harmful effect of multiple poles is due to the fact that as a rule they cause an abnormal chromatin complement in the daughter cells" [4].

Centrosomes in Development and Disease. Edited by Erich A. Nigg

ISBN 3-527-30980-2

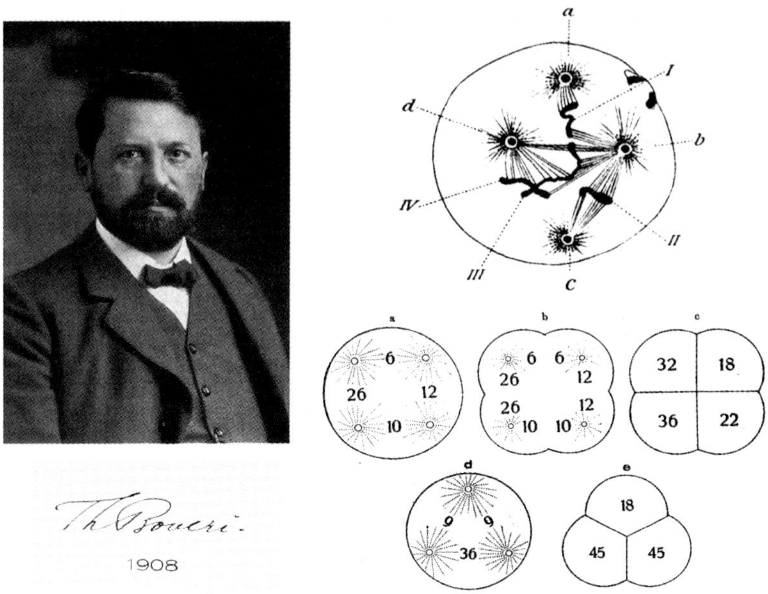

Figure 16.1 Th. Boveri (1908) and tetrapolar spindle in an egg of *Ascaris megalocephala* bivalens. Chromosomes arranged on several spindles. Top right of the figure: of the four poles, two (a,c) will receive only one chromosome, b will receive three or four; d, two [2].

Boveri continued to explore the consequences of chromosome duplication by studying the complex division of fertilized sea urchin eggs. In culture, a fertilized sea urchin egg divides to form four cells without passing through the normal two-cell stage. This cell division involves four distinct spindle poles, and each of the resulting cells, if gently separated, can develop into a normal adult. In dispermic sea urchin eggs, however, when Boveri separated daughter cells at the four-cell stage and looked at their development, he found that the individual "quarter embryos" rarely developed normally. He also observed that each of the four cells developed differently from the others. Boveri therefore hypothesized that each cell needs a full set of chromosomes for normal development [4]. If any chromosomes are missing, the cell lacks "developmental potential". Boveri's theory that chromosomes are responsible for the normal cell function and his observation that aberrant chromosome combinations in dispermic eggs cause damage to such cells, led him to postulate a link between cell division and cancer. When Boveri published the results of his experiments on the development of double-fertilized sea urchin eggs in 1902, he added the suggestion that "malignant tumors might be the result of a certain abnormal condition of the chromosomes, which may arise from multipolar mitosis". His perception that the "tumor problem is a cell problem", seems obvious today, but it represented an important step toward understanding the significance of the biology of cell division in cancer.

David von Hansemann was another early pathologist who recognized the potential role of the mitotic process in cancer. Despite von Hansemann's lesser stature in the scientific world of the 19th century, his writings significantly influenced Boveri. Von Hansemann coined the term anaplasia and defined it as:

> *A process carrying the cell in some entirely new direction – a direction, moreover, which is not the same in all tumors, nor even constant in the same tumor – the anaplastic cell is one in which, through some unknown agency, a progressive disorganization of the mitotic process occurs. [5, 6]*

Although Boveri never studied cancer *per se*, his seminal paper "Zur Frage der Entstehung maligner Tumoren" (The origin of malignant tumors), postulated (a) that the neoplastic properties of a cancer cell derive from chromosomal aberrations, and (b) that a malignant transformation results from the clonal expansion and proliferation of a single chromosomally-altered somatic cell [1].

From meticulous experimentation and tested hypotheses, Boveri concluded that for the products of cell division to develop normally, they must receive from their progenitor a full set of the structures that determine inheritance – the chromosomes. With these elegant experiments he set the agenda for cytogenetics and cell division research for the centuries to come. E. B. Wilson wrote in memory of Boveri:

> *The work was in high degree original, logical, accurate, and thorough. It enriched biological science with some of the most interesting discoveries and fruitful new conceptions of our time. [7]*

Boveri's hypothesis on the role of the centrosome in normal mitosis and its significance in cancer was an attempt to construct a unifying vision for cancer, without significant experimental verification. He acknowledged this and in 1904, concluded that:

> *...building castles in the air can provide the stimulus which is essential for carrying out painstaking experimental work. [6]*

No longer "castles in the air", Boveri's innovative hypothesis implicating centrosomes in the development of cancer has recently been resurrected [8] and re-examined [8–12], and currently finds support in several studies, using both tumor-derived cell lines and animal tumor models. During their development and progression, both primary and metastatic human tumors have been found to display prominent centrosome anomalies, especially centrosome amplification. Careful examination of human breast carcinomas and normal breast specimens revealed that structural and functional centrosome abnormalities are characteristic of cancer cells *in situ* [9, 10]. Characteristic structural alterations included increased centrosome number and volume, supernumerary centrioles, accumulation of excess pericentriolar material, and inappropriate phosphorylation of centrosomal proteins. As originally proposed by Boveri, these observations have important implications for understanding the mechanisms underlying two characteristics of high-grade tumors: genomic instability and the loss of cell polarity [11, 12]. At least one pathway to genomic instability appears to be linked to the amplification of functional centrosomes, a phenomenon that is frequently observed in tumor types [13]. A significant fraction of precursor lesions were found to display centrosome defects, including *in situ* carcinoma of the uterine cervix, prostate, and female breast [14]. Moreover, centrosome defects occurred together with mitotic spindle defects, chromosome instability, and high cytological grade.

These reports were followed by additional findings and two articles entitled "Managing the centrosome numbers game: from chaos to stability in cancer cell division" [15] and "Centrosome aberrations: cause or consequence of cancer progression" [16], that offered insights into mechanisms and molecular scenarios for the origin of supernumerary centrosomes.

16.2 Origin of Centrosome Anomalies

At least four mechanisms allow cells to acquire more than two centrosomes (Figure 16.2) and these will be discussed in Sections 16.2.1–16.2.4.

16.2.1 Deregulation of Centrosome Duplication

Two major controls regulate centrosome duplication: proper timing of the initiation of duplication, which occurs at late G1/S-phase of the cell cycle, and suppression of reduplication of duplicated centrosomes (see Chapter 9). If these controls

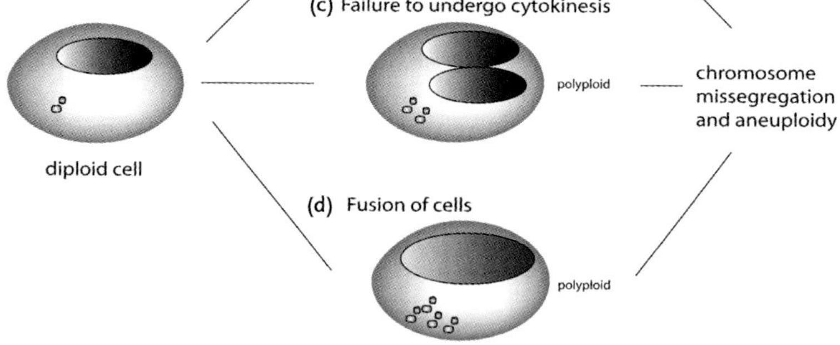

Figure 16.2 Origin of supernumerary centrosomes. Four mechanisms for the accumulation of an excess number of centrosomes are illustrated: (a) deregulation of centrosome duplication; (b) *de novo* formation of centrioles or uncontrolled separation of centriole pairs; (c) failure to undergo cytokinesis, and (d) fusion of cells.

are abrogated, centrosomes will duplicate multiple times within a single cell cycle, resulting in centrosome hyperamplification [17, 18].

By fusing cells in different cell-cycle stages, Wong et al. found that G2 centrosomes were unable to reduplicate in a cellular environment that supports centrosome duplication [19]. The block to reduplication is intrinsic to the centrosomes rather than the cytoplasm. Human primary cells exert tight control over centrosome number during prolonged S-phase arrest but partially lose this control in transformed cells. Thus, genomic stability may require control over centrosome duplication [19]. The presence of aberrant centrosome numbers in many types of cancer [10, 20, 21] make it likely that cancer cells characteristically have defects that prevent them from suppressing centrosome reduplication.

16.2.2
De Novo Formation of Centrioles or Uncontrolled Separation of Centriole Pairs

Studies on the consequences of centrosome ablation by laser microsurgery (see Chapter 10) indicate that many vertebrate cells can form centrioles *de novo*, a mechanism normally suppressed by existing centrioles [22]. Centrioles are physically partitioned from each other prior to duplication (see Chapter 9). Abrogation of this control causes the centriole pair to split apart, generating two centrosomes, each containing a single centriole and pericentriolar material, thereby causing acquisition of an extra microtubule organizing centre. If cell mutation(s) leads to overexpression of certain PCM components, the cells may form acentriolar centrosomes, as seen normally in the meiotic phases of oocytes that lack centrioles [23].

16.2.3
Failure to Undergo Cytokinesis

Cell division failure can have several primary causes, including the deregulation of pathways that coordinate mitotic progression and cytokinesis (see Chapter 8), mutational activation or inactivation of a certain protein that controls cytokinesis, the persistence of unrepaired DNA damage, or malfunction of the spindle-assembly checkpoint (see Chapter 11), or adaptation to a prolonged checkpoint. If these aberrations accompany cells that enter the next cell cycle, they will reduplicate the centrosomes, resulting in an excess number of centrosomes and twice the normal amount of DNA [24].

16.2.4
Fusion of Cells

Fusion-induced centrosome amplification has been observed following ectopic expression of the RAD6 ubiquitin-conjugating enzyme in human breast epithelial cells [25]. Adjacent nuclei of Rad6-overexpressing cells frequently polarized toward one another, promoting cell–cell fusion due to the fusion of the cell membranes separating them. These cells contained supernumerary centrosomes and either single giant nuclei or multiple multi-lobed nuclei.

The mechanisms described for generating centrosome anomalies are not mutually exclusive, and insufficient experimental and observational evidence exists to favor one mechanism over another.

Cell cycle regulatory proteins can modulate the numerical homeostasis of centrosomes. Thus, activation or inactivation of these proteins, which is commonly seen in cancer cells can disrupt centrosome biology and promote genomic instability. Specifically, genes implicated in centrosome amplification affect the following processes: the p53 pathway including p53, WAF1, Gadd445 and MDM2; the DNA-repair pathway including ATR, Brca1, Brca2, and XRCC2/3; protein degradation including Tsg 101, Skp2, and RAD5; and mitosis including Aurora A and survivin (for a review, see [16]).

16.3
Animal Models

Researchers have traced the pathways that cause centrosome aberrations in animal models by mimicking the deregulation of gene products of cell-cycle regulatory proteins in animal models. The remainder of this chapter will focus on recent findings regarding three such proteins, the tumor suppressor p53, the DNA damage repair protein BRCA1, and the ser/thr kinase Aurora A, and their roles in tumorigenesis as regulators of centrosome function and mitotic control.

16.3.1
Centrosome Anomalies and the p53 Pathway

Early studies of embryonic fibroblasts derived from p53 −/− mice suggested that p53 loss leads to centrosome amplification [21]. In these studies, loss of p53 was described to cause multiple rounds of centrosome duplication in a single S-phase, resulting in an abnormal number of centrosomes, which in turn, frequently resulted in defective mitosis organized by multiple spindle poles associated with chromosome loss and/or gain (aneuploidy). More recent findings suggest that p53 might directly control both initiation of centrosome duplication and suppression of reduplication. In p53 +/+ MEFs (mouse embryonic fibroblasts), initiation of centrosome duplication is tightly coupled with initiation of DNA duplication. In contrast, in p53 −/− MEFs, centrosomes initiate duplication early in G1 much before S-phase entry [26, 27].

The correlation between p53 mutations and centrosome hyperamplification (28) has been shown in animal models, cultured human tumor cell lines and human tumors, including breast tumors [29, 30], squamous cell carcinomas of the head and neck [29], cerebral primitive neuroectodermal tumor cells [31], and prostate cancer cell lines [32]. Cells expressing viral oncoproteins, which inactivate p53, also accumulate abnormal numbers of centrosomes (see Chapter 18).

Earlier studies found that expression of a p53 mutant in the epidermis of transgenic mice causes chromosome amplification and accelerated chemical carcinogenesis. This model expressed a murine *p53172R* $\rightarrow$ *H* mutant (*p53*m) under the control of a human keratin-1-based vector (HK1.p53m) [33]. HK1.p53m mice, in contrast to mice with wild-type *p53* and *p53*-knockout mice, exhibit increased susceptibility to chemical carcinogenesis with greatly accelerated benign papilloma formation, malignant conversion, and metastasis. In HK1.p53m papillomas, *in situ* examination of centrosomes demonstrated abnormalities at high frequencies (75 % of cells contained more than three centrosomes per cell). Significantly, equivalent *p53*-null tumors exhibited few abnormal centrosomes (4 % of cells contained more than three centrosomes per cell) [34]. These data suggest that the *p53* mutant accelerates tumorigenesis by exerting a gain-of-function associated with abnormal amplification of centrosomes.

In some cases, however, p53 fluctuation and centrosome anomalies do not correlate. For example, inactivation of p53 in human diploid cell lines induced neither centrosome hyperamplification nor chromosome instability [35, 36]. Analyses of breast carcinomas and tumor-derived cell lines indicate that centrosome amplification can arise independent of estrogen receptor (ER) or p53 status and is a common feature of aneuploid breast tumors [37]. Because most pre-invasive lesions are not uniformly mutant for p53, the development of centrosome defects does not appear to require abrogation of p53 function [14].

In conclusion, recent studies challenge the theory that p53 loss directly affects centrosome duplication [28]. Instead, centrosome anomalies may arise when the missing checkpoint function, caused by a loss of p53, leads to survival of cells that have undergone errors in mitosis.

16.3.2
Centrosome Anomalies and BRCA 1

Centrosome duplication is tightly controlled and occurs only once in a normal mammalian cell cycle (see Chapter 9). Interference with this process, as discussed above, can cause centrosome amplification, multipolar spindles, unequal chromosome segregation, and tumorigenesis. The breast cancer suppressor gene 1 (BRCA1) appears to contribute to the regulation of centrosome duplication. Many other biological processes also involve BRCA1, including transcriptional activation and repression, cell cycle regulation, chromatin remodeling, and DNA damage repair (see [38] for a review). BRCA1 has been localized at the centrosome [39] and seems to interact with a variety of proteins, including BRCA2, CDK2-cyclin A, CDK2-cyclin E, Gadd45, p21, p53, and Rb.

Targeted deletion of BRCA1 exon 11 has been shown to cause centrosome amplification [40]. Mouse embryonic fibroblasts derived from BRCA1 $\Delta 11/\Delta 11$ embryos contained the short isoform of BRCA1 and could grow in culture, unlike BRCA1-null cells. The BRCA1 $\Delta 11/\Delta 11$ cells grew slower in culture and exhibited more than two centrosomes per cell in 25 % of the cell population. In addition, the mutant cells were aneuploid, forming multipolar spindles during mitosis that resulted in unequal chromosome segregation and micronuclei formation.

Some evidence suggests, however, that BRCA1, like p53, may only play an intermediary role in centrosome amplification, acting as a checkpoint to prevent errors in mitosis. For example, BRCA1 mutant cells proliferate at a slower rate and have longer G1 phases. This could trigger an extra cycle of centrosome duplication thereby generating aberrant centrosome numbers [41]. In mouse embryonic fibroblast (MEF) carrying a targeted deletion of exon 11 of the BRCA1 gene, the G2/M checkpoint is defective in BRCA1 −/− cells [41]. Extensive chromosomal abnormalities accompany this G2/M checkpoint defect [41], which can be restored by the reconstitution of wild-type BRCA1 [42]. Thus, BRCA1 functions in G2/M checkpoint control, and supernumerary centrosomes might arise from the abrogation of that checkpoint function.

16.3.3
Centrosome Anomalies and Aurora A

The recent discovery of the Aurora kinases, a family of ser/thr kinases in mammalian cells, may cast new light on the mechanism of centrosome amplification. Aurora kinases are key regulators of cell division, and they also affect centrosome function, bipolar spindle assembly, and chromosome segregation processes (reviewed in [43–46]). Although many variants from the Aurora kinase family can be found in eukaryotic cells, mammalian cells generally express three Aurora variants, Aurora A, B, and C [45, 46]. All share amino-terminal regulatory domains and conserved catalytic domains [46]. Aurora A appears to be critical to the control of numerous mitotic events and the mRNA level varies throughout the cell cycle and peaks during G2/M. Aurora A localizes at the centrosome, whereas Aurora B appears to be associated with chromosomes and behaves as a "passenger protein" on the mitotic spindle.

Numerous studies have explored the diverse roles Aurora A plays in regulating centrosome function and mitotic control. Under mutational analysis, specific Aurora A domains target the centrosome and the mitotic spindle [47]. Recent work has demonstrated that Aurora A plays an active function in promoting entry into mitosis [48] by regulating local translation of mRNA stored in the cytoplasm, to produce cyclin B1 for example [49]. Substrates for the kinase Aurora A may also involve centrosomal proteins that interact with the mitotic machinery. Proteins like TACCs interact with XMAP215 to regulate microtubule behavior [50]. TPX2, a component of the spindle apparatus, seems to be required for targeting Aurora-A kinase to the spindle apparatus. Aurora-A, in turn, may regulate the function of TPX2 during spindle assembly [51, 52]. A model for Ran-stimulated spindle assembly proposes the activation of Aurora A by inhibiting PPI through TPX2 on microtubules. The Ran-GTP gradient established by the condensed chromosomes is translated into the Aurora A kinase gradient on the microtubules to regulate spindle assembly and dynamics [53].

Further research has examined the role of Aurora A in tumor formation and development. Aurora A may contribute to tumorigenesis when ectopic expression causes an increase in centrosome numbers, leading to aneuploidy and genomic instability [54, 55]. In mice and humans, Aurora A was identified as a candidate low-penetrative tumor-susceptibility gene [56].

Recent studies established centrosome amplification and overexpression of Aurora A as early events in rat mammary carcinogenesis. A model for rat mammary carcinogenesis provided evidence that Aurora A mRNA overexpression and centrosome amplification are linked to tumor development and progression [57]. Initially, this model had been established to investigate the resistance to tumorigenesis conferred when the normal hormonal milieu present during pregnancy results in persistent changes in the mammary gland [58]. Animals exposed to hormones failed to develop mammary tumors when subjected to carcinogens such as NMU (methylnitrosourea). Moreover, they displayed normal centrosome profiles with one or two centrosomes per cell (see Figure 16.3). Animals given NMU without prior hormonal treatment were unprotected and developed mammary tumors [58]. In rat mammary gland, as early as 40 days after NMU treatment, Aurora A expression levels were upregulated and peaked in tumors [57]. At the same time, small foci with higher centrosome numbers were detected (see Figure 16.3, 40 days NMU). Thus, higher Aurora A expression levels coincided with the appearance of supernumerary centrosomes. Centrosome counts showing elevated numbers were detected exclusively in epithelial cells. Adjacent stromal cells always displayed normal numbers with one or two centrosomes per cell. The abnormal number of centrosomes varied in early lesions of mammary tumors to a degree that was commensurate with the neoplastic state (Figure 16.3) [57].

Electron micrographs of mammary tumors demonstrate centrosomes with aberrant morphologies (unpublished data). Important questions for future research involving this mammary gland model include how the timing of the commencement of Aurora A overexpression after carcinogen treatment results in malfunctioning centrosomes, genomic instability and tumor formation.

In a study of human bladder cancer, elevated copy numbers of the Aurora A gene and overexpression of the kinase coincided precisely with the onset of aneuploidy and the clinical aggressiveness of these tumors. The study examined 205 patients and included lesions that were progressing from low-grade papillary urothelial carcinomas to high-grade invasive non-papillary urothelial carcinomas [54]. Centrosome amplification is currently being investigated in this human tumor model.

In summary, Aurora A is one example of a centrosome-associated protein that is apparently involved in tumorigenesis as a regulator of both centrosome function and direct mitotic control. Overexpression of Aurora A causes centrosome amplification and the formation of multipolar mitotic spindles, which leads to aneuploidy and tumorigenesis. Inhibition of Aurora A results in cell cycle arrest and apoptosis. For these reasons, Aurora A is of considerable interest as a potential drug target. Completion of the crystal structure of Aurora A [59, 60] will facilitate the development of inhibitors of the aurora kinases, which, in turn, will represent new entry points for therapies targeting abnormal centrosome function in cancer.

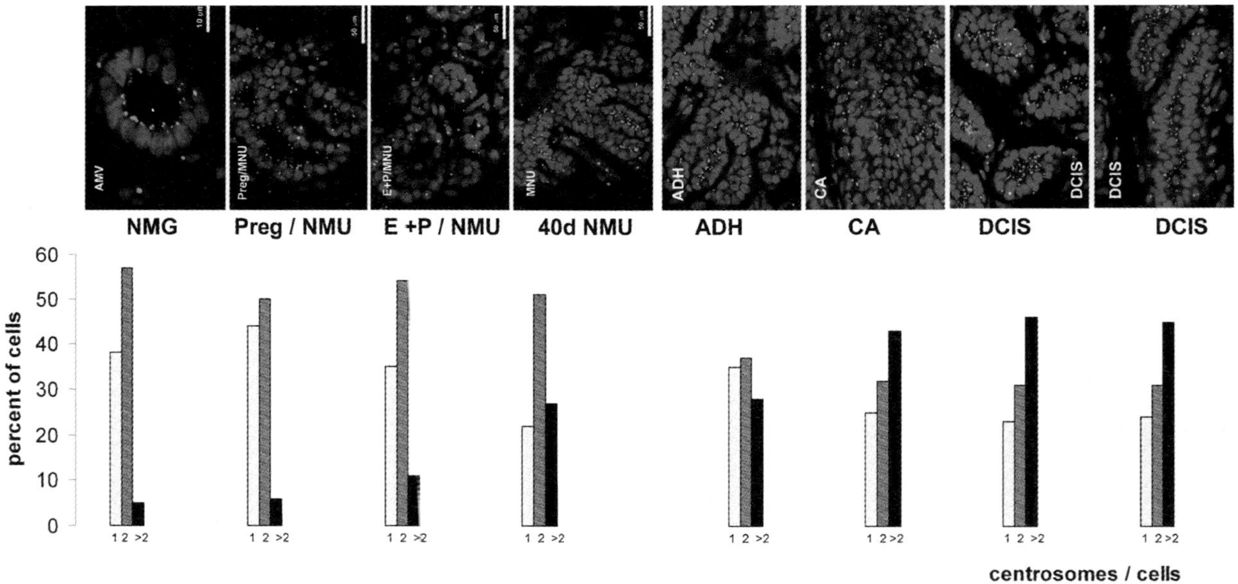

Figure 16.3 Centrosome amplification correlates with tumorigenesis in rat mammary glands. Confocal microscopy optical sections of tissue from different regimens and centrosome plots. NMG, normal mammary gland; Preg/NMU, pregnancy and involution prior to NMU; E + P/NMU, hormone treatment prior to NMU; 40 days NMU: 40 days post treatment with NMU. Pre-malignant lesions, e. g. ADH, atypical ductal hyperplasia; CA, cancer; DCIS, ductal carcinoma *in situ* (see Color Plates page XXXVI).

In *Zur Frage der Entstehung maligner Tumoren* (*The Origin of Malignant Tumors*) Boveri concludes with the following statement:

> *I may in conclusion send a wish along with this book, it is this: that my arguments may induce active investigators of the tumor problem to consider their work from the standpoint presented here and to ask in their future studies, whether what they find, contradicts or supports the theory I have here set forth. [1]*

Boveri's early vision proved to be prophetic, and we have only just begun to realize his aspirations. Centrosome anomalies and associated regulatory pathways in cancer will likely be the subject of many future studies as investigations progress from *in vitro* and *in situ* animal models to human malignancies.

Acknowledgments

We thank Melissa M. Munoz, David G. Edwards, Joshua M. Newton and Claire M. Haueter for technical assistance, and Michael Wise and Kevin Brinkley for editorial assistance. Thanks are extended to Daniel Medina, Anand Immaneni and David Stenoien for help and discussion.

This work was supported by grant number CA-64255-07 from the National Cancer Institute.

References

1. Boveri, T. H. **1914**. *Zur Frage der Entstehung maligner Tumoren*, 1–64.
2. Boveri, T. H. **1887**. Über die Befruchtung der Eier von Ascaris megalocephala. Sitzber. Ges. Morph. Phys. München.
3. Wilson, E. B. **1896**. *The Cell in Development and Inheritance*. Macmillan, New York, p. 4.
4. Boveri, T. H. **1902**. Über mehrpolige Mitosen als Mittel zur Analyse des Zellkerns. Verh D Phys Med Ges Würzburg NF, *35:* 67–90.
5. Hansemann, D. v. **1890**. Über asymmetrische Zellteilung in Epithelkrebsen und deren biologische Bedeutung. Virch Arch Path Anat, *119:* 299–326.
6. Boveri, T. H. **1904**. *Ergebnisse uber die Konstitution der chromatischen Substanz des Zellkerns*. Jena: G. Fischer.
7. Wilson, E. B. **1918**. *Erinnerung an Theodor Boveri*, W. C. Roentgen (Ed.), Mohr, Tübingen.
8. Brinkley, B. R. and Goepfert, T. M. **1998**. Supernumerary centrosomes and cancer: Boveri's hypothesis resurrected. Cell Motil Cytoskeleton, *41:* 281–288.
9. Lingle, W. L. and Salisbury, J. L. **1999**. Altered centrosome structure is associated with abnormal mitoses in human breast tumors. Am J Pathol, *155:* 1941–1951.
10. Lingle, W. L. and Salisbury, J. L. **2000**. The role of the centrosome in the development of malignant tumors. Curr Top Dev Biol, *49:* 313–329.
11. Lingle, W. L., Lutz, W. H., Ingle, J. N., Maihle, N. J., and Salisbury, J. L. **1998**. Centrosome hypertrophy in human breast tumors: implications for genomic stability and cell polarity. Proc Natl Acad Sci USA, *95:* 2950–2955.

12. Salisbury, J. L. **2001**. The contribution of epigenetic changes to abnormal centrosomes and genomic instability in breast cancer. J Mammary Gland Biol Neoplasia, *6:* 203–212.
13. Montagna, C., Andrechek, E. R., Padilla-Nash, H., Muller, W. J., and Ried, T. **2002**. Centrosome abnormalities, recurring deletions of chromosome 4, and genomic amplification of HER2/neu define mouse mammary gland adenocarcinomas induced by mutant HER2/neu. Oncogene, *21:* 890–898.
14. Pihan, G. A., Wallace, J., Zhou, Y., and Doxsey, S. J. **2003**. Centrosome abnormalities and chromosome instability occur together in pre-invasive carcinomas. Cancer Res, *63:* 1398–1404.
15. Brinkley, B. R. **2001**. Managing the centrosome numbers game: from chaos to stability in cancer cell division. Trends Cell Biol, *11:* 18–21.
16. Nigg, E. A. **2002**.Centrosome aberrations: cause or consequence of cancer progression? Nature Rev Cancer, *2:* 15–825.
17. Balczon, R., Bao, L., Zimmer, W. E., Brown, K., Zinkowski, R. P., and Brinkley, B. R. **1995**. Dissociation of centrosome replication events from cycles of DNA synthesis and mitotic division in hydroxyurea-arrested Chinese hamster ovary cells. J Cell Biol, *130:* 105–115.
18. Meraldi, P., Lukas, J., Fry, A. M., Bartek, J., and Nigg, E. A. **1999**. Centrosome duplication in mammalian somatic cells requires E2F and Cdk2-cyclin A. Nature Cell Biol, *1:* 88–93.
19. Wong, C. and Stearns, T. **2003**. Centrosome number is controlled by a centrosome-intrinsic block to reduplication. Nature Cell Biol, *5:* 539–544.
20. Pihan, G. A. D., Doxsey, S. J. **1999**. The mitotic machinery as a source of genetic instability in cancer. Semin Cancer Biol, *9:* 289–302.
21. Fukasawa, K., Choi, T., Kuriyama, R., Rulong, S., and Vande Woude, G. F. **1996**. Abnormal centrosome amplification in the absence of p53. Science, *271:* 1744–1747.
22. Khodjakov, A., Rieder, C. L., Sluder, G., Cassels, G., Sibon, O., and Wang, C. L. **2002**. *De novo* formation of centrosomes in vertebrate cells arrested during S phase. J Cell Biol, *158:* 1171–1181.
23. Schatten, H. and Schatten, G. **1986**. Motility and centrosomal organization during sea urchin and mouse fertilization. Cell Motil Cytoskeleton, *6:* 163–175.
24. Millband, D. N., Campbell, L., and Hardwick, K. G. **2002**. The awesome power of multiple model systems: interpreting the complex nature of spindle checkpoint signaling. Trends Cell Biol, *12:* 205–209.
25. Shekhar, M. P., Lyakhovich, A., Visscher, D. W., Heng, H., and Kondrat, N. **2002**. Rad6 overexpression induces multinucleation, centrosome amplification, abnormal mitosis, aneuploidy, and transformation. Cancer Res, *62:* 2115–2124.
26. Tarapore, P., Horn, H. F., Tokuyama, Y., and Fukasawa, K. **2001**. Direct regulation of the centrosome duplication cycle by the p53–p21Waf1/Cip1 pathway. Oncogene, *20:* 3173–3184.
27. Tarapore, P. and Fukasawa, K. **2002**. Loss of p53 and centrosome hyperamplification. Oncogene, *21:* 6234–6240.
28. Meraldi, P., Honda, R., and Nigg, E. A. **2002**. Aurora-A overexpression reveals tetraploidization as a major route to centrosome amplification in p53−/− cells. EMBO J, *21:* 483–492.
29. Carroll, P. E., Okuda, M., Horn, H. F., Biddinger, P., Stambrook, P. J., Gleich, L. L., Li, Y. Q., Tarapore, P., and Fukasawa, K. **1999**. Centrosome hyperamplification in human cancer: chromosome instability induced by p53 mutation and/or Mdm2 overexpression. Oncogene, *18:* 1935–1944.
30. Lingle, W. L., Barrett, S. L., Negron, V. C., D'Assoro, A. B., Boeneman, K., Liu, W., Whitehead, C. M., Reynolds, C., and Salisbury, J. L. **2002**. Centrosome amplification drives chromosomal instability in breast tumor development. Proc Natl Acad Sci USA, *99:* 1978–1983.

31. Weber, R. G., Bridger, J. M., Benner, A., Weisenberger, D., Ehemann, V., Reifenberger, G., and Lichter, P. **1998**. Centrosome amplification as a possible mechanism for numerical chromosome aberrations in cerebral primitive neuroectodermal tumors with TP53 mutations. Cytogenet Cell Genet, *83:* 266–269.
32. Ouyang, X., Wang, X., Xu, K., Jin, D. Y., Cheung, A. L., Tsao, S. W., and Wong, Y. C. **2001**. Effect of p53 on centrosome amplification in prostate cancer cells. Biochim Biophys Acta, *1541:* 212–220.
33. Wang, X. J., Greenhalgh, D. A., Jiang, A., He, D., Zhong, L., Medina, D., Brinkley, B. R., and Roop, D. R. **1998**. Expression of a p53 mutant in the epidermis of transgenic mice accelerates chemical carcinogenesis. Oncogene, *17:* 35–45.
34. Wang, X. J., Greenhalgh, D. A., Jiang, A., He, D., Zhong, L., Brinkley, B. R., and Roop, D. R. **1998**. Analysis of centrosome abnormalities and angiogenesis in epidermal-targeted p53172H mutant and p53-knockout mice after chemical carcinogenesis: evidence for a gain of function. Mol Carcinogenesis, *23:* 185–192.
35. Lengauer, C., Kinzler, K. W., and Vogelstein, B. **1997**. Genetic instability in colorectal cancers. Nature, *386:* 623–627.
36. Bunz, F., Fauth, C., Speicher, M. R., Dutriaux, A., Sedivy, J. M., Kinzler, K. W., Vogelstein, B., and Lengauer, C. **2002**. Targeted inactivation of p53 in human cells does not result in aneuploidy. Cancer Res, *62:* 1129–1133.
37. D'Assoro, A. B., Barrett, S. L., Folk, C., Negron, V. C., Boeneman, K., Busby, R., Whitehead, C., Stivala, F., Lingle, W. L., and Salisbury, J. L. **2002**. Amplified centrosomes in breast cancer: a potential indicator of tumor aggressiveness. Breast Cancer Res Treat, *75:* 25–34.
38. Deng, C. X. **2002**.Roles of BRCA1 in centrosome duplication. Oncogene, *21:* 6222–6227.
39. Hsu, L. C., Doan, T. P., and White, R. L. **2001**. Identification of a gamma-tubulin-binding domain in BRCA1. Cancer Res, *61:* 7713–7718.
40. Xu, X., Wagner, K. U., Larson, D., Weaver, Z., Li, C., Ried, T., Hennighausen, L., Wynshaw-Boris, A., and Deng, C. X. **1999**. Conditional mutation of Brca1 in mammary epithelial cells results in blunted ductal morphogenesis and tumour formation. Nature Genet, *22:* 37–43.
41. Xu, X., Weaver, Z., Linke, S. P., Li, C., Gotay, J., Wang, X. W., Harris, C. C., Ried, T., and Deng, C. X. **1999**. Centrosome amplification and a defective G2–M cell cycle checkpoint induce genetic instability in BRCA1 exon 11 isoform-deficient cells. Mol Cell, *3:* 389–395.
42. Xu, B., Kim, S., and Kastan, M. B. **2001**. Involvement of Brca1 in S-phase and G(2)-phase checkpoints after ionizing irradiation. Mol Cell Biol, *21:* 3445–3450.
43. Bischoff, J. R. and Plowman, G. D. **1999**.The Aurora/Ipl1p kinase family: regulators of chromosome segregation and cytokinesis. Trends Cell Biol, *9:* 454–459.
44. Goepfert, T. M. and Brinkley, B. R. **2000**. The centrosome-associated Aurora/Ipl-like kinase family. Curr Top Dev Biol, *49:* 331–342.
45. Nigg, E. A. **2001**. Mitotic kinases as regulators of cell division and its checkpoints. Nature Rev Mol Cell Biol, *2:* 21–32.
46. Katayama, H., Brinkley, W. R., and Sen, S. **2003**. The Aurora kinases: role in cell transformation and tumorigenesis. Cancer Metastasis Rev, *22:* 451–464.
47. Stenoien, D. L., Sen, S., Mancini, M. A., and Brinkley, B. R. **2003**. Dynamic association of a tumor amplified kinase, Aurora-A, with the centrosome and mitotic spindle. Cell Motil Cytoskeleton, *55:* 134–146.
48. Hirota T., Kunitoku N., Sasayama T., Marumoto T., Zhang D., Nitta M., Hatakeyama K., and H., S. **2003**. Aurora-A and an interacting activator, the LIM protein Ajuba, are required for

mitotic commitment in human cells. Cell, *114:* 585–598.

49. Warner, S. L., Bearss, D. J., Han, H., and Von Hoff, D. D. **2003**. Targeting Aurora-2 kinase in cancer. Mol Cancer Ther, *2:* 589–595.
50. Raff, J. W. **2002**.Centrosomes and cancer: lessons from a TACC. Trends Cell Biol, *12:* 222–225.
51. Kufer, T. A., Sillje, H. H., Korner, R., Gruss, O. J., Meraldi, P., and Nigg, E. A. **2002**. Human TPX2 is required for targeting Aurora-A kinase to the spindle. J Cell Biol, *158:* 617–623.
52. Eyers, P. A., Erikson, E., Chen, L. G., and Maller, J. L. **2003**. A novel mechanism for activation of the protein kinase aurora a. Curr Biol, *13:* 691–697.
53. Tsai, M. Y., Wiese, C., Cao, K., Martin, O., Donovan, P., Ruderman, J., Prigent, C., and Zheng, Y. **2003**. A Ran signalling pathway mediated by the mitotic kinase Aurora A in spindle assembly. Nature Cell Biol, *5:* 242–248.
54. Bischoff, J. R., Anderson, L., Zhu, Y., Mossie, K., Ng, L., Souza, B., Schryver, B., Flanagan, P., Clairvoyant, F., Ginther, C., Chan, C. S., Novotny, M., Slamon, D. J., and Plowman, G. D. **1998**. A homologue of *Drosophila* aurora kinase is oncogenic and amplified in human colorectal cancers. EMBO J, *17:* 3052–3065.
55. Zhou, H., Kuang, J., Zhong, L., Kuo, W. L., Gray, J. W., Sahin, A., Brinkley, B. R., and Sen, S. **1998**. Tumour amplified kinase STK15/BTAK induces centrosome amplification, aneuploidy and transformation. Nature Genet, *20:* 189–193.
56. Ewart-Toland, A., Briassouli, P., de Koning, J. P., Mao, J. H., Yuan, J., Chan, F., MacCarthy-Morrogh, L., Ponder, B. A., Nagase, H., Burn, J., Ball, S., Almeida, M., Linardopoulos, S., and Balmain, A. **2003**. Identification of Stk6/STK15 as a candidate low-penetrance tumor-susceptibility gene in mouse and human. Nature Genet, *34:* 403–412.
57. Goepfert, T. M., Adigun, Y. E., Zhong, L., Gay, J., Medina, D., and Brinkley, W. R. **2002**. Centrosome amplification and overexpression of aurora A are early events in rat mammary carcinogenesis. Cancer Res, *62:* 4115–4122.
58. Sivaraman, L., Stephens, L. C., Markaverich, B. M., Clark, J. A., Krnacik, S., Conneely, O. M., O'Malley, B. W., and Medina, D. **1998**. Hormone-induced refractoriness to mammary carcinogenesis in Wistar–Furth rats. Carcinogenesis, *19:* 1573–1581.
59. Littlepage, L. E., Wu, H., Andresson, T., Deanehan, J. K., Amundadottir, L. T., and Ruderman, J. V. **2002**. Identification of phosphorylated residues that affect the activity of the mitotic kinase Aurora-A. Proc Natl Acad Sci USA, *99:* 15440–15445.
60. Cheetham, G. M., Knegtel, R. M., Coll, J. T., Renwick, S. B., Swenson, L., Weber, P., Lippke, J. A., and Austen, D. A. **2002**. Crystal structure of aurora-2, an oncogenic serine/threonine kinase. J Biol Chem, *277:* 42419–42422.

17
Radiation Therapy and Centrosome Anomalies in Pancreatic Cancer

Norihiro Sato, Kazuhiro Mizumoto, and Masao Tanaka

Abstract

Radiation therapy has been widely used as a major treatment option for patients with malignant tumors, including pancreatic adenocarcinoma. Nevertheless, the precise mechanism by which radiation-induced cellular damage leads to cell death or growth arrest is not yet fully understood. Ionizing radiation has been shown to induce extensive induction of aberrations in centrosome number in various solid tumor cell lines. Importantly, such supernumerary centrosomes have been associated with the formation of multipolar spindles and chromosome missegregation. Possible mechanisms leading to abnormal centrosome numbers in response to radiation include dissociation between centrosome duplication and DNA replication cycles coupled with loss of a centrosome-intrinsic mechanism to block reduplication, failure of cell division (cytokinesis) associated with polyploidization, and centrosome fragmentation. Integrity of DNA damage checkpoint pathways as well as tumor microenvironment (extracellular factors) may determine the extent of centrosome aberrations after irradiation. The formation of multiple centrosomes that retain microtubule-nucleating activity provides an attractive model for radiation-induced nuclear fragmentation and subsequent cell death, although the causality remains to be established. A better understanding of the origins and consequences of centrosome anomalies after irradiation could lead to the development of novel prognostic markers with respect to radiation sensitivity and therapeutic modalities to improve local control and outcomes of current radiation therapy.

17.1
Introduction

The centrosome is a tiny organelle but is crucial for a variety of cellular processes, including cell motility, intracellular transport, cell cycle progression, chromosome segregation, cytokinesis, and the establishment of cell shape and cell polarity [1–4].

Centrosomes in Development and Disease. Edited by Erich A. Nigg

ISBN 3-527-30980-2

Considering the multitude of essential cellular events that require proper centrosome function, it is not surprising that defects in centrosome function have been associated with various human diseases [5, 6]. One of the most important functions of the centrosome is to contribute to the establishment of bipolar mitotic spindles that orchestrate the balanced segregation of chromosomes. To maintain mitotic fidelity and cell viability, the centrosome duplication cycle should be strictly coordinated with DNA replication and cell division [7–10]. Numerical and functional abnormalities of the centrosome can result in chromosome segregation errors through the formation of defective spindles, possibly leading to genetic instability (aneuploidy) or lethal mitotic events [11–13]. Identification of centrosome abnormalities in virtually all common types of human malignant tumors has suggested a potential role of this organelle in the pathogenesis of cancer [6 ,11, 12 ,14–18]. It has been proposed that an abnormal centrosome phenotype accelerates genetic instability and thus causes the tumor to progress to a more advanced stage [6, 12, 14, 15, 17, 19–25]. On the other hand, abnormal centrosomes and the resultant multipolar spindles have been considered apparently disadvantageous for continued cell divisions and cell viability [6, 13, 19].

Centrosome anomalies have also been shown to arise as a response to DNA damage or other genotoxic stimuli. For example, exposure of human cell lines derived from different solid tumors to γ-radiation results in a massive induction of aberrations in centrosome number, although the underlying mechanism is unknown [26, 27]. This finding is potentially of major importance because supernumerary centrosomes and the resultant multipolar spindles might contribute, at least in part, to nuclear fragmentation and subsequent cell death after irradiation. Detailed understanding of the relationship between centrosome anomalies and radiation-induced cell death will provide further insights into the role of centrosomes in the cellular response to DNA damage and in the maintenance of cell viability.

17.2 Radiation-induced Cell Death: Apoptosis or Mitotic Cell Death?

Radiation therapy, either alone or in combination with surgery and chemotherapy, has long been used in the management of a majority of patients with different types of solid tumors. It has been generally accepted that ionizing radiation induces two types of cell death that are routinely referred to as apoptosis and mitotic cell death [28, 29]. Apoptosis (or programmed cell death) is an active form of cell death which appears to occur preferentially in lymphoma and leukemia cells after irradiation [30, 31]. In most solid tumor cell types, however, cell death after radiation occurs predominantly as a result of aberrant mitotic events, namely, mitotic cell death or mitotic catastrophe [28, 32–34]. This form of cell death has been defined as loss of reproductive integrity after inappropriate entry into mitosis, and is characterized by the emergence of large nonviable cells containing multiple nuclear fragments or "micronuclei" [28, 32, 35]. In our study, a panel of 10 solid tumor cell lines treated with γ-radiation (10 Gy) showed predominantly mitotic cell death, as evidenced by the emergence of multi- or micronulceated cells [27]. Mitotic cell death can also be induced by heat shock [36] or various chemotherapeutic agents including etoposide, doxorubicin, cisplatin, bleomycin, and taxol [29]. Furthermore, mitotic cell death may occur as a consequence of prolonged growth arrest. Chang and colleagues have shown that after release from growth arrest induced by the Cdk (cyclin-dependent kinase) inhibitor $p21^{Waf1/Cip1}$, human fibrosarcoma cells re-enter the cell cycle but display multiple mitotic abnormalities characteristic of mitotic cell death [37]. Importantly, it has been suggested that apoptosis and the genes controlling it, such as *p53* and *Bcl2*, play little or no role in the sensitivity of most solid tumor cell types to overall killing by anticancer drugs and radiation [30]. Furthermore, at least in some experimental models, inhibition of apoptosis results in a subsequent increase in the proportion of cells undergoing mitotic cell death [29]. Lock and Stribinskiene have shown that overexpression of Bcl2 inhibits apoptosis of HeLa cells treated with etoposide but enhances the induction of mitotic cell death, resulting in no change in the clonogenic survival [38]. Similarly, inhibition of apoptosis by inducible expression of MDR1 does not protect tumor cells from proliferative death, which may occur through mitotic cell death or senescence-like terminal growth arrest [39]. These findings highlight the predominant role of mitotic cell death rather than apoptosis in the overall killing of solid tumor cells induced by DNA-damaging agents. Therefore, it is conceivable that strategies to augment mitotic cell death could improve the efficacy of radiation therapy for cancer patients. Despite this, little is known about the molecular or cellular pathways leading to mitotic cell death.

17.3
Centrosome Anomalies Induced by Radiation

Because centrosomes are essential for the establishment of bipolar spindles and proper partitioning of chromosomes during mitosis, it could be hypothesized that defects in centrosome function play a role in the lethal chromosomal fragmentation and catastrophic mitosis following irradiation. So far, relatively few studies have addressed the effects of radiation on centrosome dynamics. In 1983, the first observation of centrosomes in cells exposed to radiation was described by Sato and coworkers showing that X-irradiation of murine cells resulted in centrosomes that were atypical in number, structure, and microtubule nucleating activity [40]. Bunz and co-workers studied the role of p53 and p21 in the G2 cell-cycle checkpoint after DNA damage and found abnormally high numbers of centrosomes in p53- or p21-defective cancer cells undergoing abnormal mitosis after γ-radiation, although the frequency of these centrosome abnormalities was not described [41]. Using immunofluorescence microscopy, we have examined centrosome profiles in U2-OS osteosarcoma cells following γ-irradiation [26]. Although abnormal cells containing three or more centrosomes occurred with frequencies of less than 2% in untreated cultures, this population markedly increased to ~ 60% at 72 h after 10 Gy irradiation (Figure 17.1). In addition, the fraction of

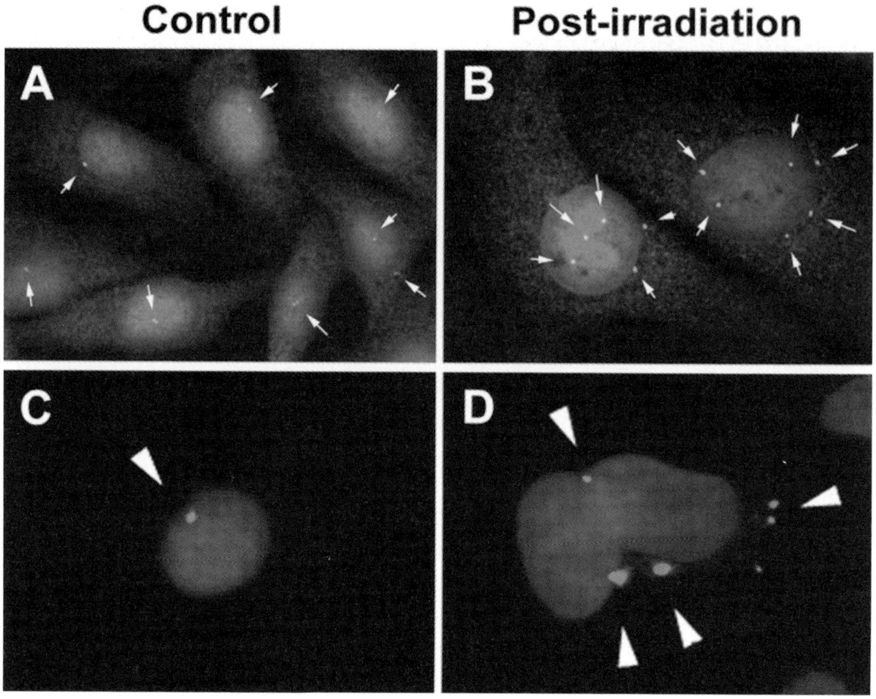

Figure 17.1 Ionizing radiation induces numerical aberrations of centrosomes in human tumor cells. Cells were stained with an antibody to γ-tubulin (green, A and B) or pericentrin (red, C and D) for centrosomes and counterstained with propidium iodide (red, A and B) or Hoechst 33258 (blue, C and D) for nuclear labeling. U2-OS osteosarcoma cells before (A) and 48 h after (B) γ-radiation at 10 Gy. Mia PaCa2 pancreatic cancer cells before (C) and 48 h after (D) γ-radiation at 10 Gy (see Color Plates page XXXVII).

mitotic cells that display multipolar spindles increased after irradiation along with the increased fraction of cells with an abnormal centrosome phenotype. Remarkably, the cells containing multiple centrosomes frequently exhibited abnormal nuclear morphology, including micronuclei. A more detailed study investigating the effect of radiation on centrosome dynamics was performed using a large panel of 10 cell lines derived from different types of human solid tumors (including pancreatic, breast, colorectal, and cervical carcinoma) [27]. Exposure to γ-irradiation at a single dose of 10 Gy led to a marked increase in the fraction of cells containing abnormally high numbers of centrosomes in all the cell lines tested, suggesting that numerical centrosome aberration is a common response of tumor cells to radiation (Figure 17.1). The extent of centrosome anomalies, however, varied among cell lines and the percentage of abnormal cells with multiple centrosomes ranged from ~ 20 to ~ 60% at 48 h after irradiation. Importantly, double-staining of irradiated cells with antibodies to α-tubulin and pericentrin have revealed that these extra copies of centrosomes have the potential to nucleate microtubules and form multipolar spindles at mitosis (Figure 17.2).

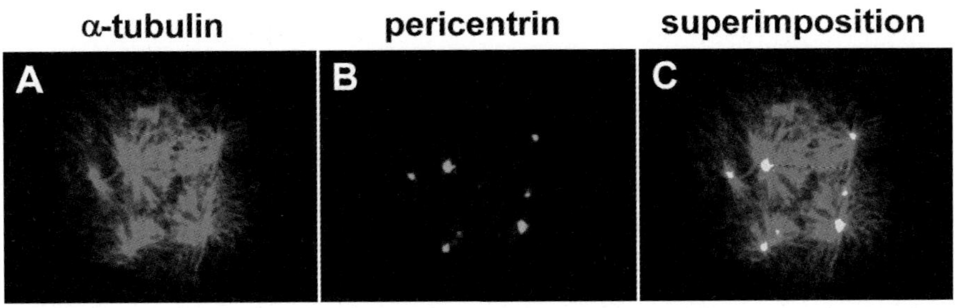

Figure 17.2 Aberrant mitotic cell with multipolar spindles after γ-radiation at 10 Gy. Cells were double-stained with α-tubulin (red) and pericentrin (green) antibodies (see Color Plates page XXXVII).

Similar centrosome aberrations have also been observed in response to other DNA-damaging agents or stress situations. Augustin and coworkers have shown that multiple centrosome formation is induced not only by γ-radiation but also by other DNA-damaging agents including N-methyl-N-nitrosourea and H_2O_2 [42]. Furthermore, increased numbers of centrosomes and spindle abnormalities have been reported in Chinese hamster ovary (CHO) cells treated with fostriecin, a phosphate ester known to inhibit DNA topoisomerase II and protein phosphatase activity [43]. The authors demonstrated that treatment with okadaic acid and cantharidin, two established protein phosphatase inhibitors, also induces similar centrosome aberrations, implying that these effects on centrosomes may be related to protein phosphatase inhibition. Notably, Meraldi and Nigg have also shown that phosphatase inhibition leads to centrosome splitting in U2-OS cells [44]. In addition, it has been documented that heat shock and bacterial cytolethal distending toxin (CDT) induces multiple centrosomes and multipolar spindle formation that closely resemble those seen in irradiated cells [45–47]. Thus, numerical centrosome aberrations may be a common cellular response to genotoxic damage or other stress situations, although it is unclear whether they arise through the same or different mechanism(s).

17.4
The Mechanism(s) Leading to Centrosome Anomalies after Radiation Treatment

The mechanism underlying the centrosome anomalies which occur after irradiation is unknown. Over the past several years, a substantial amount of research has been devoted to exploring the origin(s) of centrosome abnormalities in cancer and these studies have provided important clues to understanding the mechanisms that potentially cause centrosome multiplicity. Although it is unlikely that centrosome abnormalities seen in cancer and those induced by radiation arise through the exact same mechanism, they might share one or more common pathway(s).

One possible mechanism for the formation of multiple centrosomes after irradiation is dissociation between the centrosome duplication cycle and the DNA replication cycle. To ensure that only one new centrosome assembles per cell cycle and to limit centrosome numbers to one or two within a cell, the centrosome duplication cycle should be precisely coordinated with other cell cycle events, particularly DNA replication and mitotic cell division [7–10, 13, 48]. It has become clear that centrosome duplication is regulated by complex mechanisms, including phosphorylation [8, 9, 49] and ubiquitin-dependent proteolysis of key proteins [50–52]. It has been shown that both centrosome duplication and DNA replication require the hyperphosphorylation of the retinoblastoma (RB) protein and the activation of cyclin-dependent kinase 2 (Cdk2), suggesting that the two cycles are synchronized, in part, by sharing a common regulatory pathway [53–56]. In some experimental models, however, the centrosome cycle can be dissociated from other cell-cycle events. For example, treatment of *Xenopus* embryos with cycloheximide allows the continuation of centrosome duplication in the absence of a detectable cell cycle, resulting in numerous centrosomes [57]. Arresting DNA replication at the G_1/S boundary using hydroxyurea (HU) or aphidicolin, results in the continuation of centrosome duplication in CHO cells, producing cells with multiple centrosomes [58]. We have observed that after γ-irradiation of U2-OS cells, the fraction of cells with multiple centrosomes continues to increase during a transient cell-cycle delay in S-phase and prolonged G2 arrest [27]. Although there might be differences in centrosome behavior between inhibition of DNA replication and DNA damage, it is a plausible scenario that centrosomes accumulate in these cells through a process of repeated duplication despite radiation-triggered arrest or delay of DNA replication, thereby producing cells with multiple centrosomes. In specific cell types, however, centrosome reduplication during the HU-induced S-phase arrest does not occur [59], leading to the hypothesis that there is a mechanism that limits centrosome duplication to only once per cell cycle. In support of this hypothesis, there is emerging evidence for the existence of a centrosome-inactivation pathway in response to DNA damage [60]. It has been shown that in *Drosophila melanogaster* embryos DNA damage or incomplete DNA replication at the onset of mitosis triggers centrosome inactivation, leading to the failure of the damaged chromosomes to segregate [61, 62]. Takada and colleagues have shown that checkpoint kinase 2 (Chk2) is essential for the signaling pathway that leads to disruption of centrosome function in response to genotoxic stress [62]. This centrosome inactivation pathway is one of the damage-control systems which ensure genomic integrity, and a comparable mechanism for controlling centrosome number has been identified in higher eukaryotes [63]. Using a cell fusion assay, Wong and Stearns have recently shown that in normal somatic mammalian cells centrosome numbers are regulated by a centrosome-intrinsic mechanism that limits centrosome duplication to one round per cell cycle [63]. In normal human cells (primary diploid fibroblasts), centrosome reduplication is blocked during prolonged S-phase arrest induced by treatment with HU. By contrast, cells containing more than two centrosomes accumulate during the S-phase arrest in HCT116 colorectal cancer cells in response to the same drug, suggesting that this control to block centrosome reduplication is likely to be abrogated in cancer cells. Thus, loss of the block to centrosome reduplication in tumor cells could contribute to numerical aberrations of the centrosome after irradiation.

Another possible mechanism of centrosome multiplication in response to radiation is failures in cell division (cytokinesis) and subsequent polyploidization, which is, so far, the most likely mechanism for the abnormal centrosome phenotype seen in cancer cells [69, 64]. In support of this model, we have shown that flow cytometric analysis of U2-OS cells after irradiation revealed the appearance of tetraploid cells [27], and that most of the cells harboring abnormal numbers of centrosomes were multinucleated [26 ,27]. Induction of polyploidization after irradiation has been well documented in various tumor cells, especially in the absence of functional p53 or p21 [65–68]. It has been reported that p53−/− and p21−/− cancer cells never complete cytokinesis after γ-radiation, resulting in cells containing multilobulated nuclei and an abnormally high number of centrosomes [41]. In these cells, abnormal centrosomes were located in a cleft that likely was a remnant of the cleavage furrow associated with the failure of cytokinesis, supporting the role of cytokinesis failure in the synthesis of multiple centrosomes. Based on these findings, cytokinesis failure coupled with polyploidization is a plausible explanation for multiple centrosome formation in tumor cells after irradiation, although it is unlikely the only mechanism.

Finally, Hut and coworkers treated CHO cells with HU or mitomycin C to demonstrate that in the presence of incompletely replicated or damaged DNA, centrosomes split into fragments containing only one centriole, leading to the formation of multipolar spindles during mitosis [69]. Similar to the findings observed in irradiated cultures, the cells with abnormal spindles subsequently exit from mitosis, producing multinucleated or aneuploid cells. This centrosome splitting occurs only when mitosis is initiated, suggesting that the formation of multiple centrosome-like structures is not the result of multiple rounds of centrosome duplication but the result of centrosome fragmentation which is triggered by entry into mitosis in the presence of impaired DNA integrity. These findings provide evidence for a novel mechanism by which DNA damage induces multiple centrosome-like structures (fragments), multipolar spindles, and cells with multiple nuclei. Careful analysis with electron microscopy is needed to generalize this model to other cell types treated with different types of DNA damage-inducing agents, including radiation.

17.5
The Consequence of Centrosome Anomalies after Irradiation

Increased numbers of centrosomes after irradiation can contribute to the formation of multipolar spindles [26, 27]. It is likely that multipolar spindles may lead, in most cases, to loss of cell viability through lethal mitotic events. However, the exact fate of cells containing multipolar spindles after irradiation is uncertain. To address this issue, we treated MIA PaCa-2 pancreatic cancer cells with 10 Gy γ-radiation and the mitotic cells were collected after 24 h by the method of mitotic shake-off [27]. Staining of these mitotic cells with an anti-γ-tubulin antibody revealed that over 60 % of them displayed multiple centrosomes which were dispersed chaotically throughout the condensed chromosomes (Figure 17.3A). The cells were then re-plated in culture dishes and incubated with complete medium. After 24 h, we found that ~ 80 % of the reseeded cells turned into giant cells containing multiple nuclear fragments of various sizes and shapes (Figure 17.3B and C). A prolonged culture of these multinucleated cells for 7 days resulted in cell swelling, membrane blebbing, and detachment from the culture dishes in most cells. We have also observed a positive correlation between the fraction of cells with multiple centrosomes and the fraction with mitotic cell death characterized by the appearance of multi- or micronucleated cells after irradiation. To determine the effect of centrosome aberrations on radiation-induced cell killing we transfected p21, a Cdk inhibitor known to block centrosome duplication [55], into cells before irradiation. Forced expression of p21 by adenovirus gene transfer inhibited the induction of multiple centrosomes and partially prevented cells from undergoing cell death after irradiation. Collectively, these findings indicate that radiation-induced centrosome anomalies may give rise to the multinucleated phenotype, which eventually results in cell death.

Alternatively, it is possible that cells containing abnormally high number of centrosomes could be eliminated through the apoptotic pathway. Because the centrosome is involved in multiple signal transduction pathways and interacts with a number of regulatory proteins [70], it could play an important role in apoptosis and its dysregulation could trigger apoptotic signaling [71]. Evidence supporting a link between centrosomes and apoptosis comes from studies that report the localization of several proteins involved in the regulation of apoptosis, including p53 and poly(ADP-ribose) polymerase (PARP), to centrosomes [42, 72–74]. Sandal and coworkers have further demonstrated that Irod/Ian5, a protein also localized to centrosomes, functions as an inhibitor of γ-radiation- and okadaic acid-induced

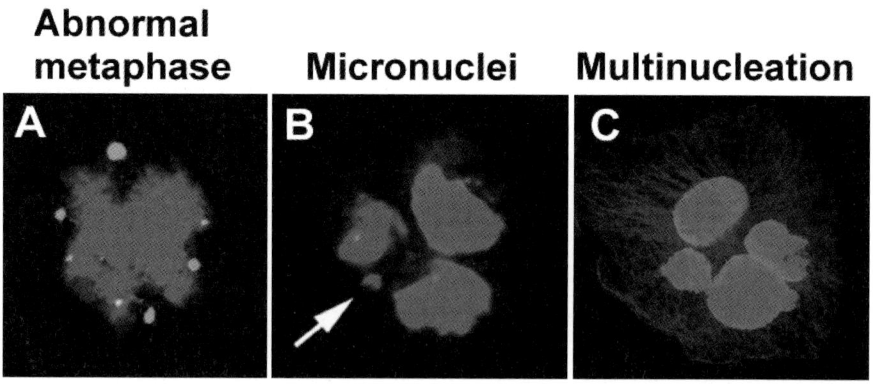

Figure 17.3 MiaPaCa2 pancreatic cancer cells after 10 Gy radiation showing abnormal mitosis (metaphase) with multiple centrosomes (green, pericentrin labeling) dispersed chaotically throughout the condensed chromosomes (A), followed by the appearance of the micro- (B) and multinucleated (C) phenotype, characteristic of mitotic cell death (see Color Plates page XXXVIII).

apoptosis [75]. These authors have also shown that calmodulin-dependent protein kinase II (CaMKII), which is known to be associated with centrosomes [76, 77], is involved in apoptosis induced by ionizing radiation. We have observed that a subset of irradiated cells with multiple centrosomes appear to undergo apoptosis as determined by TUNEL staining [27]. Bunz and coworkers also noted apoptotic cell death after γ-radiation in a subset of p53−/− or p21−/− cells containing multiple centrosomes [41]. These studies, however, did not establish causality between centrosome defects and apoptosis. It is also possible that nuclear fragmentation associated with centrosome anomalies could trigger apoptosis. Merritt and colleagues described a p53-independent delayed apoptosis in small intestinal epithelia after irradiation, which was associated with the appearance of giant cells with fragments of condensed chromatin in their nucleus [78]. In addition, several investigators have shown that p53 protein is upregulated in micronucleated cells, thereby directing these cells toward apoptosis [79 ,80]. Together, these observations raise the possibility that numerical aberrations of centrosomes and/or the resulting nuclear fragmentation may be critical events leading to apoptosis in irradiated cells. The final determination of whether cells with multiple centrosomes die by apoptosis or mitotic death may depend on the integrity of the checkpoint and/or apoptotic pathways.

17.6
Factors Affecting Centrosome Anomalies after Irradiation

DNA damage after exposure to ionizing radiation activates checkpoint pathways that inhibit progression of cells through the G1 and G2 phases [81]. In many cancer cells, these checkpoint pathways are inactivated by different mechanisms. Importantly, the integrity of the DNA damage checkpoint pathways could affect centrosome anomalies after irradiation for the following reasons. First, in the presence of intact checkpoint pathways cells with heavily damaged DNA are likely to be arrested and eliminated by apoptosis before acquisition of abnormal centrosome phenotypes. Second, considering the time required for centrosome duplication, the extent of abnormal centrosome duplication may be dependent on the duration of G2 arrest, which in turn is determined by checkpoint mechanisms. Among the genes involved in DNA checkpoint pathways, the p53 tumor suppressor is frequently mutated in human cancers [82]. In addition to its established role in the G1 checkpoint and apoptosis, p53 also blocks entry into mitosis when cells enter G2 with damaged DNA partly through inhibition of Cdk1, the cyclin-dependent kinase required for entry into mitosis [83]. Cdk1 is inhibited simultaneously by three transcriptional targets of p53, Gadd45, p21, and 14-3-3σ [83]. Notably, loss of the p53 as well as loss of its downstream targets, p21 and Gadd45, has been shown to result in supernumerary centrosomes [84–87], although the origin of this phenotype remains controversial [6]. In addition, 14-3-3σ has been reported to be frequently inactivated by aberrant DNA hypermethylation in certain cancer types, including breast carcinoma [88–91]. 14-3-3σ, which can bind to Cdk1/cyclin

B1 and sequester it in the cytoplasm, controls the duration of G2 arrest in response to DNA damage [92, 93]. HCT116 colorectal cancer cells in which 14-3-3σ was inactivated by homologous recombination were unable to sustain a G2 arrest in response to DNA damage and underwent mitotic catastrophe [92]. Interestingly, aberrant hypomethylation of 14-3-3σ and its overexpression in pancreatic cancer have also been reported [94, 95], although the biological significance of these findings remains unknown. Finally, *reprimo*, a gene involved in the p53-mediated G2 cell cycle arrest [96], has been recently identified as a target for aberrant hypermethylation in pancreatic cancer [97]. Thus, genetic and epigenetic inactivation of checkpoint genes may influence the extent of centrosome anomalies and determine the entry of cells into mitosis after irradiation.

The centrosome anomalies which occur in response to DNA damage can also be affected by tumor microenvironment (extracellular factors), such as hypoxia, acidosis, and nutritional deprivation. For example, we have shown that serum depletion of irradiated pancreatic cancer cells drastically accelerated the emergence of numerical aberrations of the centrosomes and the formation of multipolar spindles, resulting in increased nuclear fragmentation and cell death [98]. The effects of other extracellular factors (e. g. hypoxia), that are known to modulate the sensitivity of cells to radiation [99], have not yet been studied.

17.7 Conclusions and Future Directions

In conclusion, exposure of human solid tumor cells to ionizing radiation induces numerical aberrations of the centrosomes, multipolar spindle formation, and micro- or multinucleated phenotypes characteristic of mitotic cell death. The origin of supernumerary centrosomes in response to radiation is unknown, but possible mechanisms include dissociation between centrosome duplication and DNA replication cycles coupled with the loss of a centrosome-intrinsic mechanism to block reduplication, failure of cytokinesis associated with polyploidization, and centrosome fragmentation. Although there is no direct evidence to indicate that centrosome anomalies contribute to the killing of tumor cells after irradiation, supernumerary centrosomes or centrosome-like structures that retain microtubule-nucleating activity provide an attractive model for radiation-induced nuclear fragmentation and subsequent cell death. A better understanding of the origins and consequences of centrosome anomalies in response to radiation may have important clinical implications. For example, if such centrosome phenotypes correlate with overall cell killing after radiation treatment, the extent of centrosome aberrations may be used as an indicator of treatment response and as a prognostic marker. It is also important to identify and characterize the molecular pathways associated with the induction of abnormal centrosome phenotypes after irradiation. Identification of such pathways could lead to the development of novel centrosome-related strategies to enhance the efficacy of current treatment regimens for cancer patients.

References

1. Rieder CL, Faruki S, Khodjakov A. **2001**.The centrosome in vertebrates: more than a microtubule-organizing center. Trends Cell Biol 11:413–419.
2. Kellogg DR, Moritz M, Alberts BM. **1994**. The centrosome and cellular organization. Annu Rev Biochem 63:639–764.
3. Zimmerman W, Sparks CA, Doxsey SJ. **1999**. Amorphous no longer: the centrosome comes into focus. Curr Opin Cell Biol 11:122–128.
4. Doxsey S. **2001**. Re-evaluating centrosome function. Nature Rev Mol Cell Biol 2:688–698.
5. Sathasivam K, Woodman B, Mahal A, Bertaux F, Wanker EE, Shima DT, Bates GP. **2001**. Centrosome disorganization in fibroblast cultures derived from R6/2 Huntington's disease (HD) transgenic mice and HD patients. Hum Mol Genet 10:2425–2435.
6. Nigg EA. **2002**. Centrosome aberrations: cause or consequence of cancer progression? Nature Rev Cancer 2:815–825.
7. Winey M. **1996**. Keeping the centrosome cycle on track. Genome stability. Curr Biol 6:962–964.
8. Meraldi P, Nigg EA. **2002**. The centrosome cycle. FEBS Lett 521:9–13.
9. Hinchcliffe EH, Sluder G. **2001**. "It takes two to tango": understanding how centrosome duplication is regulated throughout the cell cycle. Genes Dev 15:1167–1181.
10. Sluder G, Hinchcliffe EH. **2000**. The coordination of centrosome reproduction with nuclear events during the cell cycle. Curr Top Dev Biol 49:267–289.
11. Marx J. **2001**. Cell biology. Do centrosome abnormalities lead to cancer? Science 292:426–429.
12. Doxsey S. **1998**. The centrosome – a tiny organelle with big potential. Nature Genet 20:104–106.
13. Brinkley BR. **2001**. Managing the centrosome numbers game: from chaos to stability in cancer cell division. Trends Cell Biol 11:18–21.
14. Brinkley BR, Goepfert TM. **1998**. Supernumerary centrosomes and cancer: Boveri's hypothesis resurrected. Cell Motil Cytoskeleton 41:281–288.
15. D'Assoro AB, Lingle WL, Salisbury JL. **2002**. Centrosome amplification and the development of cancer. Oncogene 21:6146–6153.
16. Lingle WL, Salisbury JL. **2000**. The role of the centrosome in the development of malignant tumors. Curr Top Dev Biol 49:313–329.
17. Salisbury JL, Whitehead CM, Lingle WL, Barrett SL. **1999**. Centrosomes and cancer. Biol Cell 91:451–460.
18. Duensing S, Munger K. **2001**. Centrosome abnormalities, genomic instability and carcinogenic progression. Biochim Biophys Acta 1471:M81–M88.
19. Sato N, Mizumoto K, Nakamura M, Maehara N, Minamishima YA, Nishio S, Nagai E, Tanaka M. **2001**. Correlation between centrosome abnormalities and chromosomal instability in human pancreatic cancer cells. Cancer Genet Cytogenet 126:13–19.
20. Shono M, Sato N, Mizumoto K, Maehara N, Nakamura M, Nagai E, Tanaka M. **2001**. Stepwise progression of centrosome defects associated with local tumor growth and metastatic process of human pancreatic carcinoma cells transplanted orthotopically into nude mice. Lab Invest 81:945–952.
21. Pihan GA, Purohit A, Wallace J, Malhotra R, Liotta L, Doxsey SJ. **2001**. Centrosome defects can account for cellular and genetic changes that characterize prostate cancer progression. Cancer Res 61:2212–2219.
22. Pihan GA, Purohit A, Wallace J, Knecht H, Woda B, Quesenberry P, Doxsey SJ. **1998**. Centrosome defects and genetic instability in malignant tumors. Cancer Res 58:3974–3985.
23. Zhou H, Kuang J, Zhong L, Kuo WL, Gray JW, Sahin A, Brinkley BR, Sen S. **1998**. Tumour amplified kinase STK15/BTAK induces centrosome

amplification, aneuploidy and transformation. Nature Genet 20:189–193.

24. Lingle WL, Lutz WH, Ingle JN, Maihle NJ, Salisbury JL. **1998**. Centrosome hypertrophy in human breast tumors: implications for genomic stability and cell polarity. Proc Natl Acad Sci USA 95:2950–2955.

25. Lingle WL, Barrett SL, Negron VC, D'Assoro AB, Boeneman K, Liu W, Whitehead CM, Reynolds C, Salisbury JL. **2002**. Centrosome amplification drives chromosomal instability in breast tumor development. Proc Natl Acad Sci USA 99:1978–1983.

26. Sato N, Mizumoto K, Nakamura M, Tanaka M. **2000**. Radiation-induced centrosome overduplication and multiple mitotic spindles in human tumor cells. Exp Cell Res 255:321–326.

27. Sato N, Mizumoto K, Nakamura M, Ueno H, Minamishima YA, Farber JL, Tanaka M. **2000**. A possible role for centrosome overduplication in radiation-induced cell death. Oncogene 19:5281–5290.

28. Hendry JH, West CM. **1997**. Apoptosis and mitotic cell death: their relative contributions to normal-tissue and tumour radiation response. Int J Radiat Biol 71:709–719.

29. Roninson IB, Broude EV, Chang BD. **2001**. If not apoptosis, then what? Treatment-induced senescence and mitotic catastrophe in tumor cells. Drug Resist Update 4:303–313.

30. Brown JM, Wouters BG. **1999**. Apoptosis, p53, and tumor cell sensitivity to anticancer agents. Cancer Res 59:1391–1399.

31. Brown JM, Wouters BG. **2001**. Apoptosis: mediator or mode of cell killing by anticancer agents? Drug Resist Update 4:135–136.

32. Chang WP, Little JB. **1991**. Delayed reproductive death in X-irradiated Chinese hamster ovary cells. Int J Radiat Biol 60:483–496.

33. Dewey WC, Ling CC, Meyn RE. **1995**. Radiation-induced apoptosis: relevance to radiotherapy. Int J Radiat Oncol Biol Phys 33:781–796.

34. Jonathan EC, Bernhard EJ, McKenna WG. **1999**. How does radiation kill cells? Curr Opin Chem Biol 3:77–83.

35. Muller WU, Nusse M, Miller BM, Slavotinek A, Viaggi S, Streffer C. **1996**. Micronuclei: a biological indicator of radiation damage. Mutat Res 366:163–169.

36. Swanson PE, Carroll SB, Zhang XF, Mackey MA. **1995**. Spontaneous premature chromosome condensation, micronucleus formation, and non-apoptotic cell death in heated HeLa S3 cells. Ultrastructural observations. Am J Pathol 146:963–971.

37. Chang BD, Broude EV, Fang J, Kalinichenko TV, Abdryashitov R, Poole JC, Robinson IB. **2000**. p21Waf1/Cip1/Sdi1-induced growth arrest is associated with depletion of mitosis-control proteins and leads to abnormal mitosis and endoreduplication in recovering cells. Oncogene 19:2165–2170.

38. Lock RB, Stribinskiene L. **1996**. Dual modes of death induced by etoposide in human epithelial tumor cells allow Bcl-2 to inhibit apoptosis without affecting clonogenic survival. Cancer Res 56:4006–4012.

39. Ruth AC, Roninson IB. **2000**. Effects of the multidrug transporter P-glycoprotein on cellular responses to ionizing radiation. Cancer Res 60:2576–2578.

40. Sato C, Kuriyama R, Nishizawa K. **1983**. Microtubule-organizing centers abnormal in number, structure, and nucleating activity in x-irradiated mammalian cells. J Cell Biol 96:776–782.

41. Bunz F, Dutriaux A, Lengauer C, Waldman T, Zhou S, Brown JP, Sedivy JM, Kinzler KW, Vogelstein B. **1998**. Requirement for p53 and p21 to sustain G2 arrest after DNA damage. Science 282:1497–1501.

42. Augustin A, Spenlehauer C, Dumond H, Menissier-De Murcia J, Piel M, Schmit AC, Apiou F, Vonesch JL, Kock M, Borneus M, De Murcia G. **2003**. PARP-3 localizes preferentially to the daughter centriole and interferes with the G1/S cell cycle progression. J Cell Sci 116:1551–1562.

43. Cheng A, Balczon R, Zuo Z, Koons JS, Walsh AH, Honkanen RE. **1998**. Fostriecin-mediated G2-M-phase growth arrest correlates with abnormal centrosome replication, the formation of aberrant mitotic spindles, and the inhibition of serine/threonine protein phosphatase activity. Cancer Res 58:3611–3619.
44. Meraldi P, Nigg EA. **2001**. Centrosome cohesion is regulated by a balance of kinase and phosphatase activities. J Cell Sci 114:3749–3757.
45. De Rycke J, Sert V, Comayras C, Tasca C. **2000**. Sequence of lethal events in HeLa cells exposed to the G2 blocking cytolethal distending toxin. Eur J Cell Biol 79:192–201.
46. Vidair CA, Doxsey SJ, Dewey WC. **1993**. Heat shock alters centrosome organization leading to mitotic dysfunction and cell death. J Cell Physiol 154:443–455.
47. Nakahata K, Miyakoda M, Suzuki K, Kodama S, Watanabe M. **2002**. Heat shock induces centrosomal dysfunction, and causes non-apoptotic mitotic catastrophe in human tumour cells. Int J Hyperthermia 18:332–343.
48. Doxsey S. **2002**. Duplicating dangerously: linking centrosome duplication and aneuploidy. Mol Cell 10:439–440.
49. Fry AM, Mayor T, Nigg EA. **2000**. Regulating centrosomes by protein phosphorylation. Curr Top Dev Biol 49:291–312.
50. Hansen DV, Hsu JY, Kaiser BK, Jackson PK, Eldridge AG. **2002**. Control of the centriole and centrosome cycles by ubiquitination enzymes. Oncogene 21:6209–6221.
51. Freed E, Lacey KR, Huie P, Lyapina SA, Deshaies RJ, Stearns T, Jackson PK. **1999**. Components of an SCF ubiquitin ligase localize to the centrosome and regulate the centrosome duplication cycle. Genes Dev 13:2242–2257.
52. Wojcik EJ, Glover DM, Hays TS. **2000**. The SCF ubiquitin ligase protein slimb regulates centrosome duplication in *Drosophila*. Curr Biol 10:1131–1134.
53. Hinchcliffe EH, Li C, Thompson EA, Maller JL, Sluder G. **1999**. Requirement of Cdk2-cyclin E activity for repeated centrosome reproduction in *Xenopus* egg extracts. Science 283:851–854.
54. Lacey KR, Jackson PK, Stearns T. **1999**. Cyclin-dependent kinase control of centrosome duplication. Proc Natl Acad Sci USA 96:2817–2822.
55. Matsumoto Y, Hayashi K, Nishida E. **1999**. Cyclin-dependent kinase 2 (Cdk2) is required for centrosome duplication in mammalian cells. Curr Biol 9:429–432.
56. Meraldi P, Lukas J, Fry AM, Bartek J, Nigg EA. **1999**. Centrosome duplication in mammalian somatic cells requires E2F and Cdk2-cyclin A. Nature Cell Biol 1:88–93.
57. Gard DL, Hafezi S, Zhang T, Doxsey SJ. **1990**. Centrosome duplication continues in cycloheximide-treated *Xenopus* blastulae in the absence of a detectable cell cycle. J Cell Biol 110:2033–2042.
58. Balczon R, Bao L, Zimmer WE, Brown K, Zinkowski RP, Brinkley BR. **1995**. Dissociation of centrosome replication events from cycles of DNA synthesis and mitotic division in hydroxyurea-arrested Chinese hamster ovary cells. J Cell Biol 130:105–115.
59. Meraldi P, Honda R, Nigg EA. **2002**. Aurora-A overexpression reveals tetraploidization as a major route to centrosome amplification in p53−/− cells. EMBO J 21:483–92.
60. Sibon OC. **2003**. Centrosomes as DNA damage regulators. Nature Genet 34:6–7.
61. Sibon OC, Kelkar A, Lemstra W, Theurkauf WE. **2000**. DNA-replication/DNA-damage-dependent centrosome inactivation in *Drosophila* embryos. Nature Cell Biol 2:90–95.
62. Takada S, Kelkar A, Theurkauf WE. **2003**. Drosophila checkpoint kinase 2 couples centrosome function and spindle assembly to genomic integrity. Cell 113:87–99.
63. Wong C, Stearns T. **2003**. Centrosome number is controlled by a centrosome-intrinsic block to reduplication. Nature Cell Biol 5:539–544.

64. Borel F, Lohez OD, Lacroix FB, Margolis RL. **2002**. Multiple centrosomes arise from tetraploidy checkpoint failure and mitotic centrosome clusters in p53 and RB pocket protein-compromised cells. Proc Natl Acad Sci USA 99:9819–9824.

65. Olive PL, Banath JP, Durand RE. **1996**. Development of apoptosis and polyploidy in human lymphoblast cells as a function of position in the cell cycle at the time of irradiation. Radiat Res 146:595–602.

66. Waldman T, Lengauer C, Kinzler KW, Vogelstein B. **1996**. Uncoupling of S phase and mitosis induced by anticancer agents in cells lacking p21. Nature 381:713–716.

67. Bulavin DV, Tararova ND, Aksenov ND, Pospelov VA, Pospelova TV. **1999**. Deregulation of p53/p21Cip1/Waf1 pathway contributes to polyploidy and apoptosis of E1A+cHa-ras transformed cells after gamma-irradiation. Oncogene 18:5611–5619.

68. Illidge TM, Cragg MS, Fringes B, Olive P, Erenpreisa JA. **2000**. Polyploid giant cells provide a survival mechanism for p53 mutant cells after DNA damage. Cell Biol Int 24:621–633.

69. Hut HM, Lemstra W, Blaauw EH, Van Cappellen GW, Kampinga HH, Sibon OC. **2003**. Centrosomes split in the presence of impaired DNA integrity during mitosis. Mol Biol Cell 14:1993–2004.

70. Lange BM. **2002**. Integration of the centrosome in cell cycle control, stress response and signal transduction pathways. Curr Opin Cell Biol 14:35–43.

71. Kong Q. **2003**. The centrosome-centered cell-brain in apoptosis. Med Hypotheses 61:126–132.

72. Brown CR, Doxsey SJ, White E, Welch WJ. **1994**. Both viral (adenovirus E1B) and cellular (hsp 70, p53) components interact with centrosomes. J Cell Physiol 160:47–60.

73. Morris VB, Brammall J, Noble J, Reddel R. **2000**. p53 localizes to the centrosomes and spindles of mitotic cells in the embryonic chick epiblast, human cell lines, and a human primary culture: An immunofluorescence study. Exp Cell Res 256:122–130.

74. Kanai M, Uchida M, Hanai S, Uematsu N, Uchida K, Miwa M. **2000**. Poly(ADP-ribose) polymerase localizes to the centrosomes and chromosomes. Biochem Biophys Res Commun 278:385–389.

75. Sandal T, Aumo L, Hedin L, Gjertsen BT, Doskeland SO. **2003**. Irod/Ian5: an inhibitor of gamma-radiation- and okadaic acid-induced apoptosis. Mol Biol Cell 14:3292–3304.

76. Pietromonaco SF, Seluja GA, Elias L. **1995**. Identification of enzymatically active Ca^{2+}/calmodulin-dependent protein kinase in centrosomes of hemopoietic cells. Blood Cells Mol Dis 21:34–41.

77. Matsumoto Y, Maller JL. **2002**. Calcium, calmodulin, and CaMKII requirement for initiation of centrosome duplication in *Xenopus* egg extracts. Science 295:499–502.

78. Merritt AJ, Allen TD, Potten CS, Hickman JA. **1997**. Apoptosis in small intestinal epithelial from p53-null mice: evidence for a delayed, p53-independent G2/M-associated cell death after gamma-irradiation. Oncogene 14:2759–2766.

79. Schwartz D, Almog N, Peled A, Goldfinger N, Rotter V. **1997**. Role of wild type p53 in the G2 phase: regulation of the gamma-irradiation-induced delay and DNA repair. Oncogene 15:2597–2607.

80. Sablina AA, Ilyinskaya GV, Rubtsova SN, Agapova LS, Chumakov PM, Kopnin BP. **1998**. Activation of p53-mediated cell cycle checkpoint in response to micronuclei formation. J Cell Sci 111:977–984.

81. Nyberg KA, Michelson RJ, Putnam CW, Weinert TA. **2002**. Toward maintaining the genome: DNA damage and replication checkpoints. Annu Rev Genet 36:617–656.

82. Hollstein M, Sidransky D, Vogelstein B, Harris CC. **1991**. p53 mutations in human cancers. Science 253:49–53.

83. Taylor WR, Stark GR. **2001**. Regulation of the G2/M transition by p53. Oncogene 20:1803–1815.

84. Fukasawa K, Choi T, Kuriyama R, Rulong S, Vande Woude GF. **1996**. Abnormal centrosome amplification in the absence of p53. Science 271:1744–1747.

85. Hollander MC, Sheikh MS, Bulavin DV, Lundgren K, Augeri-Henmueller L, Shehee R, Molinaro TA, Kim KE, Tolosa E, Ashwell JD, Rosenberg MP, Zhan Q, Fernandez-Salguero PM, Morgan WF, Deng CX, Fornace AJ, Jr. **1999**. Genomic instability in Gadd45a-deficient mice. Nature Genet 23:176–184.

86. Mantel C, Braun SE, Reid S, Henegariu O, Liu L, Hangoc G, Broxmeyer HE. **1999**. p21(cip-1/waf-1) deficiency causes deformed nuclear architecture, centriole overduplication, polyploidy, and relaxed microtubule damage checkpoints in human hematopoietic cells. Blood 93:1390–1398.

87. Carroll PE, Okuda M, Horn HF, Biddinger P, Stambrook PJ, Gleich LL, Li YQ, Tarapore P, Fukusawa K. **1999**. Centrosome hyperamplification in human cancer: chromosome instability induced by p53 mutation and/or Mdm2 overexpression. Oncogene 18:1935–1944.

88. Osada H, Tatematsu Y, Yatabe Y, Nakagawa T, Konishi H, Harano T, Tezel E, Takada M, Takahashi T. **2002**. Frequent and histological type-specific inactivation of 14-3-3sigma in human lung cancers. Oncogene 21:2418–2424.

89. Iwata N, Yamamoto H, Sasaki S, Itoh F, Suzuki H, Kikuchi T, Kaneto H, Iku S, Ozeki I, Karino Y, Satoh T, Toyota J, Satoh M, Endo T, Imai K. **2000**. Frequent hypermethylation of CpG islands and loss of expression of the 14-3-3 sigma gene in human hepatocellular carcinoma. Oncogene 19:5298–5302.

90. Suzuki H, Itoh F, Toyota M, Kikuchi T, Kakiuchi H, Imai K. **2000**. Inactivation of the 14-3-3 sigma gene is associated with 5′ CpG island hypermethylation in human cancers. Cancer Res 60:4353–4357.

91. Ferguson AT, Evron E, Umbricht CB, Pandita TK, Chan TA, Hermeking H, Marks JR, Lambers AR, Futreal PA, Stampfer MR, Sukumar S. **2000**. High frequency of hypermethylation at the 14-3-3 sigma locus leads to gene silencing in breast cancer. Proc Natl Acad Sci USA 97:6049–6054.

92. Chan TA, Hermeking H, Lengauer C, Kinzler KW, Vogelstein B. **1999**. 14-3-3Sigma is required to prevent mitotic catastrophe after DNA damage. Nature 401:616–620.

93. Hermeking H, Lengauer C, Polyak K, He TC, Zhang L, Thiagalingam S, Kinzler KW, Vogelstein B. **1997**. 14-3-3 sigma is a p53-regulated inhibitor of G2/M progression. Mol Cell 1:3–11.

94. Sato N, Maitra A, Fukushima N, van Heek NT, Matsubayashi H, Iacobuzio-Donahue CA, Rosty C, Goggins M. **2003**. Frequent hypomethylation of multiple genes overexpressed in pancreatic ductal adenocarcinoma. Cancer Res 63:4158–4166.

95. Iacobuzio-Donahue CA, Maitra A, Olsen M, Lowe AW, van Heek NT, Rosty C, Walter K, Sato N, Parker A, Ashfaq R, Jaffee E, Ryu B, Jones J, Eshleman JR, Yeo CJ, Cameron JL, Kern SE, Hruban RH, Brown PO, Goggins M. **2003**. Exploration of global gene expression patterns in pancreatic adenocarcinoma using cDNA microarrays. Am J Pathol 162:1151–1162.

96. Ohki R, Nemoto J, Murasawa H, Oda E, Inazawa J, Tanaka N, Taniguchi T. **2000**. Reprimo, a new candidate mediator of the p53-mediated cell cycle arrest at the G2 phase. J Biol Chem 275:22627–22630.

97. Sato N, Fukushima N, Maitra A, Matsubayashi H, Yeo CJ, Cameron JL, Hruban RH, Goggins M. **2003**. Discovery of novel targets for aberrant methylation in pancreatic carcinoma using high-throughput microarrays. Cancer Res 63:3735–3742.

98. Shono M, Sato N, Mizumoto K, Minamishima YA, Nakamura M, Maehara N, Urashima T, Saimura M, Qian L, Nishio S, Nagai E, Tanaka M. **2001**. Effect of serum depletion on centrosome overduplication and death of human pancreatic cancer cells after exposure to radiation. Cancer Lett 170:81–89.

99. Brown JM. **2002**. Tumor microenvironment and the response to anticancer therapy. Cancer Biol Ther 1:453–458.

18
Human Papillomavirus Infection and Centrosome Anomalies in Cervical Cancer

Karl Münger and Stefan Duensing

18.1
Genomic Instability and Malignant Progression

Most human tumor cells exhibit evidence of genomic instability. Whether genomic destabilization constitutes a mechanistically significant component of malignant progression of an emerging neoplasm, or simply represents a manifestation of cell division abnormalities that have been indiscriminately accumulated during the carcinogenic process is a matter of ongoing debate (reviewed in [1]). It has been estimated, however, that the mutation rate of normal human cells would not suffice to permit accumulation of the multitude of genetic alterations that appear to be required for the genesis of many human cancers. Hence inactivation of pathways that control genomic stability may be necessary for malignant progression *in vivo* [2]. Tumorigenic cell populations have recently been generated *in vitro* by expression of a small number of defined genetic elements in normal human cells (reviewed in [3]). Such artificially engineered tumor cells do not exhibit overt genomic instability thus underscoring that genomic destabilization does not inevitably arise as a cellular manifestation of oncogenic insults. Hence genomic instability as observed in most naturally-occurring human tumors might indeed represent an essential prerequisite that enables an emerging tumor cell to accumulate the necessary oncogenic hits.

In principle several different and cooperating mechanisms that may contribute to genomic destabilization can be envisioned (reviewed in [4]). Dysfunction of "caretaker" functions such as cell cycle checkpoint and DNA repair pathways will enable cells that have suffered genomic aberrations to remain in the proliferative pool. On the other hand, however, there may be oncogenic insults that give rise to a higher mutation rate in emerging tumors. This concept of the "mutator" phenotype of cancer cells was proposed to explain the large number of mutations in human tumors [5, 6].

The mutation rate dramatically increases in cells that have reached replicative senescence. Telomere attrition increases the frequency of chromosome fusions (reviewed in [7]) that can initiate breakage–fusion-bridge cycles and lead to chromo-

Centrosomes in Development and Disease. Edited by Erich A. Nigg

ISBN 3-527-30980-2

somal translocations [8]. In such a scenario, developing cancers exhibit a pronounced mutator phenotype exclusively at a specific relatively late stage of cellular crisis and the small number of telomerase-positive clones that grow out would not necessarily exhibit a higher mutation rate than normal cells. Emerging cancer cells, however, are constantly subject to new selection barriers, which may require them to have a higher degree of genomic plasticity not only at the end of their replicative lifespan (reviewed in [1]). Moreover, telomerase activity can be stimulated at early stages of carcinogenic progression before telomere attrition becomes critical; during cervical carcinogenesis telomerase expression is induced at a transcriptional level as a consequence of viral E6 gene expression [9, 10].

Another mechanism that might contribute to a mutator phenotype in some cancer cells is the induction of double-strand DNA breaks by certain oncogenes such as c-myc [11, 12] or HPV-16 E7 [13], which can cause chromosomal translocations. This is consistent with results of earlier studies that have shown that ectopic c-myc expression can cause genomic instability [14–17].

In general, however, the relative importance of these and other mechanisms such as mutant forms of DNA polymerases [18], for the observed mutator phenotype of human tumors has remained controversial. This may reflect the fact that for most cancer types, the molecular nature of the oncogenic insults that fuel malignant progression are molecularly diverse and/or unknown.

In contrast, cervical carcinogenesis is almost exclusively associated with infections by high-risk human papillomaviruses (HPVs). High-risk HPV infection is a prerequisite for the development of almost all the pre-malignant cervical lesions but carcinogenic progression occurs gradually and overall represents a relatively low frequency event. Accumulation of host cellular mutations may thus be rate limiting for malignant progression of high-risk HPV-positive lesions. Consistent with this model, hallmarks of genomic instability, in particular mitotic abnormalities, are evident even in early high-risk HPV-associated pre-malignant lesions (reviewed in [19]). Since only two small HPV proteins, E6 and E7, are expressed in cervical tumors it has been feasible to investigate the molecular mechanisms that can lead to genomic destabilization in an emerging neoplasm. These studies strongly suggest that the HPV E7 oncoprotein may act as a mitotic mutator by inducing primary centrosome duplication errors, whereas the cooperating E6 oncogene subverts checkpoint controls thereby allowing for the maintenance of abnormal cells in the proliferative pool.

18.2 Human Papillomaviruses

Papillomaviruses are an extensive family of small DNA viruses that have been detected in a variety of hosts from birds to humans. They are highly species specific and in each case exclusively infect epithelial cells. HPV infections generally give rise to benign hyperplasias commonly referred to as warts. Papillomaviruses possess a double-stranded circular DNA genome of approximately 8000 base pairs in size. The entire coding information is contained on a single DNA strand. Papillomavirus genomes can be divided into three portions; the early coding region that encompasses approximately eight open reading frames (ORFs), the late coding region consisting of two ORFs, and a third "long control region" (LCR) that does not contain extensive coding information but harbors multiple cis elements that control the viral replication and transcription programs. Individual ORFs are designated "E" or "L" for early and late, respectively, followed by a numeral that denotes their relative sizes; the higher the number the smaller the encoded ORF (reviewed in [20]).

Like all viruses, papillomaviruses are obligatory intracellular parasites and the function of the viral E genes collectively is to ensure replication and propagation of viral progeny, whereas the two L genes encode the viral capsid proteins. Approximately 200 HPVs have been discovered and a subgroup of approximately 30 HPVs specifically infects the anogenital and oral tract mucosa. These HPVs are further classified as "high-risk" or low-risk" depending on whether the lesions they cause are likely to progress to malignancy. Low-risk HPVs cause benign genital warts, condyloma acuminata, whereas infections with high-risk HPVs can give rise to squamous intraepithelial lesions (SILs) that have a finite potential for malignant progression (reviewed in [21]). The vast majority ($>$ 99 %) of all cervical cancers, a leading cause of cancer death in young women worldwide are associated with infections of high-risk HPVs (reviewed in [22]). In addition, a variety of other squamous cell carcinoma of the anogenital tract and approximately 20 % of oral cancers are also caused by this same high-risk HPV group [23]. Hence, high-risk HPVs may be one of the first-ever identified, necessary and molecularly defined causative agents of almost all cases of a major human cancer [24].

18.3 Biological Activities of HPV E6/E7 Oncogenes

HPV genomes frequently integrate into a host cellular chromosome upon malignant progression. There is no evidence for insertional mutagenesis as integration is relatively non-specific with respect to the host genome [25]. The integration event follows a more specific pattern with respect to the viral genome, however, and only two HPV genes; E6 and E7 are consistently expressed in cervical cancers. Since integration disrupts expression of the viral E2 transcriptional repressor protein, HPV E6/E7 expression is controlled by cellular transcription factors after integration, and cells are endowed with an additional growth advantage [26]. The HPV E6 and E7 genes each have oncogenic activities in cell culture and transgenic mouse models, and their persistent expression is required for the maintenance of the transformed phenotype of cervical cancer cell lines (reviewed in [20]). In contrast to other viruses where integration can establish latency, HPV genome integration is an irreversible event that effectively terminates the viral life cycle. Thus HPV-associated carcinogenesis is an accident with potentially dire consequences not only for the infected host, but definitely for the intruding virus. Therefore, the biological activities of the HPV E6 and E7 "onco"proteins did not evolve to cause oncogenic transformation of the infected host cell, but to contribute to the viral life cycle by supporting HPV genome replication in the infected host cells. This is a mechanistically challenging endeavor as high-level HPV genome replication and production of viral progeny are confined to terminally differentiated epithelial cells that would have normally withdrawn permanently from the cell division cycle (reviewed in [27]). Since HPVs do not encode all the necessary replication enzymes, an important function of the E6 and E7 proteins is to activate the cellular DNA replication machinery by reprogramming critical cellular control circuits. High-risk HPV E6 and E7 encode small proteins comprised of approximately 160 and 100 amino acid residues, respectively. Like other DNA tumor virus oncoproteins, HPV E6 and E7 are unique proteins with no extensive similarities to cellular proteins. They possess no known enzymatic or DNA binding activities and chiefly function by interacting with host cellular regulatory proteins, thereby subverting their biological activities. A considerable number of putative cellular targets of the HPV E6 and E7 proteins have been identified (reviewed in [28]), and only those that bear potential relevance to genomic destabilization will be discussed here. The high-risk HPV E6 proteins interact with the ubiquitin ligase E6-AP [29] and reprogram it to target the p53 tumor suppressor for rapid proteasome-mediated degradation [30] whereas the normal degradation pathway of p53 through mdm2 is abrogated in cervical carcinoma cell lines [31]. High-risk HPV E7 oncoproteins interact with the retinoblastoma tumor suppressor pRb and the related "pocket protein" family members p107 and p130 and cause their proteolytic degradation (reviewed in [32]). Since pRb, p107 and p130 limit G1/S cell cycle transition, HPV E7-expressing cells proliferate aberrantly. Furthermore, HPV E7 proteins inactivate cyclin-dependent kinase (Cdk) inhibitors such as $p21^{Cip1}$ [33, 34] and $p27^{Kip1}$ [35] causing further dysregulation of G1/S cell cycle control (Figure 18.1).

Such aberrant proliferation normally elicits activation of the p53 tumor suppressor to enforce eradication of such deregulated cells through apoptosis (reviewed in [36]). Since p53 is rendered functionally defective as a consequence of E6-mediated degradation [37], high-risk HPV infected cells will persist in the proliferative pool. In addition, E6 also causes increased telomerase transcription [38–40], potentially through a c-myc-dependent mechanism [41, 42], and thus HPV E6- and E7-expressing cells have a greatly increased proliferative capacity (Figure 18.1). As pointed out previously, high-risk HPV infected cells are thus unlikely to undergo telomere attrition-related genome destabilization.

Consistent with this notion, ectopic expression of the HPV E6/E7 oncoproteins extends the lifespan of primary human epithelial cells and facilitates their immortalization [43, 44]. When grown under conditions that allow formation of a stratified skin-like structure, HPV E6/E7-expressing epithelial cells exhibit histo-morphological alterations that are characteristic of high-grade pre-malignant intra-epithelial lesions [45, 46]. Nevertheless, early passage HPV E6/E7-immortalized keratinocyte cultures are non-tumorigenic in standard assays [47]. Tumorigenic conversion, however, can occur when such cells are engineered to express addi-

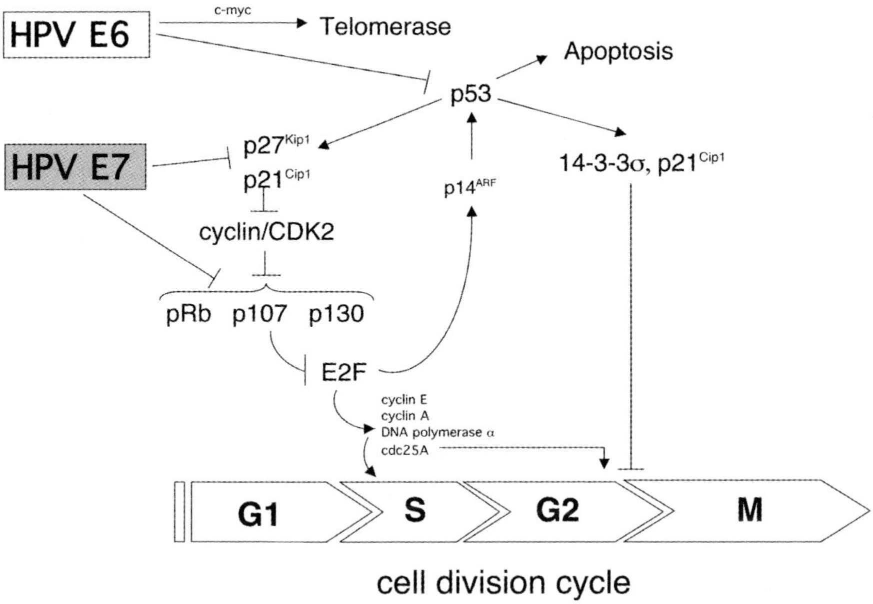

Figure 18.1 Major cellular targets of high-risk HPV E6 and E7 oncoproteins as they relate to cell cycle regulation. See text for details.

tional oncogenes such as ras or fos [48, 49]. Fully transformed clones also arise upon extended passaging *in vitro*, and malignant progression is associated with accumulation of chromosomal alterations [50]. HPV-associated cervical carcinogenesis can be recapitulated in transgenic mice when HPV E6 and E7 expression is targeted to basal epithelial cells. Malignant progression in this model, however, is dependent on chronic low-level estrogen exposure [51].

Hence, the results with each of these experimental systems suggest that HPV E6 and E7 expression is sufficient to induce cellular alterations that are reminiscent of pre-malignant lesions, but that malignant progression is a stochastic event that is reliant on additional mutations of the host genome. This clearly parallels the situation in the clinic where only a relatively small number of high-risk HPV-infected patients will eventually develop invasive cervical cancer and often only after a prolonged time after the initial infection event.

18.4 HPV-mediated Cervical Carcinogenesis as a Model System to Study Genomic Instability and Malignant Progression

Progression of a high-risk HPV-positive cervical lesion to invasive carcinoma is a multi-step process that occurs at a relatively low rate and efficiency. Some lesions regress spontaneously, most likely because the host is able to mount an effective immune response (reviewed in [52]). Nuclear alterations and mitotic abnormalities suggestive of genomic instability, however, are evident in low-grade lesions caused by high-risk HPVs, suggesting that genomic instability develops at an early stage and precedes malignant progression [53–57]. Similarly, cells engineered to express HPV oncoproteins *in vitro* rapidly exhibit hallmarks of genomic instability, even prior to immortalization [58]. Consistent with these observations, high-risk HPV E6 and E7 proteins can each individually induce genomic instability in primary human cells [59, 60]. In contrast, however, low-risk HPV E6/E7-expressing cells show no evidence of genomic instability. In these experiments, cells were selected to acquire resistance to the drug N-phosphonoacetyl-L-aspartate (PALA), a transition state analog inhibitor of aspartate transcarbamylase, one of the activities of the trifunctional carbamyl phosphate synthetase, aspartate transcarbamylase, dihydroorotase (CAD) enzyme that catalyzes the first three steps of cellular UMP synthesis. Normal cells succumb due to nutritional deprivation and do not develop PALA resistance at a high frequency [61]. When PALA-resistant clones emerge, they generally contain additional copies of the CAD gene [62, 63]. Genomic analyses of PALA resistant E6- and E7-expressing clones revealed that they had acquired PALA resistance through distinct mechanisms. HPV E6-expressing PALA-resistant clones revealed evidence for amplification of the CAD gene, whereas HPV E7-expressing PALA-resistant clones contained increased copy numbers of chromosome 2 that encodes the CAD gene [59]. The ability of the high-risk HPV E7 protein to induce aneuploidy under these conditions was consistent with earlier reports that revealed spontaneous development of aneuploidy in high-risk HPV E7-expressing keratinocytes [43, 64]. The HPV E6/E7-induced deterioration of genomic stability, in particular the ability to induce structural chromosomal alterations, might also contribute to the frequent integration of HPV genomes that accompanies malignant progression [65].

18.5
Centrosome Abnormalities and Genomic Instability: Cause or Effect?

Aneuploidy is a consequence of abnormal chromosome segregation during mitosis and represents the most frequent manifestation of genomic instability in human tumors. “Asymmetric” cell division, including multipolar mitotic figures, had already been described by Hansemann as one of the hallmarks of epithelial tumors [66], but the mechanisms that drive the development of such abnormalities remained unknown for almost a century. Based on his careful descriptive studies on abnormal multipolar cell division in polyspermic eggs [67], Theodor Boveri hypothesized in his visionary book *Zur Frage der Entstehung maligner Tumoren*, that “a single multipolar mitosis going on in a healthy tissue, caused perhaps by the simultaneous multiple division of the centrosome, might produce the primordial cell of a malignant tumor” [68]. Even though numerical and structural centrosome abnormalities have subsequently been detected in numerous human tumors and various tissue culture and transgenic mouse models of human cancer, this attractive theory still awaits experimental confirmation (reviewed in [69, 70]).

Like many other tumor types, HPV-associated lesions and cancers exhibit evidence for numerical and structural chromosome abnormalities [71]. Microsatellite instability has been observed in a subset of advanced tumors and thus may not be directly caused by HPV oncogene expression [72]. Abnormal, multipolar mitoses have long been recognized as hallmarks of high-risk HPV-associated lesions [73–75], and recent studies have revealed that these can arise as a consequence of supernumerary centrosomes [58]. A small survey of cervical neoplasms suggests that the degree of centrosome abnormalities increases in parallel with the clinical severity of the lesion [76]. Similar to reports for some other cancers [77–80], centrosome abnormalities are already detected at early stages of the disease, including premalignant high-risk HPV-associated squamous intraepithelial lesions [58, 76].

As outlined earlier, the unique advantage of studying cervical carcinogenesis is that it is almost exclusively associated with high-risk HPV infections and only two small viral oncoproteins, E6 and E7, are generally expressed in these cancers. Expression of HPV E6/E7 in epithelial cells has yielded relatively simple tissue culture and transgenic animal model systems that closely recapitulate key aspects of the clinical lesions. Hence it is possible to study the mechanisms and biological relevance of HPV oncoprotein-induced genomic instability for initiation and progression of HPV-associated tumorigenesis. It is conceivable that the results from these studies may also be more generally applicable to other, non-HPV-associated cancers.

18.6 Induction of Centrosome Abnormalities by HPV Oncoproteins: Boveri's Model Revisited

Since expression of individual high-risk HPV oncogenes in primary human cells causes genomic instability [59] this model system appeared particularly useful for studying the emergence of centrosome abnormalities and their relevance to genomic destabilization. These studies revealed that primary human foreskin keratinocytes engineered to ectopically express HPV-16 E6 or E7 each developed numerical centrosome abnormalities within a few passages in culture. Supernumerary centrosomes were detected both in interphase and metaphase cells, and were associated with the appearance of multipolar metaphases and anaphases. Interestingly, however, there was no evidence of the development of structural centrosome abnormalities as detected *in vivo* [58]. This finding suggests that structural centrosome abnormalities may not arise as a primary consequence of HPV oncogene expression. Importantly, primary human keratinocytes expressing E6 or E7 genes derived from "low-risk" genital wart-associated HPVs did not develop centrosome abnormalities [58]. Hence the ability of HPV E6/E7 proteins to induce centrosome abnormalities parallels the clinical association of high-risk or low-risk of the HPVs they are derived from. Moreover, the inability of the low-risk HPV E6/E7 proteins to induce centrosome abnormalities is consistent with the finding that expression of these proteins does not cause genomic instability [59], and may retrospectively explain Hansemann's failure to detect asymmetrical cell division in a penile wart that was likely caused by infection with a low-risk HPV [66].

Given that high-risk HPV E6 and E7 target distinct cellular pathways (reviewed in [81]), the finding that cells expressing individual HPV oncoproteins each developed centrosome anomalies was initially somewhat disconcerting. It appeared possible that centrosome abnormalities in these cell populations might have arisen as a generic consequence of subverting proliferation control mechanisms and/or cell cycle checkpoints rather than via disruption of pathways that specifically modulate centrosome duplication. Cell populations that expressed HPV E6 and E7 proteins in combination, however, contained a higher fraction of cells with supernumerary centrosomes. This suggested that the two HPV oncoproteins may have caused these defects through biochemically distinct cooperating pathways [58]. This notion was supported by additional studies that showed that acute expression of HPV E7 in primary human keratinocytes caused rapid synthesis of supernumerary centrosomes within one or two rounds of cell division. These results suggested that expression of the high-risk HPV E7 oncoprotein could rapidly uncouple centrosome duplication from the cell division cycle. In sharp contrast, acute expression of HPV E6 failed to evoke immediate subversion of centrosome homeostasis [58]. Additional evidence for the notion that centrosome abnormalities in HPV E6- or E7-expressing cells arise via fundamentally different mechanisms was obtained from experiments with a U2OS human osteosarcoma cell line engineered to express a centrin molecule fused to the green fluorescent protein (GFP) [82]. Individual centrioles are easily discerned in such cells (Figure 18.2A), and the effects of oncogenic stimuli on centriole synthesis can be directly visualized by fluorescence microscopy. Consistent with the transient transfection experiments in primary cells, acute expression of HPV E7 in these cells caused a rapid increase in GFP–centrin-positive structures (Figure 18.2A). These newly formed structures likely represent functional centrioles since they contribute to the formation of aberrant mitotic spindle poles during mitosis [83]. Hence there is compelling evidence that the HPV E7 oncoprotein can induce primary centrosome duplication errors through "the simultaneous multiple division of the centrosome" as originally postulated by Boveri [68].

HPV E7-expressing cells exhibit evidence for double-strand DNA breaks and DNA repair [13]. This may cause DNA replication forks to stall thus prolonging S-phase, which in turn may allow for aberrant centriole synthesis. Consistent with the notion that centrosome reduplication is intrinsically blocked in many cells [84], human keratinocytes that are S-phase arrested by hydroxyurea do not undergo additional rounds of centrosome duplication (S. Duensing and K. Münger, unpublished observations). Hence HPV E7 may have to overcome such a block to induce centrosome duplication errors.

One might envision at least two distinct but not mutually exclusive mechanisms by which E7 may induce aberrant centriole synthesis. Maternal centrioles in HPV E7-expressing cells might carry a persistent licensing signal, and upon abrogation of the intrinsic block to centrosome re-duplication [84], this may allow for the

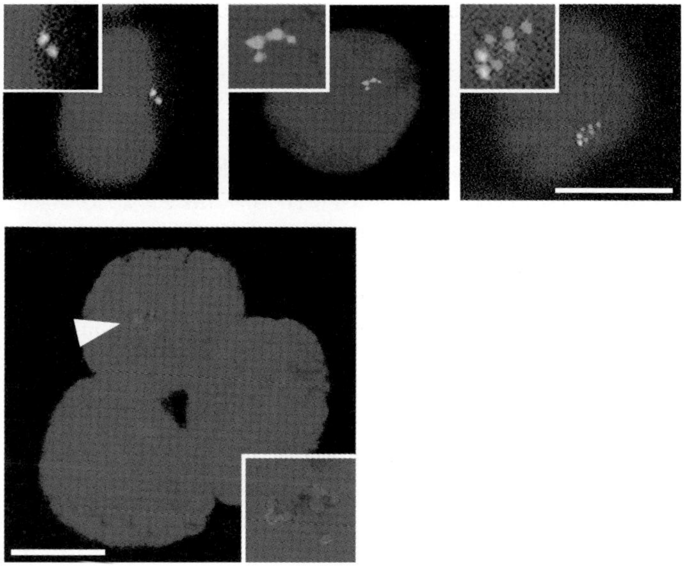

Figure 18.2 (A) Rapid induction of centriole duplication errors by the HPV-16 E7 oncoprotein. The human U-2 OS osteosarcoma cell line was manipulated to stably express centrin-GFP (kindly provided by M. Bornens, Institut Curie, Paris), thus allowing the visualization of individual centrioles as green dots by fluorescence microscopy. Prior to cell division, the normal centrosome of a cell contains two centrioles (left panel) that duplicate during S-phase and give rise to two mother–daughter centriole pairs (diplosome) in G2 (middle panel). Transient overexpression of the HPV-16 E7 oncoprotein increases the proportion of cells with uncontrolled centriole duplication (right panel). Cells were evaluated at 48 h post transfection; no drug treatment was used to induce S-phase arrest. Note the differences in fluorescence intensity between centrioles. Younger centrioles have been shown to contain less centrin than older centrioles supporting the notion that HPV E7 triggers abnormal daughter centriole formation. Nuclei stained with DAPI. Scale bar indicates 10 μm. (B) Centrosome accumulation in HPV-16 E6-expressing cells. An example of a primary human keratinocyte expressing HPV-16 E6 oncoprotein that has accumulated multiple centrosomes in parallel with nuclear atypia. Centrosomes were detected by immunofluorescence microscopy for γ-tubulin, a pericentriolar marker. The abnormal cell nucleus (stained with DAPI) is lobulated and dramatically enlarged. This nuclear morphology suggests a division failure, indicating that DNA replication may be uncoupled from cell division. Scale bar indicates 10 μm (see Color Plates page XXXVIII).

sequential or concomitant synthesis of multiple daughter centrioles from a single maternal template. Alternatively, newly formed daughter centrioles may in turn acquire licensing signals, leading to the formation of "grand-daughter" centrioles as has been recently reported in Drosophila wing discs in response to inhibition of Cdk1 [85].

Even though some cells do not engage efficient checkpoint mechanisms to thwart mitosis in the presence of aberrant multipolar spindles [86], there is an approximately 10-fold difference between the number of multipolar metaphases and multipolar anaphases in HPV E6/E7 oncogene-expressing primary human cells [13]. Hence it is conceivable that some control mechanism may be activated when a cell with supernumerary centrosomes attempts to undergo multipolar mitosis. This hypothesis is consistent with a recent analysis of a case of a malignant Burkitt's lymphoma. Despite evidence for numerical and structural centrosome abnormalities in approximately 30–50% of all cells, including some with multipolar mitoses, the tumor apparently maintained a stable pseudo-diploid karyotype [87].

A careful reevaluation of centrosome abnormalities in human keratinocytes with stable expression of HPV E6 or E7 revealed that centrosome abnormalities in E7-expressing cells were manifest in diploid cells devoid of overt nuclear anomalies. In contrast, centrosome abnormalities in HPV E6-expressing cells were mostly confined to cells that contained extensive nuclear abnormalities such as multiple nuclei, including micronuclei and/or abnormally enlarged, multilobulated nuclei and did not arise in diploid cells (Figure 18.2B). The formation of such pronounced nuclear abnormalities may be related to persistent cytokinesis defects similar to those that have been observed in cells that lack $p21^{Cip1}$ or p53 function [88]. Similar to these p53- or $p21^{Cip1}$-deficient cells, which were frequently eliminated by apoptosis [88], the HPV E6-expressing multinucleated keratinocytes often expressed markers of cellular senescence and thus they are unlikely to remain in the proliferative pool and may not give rise to viable daughters [83].

Based on these findings one has to conclude that the mere presence of centrosome abnormalities in a tumor cell does not necessarily predict that these abnormalities represent a primary defect of the centrosome duplication process but may have accumulated as a consequence of abortive mitoses. Cells with diploid or tetraploid genomes that contain centrosome abnormalities can undergo additional rounds of cell division with the potential of forming abnormal, multipolar mitotic spindles leading to chromosomal missegregation and development of aneuploidy. In contrast, cells with extensive nuclear abnormalities may be persistently defective in undergoing cytokinesis [88] and thus may be less potent in fueling malignant progression. The significance of detecting centrosome abnormalities in *in vitro*-manipulated cells and transgenic mouse models needs to be interpreted with these same caveats in mind.

18.7
Do HPV E7-induced Centrosome Anomalies Contribute to Carcinogenic Progression?

Even though HPV oncogene-expressing cells display additional signs of genomic instability that are unrelated to numerical centrosome abnormalities [13, 89, 90], there are multiple lines of evidence to support the notion that genomic destabilization caused by HPV E7-induced centriole duplication errors may contribute to carcinogenic progression. For one, centrosome abnormalities have been detected in human keratinocytes that, similar to pre-malignant lesions, retain episomal copies of HPV-16, and E6/E7 oncogene expression in these cells is controlled by viral factors. When such cells are grown under organotypic conditions that allow for the formation of a stratified skin-like structure that shows histo-morphological alterations reminiscent of dysplasia, centrosome abnormalities are observed in basal cells that contain HPV genomes at a low copy number. Centrosome abnormalities in these basal cells are associated with multipolar mitoses, and, as expected, these cells are aneuploid [91]. The incidence of centrosome abnormalities in basal cells was lower when they expressed an HPV-16 genome with a mutation in E7, lending additional support to the notion that E7 plays a direct role in inducing these abnormalities [91].

Centrosome abnormalities have also been noted in a transgenic mouse model of cervical carcinogenesis. These mice were engineered to express HPV-16 oncogenes in basal epithelial cells under the control of the human keratin K14 promoter and to develop cervical tumors when treated with low doses of estrogen (reviewed in [92]). Similar to reports for HPV-associated human cervical lesions, supernumerary centrosomes were already apparent in cervical precursor lesions, and the degree of abnormalities in this model increased in parallel with the severity of the lesions. When mice were generated that expressed HPV-16 E6 or E7 separately, the HPV E7-expressing animals developed high-grade cervical dysplasia and invasive cervical carcinomas. Expression of the HPV E6 oncoprotein, however, only induced low-grade cervical dysplasia and neoplastic progression did not occur [93]. As expected from the results with HPV E6 or E7 oncogene-expressing human keratinocytes, hyperplastic lesions that developed in HPV E6- or E7-expressing animals each displayed a similar degree of centrosome abnormalities [93]. The obvious discordance in neoplastic potency between HPV E7 and E6 in this transgenic mouse model despite similar elevations in centrosome abnormalities further reinforces the conclusion that mere detection of centrosome abnormalities in a tumor *per se* is not an appropriate predictor of neoplastic progression. The finding that lesions in mice expressing E7 can undergo carcinogenic progression, however, is consistent with the notion that primary centrosome duplication errors induced by oncogenic stimuli such as HPV E7 may have the capacity to drive genomic destabilization and thus contribute to malignant progression.

18.8 Mechanistic Considerations

The high-risk HPV E6 proteins form a complex with the ubiquitin ligase E6AP and re-target it to induce multi-ubiquitination and proteasome-mediated degradation of p53 [30]. Loss of p53 activity has been shown to result in centrosome hyperamplification [94]. This is at least in part a transcriptional effect since it can be rescued by ectopic expression of the transcriptional target $p21^{Cip1}$ [95], and cells lacking $p21^{Cip1}$ also exhibit centrosome abnormalities [96]. The detection of centrosome-associated p53 has advanced the model that p53 may have additional direct activities in modulating centrosome duplication (reviewed in [97]). In many cancers, however, centrosome abnormalities develop in pre-maligant lesions that retain bona fide wild-type p53 function [77, 78]. This suggests that loss of p53 is not a necessary, rate-limiting step for induction of centrosome anomalies in human tumors. Centrosome abnormalities in p53-deficient cells may accumulate as a consequence of mitotic errors and failure to detect and thwart such defects [98]. Since centrosome abnormalities in HPV E6-expressing cells are accompanied by nuclear abnormalities, particularly multinucleation [83] (Figure 18.2B), the accumulation of centrosome abnormalities in HPV E6-expressing cells may be related to p53 inactivation. The high-risk HPV E6 and E7 proteins can each affect aspects of G2/M checkpoint control [58, 99, 100]. In the case of E6, this is related to p53 inactivation [99] consistent with the finding that p53 is necessary to sustain G2 arrest after DNA damage through $p21^{Cip1}$- and 14-3-3σ-dependent pathways [88, 101, 102]. The mechanistic basis for the ability of HPV E7 to affect mitotic checkpoint control may be related to the ability to interfere with $p21^{Cip1}$ function [33, 34], or gradual accumulation of mdm2, which may inhibit p53 function [100].

In contrast, the ability of HPV E7 to induce centrosome duplication errors is independent of the p53 status of the cell [103], and expression of a high-risk HPV E7 protein rapidly induces centriole duplication errors in phenotypically normal, diploid human cells [83]. Hence HPV E7 may target molecule(s) that are directly involved in centrosome homeostasis. Obvious candidates are the retinoblastoma tumor suppressor pRb and the related "pocket" proteins p107 and p130 that are targeted by high-risk HPV E7 for proteasome-mediated degradation [104]. The pocket proteins control the transcriptional activity of E2F transcription factor family members (reviewed in [105]), which in turn regulate expression of transcriptional targets such as cyclin E, cyclin A, DNA polymerase α, and thymidine kinase that are rate limiting for G1/S transition as well as components of DNA repair pathways and G2/M checkpoint control [106, 107]. E2F as well as cyclin A/Cdk2 and cyclin E/Cdk2 complexes have been implicated in centrosome duplication [108–112]. Mouse embryo fibroblasts lacking E2F-3, but not those deficient in other E2F family members, accumulate centrosome abnormalities, but as in p53-deficient cells, these defects appear to arise mostly in cells that also exhibit nuclear aberrations [113]. Since cyclin A/Cdk2 and cyclin E/Cdk2 complexes can shuttle between the nucleus and cytoplasm [114] they may be directly involved in licensing centrosomes for duplication by phosphorylation of substrates such as B23 nucleophosmin [115 ,116], the Mps1p kinase [117] or CP110 [118]. Interestingly, however, suppression of Cdk2 activity in colon carcinoma cell lines did not appear to interfere with normal centrosome duplication [119] and genetic ablation of cyclin E [120, 121] and Cdk2 [122] in mice does not markedly interfere with normal cell division and development demonstrating that Cdk2 activity is not strictly required for normal centrosome duplication. This is consistent with an earlier study in Xenopus egg extracts that revealed that the initial round of centrosome duplication is Cdk2 independent but requires calcium/calmodulin-dependent protein kinase II (CaMKII) activity [123]. Alternatively, the centrosome-related function of Cdk2 in Cdk2- or cyclin E-deficient cells may be compensated by other Cdk/cyclin complexes.

Inhibition of E2F and Cdk2 activity by dominant negative mutants and/or small molecule inhibitors, however, each interfered with the ability of HPV E7 to induce abnormal centrosome duplication [58]. Additional experiments revealed that HPV E7 expression induced centrosome abnormalities in cell lines with mutated pRb and even in mouse embryo fibroblasts that lack expression of pRb, p107 and p130 altogether [103]. Even though mouse embryo fibroblasts appear to be defective in certain mitotic control mechanisms and undergo spontaneous tetraploidization [124], these results demonstrate that the capacity of HPV E7 to induce centrosome duplication errors is at least in part independent of its ability to subvert pRb and related pocket protein function [103]. This is consistent with other studies that have shown that loss of $p16^{INK4A}$ expression, which compromises pRb function through constitutive Cdk4/Cdk6-mediated hyperphosphorylation, or expression of SV40 T antigen, which interacts with pRb family members and disrupts pRb/E2F complexes, are not sufficient to rapidly induce centrosome abnormalities [124, 125]. The adenovirus E1A protein, which also interacts with pRb induces centrosome hyperamplification through a pRb-independent pathway that involves the small GTPase protein Ran [126]. The ability of HPV E7 to induce centrosome duplication errors depends on the integrity of a specific domain that in addition to serving as the pRb core binding site also mediates binding of HPV-16 E7 to the Cdk2 inhibitor $p21^{Cip1}$ and most likely other cellular proteins [33, 103]. Hence, while Cdk2 activity may not be strictly necessary for normal centrosome duplication, HPV-16 E7-induced Cdk2 hyperactivity may be critical for induction of abnormal centrosome duplication [58]. If this were indeed the case, Cdk2 inhibitors might be particularly useful as drugs to suppress genomic destabilization triggered by oncogene-induced centrosome duplication errors.

18.9 Concluding Remarks

Studies with HPV oncoproteins have yielded strong support for Boveri's hypothesis that some oncogenic stimuli can indeed give rise to aberrant centrosome synthesis, which may then drive genomic destabilization and provide genomic plasticity to an emerging neoplasm. Theses experiments also underscore, however, that in many cases, centrosome abnormalities can arise as secondary defects in cells that have undergone nuclear changes, most likely as a consequence of persistent cytokinesis defects. The mere presence of centrosome abnormalities in a tumor or tissue culture system cannot be taken as evidence that a certain oncogenic stimulus has directly caused these defects.

Virologists may raise the question whether the ability of the high-risk HPV E6 and E7 oncoproteins to induce centrosome-associated mitotic abnormalities and chromosome missegregation in their host cells may be of advantage to the viral life cycle or whether this simply represents an unavoidable corollary of the particularly nefarious viral replication strategy that these viruses have adopted. Since the low-risk HPV E6 and E7 proteins do not noticeably interfere with centrosome duplication, it is unlikely that induction of centrosome defects represents a necessary feature of the viral replication strategy. Interestingly, however, microcell-mediated transfer of chromosome 4 into the HPV-18-positive HeLa cervical carcinoma cell line led to suppression of telomerase activity [127]. Hence chromosomal gains and losses may not only alter the malignant phenotype but also augment the proliferative potential of HPV oncogene-expressing cells. Moreover, whereas many nuclei of the highly abnormal multinucleated HPV-16 E6/E7-expressing cells expressed senescence-associated β-galactosidase activity, some nuclei had retained the ability to undergo S-phase progression as evidenced by expression of the Ki67 proliferation marker [83]. High-level papillomavirus genome synthesis in differentiated keratinocytes may be confined to a relatively small number of "jackpot" cells that contain a very large number of HPV genomes [128]. Analysis of HPV genome synthesis by fluorescence *in situ* hybridization (FISH) analysis in differentiated keratinocytes grown under organotypic conditions revealed that some cells that contained high numbers of HPV-16 genomes also displayed nuclear abnormalities [91]. This indicates that, at the very least, generation of nuclear abnormalities does not interfere with the synthesis of viral progeny.

Over the years studies with small DNA tumor virus oncoproteins have yielded fundamentally relevant insights regarding the basic mechanisms of human carcinogenesis. In particular, the retinoblastoma and p53 tumor suppressors, important cellular targets of small DNA tumor virus oncoproteins, are mutated at some level in almost all human cancers and hence are likely to play a general role in tumorigenesis. Likewise, studies with HPV E6 and E7 oncoproteins may lead to the discovery of essential and generally applicable concepts and molecular pathways that can disrupt centrosome homeostasis in otherwise normal cells and may illuminate how such abnormalities might contribute to carcinogenesis. It may finally be possible to perform the critical experiment that Boveri "followed so far without success, (...) to bring about, with as slight injury as possible, multipolar mitoses in a healthy tissue and to observe whether a malignant tumor sometimes arises on such a soil" [68].

Acknowledgments

The authors' work on HPV-associated centrosome abnormalities is supported by grant CA66980 from the National Cancer Institute to K. M. S. D. was supported in part by a research fellowship from the Deutsche Forschungsgemeinschaft.

References

1. D. P. Cahill, K. W. Kinzler, B. Vogelstein, C. Lengauer. **1999**. Trends Cell Biol 9, M57–M60.
2. M. A. Nowak, N. L. Komarova, A. Sengupta, P. V. Jallepalli, M. Shih Ie, B. Vogelstein, C. Lengauer. **2002**. Proc Natl Acad Sci USA 99, 16226–16231.
3. W. C. Hahn, R. A. Weinberg. **2002**. N Engl J Med 347, 1593–1603.
4. K. W. Kinzler, B. Vogelstein. **1997**. Nature 386, 761, 763.
5. L. A. Loeb. **1991**. Cancer Res 51, 3075–3079.
6. I. P. Tomlinson, M. R. Novelli, W. F. Bodmer. **1996**. Proc Natl Acad Sci USA 93, 14800–14803.
7. S. Chang, C. Khoo, R. A. DePinho. **2001**, Semin Cancer Biol 11, 227–239.
8. B. McClintock. **1940**. Genetics 26, 234–282.
9. T. Iwasaka, P. S. Zheng, M. Yokoyama, K. Fukuda, Y. Nakao, H. Sugimori. **1998**. Obstet Gynecol 91, 260–262.
10. K. R. Shroyer, L. C. Thompson, T. Enomoto, J. L. Eskens, A. L. Shroyer, J. A. McGregor. **1998**. Am J Clin Pathol 109, 153–162.
11. O. Vafa, M. Wade, S. Kern, M. Beeche, T. K. Pandita, G. M. Hampton, G. M. Wahl. **2002**. Mol Cell 9, 1031–1044.
12. A. Karlsson, D. Deb-Basu, A. Cherry, S. Turner, J. Ford, D. W. Felsher. **2003**. Proc Natl Acad Sci USA, 100, 9974–9979.
13. S. Duensing, K. Münger. **2002**. Cancer Res 62, 7075–7082.
14. S. Mai, M. Fluri, D. Siwarski, K. Huppi. **1996**. Chromosome Res 4, 365–371.
15. S. Mai, J. Hanley-Hyde, M. Fluri. **1996**. Oncogene 12, 277–288.
16. X. Y. Yin, L. Grove, N. S. Datta, M. W. Long, E. V. Prochownik. **1999**. Oncogene 18, 1177–1184.
17. D. W. Felsher, J. M. Bishop. **1999**. Proc Natl Acad Sci USA 96, 3940–3944.
18. T. Flohr, J. C. Dai, J. Buttner, O. Popanda, E. Hagmuller, H. W. Thielmann. **1999**. Int J Cancer 80, 919–929.
19. H. zur Hausen. **2002**, Nature Rev Cancer 2, 342–350.
20. P. M. Howley, D. R. Lowy. **2001**. Papillomaviruses and their replication. In *Virology*, D. M. Knipe and P. M. Howley (Eds). Lippincott Williams & Wilkins: Philadelphia, 2197–2229.
21. D. R. Lowy, P. M. Howley. **2001**. Papillomaviruses. In *Virology*, D. M. Knipe and P. M. Howley (Eds). Lippincott Williams & Wilkins: Philadelphia, 2231–2264.
22. F. X. Bosch, N. Munoz. **2002**. Virus Res 89, 183–190.
23. M. L. Gillison, W. M. Koch, R. B. Capone, M. Spafford, W. H. Westra, L. Wu, M. L. Zahurak, R. W. Daniel, M. Viglione, D. E. Symer, K. V. Shah, D. Sidransky. **2000**. J Natl Cancer Inst 92, 709–720.
24. IARC Monographs on the evaluation of carcinogenic risks to humans: Human Papillomaviruses. **1995**. IARC Monographs on the evaluation of carcinogenic risks to humans, Vol. 64. International Agency for Research on Cancer/World Health Organization: Lyon, France, 409.
25. C. Ziegert, N. Wentzensen, S. Vinokurova, F. Kisseljov, J. Einenkel, M. Hoeckel, M. von Knebel Doeberitz. **2003**. Oncogene 22, 3977–3984.
26. S. Jeon, B. L. Allen-Hoffmann, P. F. Lambert. **1995**. J Virol 69, 2989–2997.
27. F. Stubenrauch, L. A. Laimins. **1999**. Semin Cancer Biol 9, 379–386.
28. K. Munger. **2002**. Cancer Invest 20, 71–81.
29. J. M. Huibregtse, M. Scheffner, P. M. Howley. **1991**. EMBO J 10, 4129–4135.
30. M. Scheffner, J. M. Huibregtse, R. D. Vierstra, P. M. Howley. **1993**. Cell 75, 495–505.
31. A. Hengstermann, L. K. Linares, A. Ciechanover, N. J. Whitaker, M. Scheffner. **2001**. Proc Natl Acad Sci USA 98, 1218–1223.
32. K. Munger, J. R. Basile, S. Duensing, A. Eichten, S. L. Gonzalez, M. Grace,

V. L. Zacny. **2001**. Oncogene 20, 7888–7898.
33. D. L. Jones, R. M. Alani, K. Münger. **1997**. Genes Develop 11, 2101–2111.
34. J. O. Funk, S. Waga, J. B. Harry, E. Espling, B. Stillman, D. A. Galloway. **1997**. Genes Develop 11, 2090–2100.
35. K. Zerfass-Thome, W. Zwerschke, B. Mannhardt, R. Tindle, J. W. Botz, P. Jansen-Durr. **1996**. Oncogene 13, 2323–2330.
36. K. H. Vousden, X. Lu. **2002**. Nature Rev Cancer 2, 594–604.
37. M. Scheffner, B. A. Werness, J. M. Huibregtse, A. J. Levine, P. M. Howley. **1990**. Cell 63, 1129–1136.
38. A. J. Klingelhutz, S. A. Foster, J. K. McDougall. **1996**. Nature 380, 79–82.
39. S. T. Oh, S. Kyo, L. A. Laimins. **2001**. J Virol 75, 5559–5566.
40. T. Veldman, I. Horikawa, J. C. Barrett, R. Schlegel. **2001**. J Virol 75, 4467–4472.
41. T. Veldman, X. Liu, H. Yuan, R. Schlegel. **2003**. Proc Natl Acad Sci USA 100, 8211–8216.
42. H. R. McMurray, D. J. McCance. **2003**. J Virol 77, 9852–9861.
43. P. Hawley-Nelson, K. H. Vousden, N. L. Hubbert, D. R. Lowy, J. T. Schiller. **1989**. EMBO J 8, 3905–3910.
44. K. Münger, W. C. Phelps, V. Bubb, P. M. Howley, R. Schlegel. **1989**. J Virol 63, 4417–4421.
45. D. J. McCance, R. Kopan, E. Fuchs, L. A. Laimins. **1988**. Proc Natl Acad Sci USA 85, 7169–7173.
46. J. B. Hudson, M. A. Bedell, D. J. McCance, L. A. Laimins. **1990**. J Virol64, 519–526.
47. P. Kaur, J. K. McDougall. **1988**. J Virol 62, 1917–1924.
48. M. Dürst, D. Gallahan, G. Jay, J. S. Rhim. **1989**. Virology 173, 767–771.
49. X. F. Pei, J. M. Meck, D. Greenhalgh, R. Schlegel. **1993**. Virology 196, 855–860.
50. P. J. Hurlin, P. Kaur, P. P. Smith, N. Perez-Reyes, R. A. Blanton, J. K. McDougall. **1991**. Proc Natl Acad Sci USA 88, 571–574.
51. J. M. Arbeit, P. M. Howley, D. Hanahan. Proc Natl Acad Sci USA **1996**. 93, 2930–2935.
52. R. W. Tindle. **2002**. Nature Rev Cancer 2, 59–65.
53. M. Bibbo, H. E. Dytch, E. Alenghat, P. H. Bartels, G. L. Wied. **1989**. Am J Clin Pathol 92, 261–265.
54. J. Bulten, P. J. Poddighe, J. C. M. Robben, J. H. Gemmink, P. C. M. de Wilde, A. G. J. M. Hanselaar. **1998**. Am J Pathol 152, 495–503.
55. R. Reid, C. P. Crum, B. R. Herschman, Y. S. Fu, L. Braun, K. V. Shah, S. J. Agronow, C. R. Stanhope. **1984**. Cancer 53, 943–953.
56. K. Heselmeyer, E. Schrock, S. du Manoir, H. Blegen, K. Shah, R. Steinbeck, G. Auer, T. Ried. **1996**. Proc Natl Acad Sci USA 93, 479–484.
57. R. G. Steinbeck. **1997**. Acta Oncol 36, 3–12.
58. S. Duensing, L. Y. Lee, A. Duensing, J. Basile, S. Piboonniyom, S. Gonzalez, C. P. Crum, K. Munger. **2000**. Proc Natl Acad Sci USA 97, 10002–10007.
59. A. E. White, E. M. Livanos, T. D. Tlsty. **1994**. Genes Develop 8, 666–677.
60. T. D. Tlsty, A. White, E. Livanos, M. Sage, H. Roelofs, A. Briot, B. Poulose. **1994**. Cold Spring Harb Symp Quant Biol 59, 265–275.
61. T. D. Tlsty. **1990**. Proc Natl Acad Sci USA 87, 3132–3136.
62. T. D. Kempe, E. A. Swyryd, M. Bruist, G. R. Stark. **1976**. Cell 9, 541–550.
63. G. M. Wahl, R. A. Padgett, G. R. Stark. **1979**. J Biol Chem 254, 8679–8689.
64. T. Hashida, S. Yasumoto. **1991**. J Gen Virol 72, 1569–1577.
65. T. D. Kessis, D. C. Connolly, L. Hedrick, K. R. Cho. **1996**. Oncogene 13, 427–431.
66. D. Hansemann. **1890**. Arch Pathol Anat Physiol Klin Med 119, 299–326.
67. T. Boveri. **1902**. Verh d phys-med Ges Würzburg NF 35, 67–90.
68. T. Boveri. **1914**. *Zur Frage der Entstehung Maligner Tumoren*. Jena: Fischer.
69. J. Marx. **2001**. Science 292, 426–429.
70. E. A. Nigg. **2002**. Nature Rev Cancer 2, 815–825.
71. F. Mitelman, B. Johansson, F. Mertens. Mitelman database of chromosome aberrations in cancer. http://cgap.nci.nih.gov/Chromosomes/Mitelman.

72. Y. F. Wong, T. H. Cheung, K. Y. Poon, V. W. Wang, J. C. Li, K. W. Lo, S. F. Yim, M. Y. Yu, G. Lahr, T. K. Chung. **2003**. Gynecol Oncol 89, 434–439.
73. M. Ionesco, S. Baudenon, R. Barrasso, P. Buu Hoi, T. Beuret, M. Moniot-Vittoz, G. Sadoul, J. de Brux, F. Coupez, G. Orth. **1987**. Bull Cancer 74, 397–406.
74. B. Winkler, C. P. Crum, T. Fujii, A. Ferenczy, M. Boon, L. Braun, W. D. Lancaster, R. M. Richart. **1984**. Cancer 53, 1081–1087.
75. C. P. Crum, H. Ikenberg, R. M. Richart, L. Gissman. **1984**. N Engl J Med 310, 880–883.
76. B. Skyldberg, K. Fujioka, A. C. Hellstrom, L. Sylven, B. Moberger, G. Auer. **2001**. Mod Pathol 14, 279–284.
77. W. L. Lingle, S. L. Barrett, V. C. Negron, A. B. D'Assoro, K. Boeneman, W. Liu, C. M. Whitehead, C. Reynolds, J. L. Salisbury. **2002**. Proc Natl Acad Sci USA 99, 1978–1983.
78. G. A. Pihan, J. Wallace, Y. Zhou, S. J. Doxsey. **2003**. Cancer Res 63, 1398–1404.
79. N. Sato, K. Mizumoto, M. Nakamura, K. Nakamura, M. Kusumoto, H. Niiyama, T. Ogawa, M. Tanaka. **1999**. Clin Cancer Res 5, 963–970.
80. K. K. Kuo, N. Sato, K. Mizumoto, N. Maehara, H. Yonemasu, C. G. Ker, P. C. Sheen, M. Tanaka. **2000**. Hepatology 31, 59–64.
81. K. Munger, P. M. Howley. **2002**. Virus Res 89, 213–228.
82. M. Piel, P. Meyer, A. Khodjakov, C. L. Rieder, M. Bornens. **2000**. J Cell Biol 149, 317–330.
83. S. Duensing, A. Duensing, C. P. Crum, K. Munger. **2001**. Cancer Res 61, 2356–2360.
84. C. Wong, T. Stearns. **2003**. Nature Cell Biol 5, 539–544.
85. S. J. Vidwans, M. L. Wong, P. H. O'Farrell. **2003**. J Cell Sci 116, 137–143.
86. G. Sluder, E. A. Thompson, F. J. Miller, J. Hayes, C. L. Rieder. **1997**. J Cell Sci 110(Pt 4), 421–429.
87. S. Duensing, B. H. Lee, P. Dal Cin, K. Münger. **2003**. Mol Cancer 2, 30.
88. F. Bunz, A. Dutriaux, C. Lengauer, T. Waldman, S. Zhou, J. P. Brown, J. M. Sedivy, K. W. Kinzler, B. Vogelstein. **1998**. Science 282, 1497–1501.
89. A. W. Plug-Demaggio, J. K. McDougall. **2002**. Oncogene 21, 7507–7513.
90. R. Furuta, Y. Hirai, K. Katase, S. Tate, T. Kawaguchi, F. Akiyama, Y. Kato, K. Kumada, T. Iwasaka, N. Yaegashi, K. Kanazawa, H. Yoshikawa, T. Kitagawa. **2003**. Int J Cancer 106, 167–171.
91. S. Duensing, A. Duensing, E. R. Flores, A. Do, P. F. Lambert, K. Munger. **2001**. J Virol 75, 7712–7716.
92. J. M. Arbeit. **1996**. Cancer Survey 26, 7–34.
93. R. R. Riley, S. Duensing, T. Brake, K. Munger, P. F. Lambert, J. M. Arbeit. **2003**. Cancer Res 63, 4862–4871.
94. K. Fukasawa, T. Choi, R. Kuriyama, S. Rulong, G. F. Vande Woude. **1996**. Science 271, 1744–1747.
95. P. Tarapore, H. F. Horn, Y. Tokuyama, K. Fukasawa. **2001**. Oncogene 20, 3173–3184.
96. C. Mantel, S. E. Braun, S. Reid, O. Henegariu, L. Liu, G. Hangoc, H. E. Broxmeyer. **1999**. Blood 93, 1390–1398.
97. P. Tarapore, K. Fukasawa. **2002**. Oncogene 21, 6234–6240.
98. P. Meraldi, R. Honda, E. A. Nigg. **2002**. EMBO J 21, 483–492.
99. D. A. Thompson, G. Belinsky, T. H.-T. Chang, D. L. Jones, R. Schlegel, K. Münger. **1997**. Oncogene 15, 3025–3036.
100. J. T. Thomas, L. A. Laimins. **1998**. J Virol 72, 1131–1137.
101. T. A. Chan, H. Hermeking, C. Lengauer, K. W. Kinzler, B. Vogelstein. **1999**. Nature 401, 616–620.
102. T. A. Chan, P. M. Hwang, H. Hermeking, K. W. Kinzler, B. Vogelstein. **2000**. Genes Dev 14, 1584–1588.
103. S. Duensing, K. Munger. **2003**. J Virol 77, 12331–12335.
104. S. L. Gonzalez, M. Stremlau, X. He, J. R. Basile, K. Münger. **2001**. J Virol 75, 7583–7591.
105. J. M. Trimarchi, J. A. Lees. **2002**. Nature Rev Mol Cell Biol 3, 11–20.
106. A. S. Weinmann, P. S. Yan, M. J. Oberley, T. H. Huang, P. J. Farnham. **2002**. Genes Dev 16, 235–244.
107. B. Ren, H. Cam, Y. Takahashi, T. Volkert, J. Terragni, R. A. Young, B. D.

Dynlacht. **2002**. Genes Dev 16, 245–256.

108. E. H. Hinchcliffe, C. Li, E. A. Thompson, J. L. Maller, G. Sluder. **1999**. Science 283, 851–854.

109. P. Meraldi, J. Lukas, A. M. Fry, J. Bartek, E. A. Nigg. **1999**. Nature Cell Biol 1, 88–93.

110. K. R. Lacey, P. K. Jackson, T. Stearns. **1999**. Proc Natl Acad Sci USA 96, 2817–2822.

111. Y. Matsumoto, K. Hayashi, E. Nishida. **1999**. Curr Biol 9, 429–432.

112. P. Tarapore, M. Okuda, K. Fukasawa. **2002**. Cell Cycle 1, 75–81.

113. H. I. Saavedra, B. Maiti, C. Timmers, R. Altura, Y. Tokuyama, K. Fukasawa, G. Leone. **2003**. Cancer Cell 3, 333–346.

114. M. Jackman, Y. Kubota, N. den Elzen, A. Hagting, J. Pines. **2002**. Mol Biol Cell 13, 1030–1045.

115. M. Okuda, H. F. Horn, P. Tarapore, Y. Tokuyama, A. G. Smulian, P. K. Chan, E. S. Knudsen, I. A. Hofmann, J. D. Snyder, K. E. Bove, K. Fukasawa. **2000**. Cell 103, 127–140.

116. Y. Tokuyama, H. F. Horn, K. Kawamura, P. Tarapore, K. Fukasawa. **2001**. J Biol Chem 276, 21529–21537.

117. H. A. Fisk, M. Winey. **2001**. Cell 106, 95–104.

118. Z. Chen, V. B. Indjeian, M. McManus, L. Wang, B. D. Dynlacht. **2002**. Dev Cell 3, 339–350.

119. O. Tetsu, F. McCormick. **2003**. Cancer Cell 3, 233–245.

120. T. Parisi, A. R. Beck, N. Rougier, T. McNeil, L. Lucian, Z. Werb, B. Amati. **2003**. EMBO J 22, 4794–4803.

121. Y. Geng, Q. Yu, E. Sicinska, M. Das, J. E. Schneider, S. Bhattacharya, W. M. Rideout, R. T. Bronson, H. Gardner, P. Sicinski. **2003**. Cell 114, 431–443.

122. S. Ortega, I. Prieto, J. Odajima, A. Martin, P. Dubus, R. Sotillo, J. L. Barbero, M. Malumbres, M. Barbacid. **2003**. Nature Genet 35, 25–31.

123. Y. Matsumoto, J. L. Maller. **2002**. Science 295, 499–502.

124. F. Borel, O. D. Lohez, F. B. Lacroix, R. L. Margolis. **2002**. Proc Natl Acad Sci USA 99, 9819–9824.

125. S. Piboonniyom, S. Duensing, N. W. Swilling, P. W. Hinds, K. Münger. **2003**. Cancer Res 63, 476–483.

126. A. De Luca, R. Mangiacasale, A. Severino, L. Malquori, A. Baldi, A. Palena, A. M. Mileo, P. Lavia, M. G. Paggi. **2003**. Cancer Res 63, 1430–1437.

127. C. Backsch, N. Wagenbach, M. Nonn, S. Leistritz, E. Stanbridge, A. Schneider, M. Durst. **2001**. Genes Chromosomes Cancer 31, 196–198.

128. J. B. Ravnan, D. M. Gilbert, K. G. Ten Hagen, S. N. Cohen. **1992**. J Virol 66, 6946–6952.

19
Manipulation of Centrosomes and the Microtubule Cytoskeleton during Infection by Intracellular Pathogens

Niki Scaplehorn and Michael Way

19.1
Introduction

The evolution of pathogens is shaped by a genetic arms race between invading microorganisms and the defense mechanisms of their unwilling hosts. From infecting and evading detection, to replicating and spreading, bacteria and viruses have evolved to manipulate the molecular machinery of the cell by intercepting or mimicking the signals that control normal cellular functions. These genetic weapons provide us with a rich resource of clues to establish how cellular functions are regulated. For example, the discovery of viral oncogenes such as v-src in the 1970s, and the revelation that they represent hijacked copies of host genes, completely revolutionized our understanding of signal transduction [1]. More recently, our knowledge of the regulation of actin dynamics has been transformed by work with bacteria such as *Listeria* and *Shigella*, and vaccinia virus, all of which can induce actin polymerization to enhance their spread from cell to cell [2].

Interactions between invading pathogens and the microtubule cytoskeleton have also been observed. Ever since the widespread use of electron microscopy, close associations between microtubules and virus particles have been documented. In some cases, infections with bacteria or viruses have been shown to induce dramatic reorganization of the microtubule network, but the molecular basis for these changes was not clear. However, dramatic improvements in live-cell imaging techniques, ever-increasing access to pathogen genomes, and a deeper understanding of cell regulation have all contributed to an explosion of recent advances in the study of pathogen–microtubule interactions. It has become clear that the microtubule cytoskeleton plays a fundamental role in the life cycle of many intracellular pathogens. The radial array of polarized microtubules, together with the wealth of motor proteins that carry cellular cargoes along them, forms an ideal system to facilitate the transport of viruses and bacteria around the cell. Many pathogens are able to rapidly subvert this system to promote their own spread through the cytoplasm. Much work is currently underway to determine how these pathogens are able to hijack motor proteins and microtubules so efficiently [3–5].

Centrosomes in Development and Disease. Edited by Erich A. Nigg

ISBN 3-527-30980-2

The centrosome lies at the heart of this transport network, by both organizing and maintaining the polarity of the microtubule cytoskeleton. This makes it a common site for the accumulation of many invading pathogens. This accumulation appears to play a vital role during infection, either by enhancing the targeting of virus particles to the nucleus, or by concentrating viral and host factors required to form a site of virus assembly or bacterial replication. The response of the host cell to infection is also coordinated at the centrosome, since many protein-degradation and antigen-processing activities are concentrated there [6]. The centrosome therefore promotes not only pathogen assembly and maturation, but also pathogen degradation [7].

The centrosome can also be a casualty of infection. Many viruses disrupt control of centrosome duplication, causing the cell to accumulate multiple centrosomes. This leads to the formation of abnormal mitotic spindles, which fail to accurately segregate chromosomes during cell division. The resulting genomic instability can cause the cell to acquire a transformed phenotype, suggesting that centrosome deregulation is an important part of the mechanism of virus-induced tumor progression [8] (see also Chapter 18). Recent work has also shown that some viruses are able to damage the centrosome, severely limiting its ability to nucleate microtubules [9].

Viruses and bacteria are clearly able to manipulate the microtubule network, centrosome stability, and centrosome duplication using a variety of different genetic tools. This chapter brings together these diverse mechanisms, and discusses what we might learn about the normal functions of the centrosome and microtubules by studying the interactions between intracellular pathogens and their hosts.

19.2
Microtubule-directed Movement of Viruses and Membrane Compartments during Viral Infection

As the principal microtubule organizing centre (MTOC) of the metazoan cell, the centrosome forms the core of an enormously complex transport system which spans the entire cell. By anchoring microtubule minus-ends at a single focus, close to the surface of the nucleus, the centrosome marks a reference point from which the entire system is directed [10]. The microtubule network provides the tracks for the many different motor proteins which carry cellular cargoes along the microtubules, moving either towards or away from the centrosome [11]. The localization within the cell of vesicles, the endoplasmic reticulum, the Golgi complex, mitochondria, and even the nucleus are all dependent on the microtubule–motor network.

Unfortunately, this powerful system is also an ideal target for invading microorganisms. Most intracellular pathogens need to spread through the cytoplasm. For example, many viruses are dependent on proteins of the host nucleus to replicate their genomes, and so need a fast and efficient means to locate to their site of replication. Other viruses and bacteria replicate in the cytoplasm, but need to concentrate the proteins and membranes required for their assembly and maturation in one location within the cell. These "assembly points" can be in any number of locations, from the centrosome to the plasma membrane. Finally, most bacteria and viruses that do not induce cell lysis need to move away from their site of replication and assembly, to spread infection to neighboring cells. All of these requirements are met by the microtubule cytoskeleton, and so it is unsurprising that a huge number of pathogenic organisms have evolved molecular tools to hijack this network [3].

19.2.1
Targeting the Nucleus using Motor-proteins and the Microtubule Network: *Herpes Simplex* Virus, Poliovirus and Retroviruses

Viruses which infect the large, polarized cells of the peripheral nervous system have the most dramatic need for a means of spreading over long distances, since the distance between their site of entry and their site of replication can reach several centimeters in length [12]. *Herpes simplex* virus type 1 (HSV-1) the causative agent of cold sores infects sensory neurons and spreads via synapses. Within a neuron, viral capsids must first move from the synapse, along the length of the axon to the cell body, where replication is possible. Progeny virus must then make the return journey, along the axon to the synapse, where they emerge in order to spread to neighboring cells (Figure 19.1). It has been calculated that were this movement to occur by diffusion alone, a journey of a single centimeter would take 231 years, and so it is essential that herpes virus moves by an active transport mechanism. HSV-1 capsids have long been observed to be aligned with axonal microtubules using the electron microscope, and it was suggested that capsids use this association to move around the cell [13].

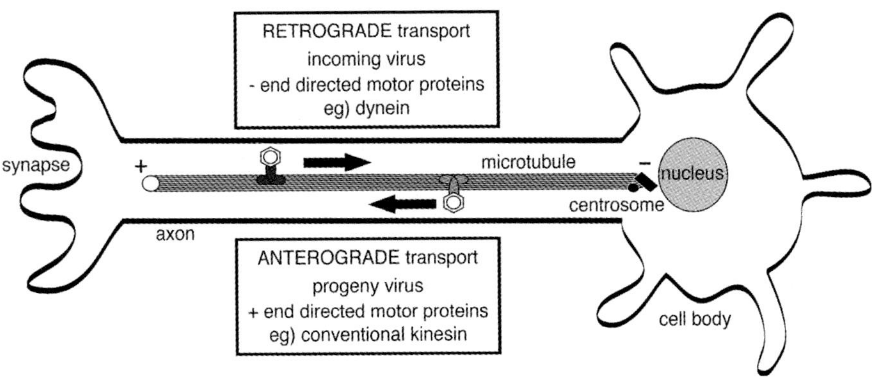

Figure 19.1 Retrograde and anterograde transport of viruses along the axon. Neurotropic viruses such as herpes simplex virus-1 subvert host motor proteins to aid their spread over long distances. Incoming virus particles move from the synapse to the cell nucleus using minus end-directed dynein-family motors. Progeny virus returns along microtubules in the opposite direction, possibly using kinesin-family motors.

More recently, the motor proteins responsible for this movement have been identified. In cultured epithelial cells, incoming nucleocapsids recruit dynein and the dynactin complex, taking full advantage of the cellular machinery normally used by the host to speed transit through the axoplasm. Retrograde transport of capsids towards the centrosome depends upon the integrity of this motor protein complex, as overexpression of a dynactin subunit, p50/dynamitin, blocks accumulation of virus near the nucleus (Figure 19.2) [14]. In contrast, anterograde transport of progeny virus in cultured neurons, away from the centrosome, is thought to involve conventional kinesin, based on its recruitment to progeny virus particles [15]. The means by which HSV-1 interacts with dynein/dynactin and kinesin complexes is yet to be fully resolved, as is the question of how this powerful subversion of host motor proteins is regulated.

While related neurotropic viruses such as poliovirus also use dynein/dynactin to locate to the nucleus, the mechanisms by which motor complexes are recruited appear to vary significantly [5]. Poliovirus enters target cells by binding a nectin-like receptor, CD155, and undergoing receptor-mediated endocytosis. A direct interaction between the cytoplasmic tail of CD155 and a dynein light chain is proposed to mediate the recruitment of virus-containing endocytic vesicles to microtubules [16]. In this way, poliovirus appears to perfectly mimic its host by subverting endogenous proteins for the task of docking with motor complexes.

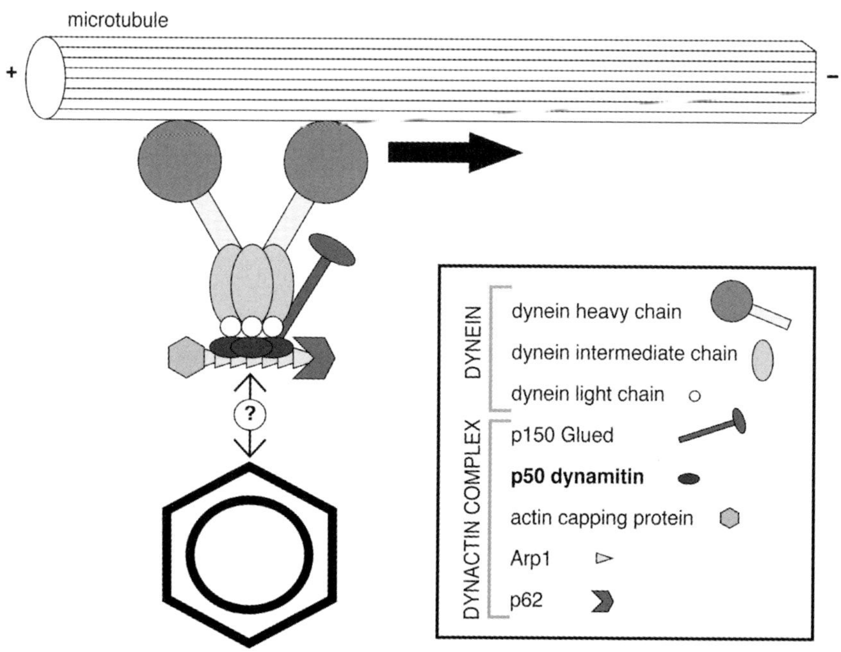

Figure 19.2 The dynein–dynactin complex and viral motility. The dynactin complex acts as an interface between dynein motors and their cargoes. Overexpression of the p50 dynamitin subunit of the dynactin complex inhibits dynein motor activity.

The journey of HIV and other retroviruses towards the nucleus, where they integrate with the host genome, has also been characterized in some detail. The virus enters the cell by engaging host receptors, and fusing with the plasma membrane. Once inside, the RNA genome is reverse transcribed to generate a DNA copy, forming a triple-helical structure which together with viral proteins comprises the reverse transcription complex (RTC). RTCs were filmed live in motion in infected cells by labeling viral proteins, newly transcribed DNA, and viral membranes (so that incomplete RTCs could be identified and excluded from analysis). RTCs move towards the centrosome with dynamics typical of microtubule-based movement, and accumulate there.

To enable the virus to move both towards and away from the centrosome, two different motor proteins must be employed. For example, if dynein-based motility in HIV infected cells is blocked by microinjection of an inhibitory antibody, virus accumulates at the cell periphery. This suggests that not only does HIV make use of minus end-directed motor proteins to move towards the centrosome, but plus end-directed motor activities are active in keeping the virus in the cell periphery [17]. This is supported by work with other retroviruses; direct interactions between dynein light chain 8 and the viral scaffold protein Gag in foamy retroviruses, and between KIF4 kinesin and Gag in murine leukemia virus have been identified [18, 19].

For a retrovirus, the centrosome is a crucial interchange during its journey towards the nucleus. While centrosomal targeting appears to be a common feature of many retroviruses, the mechanism by which each retrovirus penetrates the nuclear membrane appears to vary [20]. Many, including human T-cell leukemia virus (HTLV), require cell proliferation for productive infection, presumably because nuclear envelope breakdown during mitosis gives the virus unhindered access to the DNA of the host. Others, such as human immunodeficiency virus (HIV), encode proteins containing nuclear localization signals, which target the viral genome for nuclear import via nuclear pore complexes (NPCs). In general however, the mechanism by which viruses make the journey from the centrosome to the nuclear membrane is poorly understood. The process must at some point involve dissociation from the microtubule network, and may require the virus to acquire motor proteins with a different directionality, to allow movement away from the centrosome and towards the nuclear membrane. Establishing the molecular events that regulate transit between the centrosome and nuclear membrane will be an important aim of future research.

19.2.2
Hijacking Motor Proteins to Promote Cytoplasmic Assembly and Spread: Vaccinia Virus and African Swine Fever Virus

Vaccinia virus, the vaccine for smallpox, faces several different transport problems during infection. This is largely due to its highly complex life-cycle, which occurs exclusively in the cytoplasm, and involves formation of multiple forms of the virus in a number of different locations (Figure 19.3) [21]. A result of this nucleus-inde-

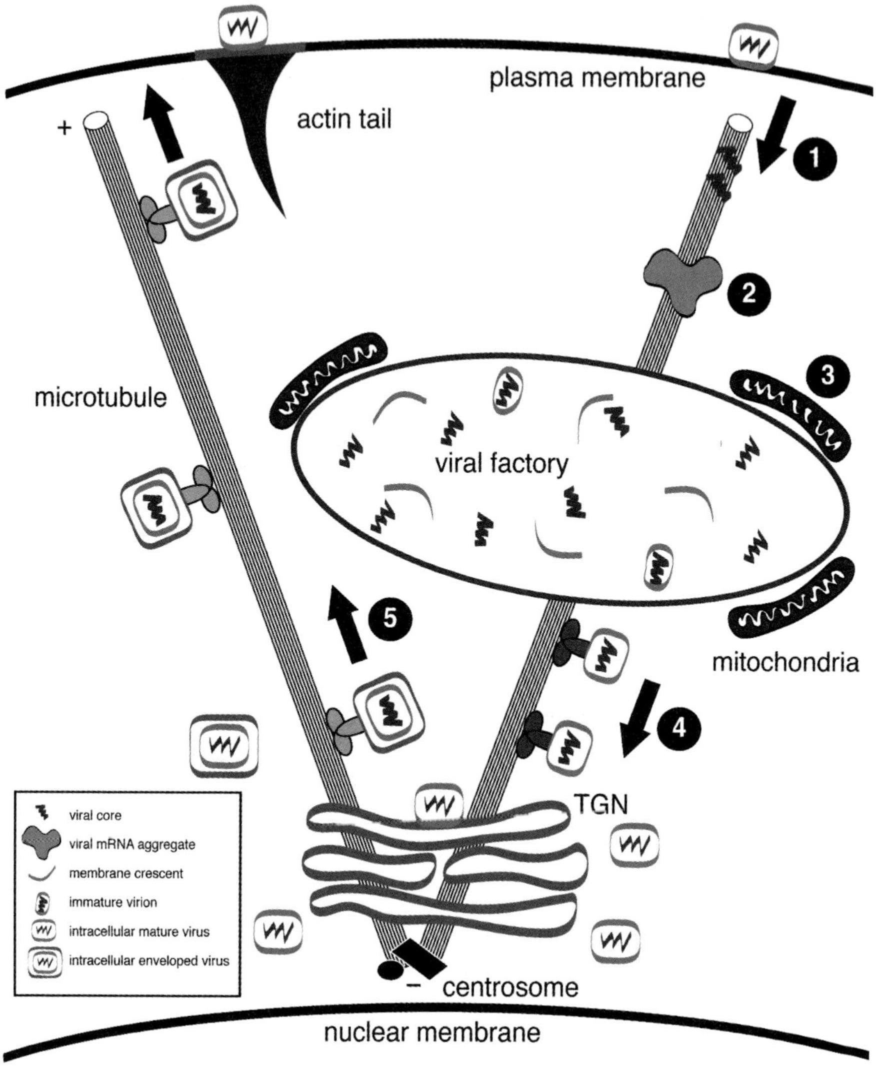

Figure 19.3 Interactions between vaccinia and microtubule cytoskeleton. Microtubules are thought to be involved in many steps of the vaccinia virus life cycle. (1) Incoming viral cores appear to associate with microtubules. (2) Viral mRNAs are organized into granules in a microtubule-dependent process. (3) Mitochondria retract from the cell periphery to surround the newly formed viral-factory region. (4) Accumulation of IMV particles at the centrosome and TGN requires intact microtubules. (5) IEV move from the TGN to the cell periphery along microtubules before leaving the cell and stimulating actin-tail formation (see Color Plates page XXXIX).

pendent existence is that vaccinia has become a masterful manipulator of intracellular membranes, essentially building for itself the structures required for viral gene expression, DNA replication, morphogenesis and spread. Microtubules play a role in coordinating these processes from the earliest stages of infection.

Almost immediately after entering a cell, vaccinia initiates transcription of early viral genes. This occurs within the incoming viral cores, and the resulting messenger RNAs (mRNAs) are extruded into the cytoplasm [22]. Both the viral cores and the extruded mRNAs are thought to move along the microtubule network. GFP-tagged cores have been observed by video-microscopy to move with speeds and "stop–start" dynamics characteristic of microtubule-based motility [23]. Although a motor protein responsible for this movement has not been identified, several viral proteins which are components of the core have been shown to behave as microtubule-associated proteins (MAPs), and it is possible that these proteins play some role in targeting vaccinia cores to the microtubule network [9].

Following extrusion from the cores, mRNAs become organized into granular structures, in a process which also requires intact microtubules [24]. The mechanism by which these structures form is not understood, but is likely to share features in common with mRNA transport in other systems. For example the polarized distribution of mRNAs in the *Drosophila* oocyte provides a crucial source of positional information which determines the antero-posterior axis of the fly [25]. This polarity is generated by the transport of mRNAs along microtubules which span the oocyte, and is known to depend on the activity of dynein-family, minus end-directed microtubule motors [26]. The same machinery for mRNA localization appears to be used both during oogenesis and in subsequent embryogenesis, suggesting that a conserved mechanism for mRNA targeting is at work [27]. Future studies with vaccinia might therefore yield important clues as to how mRNA localization is regulated during normal development.

Vaccinia replication and maturation occurs in regions of perinuclear cytoplasm called viral factories, in which replicated viral DNA is thought to become packaged with virally-modified membranes derived from the endoplasmic reticulum (ER)–Golgi intermediate compartment [28]. From these factories, so-called intracellular mature virus (IMV) particles emerge, and many begin to accumulate close to the MTOC. During infection, the trans-Golgi network (TGN) also becomes loaded with viral proteins [29]. A fraction of the IMVs become wrapped in the modified TGN to form intracellular enveloped virus (IEV). This trafficking of IMVs from the factories to the TGN depends on the protein encoded by the viral gene A27L, since viruses which lack this gene fail to accumulate at the centrosome, and are unable to reach their wrapping site to form IEV particles [30]. However, whether A27L plays a direct role in this process remains to be established.

Like HSV, accumulation of vaccinia at the MTOC requires the activity of the dynein–dynactin complex, since overexpression of p50/dynamitin inhibits the anchoring of virus particles close the centrosome [9]. However, in the case of vaccinia, the viral proteins responsible for binding this motor complex have not been found. A more detailed molecular characterization has been achieved with a distant relative of vaccinia, African swine fever virus (ASFV), which is also a large DNA virus that replicates in the cytoplasm. The movement of ASFV towards the centrosome is also dependent on the dynein–dynactin complex. A viral protein, p54, has been identified as a direct binding partner of dynein light chain 8. p54 is an external component of the ASFV virion, and resembles the 190-kDa family of mammalian MAPs. It is proposed that this interaction might be required for minus end-directed transport [31].

While an equivalent interaction in vaccinia-infected cells has not yet been identified, the movement of vaccinia away from the centrosome has been more intensively studied, and tantalizing clues as to how viral transport might be regulated have been found. After their formation, IEV particles move out towards the plasma membrane along the microtubule network [32, 33]. This process has been visualized using video-microscopy of GFP-tagged IEVs [34, 35]. Deletion of two different IEV-specific viral genes, A36R and F12L, blocks IEV transport without affecting IEV formation. In the case of A36R, a 30 amino-acid long region of the protein has been shown to be necessary and sufficient for virus movement [34]. This region overlaps with that required for A36R to bind a third IEV-specific protein, A33R [36]. Furthermore, a virus which lacks the A36R-binding site in A33R forms IEV particles which fail to move along microtubules to the cell periphery [36]. While A33R and A36R are integral IEV membrane proteins, F12L appears to associate specifically with those IEVs which are moving on microtubules, suggesting that recruitment of F12L plays a key role in promoting viral egress [37]. Finally, microtubule-associated IEV particles also recruit conventional kinesin, and overexpression of the cargo-binding domain of kinesin light chain efficiently blocks movement of IEV particles away from the centrosome [34]. What remains to be established is how these proteins interact to form a complex, and how the formation of this complex is regulated to generate movement of IEVs to the cell periphery.

Manipulation of the microtubule–motor protein network is not limited to the movement of virus particles, but also appears to be involved in the reorganization of intracellular membrane compartments during viral factory formation. Like vaccinia, both ASFV and iridoviruses depend on factories for their replication. In general, viral factories tend to form close to the MTOC, dramatically exclude host protein markers for different membrane compartments, and strongly recruit host chaperone proteins [38]. Additionally, mitochondria collapse from the cell periphery to surround the factory, presumably to supply the region with the vast quantities of ATP required for vigorous virus replication and morphogenesis [39]. This retraction could be a result of viral disruption of the host motor proteins which are required for normal mitochondrial distribution. Finally, factories become wrapped in a cage-like structure of vimentin intermediate filaments [40].

Such a vast and orchestrated restructuring of the contents of the cell might be expected to depend on the action of a large number of viral genes, and complex interactions with the machinery and signaling pathways of the host. However, although factory formation is undoubtedly an intricate process, there remains a strong possibility that the program of reorganization is largely an intrinsic protective reaction of the host cell, being strikingly similar to the cellular response to aggregation of unfolded proteins [41]. This response has been named the "aggresome" reaction, as it results in the formation of a large inclusion body around the MTOC, which contains aggregated, ubiquitinated proteins. The aggresome reaction can be induced by expressing proteins with mutations which prevent folding and promote aggregation [42]. Chaperones and proteasomal machinery are recruited to the aggresome structure, which becomes surrounded by mitochondria and a vimentin cage – mimicking almost perfectly the appearance and consequences of viral factory formation (Figure 19.3).

The centrosome and its accompanying microtubule network play at least two roles in this process. Firstly, the inward trafficking of aggregated proteins requires the activity of microtubules and the dynein–dynactin complex, as it is sensitive to nocodazole and expression of p50–dynamitin respectively [43]. For this reason, the natural point of accumulation for aggregated protein is the centrosome. Secondly, a significant fraction of active chaperones and proteasomal machinery of the cell is associated with the centrosome [44]. Viruses could take advantage of this rich supply of protein-processing enzymes to facilitate their assembly. For example, the centrosomal chaperone Hsp90 is known to be important for efficient replication of vaccinia virus [45].

However, the proximity of viral factories to centrosomal proteasome activity may also benefit the host cell. During infection, viral proteins are targeted by the proteasome, generating antigenic peptides which are presented to the immune system via major histocompatibility complex (MHC) class I proteins [46]. In the case of a misfolded influenza virus antigen, this process appears to occur at the MTOC, and in nuclear substructures called promyelocytic leukemia oncogenic domains (PODs) [47]. Processing of ubiquitinated HIV Nef antigen also occurs at the centrosome, and the microtubule-dependent recruitment of antigen to this site is necessary for processing and presentation of Nef peptides [48]. It will be interesting to see whether viral antigen processing at the centrosome is established as a general theme of the anti-viral response, or whether it only occurs in specific circumstances.

Studies of virus-infected cells have highlighted the dynamic nature of the protein-processing machinery of the cell. The observation that misfolded proteins are transported by the microtubule–motor protein network suggests that like protein synthesis, protein degradation is a spatially regulated process, which is organized by the cytoskeleton. Ironically, it appears that viruses can take advantage of this system to promote their assembly in the cytoplasm – despite the fact that an important function of the degradation machinery is to alert the immune response to intracellular infection. Both viral assembly and viral degradation are therefore focused around the centrosome.

19.2.3 Conclusion

Viruses intercept motor proteins and the microtubule network not only to accelerate their spread through the cytoplasm, but also to enhance their assembly by triggering the reorganization of cellular membrane compartments. The means by which they manage to do this, by hijacking and mimicking motor proteins and other MAPs at multiple levels, promises to highlight exciting facets of host biology. On a technical level, viruses are ideal tools to probe motor protein function, since they can potently concentrate motor proteins, making them highly amenable to visual analysis.

It appears that viruses are also able to efficiently regulate their direction of movement. Directionality is likely to be controlled by a number of factors, including post-translational modification of "cargo" proteins on the virus surface, by acquisition of host or viral adaptor proteins, or by a combination of the two processes. In particular, it will be interesting to see whether viruses, like other cellular cargoes, can simultaneously recruit a number of different motor protein complexes. The mechanism by which either retrograde or anterograde transport is selected by the virus may offer fascinating parallels with the control of vesicular and organelle trafficking in normal, uninfected cells. The centrosome itself is such an organelle, being strongly dependent on the activity of motor proteins both to maintain its position relative to the nucleus in non-dividing cells, and to migrate around the nucleus following centrosome separation [49, 50].

Viruses which take control over the motor proteins required for normal centrosome function might therefore be expected to have a dramatic impact on microtubule organization and cell division. However, finding a clear link in infected cells between viral motor-abuse and microtubule reorganization is hindered by the sheer number of different viral factors and motor proteins that might play a part in this process. The following section describes some of the many ways in which viral infection can result in disruption of the microtubule network and centrosome function.

19.3 Virus-mediated Damage to the Centrosome and Microtubule Network

All pathogens tread a fine line between manipulating the systems of their hosts, and destroying them. While cell lysis may be a useful means by which viruses promote their release and spread, it is possible that in many cases cellular damage is simply a result of an intolerable level of viral abuse of the cell's signals and resources. Bearing this in mind, together with the fact that the microtubule cytoskeleton and centrosome are frequently hijacked by intracellular pathogens, it is unsurprising that both are frequently reorganized or even destroyed during many different infections.

19.3.1 Viral Disruption of Microtubule Organization

Microtubule reorganization is presumably a result of viral manipulation of the proteins which normally regulate the stability and localization of microtubule filaments. The hijacking of motor proteins has been discussed, but other components of the microtubule network are also targeted by viruses [3]. In many cases, the mechanism by which microtubules are reorganized during infection is completely unknown.

As discussed above, both vaccinia and ASFV encode proteins which have MAP-like properties, and so may be partly responsible for microtubule reorganization during infection [9, 31]. MAPs are found in many other viruses – for example, the "movement protein" (MP) of tobacco mosaic virus (TMV) is an endoplasmic reticulum-resident protein which binds microtubules, and is required for transmission of virus from cell to cell through plasmodesmata [51]. MP has a potent ability to disrupt the microtubule cytoskeleton: its expression in uninfected mammalian cells causes microtubules to detach from the centrosome, by an unknown mechanism [51]. The HSV-1 tegument protein, VP22 also behaves as a MAP, and is an aggressive manipulator of the microtubule cytoskeleton, inducing microtubule bundling when expressed in uninfected cells [52]. Like TMV movement protein, VP22 is required for the efficient spread of virus from cell to cell [53].

Other viruses affect microtubule organization by disrupting the function of host MAPs. Poliovirus, for example, encodes a protease that cleaves MAP4. This cleavage correlates with collapse of microtubules during infection [54]. HIV also expresses a protease capable of cleaving MAP2, although the significance of this is unknown [55]. Finally, flaviviruses such as Kunjin virus and hepatitis C virus have long been known to reorganize microtubules into large crystalline structures [56, 57]. The function of these paracrystals is unclear, as is the molecular basis of their origin.

It is likely that viral disruption of the microtubule network, either by expression of viral MAPs, by cleavage of host MAPs, or by other mechanisms, enhances the ability of the virus to interact with microtubules or indeed promotes infection in some other way. However, it is equally possible that many disruptive effects of virus infection are simply tolerable side-effects of other viral manipulations or host responses to invasion. Despite this, such phenomena may still be of significant interest to those studying regulation of the microtubule network, since they may shed light on novel ways in which the cytoskeleton may be controlled in uninfected cells.

19.3.2
Virus-mediated Centrosomal Damage

In a small number of cases, virus infection has been shown to result in almost complete ablation of the centrosome. In vaccinia-infected cells, centrosome destruction is characterized by a rapid loss of centrosomal proteins from the MTOC as determined by immunofluorescence, and starts from 2 h post-infection (Figure 19.4). This effect appears to be due to a redistribution of proteins within the cell, since levels of expressed centrosomal proteins, measured by Western blot, are not significantly reduced over the same period. The ability of the centrosome to re-nucleate microtubules is also greatly diminished by vaccinia infection.

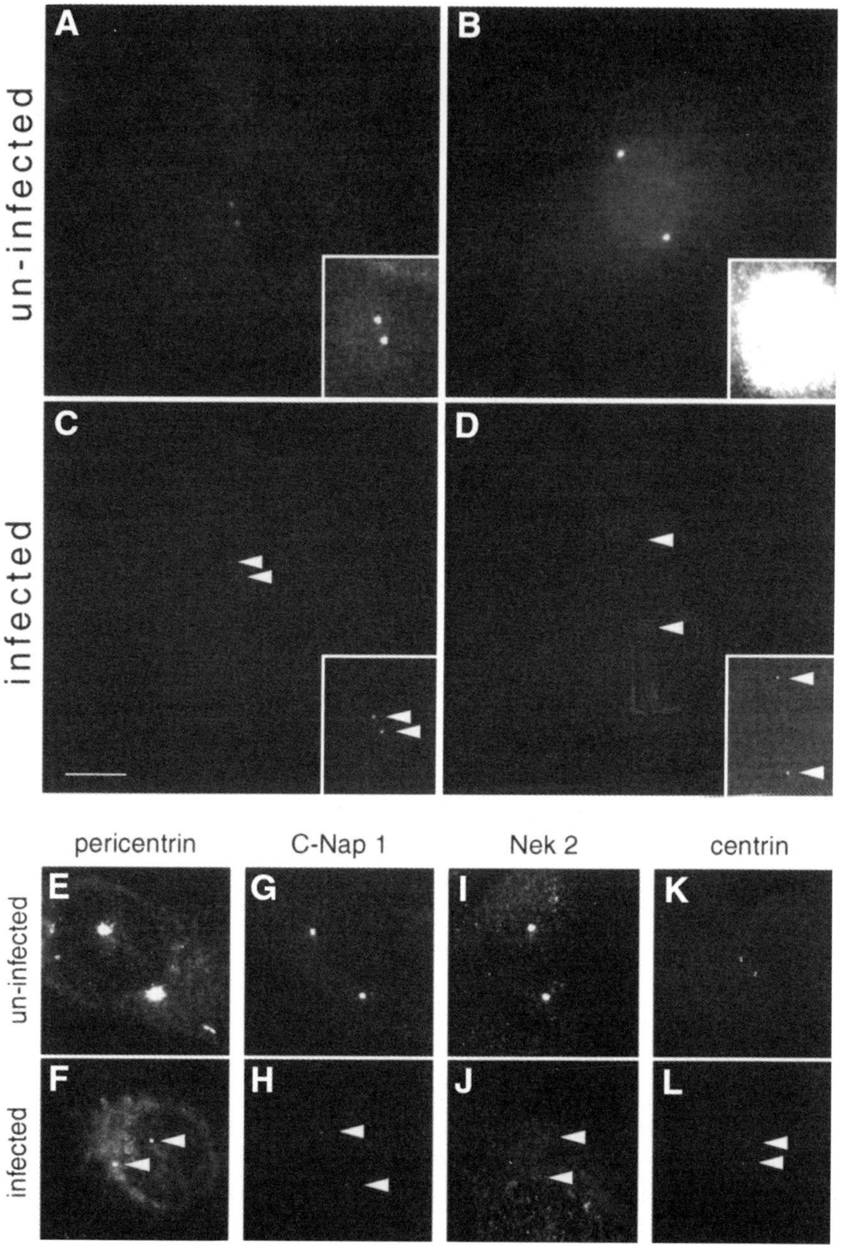

Figure 19.4 Vaccinia infection damages the centrosome. Centrosomes in vaccinia-infected cells show dramatically reduced labelling with antibodies against centrosomal proteins including gamma-tubulin (A–D), pericentrin (E, F), C-Nap1 (G, H) and centrin (K, L). Reproduced from [9] with permission from Oxford University Press.

Centrosome destruction also appears to require the expression of viral genes, although the gene responsible has not been identified [9].

More recently, centrosome destruction was also observed in HSV-1-infected cells, where VP22-dependent microtubule reorganization may be involved. In this case, MTOCs have been shown to return after prolonged infection, suggesting that the process of destruction is reversible [58]. Centrosomal defects have also been described by electron microscopy in cytomegalovirus (CMV)-infected cells, where disruption of centriole structure and detachment of fibrillar material occur [59]. Whether these changes are also seen during vaccinia- or HSV-1-mediated centrosome destruction, and so represent a general response to a variety of different viral infections, remains to be seen.

Centrosome destruction was first observed many years ago, as a cellular response to heat shock [60]. Both viral infection and heat shock are potent activators of stress pathways, raising the intriguing possibility that centrosome stability is under the control of stress-activated signaling [61]. Centrosomal damage may therefore be an underlying cause of microtubule reorganization in many other viral infections, since in many cases the centrosome has been overlooked when characterizing virus-induced alterations in the microtubule network.

19.3.3
Summary

Identifying the mechanisms by which viral infections cause damage to cellular organization may be a fruitful area of research in the future. Such work may draw our attention to new ways in which the stability of microtubules and the cytoskeleton is regulated in uninfected cells. This has certainly been the case in the cell cycle field, where the deregulation of centrosome duplication and checkpoint signaling during virus infection has been intensively studied. Insights from viral proteins which disrupt these centrosomal functions are described in the next section.

19.4
Viral Disruption of the Centrosome Duplication Cycle and Spindle Checkpoints

The centrosome is the only organelle outside the nucleus which is known to duplicate precisely once during each round of the cell cycle. How this remarkable level of control is possible for a structure consisting entirely of proteins is not yet fully understood, although pathogens may offer some insights. Many viruses have been found to cause uncontrolled duplication of the centrosome. These viruses somehow uncouple the centrosome from the mechanisms which link it to the rest of the cell cycle – the system which normally ensures that centrosomal duplication occurs in concert with replication of the host genome in S-phase [62] (see Chapter 9).

Cells are normally protected from the consequences of centrosome reduplication by mitotic checkpoints: signals which become active if the mitotic apparatus fails to be assembled correctly, and which act to arrest cell cycle progression to prevent the otherwise inevitable onset of genomic instability. However, several viruses have been identified which clearly interfere with this checkpoint system, allowing cells with multiple centrosomes to proceed through the next round of DNA replication having failed to complete mitosis. By disrupting both centrosome duplication and the protective checkpoints that buffer the cell from such damage, some viruses can be potently oncogenic. Identifying viral genes which are involved in this disruption has shed new light on how the intricacies of the host cell cycle are coordinated.

Perhaps the best characterized host–pathogen interaction involving centrosome cycle regulation is that of papilloma virus. The role of papilloma viral genes E6 and E7 in promoting ectopic centrosome duplication is covered extensively elsewhere in this book (see Chapter 18), and so will not be described here. However, studies with many other viruses are now following in the wake of papilloma virus, and the variety of different ways in which cell-cycle disruption may occur is proving to be particularly informative.

19.4.1 Early Studies on Centrosome Number: Paramyxoviral Syncytia

Some of the very first experiments to address the question of how centrosome number is regulated used viruses as an experimental tool. Infection with paramyxoviruses such as Sendai virus and simian virus 5 (SV5) causes cells to fuse with one another. In SV5-infected cell monolayers for example, large syncytia are formed in which nuclei align with each other in ordered rows. In these cells, centrosomes were observed to aggregate, forming a single, functional MTOC, from which bundles of microtubules emerge. These bundles were shown to play a role in guiding the nuclei into such organized arrays, providing us with one of the first clear examples of how microtubules can manipulate membrane-bound organelles [63]. Using Sendai virus to generate binuclear cells, MTOCs were also found to merge, to form a single, functional center [64]. However, in both cases, centrosome number is not modified directly through the signaling pathways of the host, but is increased simply through cell–cell fusion. Interestingly, centrosome aggregation is not a direct consequence of cell fusion, since other fusion techniques do not appear to result in the same outcome. Whether the aggregation response is a regulatory reaction to increased centrosome number, a function of the activity of the virus or an intrinsic property of centrosomes is not clear. In Sendai virus-infected cells, multipolar spindles are often seen, suggesting that centrosome aggregation does not restore their ability to reliably assemble a normal, bipolar spindle.

19.4.2
Multiple Centrosomes: Human Immunodeficiency Virus and the DNA Damage Checkpoint

One of the hallmarks of the late stages of AIDS is a dramatic and acute decline of $CD4^+$ T-cells in the HIV-infected patient. This is thought to be at least partially a result of cytotoxicity associated with the expression of viral proteins. One such viral component, the accessory protein Vpr, is known to have such an impact, since its expression in many cell types results in prolonged cell-cycle arrest and apoptosis. Importantly, Vpr is both necessary and sufficient for these effects, since viruses lacking Vpr fail to induce cell-cycle arrest, while ectopic expression of Vpr reproduces the arrest phenotype [65, 66]. In particular, Vpr expression induces a long arrest in the G2 phase, during which multiple centrosomes accumulate, followed by aberrant entry into mitosis, generation of abnormal spindles, and induction of apoptosis. Inappropriate centrosome reduplication is also observed in HIV-infected cells [67, 68].

Several signaling pathways that are known to be involved in regulation of the centrosome cycle have also been shown to be affected by Vpr expression. Most significantly, Cdc2 (also known as Cdk1) becomes phosphorylated on tyrosine 15 in Vpr-expressing cells, resulting in suppression of the Cdc2–cyclin B kinase activity required for G2–M transition. This effect is conserved in fission yeast, where genetic studies have pointed to a role for the Cdc2 phosphatase Cdc25 and the Cdc2 kinase Wee1 in mediating the impact of Vpr on Cdc2 phosphorylation [69]. Protein phosphatase 2A (PP2A) is also proposed to function in this cascade, although its relationship with Vpr and the significance of its activity remains controversial [70]. More recently, proteins of the DNA-damage checkpoint pathway, ATR and Chk1 have been shown to act downstream of Vpr and upstream of Cdc25 in mammalian cells [71]. Finally, while none of the above processes appear to be dependent on the tumor suppressor, p53, an additional effect of Vpr is to upregulate expression of the cyclin-dependent kinase inhibitor, $p21^{Waf1/Cip1}$, and this particular pathway requires p53 activity (Figure 19.5) [72].

How does viral manipulation of this intricate web of interactions result in centrosome reduplication? The identification of DNA-damage checkpoints as a major target of Vpr may be particularly revealing. The ATR/Chk2 signal is activated in response to treatments which disrupt accurate DNA replication during S-phase, while a sister pathway involving ATM and Chk2 are regulated by other genotoxic stresses, such as gamma-irradiation [73]. Both types of stress are known to promote the formation of ectopic centrosomes under some circumstances [74, 75] (see Chapter 17). Together, these results suggest that activation of DNA-damage checkpoints is often sufficient to inhibit the onset of mitosis, and yet insufficient to inhibit centrosome duplication. In addition, unexpected connections between the machinery that monitors genome integrity and the centrosome have been identified (see Chapter 11). For example, mutations in several genes required for DNA integrity, including Brca1 and Brca2, result in centrosome-number abnormalities [76, 77]. By what mechanisms these centrosome amplifications come about is un-

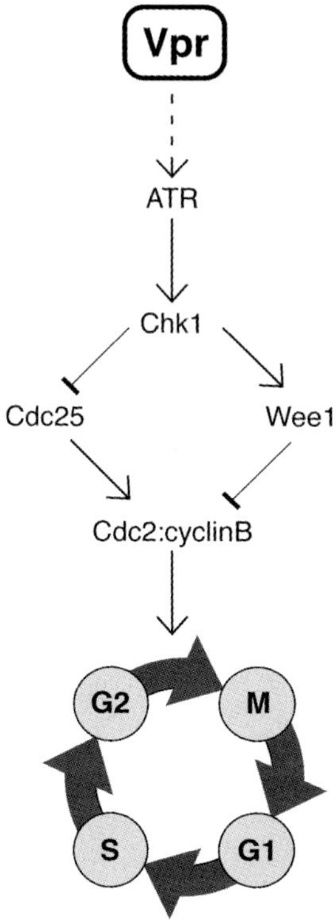

Figure 19.5 HIV Vpr inhibits progression from G2 to M phases of the cell cycle. HIV Vpr expression activates the ATR/Chk1 DNA-damage checkpoint, which acts on Cdc2 : cyclin B to inhibit the onset of mitosis. Cells stalled in this way accumulate multiple centrosomes.

clear, but aborted cell division constitutes one possible mechanism [78]. Clearly, the system which licenses centrosome replication is somehow integrated with the system which licenses the G2–M transition, although unsurprisingly, this integration is far from perfect. Such inconsistencies are exploited to great effect by viral genes such as Vpr.

Given the importance of cell-death in the pathology of AIDS, a key point which must be resolved is whether centrosomal abnormalities induced by Vpr represent a mechanism by which apoptosis is induced. Several lines of evidence, and especially the results of mutational analyses of Vpr, suggest that the cell-cycle arrest and apoptotic functions of Vpr may be separated [79, 80]. However, the interdependence between cell-cycle arrest and centrosome dysfunction is clearly highly complicated, and so a direct connection between such abnormalities and apoptosis may not be evident. What is clear is that future work in this field promises to help untangle the complicated triangular relationship between centrosome duplication, DNA replication checkpoints, and apoptosis.

19.4.3
Multiple Centrosomes: DNA Tumor Viruses, Retinoblastoma and Ran GTPase

Like HIV, DNA tumor viruses such as adenovirus and SV40, also induce centrosome amplification in infected cells. However, while HIV interrupts the cell cycle at the G2–M transition, DNA tumor viruses exert an effect on progression between G1- and S-phase, forcing infected cells to begin DNA replication independently of the permissive signals normally required for this to occur [8]. In the case of adenovirus, the oncoprotein E1A is sufficient to mediate this effect – in fact, a key regulator of G1–S transition, the retinoblastoma protein pRB, was originally identified as a ligand of E1A [81]. One of the many functions of pRB is to act as a tumor suppressor, by sequestering transcription factors such as E2F which when released, upregulate expression of genes required for S-phase entry [82]. E1A acts by binding to pRB, forcing it to release its other binding partners, and driving the cell into S-phase prematurely [83]. SV40 T antigen behaves in an identical manner [84].

As a result of pRB inactivation, there is a dramatic upregulation of E- and A-type cyclins, which together with the cyclin-dependent kinase Cdk2 form complexes capable of phosphorylating and activating key S-phase targets. These are not only involved in DNA replication, but also in the initiation of centrosome duplication [85]. In this way, E1A and SV40 T antigen could rapidly bring about premature centrosome duplication. This does not however explain how these oncoproteins disrupt the controls which normally limit this duplication to a single event in each round of the cell cycle. Inactivation of pRB by alternative routes, by overexpressing Cdk4 or impairing cyclin-dependent kinase inhibitor function, both of which lead to inhibitory phosphorylation of pRB, does not result in centrosome amplification [86]. This suggests that another function of viral oncoproteins is required for full onset of genomic instability. The identification of Ran GTPase as an additional binding partner of E1A is particularly exciting in this regard.

Ran interacts directly with E1A both *in vitro* and *in vivo*. While the meaning of this association is not understood, E1A acts to inhibit the exchange of GDP for GTP on Ran by the guanine nucleotide exchange factor RCC1 *in vitro*. Most importantly, induction of supernumerary centrosomes by E1A expression depends on the activity of the Ran signaling network, since centrosome amplification fails to occur in a cell line in which RCC1 is inactivated, even though premature S-phase entry still occurs [87].

Ran was first identified as an important regulator of the directionality of nucleocytoplasmic transport. The exchange factor activity of RCC1 is concentrated in the nucleus via its association with chromatin, whereas Ran GTPase activating proteins are found largely in the cytoplasm. This generates a RanGTP–RanGDP gradient which is used as a source of positional information by proteins passing through the nuclear pore complex [88]. More recently, Ran was also found to regulate the assembly and disassembly of microtubules. Upon the breakdown of the nuclear envelope, the Ran gradient was proposed to be critical in establishing the organization of the mitotic spindle [89]. A fraction of cellular Ran is found at centrosomes, bound to the A-kinase anchoring protein (AKAP450) [90]. An important regulator of Ran activity, RanBP1, also localizes to the centrosome, and its overexpression leads to premature centrosome splitting, to form multipolar spindles from the resulting, partially functional organizing centers [91].

However, despite this progress, the mechanism by which Ran is involved in the regulation of the centrosome cycle has not been clarified. One distinct possibility is that Ran activity is required for the shuttling of essential centrosomal components between the nucleus and the centrosome in interphase. This would suggest that E1A causes the formation of supernumerary centrosomes because it disrupts the flow of regulators between these two organelles. This is supported by recent findings with the unrelated, non-tumorigenic, hepatitis B virus, which also causes aberrant centriole replication. This effect has been mapped to the viral HBx gene, which is necessary and sufficient for the phenotype. HBx sequesters a nuclear export receptor, Crm1 in the cytoplasm, preventing export of nuclear proteins such as NF-κB. Crm1 is a binding partner of Ran, and in common with Ran, a fraction of the protein is centrosome associated [92].

Using Ran to hijack the nucleocytoplasmic transport system might pay dividends for the virus, by enhancing its access to the nucleus. Alternatively, Ran might directly affect the assembly, stability and replication of the centrosome. Adenovirus E1A and its equivalents in other viruses will be important tools to establish which of these models is correct. What is clear is that the Ran GTPase network is a common target for viruses that deregulate the centrosome cycle.

19.4.4
Targeting the Spindle Assembly Checkpoint: Human T-Cell Leukemia Virus-1

The centrosome has long been known to act as an organizing center for the assembly of the mitotic spindle. However, it is now becoming apparent that the centrosome also has a vital role in regulating spindle function during the transition from metaphase to anaphase. It does this by playing host to signaling pathways which ensure that chromosomes only separate when each has a bipolar attachment to the spindle [93]. This so-called spindle-assembly checkpoint is crucial for the maintenance of genomic stability. Viruses such as human T-cell leukemia virus-1 (HTLV-1) are capable of interfering with the spindle-assembly checkpoint, by targeting proteins which control the onset of anaphase.

HTLV-1 is the transforming retrovirus responsible for adult T-cell leukemia (ATL), a disorder characterized by immortalization of T-lymphocytes, which tend to carry severe cytogenetic defects [94]. Such cells are often multinuclear, containing micronuclei and both centric and acentric chromosome fragments. Similar cytogenetic defects are also seen in uninfected cells expressing a single HTLV-1 gene, Tax [95]. Tax is a multifunctional oncoprotein, acting on at least two different levels to deregulate cell-cycle progression. Firstly, Tax acts on a transcriptional level, both promoting viral gene expression and interacting with the transcription factors that control the essential cell cycle regulatory genes of the host. Secondly, Tax is itself able to directly suppress the activity of proteins directing cell-cycle progression, as well as the detection and repair of DNA damage [96].

One way in which Tax interferes with chromosome segregation is by targeting a component of the spindle-assembly checkpoint, MAD1. MAD1 was identified in a screen for mitotic arrest defects in budding yeast. The human homolog, HsMAD1 was subsequently identified by virtue of its affinity for HTLV-1 Tax [97, 98]. HsMAD1 operates as a homodimer, and heterodimerizes with another regulator, HsMAD2 [99]. This heterodimeric MAD complex is highly dynamic during the cell cycle; HsMAD2 is recruited to unattached kinetochores via its interaction with HsMAD1, where it acts in a signaling pathway that ultimately controls the proteolytic cleavage of proteins which maintain the physical connection between sister chromatids. Once the kinetochores have stably captured spindle microtubules emanating from both poles, the MAD complex dissociates from the kinetochores, releasing its inhibitory activity towards chromatid separation. As mitosis proceeds, HsMAD1 relocalizes to the spindle pole, the spindle mid-zone, and finally during telophase, to the midbody [100].

Tax interferes with MAD complex formation by stably associating with HsMAD1 and preventing its recruitment of HsMAD2. This results in compromised cohesion between sister chromatids, and anaphase takes place in the absence of proper chromosome alignment. For reasons that are not clearly understood, subsequent cytokinesis fails, resulting in the formation of multinuclear giant cells with increased centrosome number. Disruption of HsMAD1 by overexpression of dominant negative forms of the protein is sufficient to generate multinuclear cells, suggesting that it is the most significant target of Tax, at least with respect to generating cytogenetic abnormalities and centrosome amplification. Suppression of the spindle assembly checkpoint by Tax is also relevant to the pathology of HTLV-1, since the checkpoint appears to be non-functional in cells from ATL patients [97]. Studies with a viral oncoprotein have therefore not only revealed novel insights into the basic functions of the mitotic spindle, but have also identified a molecular basis for the pathology of a devastating human disease.

19.4.5 Summary

The above examples represent just four ways in which virus infection can disrupt centrosome number. Each one represents an opportunity to understand how the centrosome duplication cycle is integrated with that of the cell as a whole. What is perhaps most astonishing is that single, viral proteins can inflict such severe damage on the host cell. Such factors act either by stalling the cell cycle without halting the centrosome cycle, by stimulating the centrosome cycle inappropriately, or by silencing the controls which prevent the amplification of mitotic errors. By uncovering their mode of action, new connections between the core cell-cycle apparatus, the DNA-repair machinery, the nucleocytoplasmic transport system, and the mitotic spindle have been found. These functions are all organized around or dependent upon the centrosome.

In many cases, interfering with the host centrosome in this way offers obvious advantages to a virus. By promoting cell division, viruses may gain easy access to the nucleus. Alternatively, maintaining the cell in a particular cell-cycle phase might promote an environment that is particularly conducive to viral replication or morphogenesis. Formation of syncytia may be an efficient means for a virus to spread from cell to cell without alerting the attention of the host immune system, while stimulating uncontrolled proliferation achieves a similar aim, promoting the spread of infection throughout the body of the host, and generating a plentiful supply of appropriate cells in which to replicate.

From hijacking the microtubule–motor network and destabilizing the centrosome, to disrupting cell-cycle progression and checkpoint controls, viruses are clearly adept at infiltrating and reprogramming their hosts to promote their survival and spread. However, compared with viruses, interactions between bacteria and the microtubule cytoskeleton have been less commonly observed, presumably because bacteria are intrinsically less dependent on the machinery of their host for replication. The final section of this chapter reveals that a remarkably sophisticated degree of bacterial intervention in host signaling networks has become apparent.

19.5
Bacterial Manipulation of the Centrosome and Microtubules

Intracellular bacteria are also largely dependent on active transport for their passage through the cytoplasm. Actin-based motility, achieved by stimulating actin-nucleation on the bacterial surface to generate rocket-like propulsion, is one solution to this problem, but many bacteria take advantage of the microtubule network for their spread. Moreover, microtubule manipulation is also successfully used by bacteria to promote their entry, and in an analogous manner to viral-factory formation, to generate specialized sites of replication within the cell. In this respect, bacteria and viruses appear to use very similar molecular strategies to take control of cell regulation. However, unlike viruses, interactions between bacteria and centrosome have rarely been documented, but are likely to emerge.

19.5.1
Bacterial Manipulation of the Microtubule Network

Microtubules seem to be involved in the very earliest stages of infection, during which bacteria are internalized by the cell. Unfortunately, in most cases our knowledge of this involvement is limited to an observed sensitivity of bacterial uptake to microtubule destabilizing drugs such as nocodazole [101]. Furthermore, some species appear to invoke an opposite mechanism, by destabilizing microtubules upon entry. One example is that of *Shigella*, which locally disrupts the microtubule network beneath its site of attachment. This is thought to involve the virulence gene VirA, which is secreted by the bacterial type III secretion system, and appears to act directly on the microtubule network by binding to tubulin heterodimers, preventing their incorporation into microtubules. By inducing microtubule instability, it is proposed that *Shigella* activates the Rho GTPase Rac1, which induces actin-mediated membrane ruffling and stimulates bacterial uptake [102]. In this way, VirA makes a dramatic impact on both the actin and microtubule cytoskeletons, even before the bacterium has entered the cell. A similar mechanism may also operate during internalization of *Yersinia pseudotuberculosis*, since microtubule-dependent Rho GTPase signaling is also required for bacterial uptake in this system [103, 104].

Following phagocytosis or endocytosis, bacteria face the impending threat of being passed into the lysosome for enzymatic destruction. Some, including the actin-tail forming bacteria *Shigella* and *Listeria*, escape the endosome entirely, while others have evolved a means of manipulating both the endocytic pathway and the microtubule network to such an extent that their passage to the lysosome is blocked [105]. One of the most ingenious members of this group is *Chlamydia*, the bacterial genus which is both responsible for a sexually-transmitted disease that constitutes the most common reportable infection in the United States, and is also the world's leading cause of infectious blindness.

Chlamydia exist in two morphogenetic forms: the infectious elementary bodies (EBs) which are compact and resistant to environmental stresses, and the intracellular reticulate bodies (RBs), which are metabolically active, and replication competent. EBs are phagocytosed, and become surrounded by endosomal membranes, forming an "inclusion", in which differentiation to the RB form occurs. The inclusion migrates from the periphery of the cell to the MTOC, in close proximity to the Golgi membranes from which exocytic vesicles containing sphingomyelin and cholesterol are captured and incorporated (Figure 19.6). By hijacking the exocytic pathway in this manner, the chlamydial inclusion escapes from the endocytic pathway and avoids destruction in the lysosome [106].

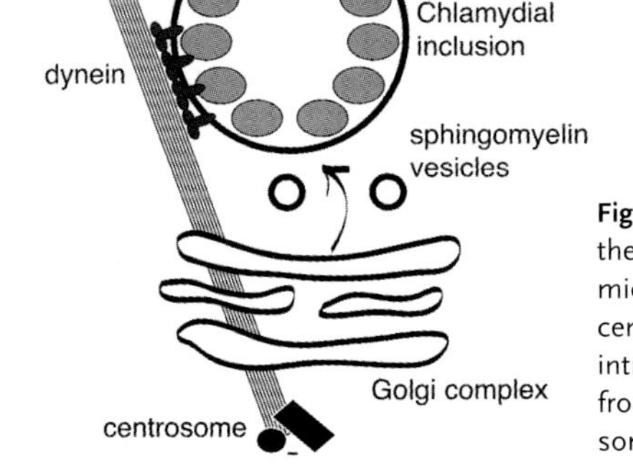

Figure 19.6 Chlamydial inclusions accumulate at the centrosome. Chlamydial inclusions move along microtubules, using dynein to locate to the centrosome. The inclusion becomes a protective intracellular niche, incorporating sphingomyelin from the Golgi and escaping passage to the lysosome.

The targeting of the inclusion to the centrosome appears to depend on the expression of bacterial proteins, which are secreted and incorporated into the endosomal membrane [107]. In a similar fashion to retrograde transport of many viruses, movement is dependent on microtubules. Moreover, inclusions recruit both the motor protein dynein, and a component of the dynactin complex, p150/Glued. However, surprisingly, inward migration of the inclusion is insensitive to overexpression of the dynactin subunit p50–dynamitin, suggesting that this cargo-binding protein is not involved in binding to the inclusion [108]. While the significance of these findings is not yet clear, it is tempting to speculate that bacterial genes which encode components of the inclusion membrane may be able to partially substitute for components of the dynactin complex. It remains an exciting possibility that this mimicry of motor-protein components may also be a strategy by which viruses facilitate their movement.

In a further parallel with viral-factory formation, at least one of the chlamydial species, *C. psittaci*, causes mitochondria to retract towards the region surrounding the inclusion, presumably to enhance the supply of ATP for bacterial replication. This reorganization seems to require the activity of kinesin motors, implying that this species is able to recruit both dynein and kinesin protein families to construct a structure in which RBs can multiply [109]. An understanding of the bacterial proteins involved in these processes will offer exciting insights into the regulation of motor proteins in uninfected cells.

The typhoid bacterium *Orientia tsutsugamushi* also utilizes the dynein motor network to accumulate near the centrosome. *Orientia* moves along microtubules, in a nocodazole- and p50–dynamitin-sensitive manner. However, in contrast to *Chlamydia*, *Orientia* is thought to escape the endosomal compartment early during infection. This would suggest that *Orientia* can interact with motor proteins in the absence of intervening cellular membranes. It is unclear whether escape from the endosome is a prerequisite for microtubule-dependent movement, or whether both endosomal and non-endosomal transport is possible. The reason for this perinuclear aggregation is also unknown, as is the case for several other species which gather close to the centrosome during infection. One possibility is that centrosomal localization is necessary for efficient replication of the bacterium.

As has been shown to be true for viruses such as HIV, motor proteins can also be recruited to transport bacterial virulence factors, rather than the bacteria themselves. In this scenario, the centrosome plays its role as a nexus between microtubules and the nuclear import system. Two bacterial proteins in particular are known to use the centrosome in this way: the YopM protein of *Yersinia pestis*, and $IpaH_{9.8}$ of *Shigella flexneri* [110, 111]. Both are virulence factors which belong to a family of leucine-rich repeat-containing proteins that are secreted into the host cytoplasm during infection. Both proteins accumulate at the centrosome and are subsequently targeted to the nucleus. Their functions in the nucleus are unknown, although both are suggested to modulate the inflammatory or immune responses of the host, presumably by disrupting or stimulating programs of host gene expression. The extent to which the transport of these proteins involves the microtubule–

motor protein network is also unclear, although it is possible that their route to the nucleus mirrors that of viruses such as HSV and HIV.

19.5.2 Interactions between Bacteria and the Centrosome

Finally, it remains possible that bacterial proteins might act on the centrosome directly to carry out their functions. Perhaps the most dramatic example of a bacterium which can interfere with centrosome function is *Wolbachia*, an endosymbiont which infects arthropods, and causes several reproductive defects in their hosts [112]. *Wolbachia* is also responsible for an unusual genetic puzzle which has been described in several different insects, that of cytoplasmic incompatibility (CI). This arises when sperm from infected male insects fertilize eggs from uninfected females. The resulting embryos fail to develop normally, having severe defects in metaphase chromosome alignment which prevent the segregation of paternal chromosomes.

The most enigmatic feature of CI is the fact that crosses between infected males and infected females, and also those between uninfected males and infected females, are all viable. Early models suggested that *Wolbachia* somehow disrupts chromatin condensation or the correct formation of the mitotic spindle, and such arguments were favored by findings which showed that *Wolbachia*, like *Chlamydia* and *Orientia*, cluster around the centrosome in infected embryos.

Dramatic insights into the mechanism of CI have been obtained by following in real-time the dynamics of centrosomes and pronuclei during fertilization in the parasitoid wasp, *Nasonia vitripennis*. This showed that *Wolbachia* does not prevent transmission of paternal centrosomes to the embryo, and that centrosomes from infected males separate normally in preparation for the first mitotic division. However, the usual pattern of centrosomal inheritance was disrupted. In *Nasonia*, centrosomes are transmitted reciprocally, such that female embryos inherit the paternal centrosomes whereas male embryos inherit maternal centrosomes. In contrast, centrosomes derived from the sperm of *Wolbachia*-infected males were seen to dominate, even in male embryos in which the chromosomes which prevailed were all female. Also, the envelope of the pronucleus derived from infected males fails to break down in synchrony with the pronucleus of uninfected females, causing chromosome condensation and alignment on the metaphase plate to occur too late for successful, diploid mitosis [113].

Delayed nuclear envelope breakdown neatly explains why crosses involving infected females are always viable. *Wolbachia* are present during spermatogenesis, but the bacteria are shed during its final stages. In the egg, however, *Wolbachia* persists. If the bacterium acts by delaying the breakdown of the pronucleus, then bacteria in an infected egg may affect both the male and female pronuclei, and so both are delayed and asynchrony is avoided. Only if the male pronucleus is delayed by exposure to the bacterium, while the female pronucleus is unexposed, will asynchrony and CI follow. How *Wolbachia* disrupts the timing of nuclear envelope breakdown and the inheritance of centrosomes remains to be seen. The mitotic cyclin-dependent kinase activity of Cdk1/cyclin B has been shown to play an important role in the onset of nuclear envelope breakdown, and so is a possible target [114]. Interestingly, cyclin B associates with the centrosome during interphase, and so would be in the vicinity of bacteria which accumulate at the MTOC [115]. If mitotic cyclins are indeed targeted by *Wolbachia*, this pathogen would join the ever-growing list of those which are able to re-engineer progression of the cell cycle.

19.5.3
Summary

These examples demonstrate that bacteria, like viruses, are capable of sophisticated manipulation of the microtubule network. While the identification of bacterial genes responsible for these interactions lags behind work with viruses, their distinctive evolutionary heritage means that bacteria promise to be an exciting new source of clues for those investigating the regulation of microtubules and centrosomes.

19.6
Conclusion

For an invading pathogen, the inside of the cell is an enormous and hostile place. Bacteria and viruses are able to manipulate this harsh environment with astonishing ease, using a powerful and varied genetic armory. One of the most exciting discoveries that has been made by those studying this "genetic terrorism", is that most of the weapons identified so far appear to act on a common set of host targets. For example, the fact that organisms as diverse as vaccinia and *Orientia* both target the dynein–dynactin complex to accumulate near the centrosome highlights the fundamental importance these proteins must play in the organization of the cell.

The breadth and diversity of interactions between viral and bacterial pathogens and the centrosome, which are only now becoming apparent, reveals the central role this tiny organelle plays in the organization and regulation of the cell. Previously unforeseen connections between the microtubule network, motor proteins, the machinery of nucleocytoplasmic transport, the activities of chaperones and the proteasome, the controlled progression of the cell cycle, and at the center of this web, the centrosome, are all currently emerging from the study of the host–pathogen interface. Our efforts to catch up with these compulsive cellular engineers will not only further our ability to combat infection and disease, but will also deepen our understanding of fundamental processes in cell biology, including the many functions of the mysterious centrosome.

Acknowledgments

We would like to thank Drs Tim P. Newsome and Ferran Valderrama for comments on the text.

References

1. Oppermann, H., A. D. Levinson, H. E. Varmus, L. Levintow, and J. M. Bishop. **1979**. Uninfected vertebrate cells contain a protein that is closely related to the product of the avian sarcoma virus transforming gene (src). *Proc Natl Acad Sci USA* 76:1804–1808.
2. Frischknecht, F., and M. Way. **2001**. Surfing pathogens and the lessons learned for actin polymerization. *Trends Cell Biol* 11:30–38.
3. Ploubidou, A., and M. Way. **2001**. Viral transport and the cytoskeleton. *Curr Opin Cell Biol* 13:97–105.
4. Smith, G. A., and L. W. Enquist. **2002**. Break ins and break outs: viral interactions with the cytoskeleton of Mammalian cells. *Annu Rev Cell Dev Biol* 18:135–161.
5. Sodeik, B. **2000**. Mechanisms of viral transport in the cytoplasm. *Trends Microbiol* 8:465–472.
6. Fabunmi, R. P., W. C. Wigley, P. J. Thomas, and G. N. DeMartino. **2000**. Activity and regulation of the centrosome-associated proteasome. *J Biol Chem* 275:409–413.
7. Hung, C. F., W. F. Cheng, L. He, M. Ling, J. Juang, C. T. Lin, and T. C. Wu. **2003**. Enhancing major histocompatibility complex class I antigen presentation by targeting antigen to centrosomes. *Cancer Res* 63:2393–2398.
8. Lavia, P., A. M. Mileo, A. Giordano, and M. G. Paggi. **2003**. Emerging roles of DNA tumor viruses in cell proliferation: new insights into genomic instability. *Oncogene* 22:6508–6516.
9. Ploubidou, A., V. Moreau, K. Ashman, I. Reckmann, C. Gonzalez, and M. Way. **2000**. Vaccinia virus infection disrupts microtubule organization and centrosome function. *EMBO J* 19:3932–3944.
10. Bornens, M. **2002**. Centrosome composition and microtubule anchoring mechanisms. *Curr Opin Cell Biol* 14:25–34.
11. Vale, R. D. **2003**. The molecular motor toolbox for intracellular transport. *Cell* 112:467–480.
12. Tomishima, M. J., G. A. Smith, and L. W. Enquist. **2001**. Sorting and transport of alpha herpesviruses in axons. *Traffic* 2:429–436.
13. Schwartz, J., and T. S. Elizan. **1975**. Herpes simplex virus encephalitis in suckling mice: ultrastructural studies of virus replication. *J Neuropathol Exp Neurol* 34:359–368.
14. Dohner, K., A. Wolfstein, U. Prank, C. Echeverri, D. Dujardin, R. Vallee, and B. Sodeik. **2002**. Function of dynein and dynactin in herpes simplex virus capsid transport. *Mol Biol Cell* 13:2795–2809.
15. Diefenbach, R. J., M. Miranda-Saksena, E. Diefenbach, D. J. Holland, R. A. Boadle, P. J. Armati, and A. L. Cunningham. **2002**. *Herpes simplex* virus tegument protein US11 interacts with conventional kinesin heavy chain. *J Virol* 76:3282–3291.
16. Mueller, S., X. Cao, R. Welker, and E. Wimmer. **2002**. Interaction of the poliovirus receptor CD155 with the dynein light chain Tctex-1 and its implication for poliovirus pathogenesis. *J Biol Chem* 277:7897–7904.
17. McDonald, D., M. A. Vodicka, G. Lucero, T. M. Svitkina, G. G. Borisy, M. Emerman, and T. J. Hope. **2002**. Visualization of the intracellular behavior of HIV in living cells. *J Cell Biol* 159:441–452.
18. Kim, W., Y. Tang, Y. Okada, T. A. Torrey, S. K. Chattopadhyay, M. Pfleiderer, F. G. Falkner, F. Dorner, W. Choi, N. Hirokawa, and H. C. Morse, 3rd. **1998**. Binding of murine leukemia virus Gag polyproteins to KIF4, a microtubule-based motor protein. *J Virol* 72:6898–6901.
19. Petit, C., M. L. Giron, J. Tobaly-Tapiero, P. Bittoun, E. Real, Y. Jacob, N. Tordo, H. De The, and A. Saib. **2003**. Targeting of incoming retroviral Gag to the centrosome involves a direct interac-

tion with the dynein light chain 8. *J Cell Sci* 116:3433–3442.

20. Cullen, B. R. **2001**. Journey to the center of the cell. *Cell* 105:697–700.

21. Smith, G. L., A. Vanderplasschen, and M. Law. **2002**. The formation and function of extracellular enveloped vaccinia virus. *J Gen Virol* 83:2915–2931.

22. Pelham, H. R. **1977**. Use of coupled transcription and translation to study mRNA production by vaccinia cores. *Nature* 269:532–534.

23. Carter, G. C., G. Rodger, B. J. Murphy, M. Law, O. Krauss, M. Hollinshead, and G. L. Smith. **2003**. Vaccinia virus cores are transported on microtubules. *J Gen Virol* 84:2443–2458.

24. Mallardo, M., S. Schleich, and J. Krijnse Locker. **2001**. Microtubule-dependent organization of vaccinia virus core-derived early mRNAs into distinct cytoplasmic structures. *Mol Biol Cell* 12:3875–3891.

25. Wodarz, A. **2002**. Establishing cell polarity in development. *Nature Cell Biol* 4:E39–E44.

26. Duncan, J. E., and R. Warrior. **2002**. The cytoplasmic dynein and kinesin motors have interdependent roles in patterning the *Drosophila* oocyte. *Curr Biol* 12:1982–1991.

27. Bullock, S. L., and D. Ish-Horowicz. **2001**. Conserved signals and machinery for RNA transport in Drosophila oogenesis and embryogenesis. *Nature* 414:611–616.

28. Sodeik, B., R. W. Doms, M. Ericsson, G. Hiller, C. E. Machamer, W. van't Hof, G. van Meer, B. Moss, and G. Griffiths. **1993**. Assembly of vaccinia virus: role of the intermediate compartment between the endoplasmic reticulum and the Golgi stacks. *J Cell Biol* 121:521–541.

29. Schmelz, M., B. Sodeik, M. Ericsson, E. J. Wolffe, H. Shida, G. Hiller, and G. Griffiths. **1994**. Assembly of vaccinia virus: the second wrapping cisterna is derived from the trans Golgi network. *J Virol* 68:130–147.

30. Sanderson, C. M., M. Hollinshead, and G. L. Smith. **2000**. The vaccinia virus A27L protein is needed for the microtubule-dependent transport of intracellular mature virus particles. *J Gen Virol* 81:47–58.

31. Alonso, C., J. Miskin, B. Hernaez, P. Fernandez-Zapatero, L. Soto, C. Canto, I. Rodriguez-Crespo, L. Dixon, and J. M. Escribano. **2001**. African swine fever virus protein p54 interacts with the microtubular motor complex through direct binding to light-chain dynein. *J Virol* 75:9819–9827.

32. Hollinshead, M., G. Rodger, H. Van Eijl, M. Law, R. Hollinshead, D. J. Vaux, and G. L. Smith. **2001**. Vaccinia virus utilizes microtubules for movement to the cell surface. *J Cell Biol* 154:389–402.

33. Ward, B. M., and B. Moss. **2001**. Vaccinia virus intracellular movement is associated with microtubules and independent of actin tails. *J Virol* 75:11651–11663.

34. Rietdorf, J., A. Ploubidou, I. Reckmann, A. Holmstrom, F. Frischknecht, M. Zettl, T. Zimmermann, and M. Way. **2001**. Kinesin-dependent movement on microtubules precedes actin-based motility of vaccinia virus. *Nature Cell Biol* 3:992–1000.

35. Ward, B. M., and B. Moss. **2001**. Visualization of intracellular movement of vaccinia virus virions containing a green fluorescent protein–B5R membrane protein chimera. *J Virol* 75:4802–4813.

36. Ward, B. M., A. S. Weisberg, and B. Moss. **2003**. Mapping and functional analysis of interaction sites within the cytoplasmic domains of the vaccinia virus A33R and A36R envelope proteins. *J Virol* 77:4113–4126.

37. van Eijl, H., M. Hollinshead, G. Rodger, W. H. Zhang, and G. L. Smith. **2002**. The vaccinia virus F12L protein is associated with intracellular enveloped virus particles and is required for their egress to the cell surface. *J Gen Virol* 83:195–207.

38. Jindal, S., and R. A. Young. **1992**. Vaccinia virus infection induces a stress response that leads to association of Hsp70 with viral proteins. *J Virol* 66:5357–5362.

39. Rojo, G., M. Chamorro, M. L. Salas, E. Vinuela, J. M. Cuezva, and J. Salas. **1998**. Migration of mitochondria to viral assembly sites in African swine fever virus-infected cells. *J Virol* 72:7583–7588.

40. Ferreira, L. R., N. Moussatche, and V. Moura Neto. **1994**. Rearrangement of intermediate filament network of BHK-21 cells infected with vaccinia virus. *Arch Virol* 138:273–285.

41. Heath, C. M., M. Windsor, and T. Wileman. **2001**. Aggresomes resemble sites specialized for virus assembly. *J Cell Biol* 153:449–455.

42. Johnston, J. A., C. L. Ward, and R. R. Kopito. **1998**. Aggresomes: a cellular response to misfolded proteins. *J Cell Biol* 143:1883–1898.

43. Johnston, J. A., M. E. Illing, and R. R. Kopito. **2002**. Cytoplasmic dynein/dynactin mediates the assembly of aggresomes. *Cell Motil Cytoskeleton* 53:26–38.

44. Wigley, W. C., R. P. Fabunmi, M. G. Lee, C. R. Marino, S. Muallem, G. N. DeMartino, and P. J. Thomas. **1999**. Dynamic association of proteasomal machinery with the centrosome. *J Cell Biol* 145:481–490.

45. Hung, J. J., C. S. Chung, and W. Chang. **2002**. Molecular chaperone Hsp90 is important for vaccinia virus growth in cells. *J Virol* 76:1379–1390.

46. Hewitt, E. W. **2003**. The MHC class I antigen presentation pathway: strategies for viral immune evasion. *Immunology* 110:163–169.

47. Anton, L. C., U. Schubert, I. Bacik, M. F. Princiotta, P. A. Wearsch, J. Gibbs, P. M. Day, C. Realini, M. C. Rechsteiner, J. R. Bennink, and J. W. Yewdell. **1999**. Intracellular localization of proteasomal degradation of a viral antigen. *J Cell Biol* 146:113–124.

48. Lacaille, V. G., and M. J. Androlewicz. **2000**. Targeting of HIV-1 Nef to the centrosome: implications for antigen processing. *Traffic* 1:884–891.

49. Koonce, M. P., J. Kohler, R. Neujahr, J. M. Schwartz, I. Tikhonenko, and G. Gerisch. **1999**. Dynein motor regulation stabilizes interphase microtubule arrays and determines centrosome position. *EMBO J* 18:6786–6792.

50. Robinson, J. T., E. J. Wojcik, M. A. Sanders, M. McGrail, and T. S. Hays. **1999**. Cytoplasmic dynein is required for the nuclear attachment and migration of centrosomes during mitosis in Drosophila. *J Cell Biol* 146:597–608.

51. Heinlein, M. **2002**. The spread of tobacco mosaic virus infection: insights into the cellular mechanism of RNA transport. *Cell Mol Life Sci* 59:58–82.

52. Elliott, G., and P. O'Hare. **1998**. Herpes simplex virus type 1 tegument protein VP22 induces the stabilization and hyperacetylation of microtubules. *J Virol* 72:6448–6455.

53. Pomeranz, L. E., and J. A. Blaho. **2000**. Assembly of infectious Herpes simplex virus type 1 virions in the absence of full-length VP22. *J Virol* 74:10041–10054.

54. Joachims, M., K. S. Harris, and D. Etchison. **1995**. Poliovirus protease 3C mediates cleavage of microtubule-associated protein 4. *Virology* 211:451–461.

55. Ainsztein, A. M., and D. L. Purich. **1992**. Cleavage of bovine brain microtubule-associated protein-2 by human immunodeficiency virus proteinase. *J Neurochem* 59:874–880.

56. Ng, M. L., and S. S. Hong. **1989**. Flavivirus infection: essential ultrastructural changes and association of Kunjin virus NS3 protein with microtubules. *Arch Virol* 106:103–120.

57. Pfeifer, U., R. Thomssen, K. Legler, U. Bottcher, W. Gerlich, E. Weinmann, and O. Klinge. **1980**. Experimental non-A, non-B hepatitis: four types of cytoplasmic alteration in hepatocytes of infected chimpanzees. *Virch Arch B Cell Pathol Incl Mol Pathol.* 33:233–243.

58. Kotsakis, A., L. E. Pomeranz, A. Blouin, and J. A. Blaho. **2001**. Microtubule reorganization during herpes simplex virus type 1 infection facilitates the nuclear localization of VP22, a major virion tegument protein. *J Virol* 75:8697–8711.

59. Bystrevskaya, V. B., T. V. Lobova, V. N. Smirnov, N. E. Makarova, and A. A. Kushch. **1997**. Centrosome injury in

cells infected with human cytomegalovirus. *J Struct Biol* 120:52–60.

60. Debec, A., A. M. Courgeon, M. Maingourd, and C. Maisonhaute. **1990**. The response of the centrosome to heat shock and related stresses in a *Drosophila* cell line. *J Cell Sci* 96(Pt 3):403–412.

61. Glotzer, J. B., M. Saltik, S. Chiocca, A. I. Michou, P. Moseley, and M. Cotten. **2000**. Activation of heat-shock response by an adenovirus is essential for virus replication. *Nature* 407:207–211.

62. Stearns, T. **2001**. Centrosome duplication. a centriolar pas de deux. *Cell* 105:417–420.

63. Wang, E., J. A. Connolly, V. I. Kalnins, and P. W. Choppin. **1979**. Relationship between movement and aggregation of centrioles in syncytia and formation of microtubule bundles. *Proc Natl Acad Sci USA*. 76:5719–5723.

64. Watt, F. M., E. Sidebottom, and H. Harris. **1980**. Microtubule-organizing centres in binucleate cells and homosynkaryons. *J Cell Sci* 44:123–133.

65. Rogel, M. E., L. I. Wu, and M. Emerman. **1995**. The human immunodeficiency virus type 1 vpr gene prevents cell proliferation during chronic infection. *J Virol* 69:882–888.

66. Stewart, S. A., B. Poon, J. B. Jowett, and I. S. Chen. **1997**. Human immunodeficiency virus type 1 Vpr induces apoptosis following cell cycle arrest. *J Virol* 71:5579–5592.

67. Minemoto, Y., M. Shimura, Y. Ishizaka, Y. Masamune, and K. Yamashita. **1999**. Multiple centrosome formation induced by the expression of Vpr gene of human immunodeficiency virus. *Biochem Biophys Res Commun* 258:379–384.

68. Watanabe, N., T. Yamaguchi, Y. Akimoto, J. B. Rattner, H. Hirano, and H. Nakauchi. **2000**. Induction of M-phase arrest and apoptosis after HIV-1 Vpr expression through uncoupling of nuclear and centrosomal cycle in HeLa cells. *Exp Cell Res* 258:261–269.

69. Masuda, M., Y. Nagai, N. Oshima, K. Tanaka, H. Murakami, H. Igarashi, and H. Okayama. **2000**. Genetic studies with the fission yeast *Schizosaccharomyces pombe* suggest involvement of wee1, ppa2, and rad24 in induction of cell cycle arrest by human immunodeficiency virus type 1 Vpr. *J Virol* 74:2636–2646.

70. Elder, R. T., M. Yu, M. Chen, X. Zhu, M. Yanagida, and Y. Zhao. **2001**. HIV-1 Vpr induces cell cycle G2 arrest in fission yeast (*Schizosaccharomyces pombe*) through a pathway involving regulatory and catalytic subunits of PP2A and acting on both Wee1 and Cdc25. *Virology* 287:359–370.

71. Roshal, M., B. Kim, Y. Zhu, P. Nghiem, and V. Planelles. **2003**. Activation of the ATR-mediated DNA damage response by the HIV-1 viral protein R. *J Biol Chem* 278:25879–25886.

72. Chowdhury, I. H., X. F. Wang, N. R. Landau, M. L. Robb, V. R. Polonis, D. L. Birx, and J. H. Kim. **2003**. HIV-1 Vpr activates cell cycle inhibitor p21/Waf1/Cip1: a potential mechanism of G2/M cell cycle arrest. *Virology* 305:371–377.

73. Abraham, R. T. **2001**. Cell cycle checkpoint signaling through the ATM and ATR kinases. *Gene Dev* 15:2177–2196.

74. Cheng, A., R. Balczon, Z. Zuo, J. S. Koons, A. H. Walsh, and R. E. Honkanen. **1998**. Fostriecin-mediated G2–M-phase growth arrest correlates with abnormal centrosome replication, the formation of aberrant mitotic spindles, and the inhibition of serine/threonine protein phosphatase activity. *Cancer Res* 58:3611–3619.

75. Sato, N., K. Mizumoto, M. Nakamura, H. Ueno, Y. A. Minamishima, J. L. Farber, and M. Tanaka. **2000**. A possible role for centrosome overduplication in radiation-induced cell death. *Oncogene* 19:5281–5290.

76. Ban, S., T. Shinohara, Y. Hirai, Y. Moritaku, J. B. Cologne, and D. G. MacPhee. **2001**. Chromosomal instability in BRCA1- or BRCA2-defective human cancer cells detected by spontaneous micronucleus assay. *Mutat Res* 474:15–23.

77. Barthelmes, H. U., P. Grue, S. Feineis, T. Straub, and F. Boege. **2000**. Active DNA topoisomerase IIalpha is a com-

ponent of the salt-stable centrosome core. *J Biol Chem* 275:38823–38830.
78. Nigg, E. A. **2002**. Centrosome aberrations: cause or consequence of cancer progression? *Nature Rev Cancer* 2:815–825.
79. Nishizawa, M., M. Kamata, R. Katsumata, and Y. Aida. **2000**. A carboxy-terminally truncated form of the human immunodeficiency virus type 1 Vpr protein induces apoptosis via G(1) cell cycle arrest. *J Virol* 74: 6058–6067.
80. Waldhuber, M. G., M. Bateson, J. Tan, A. L. Greenway, and D. A. McPhee. **2003**. Studies with GFP-Vpr fusion proteins: induction of apoptosis but ablation of cell-cycle arrest despite nuclear membrane or nuclear localization. *Virology* 313:91–104.
81. Whyte, P., K. J. Buchkovich, J. M. Horowitz, S. H. Friend, M. Raybuck, R. A. Weinberg, and E. Harlow. **1988**. Association between an oncogene and an anti-oncogene: the adenovirus E1A proteins bind to the retinoblastoma gene product. *Nature* 334: 124–129.
82. Stevaux, O., and N. J. Dyson. **2002**. A revised picture of the E2F transcriptional network and RB function. *Curr Opin Cell Biol* 14:684–691.
83. Helt, A. M., and D. A. Galloway. **2003**. Mechanisms by which DNA tumor virus oncoproteins target the Rb family of pocket proteins. *Carcinogenesis* 24:159–169.
84. Ludlow, J. W., J. A. DeCaprio, C. M. Huang, W. H. Lee, E. Paucha, and D. M. Livingston. **1989**. SV40 large T antigen binds preferentially to an underphosphorylated member of the retinoblastoma susceptibility gene product family. *Cell* 56:57–65.
85. Hinchcliffe, E. H., and G. Sluder. **2002**. Two for two: Cdk2 and its role in centrosome doubling. *Oncogene* 21:6154–6160.
86. Piboonniyom, S. O., S. Duensing, N. W. Swilling, J. Hasskarl, P. W. Hinds, and K. Munger. **2003**. Abrogation of the retinoblastoma tumor suppressor checkpoint during keratinocyte immortalization is not sufficient for induction of centrosome-mediated genomic instability. *Cancer Res* 63:476–483.
87. De Luca, A., R. Mangiacasale, A. Severino, L. Malquori, A. Baldi, A. Palena, A. M. Mileo, P. Lavia, and M. G. Paggi. **2003**. E1A deregulates the centrosome cycle in a Ran GTPase-dependent manner. *Cancer Res* 63:1430–1437.
88. Kuersten, S., M. Ohno, and I. W. Mattaj. **2001**. Nucleocytoplasmic transport: Ran, beta and beyond. *Trends Cell Biol* 11:497–503.
89. Hetzer, M., O. J. Gruss, and I. W. Mattaj. **2002**. The Ran GTPase as a marker of chromosome position in spindle formation and nuclear envelope assembly. *Nature Cell Biol* 4:E177–E184.
90. Keryer, G., B. Di Fiore, C. Celati, K. F. Lechtreck, M. Mogensen, A. Delouvee, P. Lavia, M. Bornens, and A. M. Tassin. **2003**. Part of Ran is associated with AKAP450 at the centrosome: involvement in microtubule-organizing activity. *Mol Biol Cell* 14:4260–4271.
91. Di Fiore, B., M. Ciciarello, R. Mangiacasale, A. Palena, A. M. Tassin, E. Cundari, and P. Lavia. **2003**. Mammalian RanBP1 regulates centrosome cohesion during mitosis. *J Cell Sci* 116:3399–3411.
92. Forgues, M., M. J. Difilippantonio, S. P. Linke, T. Ried, K. Nagashima, J. Feden, K. Valerie, K. Fukasawa, and X. W. Wang. **2003**. Involvement of Crm1 in hepatitis B virus X protein-induced aberrant centriole replication and abnormal mitotic spindles. *Mol Cell Biol* 23:5282–5292.
93. Musacchio, A., and K. G. Hardwick. **2002**. The spindle checkpoint: structural insights into dynamic signalling. *Nature Rev Mol Cell Biol* 3:731–741.
94. Mortreux, F., A. S. Gabet, and E. Wattel. **2003**. Molecular and cellular aspects of HTLV-1 associated leukemogenesis *in vivo*. *Leukemia* 17:26–38.
95. Majone, F., O. J. Semmes, and K. T. Jeang. **1993**. Induction of micronuclei by HTLV-I Tax: a cellular assay for function. *Virology* 193:456–459.

96. Gatza, M. L., J. C. Watt, and S. J. Marriott. **2003**. Cellular transformation by the HTLV-I Tax protein, a jack-of-all-trades. *Oncogene* 22:5141–5149.

97. Jin, D. Y., F. Spencer, and K. T. Jeang. **1998**. Human T cell leukemia virus type 1 oncoprotein Tax targets the human mitotic checkpoint protein MAD1. *Cell* 93:81–91.

98. Li, R., and A. W. Murray. **1991**. Feedback control of mitosis in budding yeast. *Cell* 66:519–531.

99. Chen, R. H., A. Shevchenko, M. Mann, and A. W. Murray. **1998**. Spindle checkpoint protein Xmad1 recruits Xmad2 to unattached kinetochores. *J Cell Biol* 143:283–295.

100. Howell, B. J., D. B. Hoffman, G. Fang, A. W. Murray, and E. D. Salmon. **2000**. Visualization of Mad2 dynamics at kinetochores, along spindle fibers, and at spindle poles in living cells. *J Cell Biol* 150:1233–1250.

101. Yoshida, S., and C. Sasakawa. **2003**. Exploiting host microtubule dynamics: a new aspect of bacterial invasion. *Trends Microbiol* 11:139–143.

102. Yoshida, S., E. Katayama, A. Kuwae, H. Mimuro, T. Suzuki, and C. Sasakawa. **2002**. *Shigella* delivers an effector protein to trigger host microtubule destabilization, which promotes Rac1 activity and efficient bacterial internalization. *EMBO J* 21:2923–2935.

103. Gruenheid, S., and B. B. Finlay. **2003**. Microbial pathogenesis and cytoskeletal function. *Nature* 422:775–781.

104. McGee, K., P. Holmfeldt, and M. Fallman. **2003**. Microtubule-dependent regulation of Rho GTPases during internalisation of *Yersinia pseudotuberculosis*. *FEBS Lett* 533:35–41.

105. Amer, A. O., and M. S. Swanson. **2002**. A phagosome of one's own: a microbial guide to life in the macrophage. *Curr Opin Microbiol* 5:56–61.

106. Fields, K. A., and T. Hackstadt. **2002**. The chlamydial inclusion: escape from the endocytic pathway. *Annu Rev Cell Dev Biol* 18:221–245.

107. Scidmore, M. A., D. D. Rockey, E. R. Fischer, R. A. Heinzen, and T. Hackstadt. **1996**. Vesicular interactions of the *Chlamydia trachomatis* inclusion are determined by chlamydial early protein synthesis rather than route of entry. *Infect Immun* 64:5366–5372.

108. Grieshaber, S. S., N. A. Grieshaber, and T. Hackstadt. **2003**. *Chlamydia trachomatis* uses host cell dynein to traffic to the microtubule-organizing center in a p50 dynamitin-independent process. *J Cell Sci* 116:3793–3802.

109. Escalante-Ochoa, C., R. Ducatelle, G. Charlier, K. De Vos, and F. Haesebrouck. **1999**. Significance of host cell kinesin in the development of *Chlamydia psittaci*. *Infect Immun* 67:5441–5446.

110. Skrzypek, E., C. Cowan, and S. C. Straley. **1998**. Targeting of the *Yersinia pestis* YopM protein into HeLa cells and intracellular trafficking to the nucleus. *Mol Microbiol* 30:1051–1065.

111. Toyotome, T., T. Suzuki, A. Kuwae, T. Nonaka, H. Fukuda, S. Imajoh-Ohmi, T. Toyofuku, M. Hori, and C. Sasakawa. **2001**. Shigella protein IpaH(9.8) is secreted from bacteria within mammalian cells and transported to the nucleus. *J Biol Chem* 276:32071–32079.

112. Werren, J. H. **1997**. Wolbachia run amok. *Proc Natl Acad Sci USA* 94:11154–11155.

113. Tram, U., and W. Sullivan. **2002**. Role of delayed nuclear envelope breakdown and mitosis in *Wolbachia*-induced cytoplasmic incompatibility. *Science* 296:1124–1126.

114. Dessev, G., C. Iovcheva-Dessev, J. R. Bischoff, D. Beach, and R. Goldman. **1991**. A complex containing p34cdc2 and cyclin B phosphorylates the nuclear lamin and disassembles nuclei of clam oocytes *in vitro*. *J Cell Biol* 112:523–533.

115. Jackman, M., C. Lindon, E. A. Nigg, and J. Pines. **2003**. Active cyclin B1-Cdk1 first appears on centrosomes in prophase. *Nature Cell Biol* 5:143–148.

20
Basal Bodies and Microtubule Organization in Pathogenic Protozoa

Keith Gull, Laura Briggs and Sue Vaughan

20.1
Introduction and Appreciation

The 30 years straddling the end of the 19th and the start of the 20th centuries was a period of intense microscopical and microbiological investigation. This period brought us insights into the biology of the protozoa discovered then to be the etiological agents of major parasitic diseases of man or animals such as Malaria, Leishmaniasis, East Coast fever, Texas or red-water fever and African trypanosomiasis. The sub-structure and activity of living cells was also being described in detail and the protozoa were extremely useful in this endeavor since their unicellularity, coupled with cellular diversity, provided superb experimental material. It is not surprising therefore that in one of the earliest textbooks written at the end of this period Wenyon [1] was able to describe not only the biology of most of the parasitic protozoa that we know today but also many of the features of their cell biology. There was a clear recognition that protozoa exhibited a vast array of filamentous structures in their cytoplasm, dividing nuclei as well as in flagella and other appendages. Thus, well before microtubules were defined by the electron microscope their identity and some of their properties were well recognized. Given the precision of cellular form in parasitic protozoa and the precise control over positioning and number of flagella and other organelles it was natural that cytologists were interested in the ontogeny of these filaments and cell projections. Their polarity, association with and growth from defined centers was well recognized. The terms centriole and centrosome were in common use at that time, however they were mainly applied to the structure associated with the cell center or spindle poles. The structure at the base of the flagellum was then often termed the blepharoblast. Wenyon writes on page 31 of his textbook: "A flagellum, as pointed out by Alexeieff (1911) consists of an axial filament for which the term *axoneme*, suggested to the writer by Colonel A. Alcock, will be employed, and a thin sheath of cytoplasm. The axoneme itself takes origin in a minute granule, the *blepharoblast*, which is situated in the cytoplasm, and sometimes upon the surface of the nuclear membrane".

Centrosomes in Development and Disease. Edited by Erich A. Nigg

ISBN 3-527-30980-2

The centrosome cycle in interphase and mitosis was well established and Wenyon rehearsed the form of nuclear divisions in a number of protozoa describing those with a discrete centrosome and those without. He comments: "That the formation of a spindle may occur without a definite centrosome being identifiable has long been recognized in higher plants, so there is no reason to suppose that this may not happen amongst the protozoa". However, Wenyon goes on to describe in detail the debates amongst researchers as to the existence of a centriole rather than a centrosome in certain division types. He urges the use of good cytological technique to define the centrosome and its radiating filaments as in his views of mitosis in *Hartmanella*. However, again these structures were not confused with the structure at the base of the flagellum, which was then termed the blepharoblast. Wenyon is very clear about the fact that the growth and formation of new flagella are intimately bound up with the activities of the blepharoplast: "When the blepharoblast of a flagellate divides, the axoneme which arose from it remains attached to one daughter blepharoblast, while a new axoneme grows out from the other to form a new flagellum". In these and other statements Wenyon emphatically sets out the perceived view of these structures as the organizing centers of the various filamentous elements of pathogenic protozoan cells that we now recognize as microtubule in nature. The intervening 80 years have filled in much of the detailed biology and, in this new post-genomic era in molecular parasitology we are, at the start of this century, only just starting to understand the function of some of these structures.

It is worth reminding oneself that virtually the whole of the taxonomy, classification and identification of the then newly described parasitic protozoa in Wenyon's 1926 textbook [1] relied entirely on cytology. This precisely defined morphology of cell types in the protozoa, replicated with fidelity at each division or differentiation, reflects an internal, highly defined, microtubule cytoskeleton. The advent of the electron microscope revealed the beautiful, highly structured microtubule arrays inside the cytoplasm of protozoa as well as the (in general) 9 + 2 microtubule arrangement of the axoneme. The blepharoplast became a term used for a number of structures at the base of the flagella in different organisms and electron microscopy revealed it to be, in general, the now well-known 9-triplet microtubule basal body. However, in trypanosomes the densely staining structure was revealed by electron microscopy to include the kinetoplast. The term blepharoplast is now mainly used to describe the densely staining, spherical inclusion containing numerous lightly stained channels, which is a precursor organelle for the production of the 100–150 basal bodies of the motile sperm in ferns [2].

There is however, no doubt that the complex relationships between the flagellar/ciliary basal bodies and the centrosome were clearly recognized by many early scientists (see for example [3, 4]). These and other concepts were more recently brought together in the concept of the microtubule organizing center (MTOC) when the biochemistry and morphology of the microtubule became known [5].

The possession of microtubule-mediated motility, via flagella and cilia, by many protozoa obviously proved useful in their adaptation to a parasitic lifestyle. In a similar manner, microtubule-mediated functions in the definition of shape and form, intracellular locomotion of various cargoes and coordination of actomyosin motility phenomena have all become central pathogenicity factors in these parasites. Microtubule-mediated events therefore play important roles in invasion, pathogenicity and spread of many pathogenic protozoa.

20.2
The "Dispersed" MTOC Complement of Protozoal Cells

What is different about the protozoa is that they provide a counterbalance to the often too generally simplified view of a single centrosomal MTOC found in animal cells. This idea of an MTOC "singularity" is not the generality in protozoa. Rather, they display a diversity of MTOCs in both design and function within the same cell and Mignot [6] has pointed out some of these features. In general there are usually three distinct MTOCs operating for cytoplasmic, mitotic and flagellar/ciliary microtubules. Rather than being grouped together, as in many animal cells, these three functions are performed by spatially separated and morphologically distinct MTOCs. Mignot often refers to the MTOCs nucleating cytoplasmic microtubules as "interphase MTOCs". However, since they can operate throughout the cell cycle we prefer to use the term "cytoplasmic MTOCs" for those MTOCs that nucleate the cytoplasmic microtubules involved in intracellular transport, organelle positioning, cell shape and division. Thus the three basic types of MTOCs seen in most eukaryotic cells are the cytoplasmic MTOC, the mitotic MTOC and the flagellum/cilium MTOC. Whilst in animal cells these are often grouped together into the centrosome, in protozoan cells they are more usually spatially separated. The mitotic MTOC of protozoa undergoes the normal segregation of one to two units to set up the bipolar spindle. However, in contrast to many metazoan cells the cytoplasmic MTOCs and the flagellum/cilium MTOCs of protozoa may be completely absent or present in multiples. In the latter case the multiple cytoplasmic MTOCs may then nucleate distinct sets of microtubules, usually with precise number control. This characteristic cell biology of such distinct sets of MTOCs may be easily observed in pathogenic protozoa such as *Giardia, Trichomonas, Trypanosoma, Plasmodium and Toxoplasma* [7]. A useful illustration of this phenomenon is the African trypanosome and the *Apicomplexa.*

20.3
The *Trypanosoma brucei* Microtubule Biology

In *Trypanosoma brucei* there are distinct MTOCs for the flagellum, the mitotic spindle and the cytoplasmic microtubules [8]. Whilst there is one flagellum subtended from a single basal body (and a pro-basal body destined to nucleate the new flagellum of the next cell cycle), there are two sets of cytoplasmic microtubules nucleated in distinct manners. The major cytoplasmic microtubule complement is in the form of a sub-pellicular array forming a corset that is maintained through all stages of the cell cycle. The sub-pellicular array underlies the plasma membrane and is responsible for determining the shape of the cell body. At the posterior ends of the cell, a single flagellum emerges from a flagellar pocket that is devoid of sub-pellicular microtubules and is the sole site of endocytotic and exocytotic membrane traffic at the cell surface in the trypanosome [9, 10]. After its emergence from the flagellar pocket, the flagellum acquires a structure additional to the axoneme, a large lattice-like complex termed the paraflagellar rod (PFR) [11–13]. Flagellum attachment to the cell body is mediated through connections between the flagellum membranes and the flagellum attachment zone (FAZ) in the cell body. The FAZ consists of a series of junctional complexes positioned along a filament integrated into the sub-pellicular array. Accompanying this FAZ filament are four specialized microtubules, nucleated from the basal body region [18]. This nucleation site therefore represents a distinctly different cytoplasmic MTOC from that used for the sub-pellicular microtubules, illustrating the potential for multiplicity of cytoplasmic MTOCs in protozoa.

Positioning of these MTOCs obviously influences nucleation events and therefore microtubule polarity. The flagellum is attached along the length of cell body up to the anterior end, where there is a short overhang. The anterior and posterior ends of the trypanosome are defined according to the direction of the cell during movement and, influenced by the distinct flagellum motility, the cell anterior is at the distal end of the flagellum. Thus, the flagellar microtubules have their plus ends at the anterior end of the cell. However, the microtubules of the sub-pellicular array have their plus ends at the posterior end of the trypanosome cell [8, 15]. Since the four specialized microtubules are nucleated close to the basal bodies before they join the sub-pellicular array the best evidence suggests that they have their plus ends at the anterior end of the trypanosome. Thus, in effect, the sub-pellicular array is a corset of microtubules of one polarity with a "FAZ seam" of four microtubules of opposite polarity within it. Evidence suggests that this antiparallel arrangement of the FAZ microtubules within the corset is critical for events such as cytokinesis and intracellular membrane traffic to and from the flagellar pocket [15–17].

20.3.1
The Spindle and Cell Division

Although there is some remodeling of the cytoplasmic microtubules during cell division in protozoa it is often the case that there is a general maintenance of many arrays. This is also true for the flagella. Thus, this provides the opportunity to study how the cytoskeleton is duplicated and then distributed to the two daughter cells. In trypanosome division the cytoskeleton is maintained and all its components are duplicated to provide each daughter cell with a complete sub-pellicular array and flagellum. New microtubules are added to the main microtubules of the sub-pellicular array by insertion between the old microtubules and hence this group is distributed to the two daughters in a semi-conservative manner [14]. In contrast, a new flagellum is formed during the division process as are a

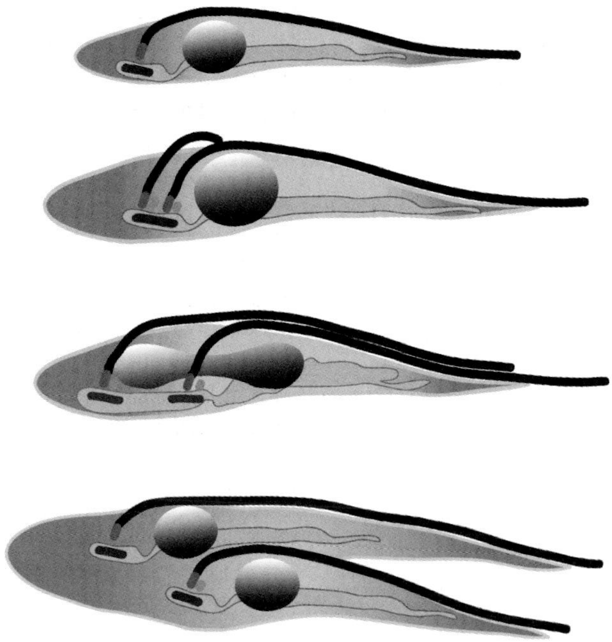

Figure 20.1 A schematic representation of the cell division cycle of pro-cyclic forms of *T. brucei.* (A) The G1 cell contains a single flagellum with a basal body and pro-basal body. The basal bodies are connected via the Tripartite Attachment Complex to the kinetoplast (dark blue) inside the mitochondrion. (B) The pro-basal body matures, nucleates a new flagellum and two new pro-basal bodies are formed. (C) Mitosis occurs across the axis of the old flagellum. (D) Cytokinesis occurs from the anterior end of the trypanosome (see Color Plates page XL).

new FAZ filament and four specialized microtubules [13, 19, 20]. Thus, these cytoskeleton elements are distributed to the daughters in a conservative manner. The initiation of a new flagellum assembly defines the very beginning of the cell duplication cycle in trypanosomes [18, 21]. The first event being the elongation and maturation of the pro-basal body, positioned next to the basal body of the existing flagellum, and the subsequent nucleation of the new flagellar axoneme (Figure 20.1). As the new axoneme elongates during the cell cycle the trypanosome cell becomes biflagellated, with both an elongating new flagellum and the existing old flagellum. Figure 20.2 shows the arrangement of microtubules and axonemes at two points in the trypanosome at this time, illustrating the conservative and semi-conservative construction and pattern of inheritance of different cytoskeletal components to the two daughter cells.

The mitotic spindle is intranuclear in trypanosomes but it has been difficult to visualize a discrete spindle pole structure analogous to the spindle pole body of the yeasts or the plaques at the poles of the malaria parasite spindle (see below). Electron microscopy does reveal that the spindle microtubules end in a discrete area of nucleoplasm [22] and imunofluorescence microscopy has shown that in

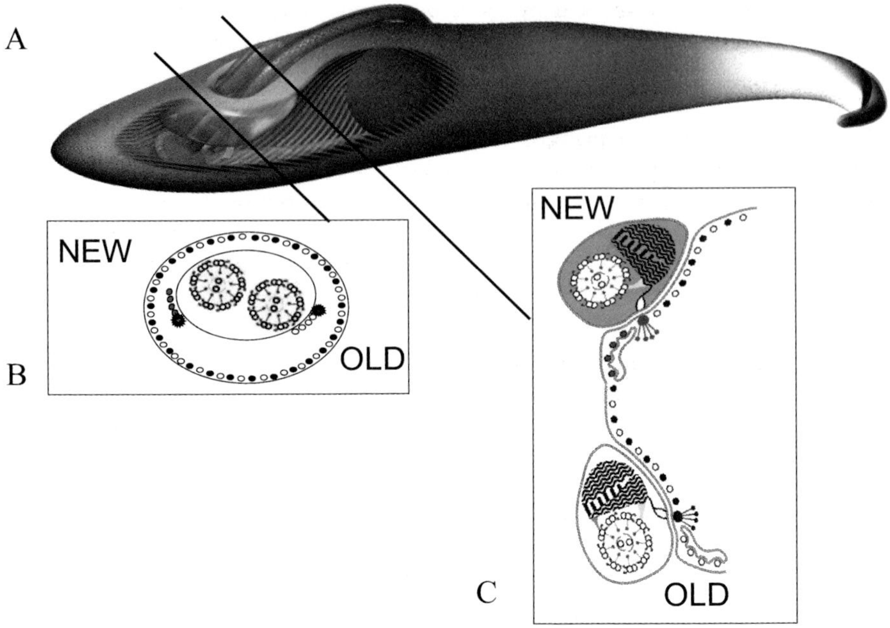

Figure 20.2 (A) A diagram of an early stage in the division of a pro-cyclic *T. brucei* [13] depicting the three-dimensional arrangement of the flagella and flagellar pocket in a cell soon after the initiation of the formation of a new flagellum. The specialized set of four FAZ microtubules (white) are nucleated near the basal bodies. The cartoon shows two theoretical slices through the trypanosome at the level of the flagellar pocket and at a position through the cell body more to the anterior of the cell. The diagrams (B) and (C) illustrate the semi-conservative inheritance of the sub-pellicular microtubules (shown as black/white for old/new) and the formation and conservative inheritance of a completely new flagellum (orange) and completely new FAZ filament and four microtubules (both red) (see Color Plates page XLI).

pre-mitotic cells a gamma tubulin dot occurs in the nucleus which duplicates in early mitotic cells indicative of an intranuclear spindle MTOC [23]. The spindle contains less than 100 microtubules and the dynamics of segregation of the diverse sets of chromosomes has been documented [24, 25].

The position of the nucleus and spindle during mitosis in procyclic forms of *T. brucei* is interesting (see Figure 20.1). In trypanosomes, there is a single mitochondrion that contains a single mitochondrial DNA structure, the kinetoplast, containing thousands of mini-circles and tens of copies of maxi-circle DNA in a catenated mass [26]. This mitochondrial genome is replicated in a periodic S-phase in the cell cycle [21] and segregated by movement apart of the flagellar basal bodies (see below). This segregation occurs before mitosis begins and so in trypanosomes the basal bodies are not located at the poles of the intranuclear spindle [18]. They are however a critical feature of this interesting phenomenon whereby a unit mitochondrial genome is segregated with fidelity after a periodic S-phase. We will discuss this in more detail later in the context of other segregation roles that have been adopted by basal bodies in parasitic protozoa in general. Whilst in *T. brucei* the segregation of basal bodies, kinetoplasts and nuclei occur towards the posterior end of the cell, the initiation of the cytokinetic process occurs at the anterior end. Work from our laboratory using cell-cycle and microtubule inhibitors and RNAi silencing of flagellar attachment zone proteins has shown that in *T. brucei* pro-cyclic forms, whose flagellum is attached for most of its length, there is interplay between flagellum and flagellum attachment zone morphogenesis, organelle positioning and the capacity for coherent cell division [15–17]. This work revealed the important contribution that the FAZ filament and the four specialized microtubules referred to earlier, play in setting up the process of cytokinesis. Compromising flagellum morphogenesis by various means leads to abnormalities in the construction of internal cytoskeletal elements, particularly in the FAZ, and subsequent failures in division.

20.3.2 Basal Bodies

T. brucei basal bodies exhibit the canonical 9 + 0 configuration of 9-triplet microtubules. The mature basal body subtends the 9 + 2 axoneme of the single flagellum and recent evidence shows that the outer doublets of the axoneme are directly nucleated onto the triplet microtubules of the basal body but that a gamma tubulin-dependent nucleation of the central pair of microtubules occurs from the distal transition zone of the basal body. Induction of a gamma tubulin RNAi cell line of *T. brucei* produced an early phenotype where cells in division with two flagella possessed an older, motile flagellum and a new immotile flagellum. The immotile flagellum lacked the central pair of doublets, but still possessed the outer doublets, clearly demonstrating that, although nucleation of new microtubules (the central pair) are gamma tubulin dependent, extension of the outer doublet microtubules is a gamma tubulin-independent process [27]. The trypanosomes possess delta tubulin and we discovered both the more evolutionarily widespread epsilon tubulin and the more restricted zeta in these organisms [28]. Our initial observation that the evolutionary distribution of these tubulins strongly suggested their role in basal body structure, function or regulation has proven correct. The new tubulins have been extensively reviewed both in this volume and in previous discussions [29, 30].

A number of general points emerge from the above analysis. First cytoplasmic microtubule arrays in protozoa are nucleated from spatially specific and functionally distinct MTOCs. Second, basal bodies are often not directly concerned with the bipolarity of the spindle. Thirdly, some events in duplication of the cytoskeleton are spread, as in mammalian cells, over more than one cell cycle (basal body formation, maturation and axonemal nucleation) whilst others occur within one cell cycle (formation of the axoneme, paraflagellar rod and the FAZ filament/four microtubules).

20.4 The Microtubule Biology of the Apicomplexa

The apicomplexan protozoa are the etiological agents of a series of devastating diseases of man and domestic animals [7]. The group includes parasites ranging from *Plasmodium* which causes malaria, *Toxoplasma* and *Cryptosporidium* which are opportunistic pathogens of humans and *Theileria*, a tick-borne parasite of cattle in Africa. Each organism exhibits a unique life cycle reflecting its biology in the host and vector. There are certain core features to all the life cycles. The first is an infective and proliferative cycle stage(s) in the mammalian host (merozoites and intra-erythrocyte stages in *Plasmodium*; Tachyzoites and Bradyzoites in *Toxoplasma*). Then follows the production of gametes followed by fertilization and zygote production (these events occurring often in alternative hosts or the vector). Subsequent divisions produce the sporozoite which is often the stage in the life cycles that re-infects the primary mammalian host [31–38]. Again, as in the trypanosome, we see that the cells of these parasites show a dispersed set of the three types of MTOC. They express the flagellum MTOC (centriole/basal body), the mitotic MTOC and diverse cytoplasmic MTOCs in a pattern that differs with organism and life cycle stage [34].

20.4.1 The Apical Polar Ring: A Unique Cytoplasmic MTOC

Apicomplexan parasite cells often exhibit a stable sub-pellicular array of microtubules that underlie and link to the pellicular membrane (Figure 20.3). A unique MTOC, the apical polar ring, in the form of a circle at one end of the cell organizes these evenly spaced, sub-pellicular microtubules which spiral down to about two-thirds of the length of the cell body [34, 39, 40]. If two polar rings are present then the microtubules are associated with the outer one. There are differences in the number and arrangement of microtubules between organisms and at particular life cycle stages. However, unlike the trypanosome sub-pellicular array, the inter-microtubule distance increases as the diameter of the cell increases and they appear not to be linked to each other. Whilst the sub-pellicular microtubules are proposed to be important for the elongated cell shape and form and daughter cell budding, [41] an interwoven mesh of 8–10-nm intermediate filament-like components associated with the pellicle, composed of two major proteins TgIMC1 and 2, is probably a major contributor to mechanical strength [42]. Although characterized in *Toxoplasma*, homologs of these proteins exist in many apicomplexan parasites. The cytoplasmic microtubules appear essential for shape and apical polarity since drug treatments, which compromise these cytoplasmic arrays, affect these processes.

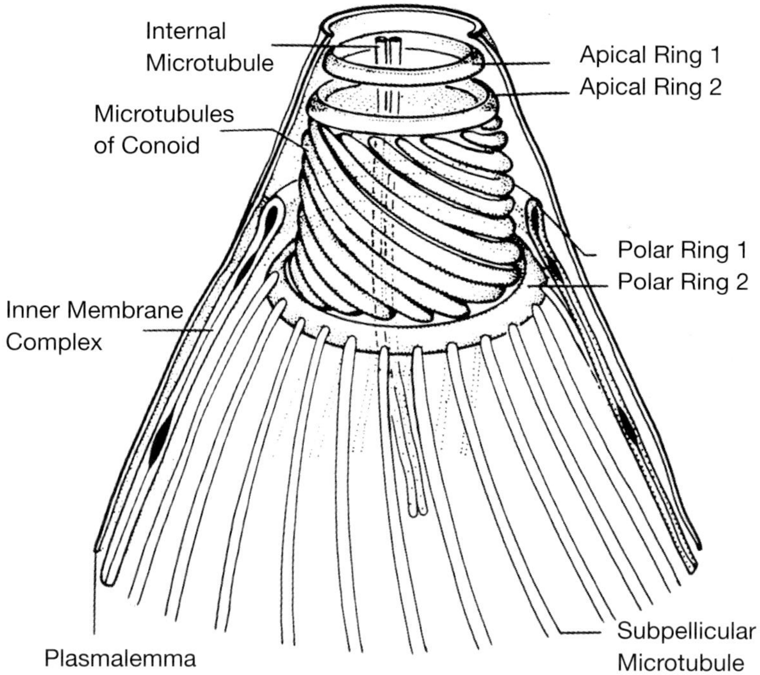

Figure 20.3 Schematic representation of the apical complex of *T. gondii* [from 7].

20.4.2
The Conoid

The conoid of some apicomplexan species (*Toxoplasma, Eimeria, Sarcocystis*) provides a beautiful example of the diversity of the tubulin molecule capabilities. This structure is suggested to be essential for invasion and pathogenicity of these intracellular parasites and may be particularly advantageous in penetration of the vertebrate gut epithelium, hence its absence in the apicomplexan that invade erthrocytes such as *Plasmodium* [32, 33, 40, 43]. Recently, evidence has been presented for a novel arrangement of tubulin molecules in this structure in *T. gondii.* Electron microscopic analysis suggests that in the mature conoid, tubulin is arranged into a novel polymer form that is quite different in form from that in typical microtubule protofilaments [44].The conoid and its set of intertwining counter-clockwise spiral filaments (Figure 20.3) can be stimulated to extrude by calcium influx [45]. Intriguingly, there are two specialized microtubules associated with the conoid, perhaps providing evidence for yet another discrete cytoplasmic MTOC in this apical area in addition to the apical ring [34].

20.4.3
Apicomplexan Basal Bodies

The apicomplexan parasites are not flagellated in their proliferative, infective stages. Flagella are produced in the male microgamete and motility is important for movement to the female macrogamete and fertilization [7, 31, 46]. The form of a typical apicomplexan male microgamete is shown in Figure 20.4 for *Eimeria maxima*. The anterior end of the microgamete contains three basal bodies and these basal bodies nucleate either two or three flagella depending on the organism. The variation comes in the fact that in some apicomplexa the third flagellum may be reduced significantly. In many Apicomplexa as in the *Eimeria* example in Figure 20.4, one of the flagellar axonemes runs inside the cell for some length before exiting. These microgametes still possess a form of cytoplasmic MTOC in that a band of cytoplasmic microtubules runs from the basal body region down past the mitochondrion and nucleus towards the posterior end.

There can be distinct differences in the presence of the basal bodies/centrioles at other stages of the life cycles in the different apicomplexan parasites. Moreover, as is the case in certain metazoans such as nematodes, the actual structure of the cen-

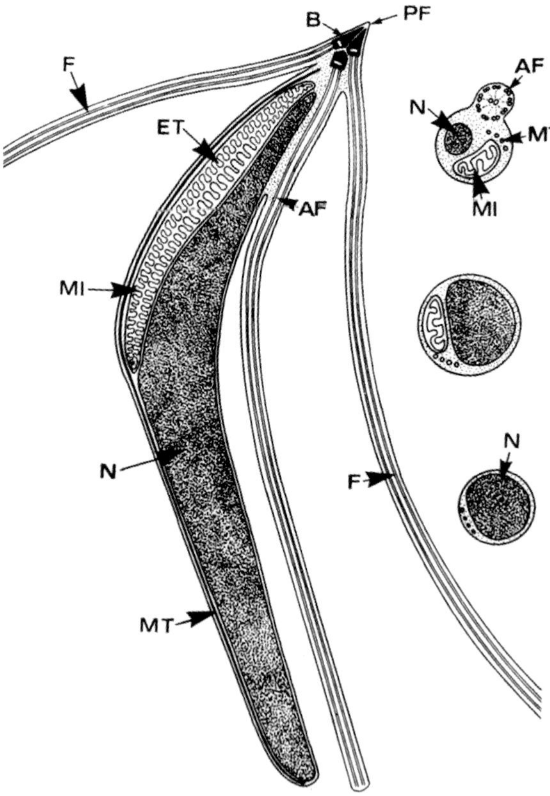

Figure 20.4 A diagrammatic representation of a microgamete of *Eimeria maxima* in longitudinal and cross section. AF, attached flagellum; B, basal body; ET, enlarged tubule; F, free flagellum; MI, mitochondrion; MT, microtubule; N, nucleus; PF, perfortorium (from [7]).

triole can vary when it is not acting as a basal body in flagellar axoneme nucleation. When the apicomplexan basal body is nucleating the 9 + 2 microtubule flagellar axoneme it exhibits the canonical 9 + 0 triplet microtubule arrangement. However, when centrioles are present in the cytoplasm of non-flagellated apicomplexan cells at other stages of the life cycles they adopt an unconventional 9 + 1 form with a single central microtubule surrounded by nine singlet microtubules [34]. This 9 + 1 configuration appears to be a conveniently "stalled" stage in the morphogenesis of the centriole/basal body providing the required configuration for the non-flagellated cell types. Such a truncation may indeed be an economical form of construction at times when additional accessory structures are unnecessary or when cell cycle times are short (cf. the early divisions in the *Drosophila* embryo). Indeed in some apicomplexan cell types such as certain of those in *Plasmodium* the centriole may not be present at all [32, 34].

20.4.4
The Spindle MTOC

Where present, the above-mentioned centrioles are located in the cytoplasm close to the nuclear envelope, and during mitosis at the spindle poles. Mitosis is intranuclear in these apicomplexan parasites. However, there is a distinct spindle MTOC that takes the form of a plaque structure, nucleates the intranuclear spindle microtubules and is referred to as a spindle pole body or a spindle pole plaque. When no centrioles are present, as in some *Plasmodium* or *Theileria* cell types these plaque structures are the only elements at the spindle poles [32, 34].

20.4.5
Apicomplexan Cell Division and Cell Morphogenesis

Parasitic protozoa exhibit an interesting set of cell division processes reflecting the architecture and karyology of particular cell types. In addition the processes often vary in the lifecycle in the context of whether the division is a purely proliferative division leading to daughter cells of the same cell type as the original cell or a differentiation division where the cell is entering a division that will lead to a new cell type. The procyclic trypanosome division referred to earlier and seen in Figure 20.1 is of a proliferative type but *T. brucei*, as other trypanosomatids such as *T. cruzi* and *Leishmania*, undergoes differentiation divisions during its life cycle [47, 48]. In each case the basal bodies and MTOCs are central to the organelle positioning and cell morphogenesis events that typify the division [15, 18, 49].

The apicomplexan parasites however, exhibit elaborate divisions that differ in phylogenetic distribution [7]. Apicomplexan parasites replicate by a variety of internal budding processes to create either two daughter cells or multiple progeny. In all cases the mitotic spindle is intranuclear. Cell division in *Toxoplasma* occurs by a process termed endodyogeny whereby the progeny are two daughter parasites. This division type is not a straightforward binary fission proliferative division as typified by the division of the *T. brucei* procyclic trypanosome where a preceding replication of the mother cell organelles and their segregation leads to two daughters. In endodyogeny, although the two daughters are ultimately the same as their mother they are formed within the intact, fully polarized mother cell (Figure 20.5). The internal daughter cells are delimited by an inner membrane complex and a new daughter cell anlage or primordium forms nucleating the associated sub-pellicular microtubules. Each daughter contains a nucleus, mitochondrion, Golgi, centriole and apicoplast (see below) as well as a complete set of apical organelles. When morphogenesis of the two daughter cells is completed the maternal apical complex is disassembled and the daughter parasites bud from the mother cell's plasma membrane [7, 34]. One thought is that this process has the attraction of allowing these tachyzoite stage cells to retain the ability to invade host cells throughout their cell cycle.

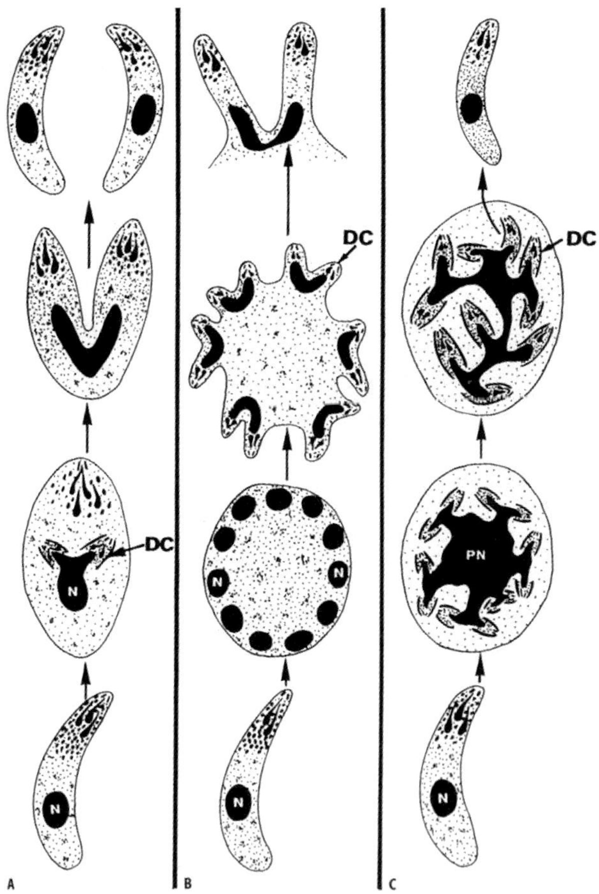

Figure 20.5 Asexual reproduction in apicomplexan protozoan parasites. (A) Endodyogeny as seen in tissue cyst stages of *Toxoplasma* and *Sarcocystis*. (B) Schizogeny (ectotype) as found in *Eimeria, Plasmodium* and *Theileria*. (C) Endopolygeny as seen in the schizonts of *Sarcocystis*. DC, the daughter cell anlage; N, nucleus; PN, polyploid nucleus (from [7]).

There are two other means of replication used by certain apicomplexan parasites that illustrate the differing ways that the MTOCs appear to act as coordinating cell centers to facilitate the formation of multiple daughter cells either after a series of mitotic divisions has produced a syncytium (Ectotype schizogeny) or by the production of many daughter cells around the periphery of a single, large, presumably endoreduplicated nucleus (Endopolygeny schizogeny) [7]. The first process of schizogeny (Figure 20.5) is seen in *Plasmodium, Theileria, Eimeria,* and *Babesia.* The second process (Figure 20.5) of endopolygeny occurs during schizogeny in *Sarcocystis* and the flagellated microgametes of *Plasmodium* spp. are produced in a similar manner.

In ectotype schizogeny the parasite sub-pellicular microtubules and apical complex are disassembled and multiple rounds of mitotic nuclear divisions occur. The formation of the classically polarized parasites occurs when the nuclei move to the cell periphery and associate with the assembling daughter cell anlagen comprising inner membrane complexes, sub-pellicular microtubules and apical organelles. The daughter cells then bud out of the mother cell as merozoites. A set of complex microtubule events accompanies each division type, which have been studied by immunofluorescence microscopy and electron microscopy. The details are fascinating but beyond the scope of this chapter. What is important is that there are unifying principles in how different MTOCs are orchestrated in each division. For instance the dominant role of the mitotic MTOC is seen in each division type acting as a cell center that then connects with the newly forming cytoplasmic MTOCs of each prospective daughter so allowing formation of the daughter cell anlagen. Further, the endopolygeny form of schizogeny implies that although segregation of genomes occurs, an individual chromosome complement is associated with each mitotic MTOC so it is maintained at the site of each forming daughter cell anlage.

In *Plasmodium*, microgametogenesis occurs in the mosquito midgut, after the ingestion of infected erythrocytes from a mammalian host. Once in the midgut the infected erythrocytes lyse to release either female macrogametocytes or male microgametocytes. These cell types are predetermined and differentiation from the merozoites is initiated while still in the mammalian host. As we have just rehearsed, the process of formation of the microgametes in *Plasmodium* occurs via the endopolygeny route whereby a single large, endoreduplicated nucleus is divided into many progeny. The microgametes of the malarial parasite are flagellated and in *Plasmodium* ultrastructural studies have shown an intriguing mechanism linking axoneme assembly to nuclear segregation [46, 50].

During microgametogenesis the genome is segregated on three successive series of intranuclear mitotic spindles. Basal bodies, which are not found in the preceding cells of the intraerythrocytic stages of the *Plasmodium* life cycle, are formed and are joined to the spindle pole plaques through electron-dense material. After the final division there is one centriole/basal body linked to one spindle pole plaque and this arrangement provides the mechanism by which each haploid set of chromosomes is eventually captured into a single microgamete equipped with a flagellum. The centriole/basal body becomes surrounded by a vesicular-tubular basket at the time of the final nuclear division. The assembly of an axoneme takes place within the cytoplasm of the gametocyte just outside the nuclear envelope and during exflagellation the axoneme, basal body and associated MTOCs and other structures are forced to the cell surface distending the plasma membrane. In the final stages of this process the prospective microgamete slides tangentially and the nuclear bud containing the now condensed haploid genome becomes detached from the main nucleus and is incorporated into the released microgamete [46, 50]. This explosive process again illustrates and emphasizes the points made in the Introduction, of the dispersed, yet distinct nature of the mitotic, cytoplasmic and flagellum MTOCs in protozoa, their independent regulation and the maintained, physical relationship of the mitotic MTOC to the segregated genome.

20.5
Basal Bodies Are More than Just Microtubule Organizers: The Hitchhiker's Guide to the Cytoskeleton!

Studies of parasitic protozoa have revealed that the precise and cell cycle-dependent duplication of basal bodies has been utilized to ensure inheritance of other structures/organelles outside the roles of microtubule organization and flagellum nucleation. Organisms of the Kinetoplastida share the fact that their mass of mitochondrial DNA is located close to a basal body and can be visualized by DAPI staining in a structure called the kinetoplast. Studies of the association between basal bodies and mitochondrial DNA in *T. brucei* have shown that it is critical for kinetoplast position and segregation [51]. The association is effected via a proximal end function of the trypanosome basal bodies. The large mitochondrial genome mass within the kinetoplast is physically connected to the flagellar basal bodies and is segregated by them during their repositioning during cell division. Recently our laboratory provided evidence for a stable transmembrane link that connects the kinetoplast DNA to the basal bodies. We showed that three specific components comprise a structure that we have termed the tripartite attachment complex (TAC) [52]. The TAC involves a set of filaments linking the basal bodies to a zone of differentiated outer and inner mitochondrial membranes and a further set of intramitochondrial filaments linking the inner face of the differentiated membrane zone to the kinetoplast (Figure 20.6). As mentioned previously, and in an event unusual for a eukaryotic cell, the kinetoplast DNA is replicated in its own periodic S-phase [26, 51]. The TAC and flagellum–kinetoplast DNA connections are sustained throughout the cell cycle and are replicated and remodeled during this periodic kinetoplast DNA S-phase. The architecture of the TAC suggests that it may also function in providing a structural and vectorial role during replication of this catenated mass of mitochondrial DNA. This understanding of the high-order transmembrane linkage provides an explanation for the spatial position of the trypanosome mitochondrial genome and its mechanism of segregation [26, 51, 52]. We have, as yet, little information on the constituents of the TAC. However, our view was that since the kinetoplast is so large and so highly organized its inter-

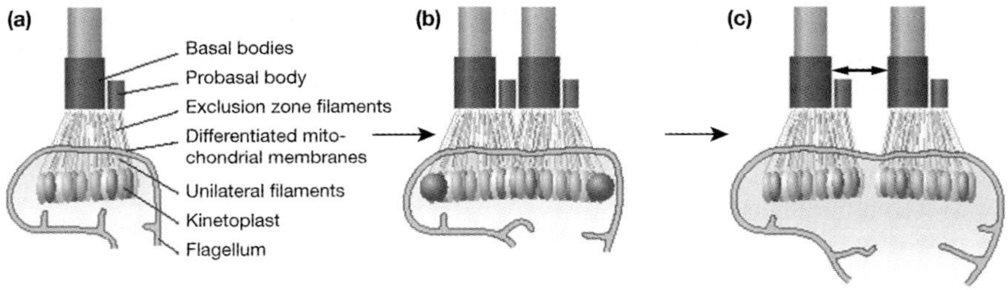

Figure 20.6 A schematic diagram of the tripartite attachment complex in Trypanosomes. Panel (a) illustrates the basal bodies, kinetoplast and the components of the TAC (exclusion zone filaments, differentiated mitochondrial membranes and unilateral filaments) in a trypanosome in G1 of the cell cycle. In this period there is a single flagellum, a basal body and a pro-basal body. Panel (b) shows the organization of the S-phase TAC. When the cell enters S-phase discrete fibrous lobes appear at the poles of the kinetoplast, the pro-basal body matures into a basal body and subtends the new flagellum and two new pro-basal bodies are formed. Two nascent TAC complexes are discernable at this period of the cell cycle. Panel (c) shows the period where movement apart of the flagella basal bodies segregates the replicated kinetoplast DNA. Note that the position and orientation of the basal bodies have been idealized in this two-dimensional diagram (from [58]) (see Color Plates page XLII).

action with the mitochondrial membrane and the transmembrane nature of the connection is consequently easily visualized. However, we suggested that this complex was likely to represent an extreme form of a more generally occurring interaction between the mitochondrion and the cytoskeleton in eukaryotic cells. This appears to be the case and recently evidence has been presented for a stable two membrane-spanning autonomous mitochondrial DNA replisome in yeast [53]. The mitochondrial DNA of yeast and other eukaryotes is organized as a complex nucleoprotein structure termed the nucleoid. Yeast has at least two populations of nucleoids that exist within the same mitochondrion and can be distinguished by their association with a discrete proteinaceous structure spanning the outer and inner mitochondrial membranes. This two membrane-spanning structure (TMS) persists and self-replicates in the absence of mitochondrial DNA but actively replicating nucleoids are associated exclusively with TMS. Thus the TAC appears to be the paradigm for a more generally occurring two membrane-spanning replisome that can provide the mechanism for physically linking mitochondrial DNA replication and inheritance.

Apicomplexan protozoan parasites such as *Plasmodium* and *Toxoplasma* possess a single non-photosynthetic plastid termed the apicoplast. The apicoplast appears to be the remnant of a eukaryotic algal plastid that was acquired at some point in evolution by secondary endosymbiosis. The presence of multiple membranes around the apicoplast is seen as an archaeological cellular signature of such an event. Although the apicoplast has only a reduced remnant genome it imports many now nuclear encoded proteins and its metabolism is essential for pathogenicity and viability [54–56]. Given this and the fact that a single apicoplast resides in the cytoplasm of *Plasmodium* and *Toxoplasma* parasites then faithful segregation and inheritance of this plastid should be important for these intracellular parasites. The mechanism for this has now become clearer and the centrosome is intimately involved in the process. In *Toxoplasma gondii*, the precise duplication and inheritance of centrioles/basal bodies is again utilized to ensure division and segregation of in this case, a single copy plastid [54]. In non-dividing cells the apicoplast is closely associated with the centrioles. During mitosis as the intranuclear spindle extends and becomes bipolar the plastid is pulled into a U-shape and its ends are associated with centrioles at each spindle pole. The plastid divides into two and each daughter remains associated with the centrioles of the daughter cell. Since these structures provide a singularity, which, along with the newly forming apical ring MTOCs, focuses and polarizes the formation of each daughter cell cytoskeleton, so then the inheritance of an apicoplast by each daughter cell is assured. Whether the plastid and its DNA and basal body are physically attached remains to be determined but appears likely. Again the trypanosome TAC described above may provide the paradigm for such a mechanism.

This hitchhiking approach that ensures fidelity of replication and segregation appears to be deeply embedded in the cell biology of organelles of ancient endosymbiotic origin. Intriguingly, this form of cellular hitchhiking involving centrioles/basal bodies or centrosomes has also been adopted by intracellular protozoan parasites of animals to ensure segregation and therefore vertical transmission to daughter host tissue cells. It is likely that *Theileria parva*, microsporidial parasites and bacteria harbored in the cells of *Drosophila* and other organisms use this mechanism [57]. We have predicted that a unifying cell biology underlies the mechanism of interaction of genome-containing organelles and intracellular parasites or symbionts with cytoplasmic or spindle microtubules, or basal bodies or centrosomes [58].

20.6
Cytoskeletal Adaptations to Parasitism

Whilst the main focus of this chapter has been on the basal bodies and MTOCs of parasitic protozoa in relation to their intrinsic cell biology it is obvious that some at least of this cytoskeletal cell biology is important as an adaptation for parasitism. The eukaryotic flagellum is an extremely versatile organelle and it has been adapted for many roles in pathogenicity.

Axoneme-associated structures are commonly found in eukaryotic flagella but one of the most dramatic is the paraflagellar rod that accompanies the axoneme in the flagella of trypanosomes. The PFR is a complex network of filaments extending the length of the flagellum after it emerges from the flagellar pocket and has been demonstrated to play an essential role in flagella motility [59, 60]. The PFR is linked directly to the axoneme through filaments and is composed primarily of two closely related proteins, PFR-1 and PFR-2. Knock down of protein levels of PFR-2 by RNAi, ablated the PFR structure with only a small part of the proximal region that connects to the axoneme remaining. Without the presence of the PFR-1 subunits, PFR-1 and other PFR constituents cannot assemble to form the PFR structure and without a complete PFR the resulting cells are viable but immotile. Similar conclusions have been reached by gene knockout approaches in *Leishmania*, which has a detached flagellum [61, 62].

Motility via a flagellum is an undoubted necessity in many pathogenic protozoa for passage through their complex life cycles. Such fast motility is no doubt essential for the African trypanosomes such as *T. brucei* during its journey in the tsetse vector from the midgut to the salivary glands where essential differentiations occur before the parasite can be successfully transmitted to another mammalian host. *T. brucei* is an extracellular parasite in the mammalian host but no doubt a flagellum is critical there for traverse of blood vessel endothelium barriers. Trypanosomatids such as *T. cruzi* and *Leishmania*, which have adapted to proliferation inside mammalian host cells such as macrophages, do so with an amastigote form that has no flagellum or just a short stub [63–65]. In the malarial parasite the flagellum, as rehearsed above, is restricted to the microgamete stage of the life cycle. This requirement for at least one of the gametes to be flagellated in order to locate the female macrogamete mirrors much of the use of the flagellum in the sexual reproduction phases of many eukaryotes.

One function of the flagellum – that it provides a cell with the ability to differentiate the environment of its surface membrane – has become of great utility in trypanosomatid parasites and is seen as a key function of the flagellum in bloodstream forms of the African trypanosome, *T. brucei* [9, 66]. Flagellum morphogenesis in *T. brucei* defines three plasma membrane domains: the surface membrane, the flagellar pocket membrane and the flagellar membrane [13]. All vesicular traffic, both into and out of the cell, passes through the flagellar pocket and it defines the dynamic portal to host or vector environment [10]. Because of the ciliary necklace connections at the basal body region of cilia and flagella, the membranes of the flagellum and the cell body are contiguous, yet able to be differentiated. A key feature is that the flagellum not only provides a mechanism for movement and attachment, it also enables the morphogenesis of a lumenal region of the plasma membrane: the flagellar pocket. This pocket provides the portal through which most of the dynamic interactions with the host occur. These interactions facilitate resistance to innate and acquired immune responses as well as acquisition of growth factors from the host [9, 66, 67]. Receptors for macromolecules such as transferrin and lipoproteins from the surrounding plasma are hidden to some extent from the immune response by being located in the flagellar pocket. The *T. brucei* bloodstream form surface is covered by a dense coat composed of a glycosylphosphatidylinositol anchored (GPI) protein, the variant surface glycoprotein (VSG). The presence of the VSG coat and its switching in individual parasites allows evasion of the host's immune response [68, 69]. Such GPI-anchored proteins are delivered and recycled to the cell surface via secretory events at the flagellar pocket.

There is evidence that in trypanosomes, as in other systems, the flagellar membrane enables the specific localization of particular signaling proteins, one example being adenylate cyclase [70]. The ability to locate a specific translocator, receptor and signaling system in a membrane domain may have particular benefits in host/parasite, parasite/parasite and parasite/environment interactions.

The flagellum allows *T. brucei* to develop these three distinct plasma membrane domains: the cell body, the flagellar pocket and the flagellum. Molecular mechanisms and signals likely to be important in targeting of particular proteins to particular regions of the three major plasma membrane domains of the trypanosome and the flagellum cytoskeleton are becoming clearer, but general paradigms have yet to emerge [9, 71].

As we commented earlier, the cytoplasmic MTOC responsible for nucleating the four specialized microtubules, which traverse a path around the flagellar pocket before joining the main sub-pellicular array in an antiparallel manner in the FAZ, is located close to the basal bodies. This configuration not only facilitates cell morphogenesis but also suggests mechanisms for setting up directed membrane trafficking and other polarity-dependent phenomena. We have suggested that apart from organelle positioning this architecture could underlie phenomena such as the observed capping and internalization of antibody molecules, receptor cycling, directed access and egress from the flagellum pocket.

The flagellum of many trypanosomatid parasites is used for attachment to host or vector surfaces. Obviously the sub-pellicular microtubules underlying the cell body plasma membrane prohibit the formation of stable complex structures being elaborated between the parasite and the host surface. In *Leishmania* spp. the flagellum facilitates attachment to various regions of the sand fly vector gut, so enabling cell type differentiations important for reinvasion of the mammalian host to occur. The *T. brucei* flagellum is used for attachment of the parasite to the tsetse mouthparts at the epimastigote stage of the life cycle. Attachment of *T. brucei* epimastigotes to the salivary gland epithelial brush border is mediated by rather elaborate, branched outgrowths of the flagellar membrane and "hemidesmosomal-type" plaques. These outgrowths diminish but the attachment plaques are maintained as the parasite differentiates to the nascent metacyclic form before the eventual release of the metacyclic form of the parasite, which is the mammal-infective form. Epimastigote forms are proliferative in the tsetse vector and therefore able to divide whilst attached. It has been suggested that this mode of attachment facilitates cell division. Electron microscopy of whole cells and detergent-extracted preparations reveals the detailed sub-structure of filaments and attachment plaques [72] but the biochemical nature of the components of both are unknown.

Parasites need to survive exposure to unstable environments, often where a fluidic motion may expel them from their host. Therefore, in order to maintain the parasites' presence in this environment it is likely that they secure an attachment to the host cells. This attachment may be a vital, long-term requirement necessary to sustain an infection, or may be a necessary but relatively transient attachment (as with trypanosomes) in being required at only a single stage in their life cycle. Perhaps no parasite shows a more extreme body plan influenced by the need to attach to host cells than *Giardia*. The water-borne parasite *Giardia* leads an exclusively extracellular existence and for the majority of its life cycle it exists as trophozoites in the small intestine of a mammalian host. Its ability to persist and colonize the small intestine of its host is essential for the maintenance of the infection and is also in part responsible for the pathology of the disease [73]. In order to maintain its presence and to prevent being swept away by peristalsis in the intestines, *Giardia* secures an attachment to the intestinal epithelium; this attachment being mediated through unique microtubular structures which dominate the parasite cytoskeleton and define its appearance [74].

The ventral disk is a concave structure covering the entire ventral surface of the parasite. It consists of a spiral array of microtubules that lie directly above the ventral membrane and are connected to it through short filaments. These microtubules originate from a set of electron-dense bands located immediately anterior to the basal bodies of the caudal and posterior-lateral flagella. Microribbons (dorsal ribbons) extend from each microtubule and project up, nearly perpendicular to the membrane. The microribbons are connected to a series of cross-linking filaments. At the edge of the disk, a network of fibers termed the lateral crest replaces the microtubules and microribbons.

While the exact method by which the ventral disk physically secures its attachment to intestinal cells has yet to be agreed, two proposed methods stand out. Firstly, that the lateral crest of the ventral disk is capable of contracting resulting in a clasping of the intestinal epithelial cells. The lateral crest is repeatedly observed in contact with the intestinal epithelial cells and after trophozoite detachment, leaves behind a footprint with sharp marks in the host cell membranes corresponding to the lateral crest region. Secondly, that the beating action of the ventral flagella during attachment generates a negative pressure under the disk.

The apicomplexan parasites such as Malaria and *Toxoplasma* have balanced their elaborate microtubule cytoskeleton with an intensive investment in acto-myosin motility systems that are critical for host cell invasion [37, 43, 75, 76]. The apicomplexan microtubule systems that have been rehearsed earlier are critical for defining a cell polarity at the apical end that is tailored towards providing anchor points and the directional secretion of specialized membranous organelles, namely micronemes, rhoptries and dense granules [77, 78]. This directed secretion of organelles is vital not only for host cell attachment, but also for motility and cell invasion.

20.7 Conclusion

There are now three completed trypanosomatid genome projects (*T. brucei, T. cruzi* and *Leishmania major*), in addition to the genomes of *Plasmodium, Toxoplasma* and *Giardia* that are complete or nearly so. In addition there is an increasing lower coverage data set available on the genomes of other parasitic protozoa. Mining of this information already shows that many of the proteins that we recognize as being associated with basal bodies and MTOCs from work with other organisms, are present in these genomes. In addition work from a number of laboratories has defined basal body and cytoskeletal proteins of parasitic protozoa by empirical experimental approaches. Given the now excellent and tractable reverse genetics of these experimental systems there will be extremely interesting opportunities for future functional analyses. These are likely to be informative for understanding fundamental components, structures, properties and functions of MTOCs, basal bodies and cilia/flagella in all eukaryotic cells, as well as revealing how these structures and organelles have been subverted by parasitic protozoa for their own pathogenicity strategies.

Acknowledgments

We thank the Wellcome Trust for support of our work. Laura Briggs is in receipt of an MRC graduate studentship.

References

1. Wenyon, C. M. **1926**. *Protozoology. A Manual for Medical Men, Veterinarians and Zoologists*, Vol. 1, Hafner, New York.
2. Tsai, C. W. W. and S. M. Wolniak **2001**. Cell cycle arrest allows centrin translation but not basal body formation during spermiogenesis in Marsilea. Jornal of Cell Science **114**: 4265–4272.
3. Henneguy, L. F. **1898**. Sur les rapportsdes cils vibratiles avec les centrosomes. Archiv fur mikroskopische Anatomie **1**: 482–496.
4. Lenhossek, M., von. **1898**. Uber Flimmerzellen. Verhandlungen der Anatomischen Gesellschaft **12**: 106–128.
5. Pickett-Heaps, J. D. **1969**. The evolution of the mitotic apparatus: an attempt at comparative ultrastructural cytology in dividing plant cells. Cytobios **3**: 257–280.
6. Mignot, J. P. **1996**. The centrosomal big bang: From a unique central organelle towards a constellation of MTOCs. Biology of the Cell **86**: 81–91.
7. Mehlhorn, H., (Ed.) **2001**. *Encyclopedic Reference of Parasitology*, Springer, Heidelberg.
8. Gull, K. **1999**. The cytoskeleton of trypanosomatid parasites. Annual Review of Microbiology **53**: 629–655.
9. Landfear, S. M. and M. Ignatushchenko **2001**. The flagellum and flagellar pocket of trypanosomatids. Molecular and Biochemical Parasitology **115**: 1–17.
10. McConville, M. J., K. A. Mullin, S.C. Ilgoutz, R.D. Teasdale **2002**. Secretory pathway of trypanosomatid parasites. **66**: 122–254.
11. Bastin, P. and K. Gull **1999**. Assembly and function of complex flagellar structures illustrated by the paraflagellar rod of trypanosomes. Protist **150**: 113–123.
12. Maga, J. A. and J. H. LeBowitz **1999**. Unravelling the kinetoplastid paraflagellar rod. Trends in Cell Biology **9**: 409–413.
13. Vaughan, S. and K. Gull **2003**. The trypanosome flagellum. Journal of Cell Science **116**: 757–759.
14. Sherwin, T. and K. Gull **1989**. Visualization of detyrosination along single microtubules reveals novel mechanisms of assembly during cytoskeletal duplication in Trypanosomes. Cell **57**: 211–221.
15. Robinson, D. R., T. Sherwin, A. Ploubidou, E. H. Byard, K. Gull **1995**. Microtubule polarity and dynamics in the control of organelle positioning, seg-

regation, and cytokinesis in the Trypanosome cell cycle. Journal of Cell Biology **128**: 1163–1172.
16. Ploubidou, A., D. R. Robinson, R. C. Docherty, E. O. Ogbadoyi, K. Gull **1999**. Evidence for novel cell cycle checkpoints in trypanosomes: kinetoplast segregation and cytokinesis in the absence of mitosis. Journal of Cell Science **112**: 4641–4650.
17. Moreira-Leite, F. F., T. Sherwin, L. Kohl, K. Gull **2001**. A trypanosome structure involved in transmitting cytoplasmic information during cell division. Science **294**: 610–612.
18. Sherwin, T. and K. Gull **1989**. The cell division cycle of *Trypanosoma brucei brucei* – timing of event markers and cytoskeletal modulations. Philosophical Transactions of the Royal Society of London Series B-Biological Sciences **323**: 573–588.
19. Sherwin, T., A. Schneider, R. Sasse, T. Seebeck, K. Gull **1987**. Distinct localization and cell cycle dependence of COOH terminally tyrosinolated alpha-tubulin in the microtubules of *Trypanosoma brucei brucei*. Journal of Cell Biology **104**: 439–446.
20. Kohl, L., T. Sherwin, K. Gull **1999**. Assembly of the paraflagellar rod and the flagellum attachment zone complex during the *Trypanosoma brucei brucei* cell cycle. Journal of Eukaryotic Microbiology **46**: 105–109.
21. Woodward, R. and K. Gull **1990**. Timing of nuclear and kinetoplast DNA replication and early morphological events in the cell cycle of *Trypanosoma brucei brucei*. Journal of Cell Science **95**: 49–57.
22. Ogbadoyi, E., K. Ersfeld, D. Robinson, T. Sherwin, K. Gull **2000**. Distinct localization and cell cycle dependence of COOH terminally tyrosinolated alpha-tubulin in the microtubules of *Trypanosoma brucei brucei*. Chromosoma **108**: 501–513.
23. Scott, V., T. Sherwin, K. Gull **1997**. Gamma-tubulin in trypanosomes: molecular characterisation and localisation to multiple and diverse microtubule organising centres. Journal of Cell Sciences. 1997 Jan; **110 (Pt 2)**: 157–168.
24. Ersfeld, K. and K. Gull **1997**. Partitioning of large and minichromosomes in *Trypanosoma brucei brucei*. Science. 1997 Apr 25; **276 (5312)**: 611–614.
25. Gull, K., S. Alsford, K. Ersfeld **1998**. Segregation of minichromosomes in trypanosomes: implications for mitotic mechanisms. Trends in Microbiology **6**: 319–323.
26. Lukes, J., D. L. Guilbride, J. Votypka, A. Zikova, R. Benne, P.T. Englund **2002**. Kinetoplast DNA network: Evolution of an improbable structure. Eukaryotic Cell **1**: 495–502.
27. McKean, P. G., A. Baines, S. Vaughan, K. Gull **2003**. Gamma-tubulin functions in the nucleation of a discrete subset of microtubules in the eukaryotic flagellum. Current Biology **13**: 598–602.
28. Vaughan, S., T. Attwood, et al. **2000**. New tubulins in protozoal parasites. Current Biology **10**: R258–R259.
29. McKean, P. G., S. Vaughan, K. Gull **2001**. The extended tubulin superfamily. Journal of Cell Science **114**: 2723–2733.
30. Dutcher, S. K. **2003**. Long-lost relatives reappear: identification of new members of the tubulin. Current Opinion in Microbiology. 2003 Dez; **6 (6)**: 634–640.
31. Baker, J. R. **1989**. Sexual processes in parasitic protozoa. International Journal for Parasitology **19**: 465–472.
32. Bannister, L. H., J. M. Hopkins, R. E. Fowler, S. Krishna, G. H. Mitchell **2000**. A brief illustrated guide to the ultrastructure of Plasmodium falciparum asexual blood stages. Parasitology Today. 2000 Oct; **16 (10)**: 427–433.
33. Carruthers, V. B. **2002**. Host cell invasion by the opportunistic pathogen *Toxoplasma gondii*. Acta Tropica **81**: 111–122.
34. Morrissette, N. S. and L. D. Sibley **2002**. Cytoskeleton of apicomplexan parasites. Microbiology and Molecular Biology Reviews **66**: 21–38.
35. Bannister, L. and G. Mitchell **2003**. The ins, outs and roundabouts of malaria. Trends in Parasitology **19**: 209–213.

36. Kappe, S. H. I., K. Kaiser, K. Matuschewski **2003**. The Plasmodium sporozoite journey: a rite of passage. Trends in Parasitology **19**: 135–143.
37. Shaw, M. K. **2003**. Cell invasion by Theileria sporozoites. Trends in Parasitology **19**: 2–6.
38. Sibley, L. D. **2003**. *Toxoplasma gondii*: perfecting an intracellular life style. Traffic **4**: 581–586.
39. Nichols, B. A. and M. L. Chiappino **1987**. Cytoskeleton of *Toxoplasma gondii*. Journal of Protozoology **34**: 217–226.
40. Dubey, J. P., D. S. Lindsay, C. A. Speer **1998**. Structures of *Toxoplasma gondii* tachyzoites, bradyzoites, and sporozoites and biology and development of tissue cysts. Clinical Microbiology Reviews **11**: 267–299.
41. Shaw, M. K., H. L. Compton, D. S. Roos, L. G. Tilney **2000**. Microtubules, but not actin filaments, drive daughter cell budding and cell division in *Toxoplasma gondii*. Journal of Cell Science **113**: 1241–1254.
42. Mann, T. and C. Beckers **2001**. Characterization of the subpellicular network, a filamentous membrane skeletal component in the parasite *Toxoplasma gondii*. Molecular and Biochemical Parasitology **115**: 257 268.
43. Barragan, A. and L. D. Sibley **2003**. Migration of *Toxoplasma gondii* across biological barriers. Trends in Microbiology **11**: 426–430.
44. Hu, K., D. S. Roos, J. M. Murray **2002**. A novel polymer of tubulin forms the conoid of *Toxoplasma gondii*. Journal of Cell Biology **156**: 1039–1050.
45. Monteiro, V. G., E. J. de Melo, M. Attias, W. de Souza **2001**. Morphological changes during conoid extrusion in *Toxoplasma gondii* tachyzoites treated with calcium ionophore. Journal of Structural Biology **136**: 181–189.
46. Sinden, R. E. **1983**. Sexual development of malarial parasites. Advances in Parasitology **22**: 153–216.
47. Matthews, K. R. and K. Gull **1994**. Cycles within cycles – the interplay between differentiation and cell division in *Trypanosoma brucei*. Parasitology Today **10**: 473–476.
48. Matthews, K. R. **1999**. Developments in the differentiation of *Trypanosoma brucei*. Parasitology Today **15**: 76–80.
49. Matthews, K. R., T. Sherwin, K. Gull **1995**. Mitochondrial genome repositioning during the differentiation of the African trypanosome between life-cycle forms is microtubule mediated. Journal of Cell Science. 1995 June; **108**: 2231–2239.
50. Sinden, R. E., E. U. Canning, B. Spain **1976**. Gametogenesis and fertilisation in *Plasmodium yoelii:* a transmission electron microscope study. Proceedings of the Royal Society. London. Series B. **193**: 55–76.
51. Robinson, D. R. and K. Gull **1991**. Basal body movements as a mechanism for mitochondrial genome segregation in the Trypanosome cell cycle. Nature **352**: 731–733.
52. Ogbadoyi, E. O., D. R. Robinson, K. Gull **2003**. A high-order trans-membrane structural linkage is responsible for mitochondrial genome positioning and segregation by flagellar basal bodies in trypanosomes. Molecular Biology of the Cell **14**: 1769–1779.
53. Meeusen, S. and J. Nunnari **2003**. Evidence for a two membrane-spanning autonomous mitochondrial DNA replisome. Journal of Cell Biology **163**: 503–510.
54. Striepen, B., M. J. Crawford, M. K. Shaw, L. G. Tilney, F. Seeber, D. S. Roos **2000**. The plastid of *Toxoplasma gondii* is divided by association with the centrosomes. Journal of Cell Biology **151**: 1423–1434.
55. Marechal, E. and M. F. Cesbron-Delauw **2001**. The apicoplast: a new member of the plastid family. Trends in Plant Science **6**: 200–205.
56. Foth, B. J. and G. I. McFdden **2003**. The apicoplast: a plastid in *Plasmodium falciparum* and other apicomplexan parasites. International Review of Cytology – a Survey of Cell Biology, **224**: 57–110.
57. Terry, R. S., A. M. Dunn, J. E. Smith **1999**. Segregation of a microsporidian parasite during host cell mitosis. Parasitology **118**: 43–48.

58. Gull, K. **2001**. Protist tubulins: new arrivals, evolutionary relationships and insights to cytoskeletal function. Current Opinion in Microbiology **4**: 427–432.

59. Bastin, P., T. Sherwin, K. Gull **1998**. Paraflagellar rod is vital for trypanosome motility. Nature **391**: 548–548.

60. Bastin, P., T. J. Pullen, F. F. Moreira-Leite, K. Gull **2000**. Inside and outside of the trypanosome flagellum: a multifunctional organelle. **2**: 1865–1874.

61. Santrich, C., L. Moore, T. Sherwin, P. Bastin, C. Brokaw, K. Gull, J. H. LeBowitz **1997**. A motility function for the paraflagellar rod of Leishmania parasites revealed by PFR-2 gene knockouts. Molecular and Biochemical Parasitology **90**: 95–109.

62. Maga, J. A., T. Sherwin, S. Francis, K. Gull, J. H. LeBowitz **1999**. Genetic dissection of the Leishmania paraflagellar rod, a unique flagellar cytoskeleton structure. Journal of Cell Science **112**: 2753–2763.

63. De Souza, W. **2002**. Basic cell biology of *Trypanosoma cruzi*. Current Pharmaceutical Design **8**: 269–285.

64. Vannier-Santos, M. A., A. Martiny, W. de Souza **2002**. Cell biology of *Leishmania* spp.: invading and evading. Current Pharmaceutical Design **8**: 297–318.

65. Barrett, M. P., R. J. Burchmore, A. Stich, J. O. Lazzari, A. C. Frasch, J. J. Cazzulo, S. Krishna **2003**. The trypanosomiases. Lancet **362**: 1469–1480.

66. Gull, K. **2003**. Host–parasite interactions and trypanosome morphogenesis: a flagellar pocketful of goodies. Current Opinions in Microbiology. 2003 Aug; **6 (4)**: 365–370.

67. Sacks, D. and A. Sher **2002**. Evasion of innate immunity by parasitic protozoa. Nature Immunology. 2002 Nov; **3 (11)**: 1041–1047.

68. Borst, P. and S. Ulbert **2001**. Control of VSG gene expression sites. Molecular and Biochemical Parasitology **114**: 17–27.

69. Vanhamme, L., E. Pays, R. McCulloch, J. D. Barry **2001**. An update on antigenic variation in African trypanosomes. Trends in Parasitology **17**: 338–343.

70. D'Angelo, M. A., A. E. Montagna, S. Sanguineti, H. N. Torres, M. M. Flawia **2002**. A novel calcium-stimulated adenylyl cyclase from *Trypanosoma cruzi*, which interacts with the structural flagellar protein paraflagellar rod. Journal of Biological Chemistry **277**: 35025–35034.

71. Bastin, P., T. H. MacRae, S. B. Francis, K. R. Matthews, K. Gull **1999**. Flagellar morphogenesis: Protein targeting and assembly in the paraflagellar rod of trypanosomes. Molecular and Cellular Biology **19**: 8191–8200.

72. Beattie, P. and K. Gull **1997**. Cytoskeletal architecture and components involved in the attachment of *Trypanosoma congolense* epimastigotes. Parasitology **115**: 47–55.

73. Adam, R. D. **2001**. Biology of *Giardia lamblia*. Clinical Microbiology Reviews **14**: 447–475.

74. Elmendorf, H. G., S. C. Dawson, J. M. McCaffery **2003**. The cytoskeleton of *Giardia lamblia*. International Journal for Parasitology **33**: 3–28.

75. Menard, R. **2001**. Gliding motility and cell invasion by Apicomplexa: insights from the Plasmodium sporozoite. Cellular Microbiology **3**: 63–73.

76. Heintzelman, M. B. **2003**. Gliding motility: The molecules behind the motion. Current Biology **13**: R57–R59.

77. Blackman, M. J. and L. H. Bannister **2001**. Apical organelles of Apicomplexa: biology and isolation by subcellular fractionation. Molecular and Biochemical Parasitology **117**: 11–25.

78. Soldati, D., J. F. Dubremetz, M. Lebrun **2001**. Microneme proteins: structural and functional requirements to promote adhesion and invasion by the apicomplexan parasite *Toxoplasma gondii*. International Journal for Parasitology **31**: 1293–1302.

Index

Centrosomes in Development and Disease. Edited by Erich A. Nigg

ISBN 3-527-30980-2

C